Soares Book on Grounding and Bonding

Soares Book on Grounding and Bonding

Tenth Edition

International Association of Electrical Inspectors
Richardson, Texas

International Association of Electrical Inspectors
901 Waterfall Way, Suite 602
Richardson, TX 75080-7702

 First Edition published 1966
Printed in the United States of America
09 08 07 06 05 04 5 4 3 2 1

ISBN: 1-890659-47-9

Photos used in this book were shot *in situ* or at tradeshows. Use of the photos does not imply endorsement by IAEI of the manufacturers or the products.

Notice to the Reader

This book has not been processed in accordance with NFPA Regulations Governing Committee Projects. Therefore, the text and commentary in it shall not be considered the official position of the NFPA or any of its committees and shall not be considered to be, nor relied upon as a formal interpretation of the meaning or intent of any specific provision or provisions of the 2008 edition of NFPA 70, *National Electrical Code*.®[1]

Publishers do not warrant or guarantee any of the products described herein or perform any independent analysis in connection with any of the product information contained herein. Publisher does not assume, and expressly disclaims, any obligation to obtain and include information referenced in this work.

The reader is expressly warned to consider carefully and adopt all safety precautions that might be indicated by the activities described herein and to avoid all potential hazards. By following the instructions contained herein, the reader willingly assumes all risks in connection with such instructions.

THE PUBLISHERS MAKE NO REPRESENTATIONS OR WARRANTIES OF ANY KIND, INCLUDING, BUT NOT LIMITED TO, THE IMPLIED WARRANTIES OF FITNESS FOR PARTICULAR PURPOSE, MERCHANTABILITY OR NON-INFRINGEMENT, NOR ARE ANY SUCH REPRESENTATIONS IMPLIED WITH RESPECT TO SUCH MATERIAL. THE PUBLISHERS SHALL NOT BE LIABLE FOR ANY SPECIAL, INCIDENTAL, CONSEQUENTIAL OR EXEMPLARY DAMAGES RESULTING, IN WHOLE OR IN PART, FROM THE READER'S USES OF OR RELIANCE UPON THIS MATERIAL.

[1]*National Electrical Code* and *NEC* are registered trademarks of the National Fire Protection Association, Inc., Quincy, MA 02269

Table of Contents

Preface

This book is dedicated to the memory of Eustace C. Soares, P.E., one of the most renowned experts in the history of the *National Electrical Code* in the area of grounding electrical systems. A wonderful teacher and man of great vision, Eustace foresaw the need for better definitions to clear up to the great mystery of grounding of electrical systems.

Eustace Soares' book, *Grounding Electrical Distribution Systems for Safety* was originally published in 1966 and was based upon the 1965 edition of the *National Electrical Code*. Over the years, this book has become a classic.

A great majority of the recommendations contained in the original edition of his book have been accepted as part of Article 250 of the *National Electrical Code*. The grounding philosophies represented in the original edition are just as relevant today as they were then. To say that Eustace contributed more than any other man to solving some of the mysteries of grounding of electrical systems would not be an overstatement of fact. Previous editions have been extensively revised both in format and in information. An effort has been made to bring this work into harmony with the 2008 edition of the *National Electrical Code* and retain the integrity of the technical information for which this work has been well known, at the same time adding additional information which may be more recent on the subject of grounding and bonding.

IAEI acquired the copyright to Soares' book in 1981 and published the second edition under the title *Soares Grounding Electrical Distribution Systems for Safety*. IAEI acknowledges the contributions of Wilford I. Summers to editions two and three and J. Philip Simmons as the principal contributor in the development of the fourth through seventh editions. IAEI acknowledges Michael J. Johnston, director of education, codes and standards, as the principal contributor in the development of the eighth, ninth, and tenth editions.

IAEI intends to revise this work to complement each new edition of the *National Electrical Code* so this will be an on-going project. Any suggestions for additional pertinent material or comments about how this work could be improved upon would be most welcome.

General Fundamentals

From the beginning, the use of electricity has presented many challenges ranging from how to install a safe electrical system to how to develop minimum *Code* requirements for safe electrical installations. These installations depend on several minimum requirements, many of which are covered in NFPA 70, *National Electrical Code*, Chapter 2, Wiring and Protection. Understanding the protection fundamentals and performance requirements in chapter 2 is essential for electrical installation, design, and inspection. To truly understand how and why things work as they do, one must always start with the basics. It is important that basic electrical circuits be understood, because grounding and bonding constitute an electrical circuit. The process of grounding and bonding creates safety circuits that work together and are associated with the electrical circuits and systems.

Objectives

To understand...

- Fundamentals and purpose of grounding of electrical systems
- Definitions relative to grounding
- Equipment from grounded and ungrounded systems
- Effects of electric shock hazards
- Purpose of grounding and bonding
- Short circuit vs. ground fault
- Basic design requirements for grounding electrical systems
- Circuit impedance and other characteristics
- Basic electrical circuit operation
- Ohm's Law

Chapter 1

The material in this book analyzes the how and why of these two functions of grounding and bonding and expresses their purpose in clear and concise language. It also examines grounding and bonding in virtually every article of the *Code* in addition to the major requirements of Article 250. Further, it provides information on grounding and bonding enhanced installations that exceed the minimum *NEC* requirements, such as for data processing facilities and sensitive electronic equipment installations. Chapter seventeen expands the information about those types of installations that are designed to exceed the *Code* requirements. It covers establishing an enhanced grounding electrode system or earthing system and installing feeders and branch circuits in a fashion that helps reduce the levels of electrical or electro-magnetic interference (EMI) noise on the grounding circuits. This is accomplished though insulation and isolation of the grounding circuit as it is routed to the original grounding point at the service or source.

Some definitions of electrical terms that should be understood as they relate to the performance of grounding and bonding circuits are also included in this first chapter. This book emphasizes the use of the defined terms of both the electrical field and the *NEC* in order to develop a common language of communication.

Taking the Mystery Out of Grounding

For many years the subjects of grounding and bonding have been considered the most controversial and misunderstood concepts in the *National Electrical Code*. Yet there is no real reason why these subjects should be treated as mysteries and given so many different interpretations. Probably the single most effective method for clearing up the confusion is for one to review and clearly understand the definitions of the various elements of the grounding system. In addition, these terms should be used correctly during all discussions and instruction on the subject so that everyone will have a common understanding. For example, using the term ground wire to mean an equipment grounding conductor does no more to help a person understand what specific conductor is referenced than does the use of the term vehicle when one specifically means a truck.

It is recommended that the reader carefully review the terms defined at the beginning of each chapter in order to develop a clear understanding of how those terms are used in regard to that particular aspect of the subject. Also, many of the terms associated with the overall grounding system are illustrated to give the reader a clearer understanding of their meaning.

This book is intended to assist the reader in establishing a strong understanding of the fundamentals of and reasons for the requirements of grounding and bonding to attain the highest level of electrical safety for persons and property. Appendix A provides information on the origin of concrete-encased electrodes. Appendix B provides a short history of the National Electrical Grounding Research Project. IAEI is committed to providing the highest quality information on grounding and bonding to the electrical industry and hopes that the reader benefits immensely from this volume.

Definitions of Electrical Terms

Current (Amperes). Current, measured in amperes, consists of the movement or flow of electricity. In most cases, the current of a circuit consists of the motion of electrons, negatively charged particles of electricity.

Capacitance. A capacitor basically consists of two conductors that are separated by an insulator. A capacitor stores electrical stress. Capacitive reactance is the opposition to current due to capacitance of the circuit. The Institute of Electrical and Electronics Engineers (IEEE) defines capacitance as, "The property of systems of conductors and dielectrics which permits the storage of electricity when potential difference exists between the conductors."

Impedance. The term resistance is often used to define the opposition to current in both ac and dc systems. The correct term for opposition to current in ac systems is impedance. Resistance, inductive reactance, and capacitive reactance all offer opposition to current in alternating-current circuits. The three elements are added together vectorially (phasorially), not directly. This results in the total impedance or opposition to current of an ac circuit.

Inductance. Inductance is the ability to store magnetic energy. Inductance is caused by the magnetic field of an alternating-current circuit as a result of the alternating current changing directions. This causes the magnetic lines of force that surround the conductor to rise and fall. Induction is measured as inductive reactance. As the magnetic lines of force rise and fall, they work to oppose the conductor and induce a voltage directly opposite the applied voltage. This induced voltage is called counter-electromotive force or counter EMF. Induction is the current effect of an ac circuit. Where there is an alternating magnetic field there will be induction. This induction will result in inductive reactance, which opposes the current. Impedance is measured in ohms

Resistance. Resistance is the name given to the opposition to current offered by the internal structure of the particular conductive material to the movement of electricity through it, i.e., to the maintenance of current in them. This opposition results in the conversion of electrical energy into heat.

Voltage (Electromotive Force). A volt is the unit of measure of electromotive force (EMF). It is the unit of measure whereby the tendency to establish and maintain electric currents can be measured. By international agreement 1 volt is the amount of EMF that will establish a current of 1 amp through a resistance of 1 ohm.

The Foundation of Grounding

The first and most vital element of a sound, safe structure is a solid footing or foundation on which to build the building. This foundation, usually consisting of concrete and reinforcing rods, must be adequate to support the weight of the building and provide a solid structural connection to the earth on which it sits. If the building or structure does not sit on a solid foundation, there can be continuous structural problems that might lead to unsafe conditions. Likewise, the grounding system serves as the foundation for an electrical service or distribution system supplying electrical energy to the structure. Often the grounding of a system or metal objects is referred to as earthing, being connected to the earth. When solidly grounded, the electrical system must be connected to a dependable grounding electrode or grounding electrode system. The grounding electrode(s) supports the entire grounding system and makes the earth connection. It must be effective and all grounding paths must be connected to it. This serves as the foundation of the electrical system. Chapter six covers the grounding electrodes, their functions, and their installations.

Electrical Circuitry Basics

Anyone who has been involved in the electrical field for any length of time has heard the phrase, "Electricity takes the path of least resistance." From the first-year apprentice to the seasoned veteran of the industry, the phrase is used to describe the path electrical current will take. The phrase is stated with pride, "Electricity takes the path of least resistance" or "Current takes the path of least resistance," and usually not much thought is given as to what is really going on. In reality, current will take all paths or circuits that are available. Where more than one path exists, current will divide among the paths (see figure 1-1). As we will review later, current will divide in opposite proportion to the impedance. The lower impedance path or circuit will carry more current than the higher impedance path(s). The study of grounding and bonding is vital to applying basic rules relative to this important safety element of the electrical circuit. It is important to review some basic principles and the fundamental elements of electricity and how current relates to electrical safety.

Ohm's Law in Review

Before we can have current, there needs to be a complete circuit (see figure 1-2). The amount of current in an electrical circuit depends on the characteristics of the circuit. Voltage or electromotive force (E) will cause

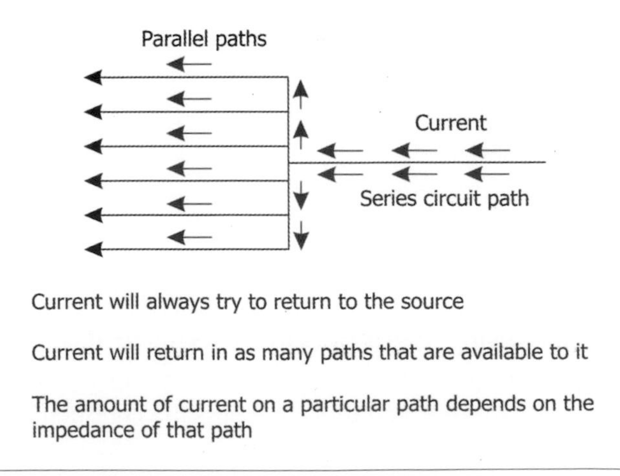

Parallel paths
Current
Series circuit path
Current will always try to return to the source
Current will return in as many paths that are available to it
The amount of current on a particular path depends on the impedance of that path

Figure 1-1. Series and parallel paths for current

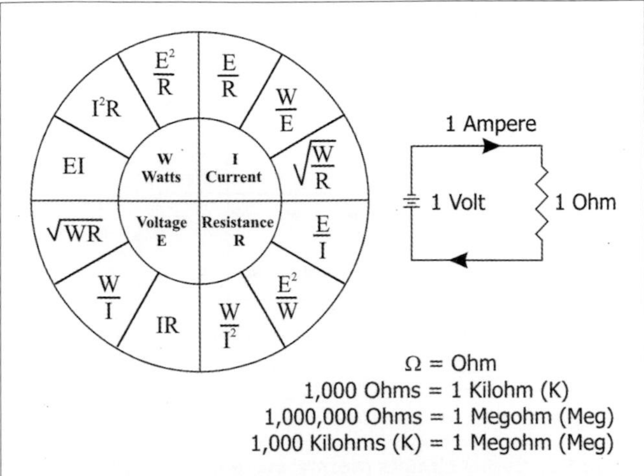

Figure 1-2. Watt's wheel—current in a circuit

(push) current or intensity (*I*) through a resistance (*R*). These are the basic components of Ohm's law (see Ohm's law and its derivatives in Watt's wheel below). Electrical current can be compared with water flowing through a water pipe. The bigger the pipe, the less the resistance is to the flow of water through the pipe. The smaller the pipe, the greater the resistance is to the flow of water through it. The same holds true for electrical current. Larger electrical conductors (paths) offer lower resistance to current. Smaller electrical conductors (paths) offer greater resistance to current. There must be a complete circuit or path and a voltage (difference of potential) or there will be no current. This is true of both normal current and fault current.

Resistance as Compared to Impedance

Understanding the differences between the pure resistance of an electric circuit and the impedance of a circuit is important in gaining a thorough understanding of the grounding or safety circuit. In Ohm's law, resistance is the total opposition to current in a dc circuit. In an alternating-current circuit, the total opposition to current is the total of three components. The impedance (*Z*) of an ac circuit is the inductance, capacitance, and the

Voltage = (E) or Pressure that pushes

Resistance = (R) or Resistance in ohms

Current = (I) or Amperes that flow

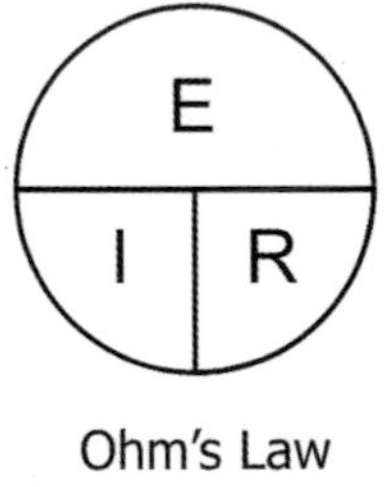

Ohm's Law

- Opposition to current in a dc circuit is *resistance*
- Opposition to current in an ac circuit is made up of three components:

R Resistance

X_L Inductive reactance

X_C Capacitive reactance

$$Z = \sqrt{R^2 + (X_L - X_C)^2}$$

Impedance formula

Impedance = Opposition (Z) or total opposition to current in an ac circuit

Figure 1-3. Basic electrical theory terms and formulas, including basic formulas for ac circuit resistance and impedance

resistance added together vectorially (phasorially) [see formula in figure 1-3]. In a 60-cycle ac circuit, alternating current changes amplitude and direction 120 times per second and develops a magnetic field that results from the inductive reactance of the circuit. Therefore, minimizing the amount of opposition to current in the grounding and bonding circuits of electrical systems is very important. These circuits can be looked upon as silent servants, just waiting to perform the important function of carrying enough current so overcurrent protective devices can operate to clear a fault.

Current in a Circuit

In any complete circuit or path that is available, current—be it normal current or fault current—will always try to return to its source. The statement on taking the path of least resistance is partially correct. Electrical current will take any and all available paths to return to its source (see figure 1-4). If several paths are available, current will divide and the resistance or the impedance of each path will determine how much current is on that particular path. It can be concluded from the above that if there is no complete circuit, there is no current. Care is given to the installation of ungrounded (phase or hot) conductors so that the circuit will be complete to provide a suitable path for current during normal operation. The same principles and fundamentals apply to the installation of grounding or safety circuits. The grounding (safety) circuit must be complete and must meet three important criteria: (1) the path for fault current must be electrically continuous; (2) it must have adequate capacity to conduct safely any fault likely to be imposed on it; and (3) it must be of low impedance

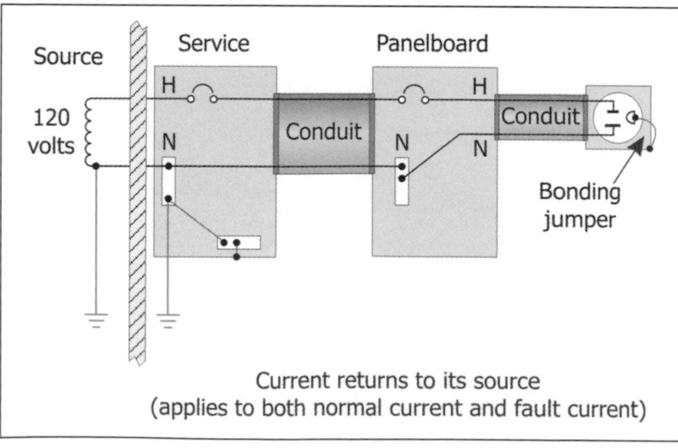

Figure 1-4. Current will try to return to its source (normal and fault current work the same way)

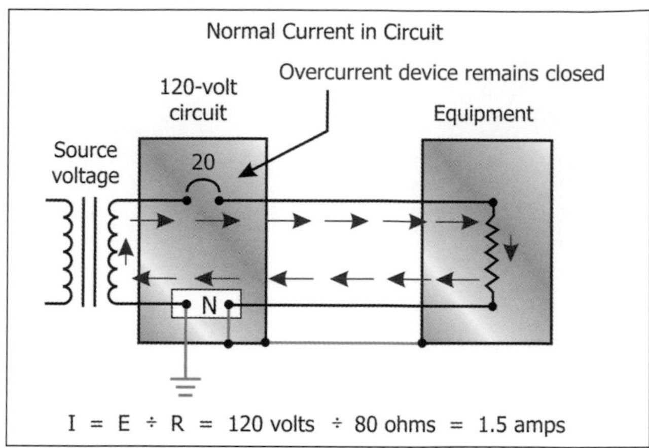

Figure 1-5. Series circuits (normal current in circuit)

(see figure 1-26 and chapter eleven for more specific information relative to clearing ground faults and short circuits).

Article 250 mentions the term low-impedance path several times. As a quick overview, the opposition to current in a dc circuit is resistance. The total opposition to current in an ac circuit is impedance, which is made up of three elements: resistance, capacitance, and inductance. When low-impedance path is used in the *Code*, it is referring to a path that offers little opposition to current whether it is normal current or fault current. The key element is low opposition or impedance.

Overcurrent Device Operation

Overcurrent devices operate because of current (amps). Generally speaking, the more current through overcurrent devices the faster they open or operate; this is because they are inverse time. The higher the impedance of the path, the lower the current through the overcurrent device. The lower the impedance of the path, the greater the current through the overcurrent device. Understanding these basic elements of electrical circuits helps one apply some important rules in Article 250. The following examples clearly demonstrate that amps operate overcurrent devices (see figures 1-5 through 1-6).

As with the electrical circuit installed for normal current, the equipment grounding (safety) circuit must also be installed for abnormal current to ensure overcurrent device operation in ground-fault conditions. The grounding or safety circuit must be complete and constructed with as little impedance as practicable for quick, sure overcurrent

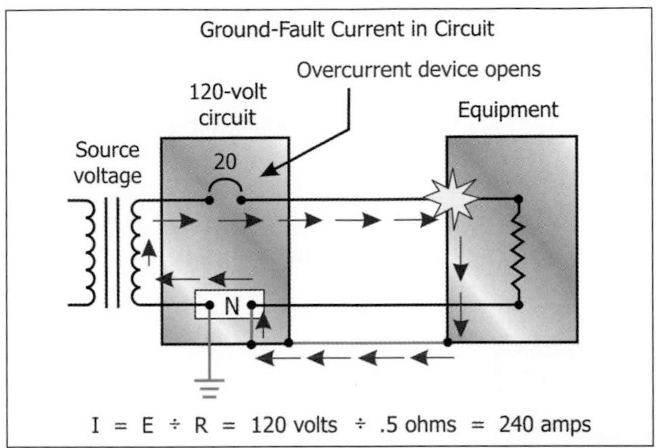

Figure 1-6. Series circuit with a high resistance in the circuit

device operation. Care must be taken when building electrical systems and circuits, including the grounding and bonding circuits of the system. Where the human body gets involved in the circuit it can, or often, results in an electrical shock or even electrocution in some cases. The human body introduces a relatively high level of resistance that impacts the overcurrent device operation. Ground-fault circuit interrupters provide protection from electrical shock, but overcurrent devices do not. Chapter 14 provides more information about ground-fault circuit interrupters.

Proper Language of Communication

A common language of communication has been established to enable one to understand the requirements of the *NEC*, in general, and of grounding and bonding, in particular. A common set of terms, defining and explaining the function of the terms as used in the *Code*, is included in Article 100 and in .2 of other articles. Two conductors of the grounding system should be mentioned and a brief story told about each: the grounded conductor and the equipment grounding conductor.

Grounded and Grounded Conductor

The grounded conductor (usually a neutral) is generally a system conductor intended to carry current during normal operation of the circuit. The grounding of the grounded (often a neutral) conductor of a system is accomplished by a connection to ground through a grounding electrode conductor either at the service or at a separately derived system. Generally, it should be understood that the grounded conductor should not be used for grounding equipment on the load side of the grounding connection at the service or source of separately derived systems. This separation between grounded conductors and equipment grounding conductors keeps the normal return current on the neutral (grounded) conductor of the system, where it belongs when returning to its source. These principles are reinforced by requirements in 110.7 and 250.24(A)(5). *Code* rules and requirements for the grounded conductors are covered in depth in chapter three of this text.

Grounding and Equipment Grounding Conductor

As used in Article 250 and other articles, grounding is a process that is ongoing. The conductor to look at is the equipment grounding conductor. The action is ongoing through every electrical enclosure it is connected to all the way to the last outlet on the branch circuit. The equipment grounding conductor provides a low-impedance path for fault-current if a ground fault should occur in the system and also connects all metal enclosures to the grounding point of the service or system.

So it is important that the equipment grounding conductor make a complete and reliable circuit back to the source. At the service is where the grounded (neutral) conductor and the equipment grounding conductor are required to be connected together through a main bonding jumper. In a separately derived system, this connection is made with a system bonding jumper installed between the grounded conductor and the equipment grounding conductor. The main bonding jumper and the system bonding jumper complete the fault-current circuit back to the source. The rules and requirements for equipment grounding conductors are covered in depth in chapter nine.

Grounding as Compared to Bonding.

Defined in Article 100, both of these functions are essential for the complete safety anticipated by the rules in Article 250 (see figure 1-9).

Ground. "The earth."

Grounded (Grounding). "Connected (connecting) to ground or to a conductive body that extends the ground connection" (see figure 1-7).

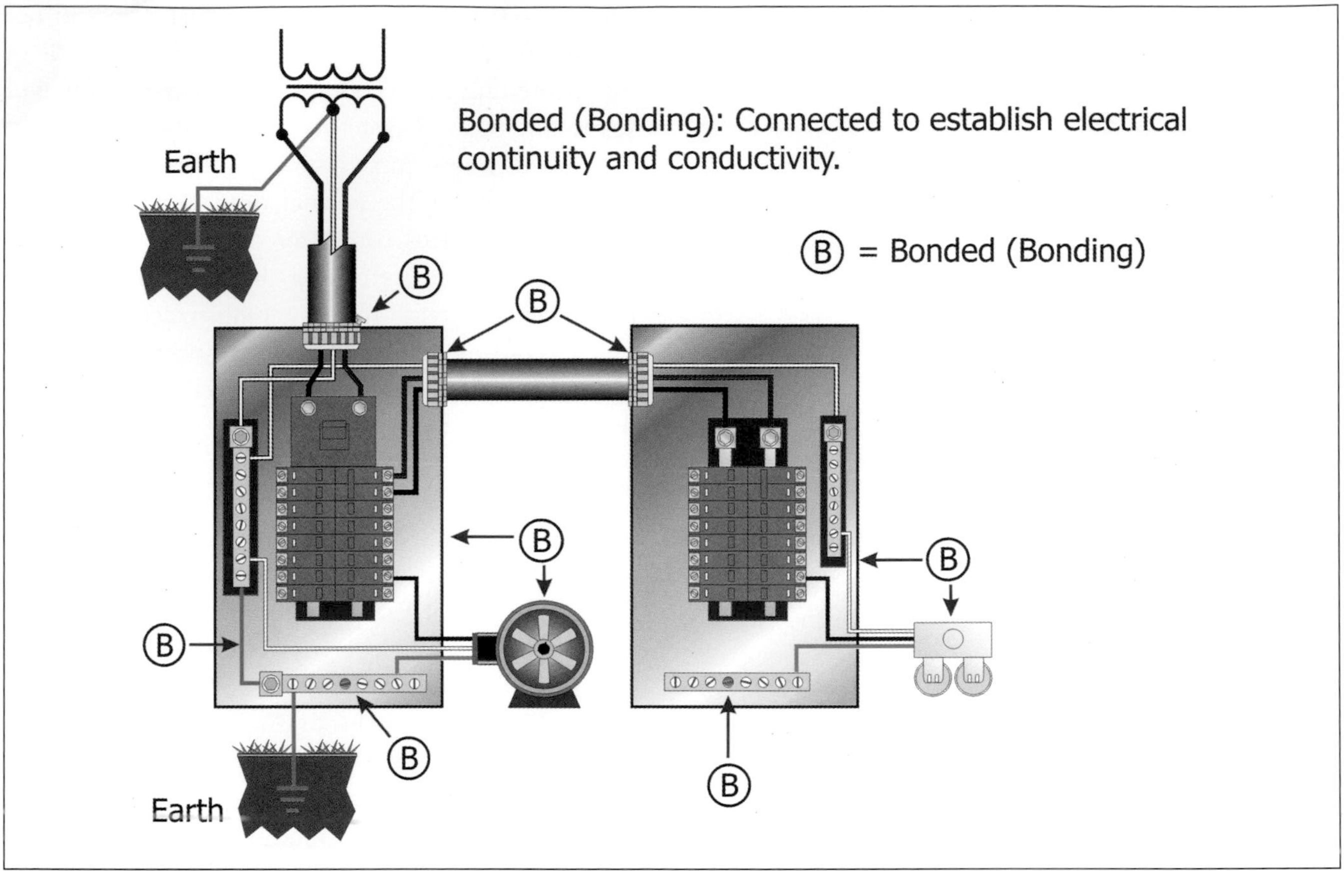

Figure 1-7. Bonding (bonded) establishes electrical continuity and conductivity.

Bonded (Bonding). "Connected to establish electrical continuity and conductivity"[1] (see figure 1-8).

These are two separate functions with two different purposes. It is important to establish a clear understanding of the grounding (earthing) circuit and its purpose as compared to the equipment grounding conductors and bonding jumpers or connections. Section 250.4 has been broken down into grounded systems and ungrounded systems.

Requirements in this section include descriptive performance requirements and establish the purposes served by each of these actions. The title of Article 250 is "Grounding and Bonding." The article contains an equally strong emphasis on bonding requirements. Chapter eight presents detailed information on these bonding requirements (see sidebar for important information about grounding and bonding terminology revisions in *NEC*-2008).

The *National Electrical Code* Trend

The *NEC* in recent cycles has been revised to reduce the use of the grounded conductor for grounding equipment downstream from the main bonding jumper in a service, or downstream from the system bonding jumper at a separately derived system. As stated earlier, the reasons are elementary. Current, be it normal current or fault current, will take all the paths available to it to try to

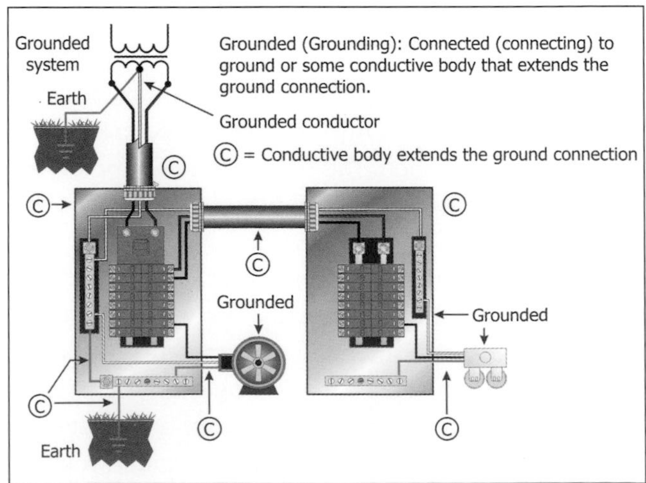

Figure 1-8. Grounding connects equipment and systems to ground (the earth).

return to its source. If the grounded conductor (neutral) and equipment grounding conductors are connected at points downstream of the service or separately derived system, such as at sub-panels, multiple paths will be available on which the current will try to return to the source. This can lead to normal neutral current on water piping systems, conduit, equipment grounding conductors, and any other electrically conductive paths and can compromise electrical safety and even overcurrent device operation in ground-fault conditions.

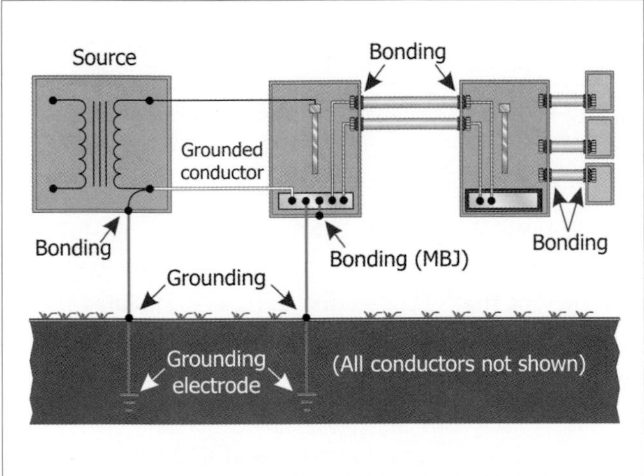

Figure 1-9. Grounding compared to bonding showing the connection to earth at the source (utility) and service and everything bonded to that point of grounding

In recent editions of the *NEC* (1996), electric range and dryer circuits were required to include an equipment grounding conductor in addition to an insulated grounded conductor. Existing range and dryer circuits are allowed to continue the use of the grounded conductor, or neutral, to ground the boxes at the outlet and the frames of the equipment. New installations, however, are required to maintain isolation (insulation) between the grounded conductor and the equipment grounding conductor.

The rules covering the use of the grounded conductor for grounding purposes at a second building or structure are provided in Section 250.32. Section 250.32(B) requires an equipment grounding conductor to be installed with the feeder supplying the second building or structure; separation between the grounded (neutral) conductor is to be maintained. There is an allowance in 250.32(B) Exception, for existing installations only, to utilize the grounded conductor of the feeder for grounding equipment under three specific

Bonding and Grounding Terminology

IAEI's *Soares Book on Grounding and Bonding* places a huge emphasis on definitions of words and terms for proper application of *Code* rules relating to the subject of grounding and bonding. Using a common language of communication is imperative to understanding this subject, and applying the *Code* to installations and systems in the field as clearly indicated in chapter one of this book. It is important that words and terms related to this subject mean what they imply by definition for all *Code* users.

NEC Grounding and Bonding Revisions

There were numerous revisions in the *NEC*-2008 development process; many related to improvements in *Code*-defined grounding and bonding terms. These revisions were the result of significant efforts of a special task group assigned by the NEC Technical Correlating Committee. The primary charter of this working group was to ensure accuracy of defined terms related to grounding and bonding, differentiate between the two concepts, and verify the use of these terms is uniform and consistent throughout the *NEC*. The work of this group resulted in changing simply the meaning of defined grounding and bonding terms to improve clarity and usability within the *NEC* requirements where they are used. *Code* rules that use defined grounding and bonding terms were revised as needed to clarify the meaning of the rule and ensure that these terms are used consistent with how they are defined in Article 100. In many instances, rules have become more prescriptive for users to provide clear direction on what is intended to be accomplished from a performance standpoint. For example, in many rules throughout the *Code* the phrase "shall be grounded" has been replaced with the phrase "shall be connected to an equipment grounding conductor." See IAEI's *Analysis of Changes NEC-2008* for more complete information and details about the changes related to defined grounding and bonding terms and the revisions to the rules in which they are used.

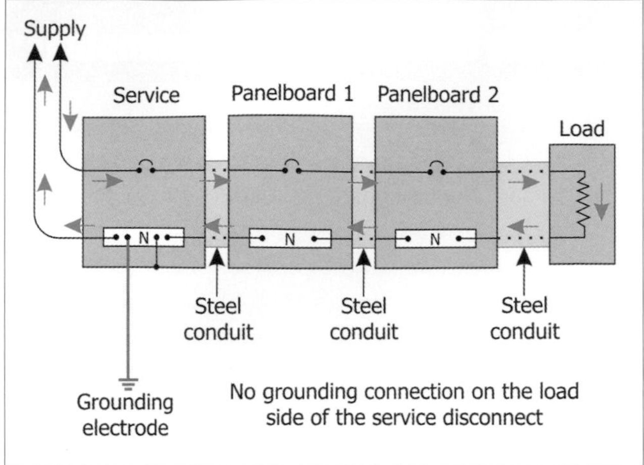

Figure 1-10. Proper path for current over grounded conductor returning to the source (correct)

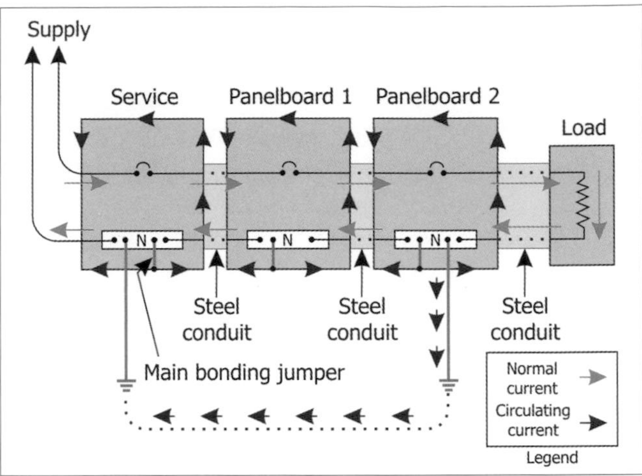

Figure 1-11. Current taking multiple paths trying to return to the source (incorrect)

and very restrictive conditions. First, an equipment grounding conductor is not included with the feeder supplying the building or structure. Second, there are no continuous metallic paths bonded to the grounding system in both buildings. Third, there is no ground-fault protection of equipment installed at the service. If all of these conditions are met at exisitng installations only, the grounded conductor may be used for grounding and must be connected to the building or structure disconnecting means.

In these cases, the grounded conductor is also required to be connected to a grounding electrode at the building or structure and installed in accordance with Part III of Article 250. This will serve as the grounding means and as the path for both normal current and fault current to clear overcurrent devices. Using the exception to 250.32(B) requires that there be no continuous metallic paths bonded to the grounding system in each structure. This requirement encompasses all paths, not just wires or conduits. These paths could include items such as metal water pipes, other metal piping, steel members, and paths such as the shielding on a communications cable or a coaxial cable installed between the structures. If this connection was made and the service was equipped with ground-fault protection in accordance with 230.95, these connections could desensitize the GFP device, possibly preventing it from operating properly when a ground-fault condition occurred because of multiple paths for current.

Current on the Proper Path

It is important that the basic elements of current be understood and carefully considered while applying the rules of the *NEC*. Section 250.24(A)(5) states that a grounding connection to any grounded circuit conductor on the load side of the service disconnecting means shall not be made, unless otherwise permitted in the article (see figure 1-10). The fine print note (FPN) refers to three situations where this is acceptable but also restrictive: for separately derived systems in 250.30(A); for separate buildings or structures in 250.32; and for grounding equipment under the limitations of 250.142.

Installers and inspectors should be watchful to ensure there are no neutral-to-ground connections on the load side of the grounding connections at either a service disconnecting means or source of a separately derived system. In other words, isolate (insulate) the neutral conductors and equipment grounding conductor connections (see figures 1-10 and 1-11). Give current, be it fault current or normal current, the low-impedance path anticipated by the requirements of the *NEC*.

Grounding and bonding must be effective and include the following vital characteristics:

1. Be a continuous path
2. Have adequate capacity for the maximum fault current likely to be imposed
3. Provide a fault-current path of low impedance

The definitions of 250.2 describe three important terms used in Article 250. These three terms are:

Effective Ground-Fault Current Path. "An intentionally

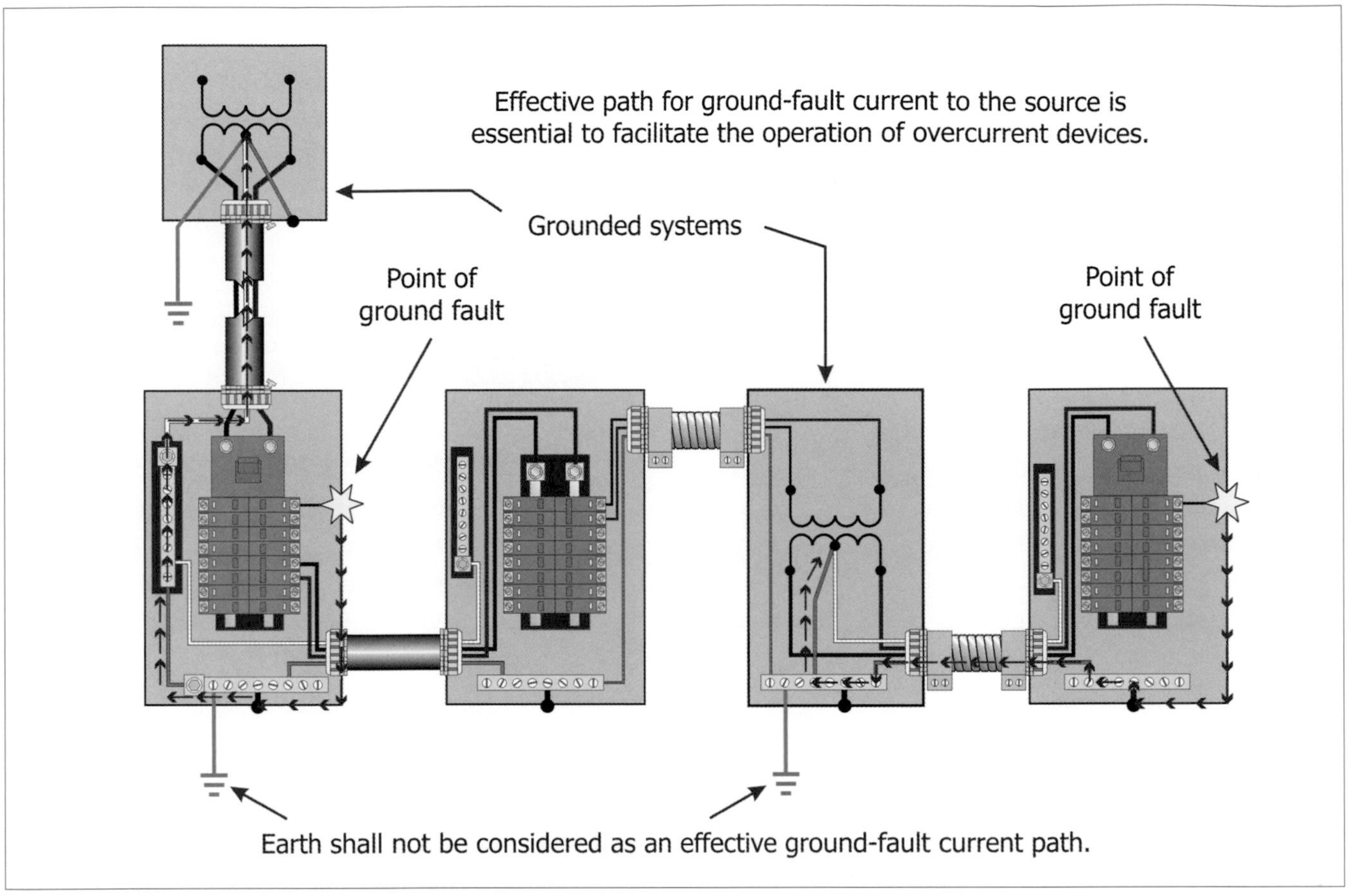

Effective path for ground-fault current to the source is essential to facilitate the operation of overcurrent devices.

Grounded systems

Point of ground fault

Point of ground fault

Earth shall not be considered as an effective ground-fault current path.

Figure 1-12. Effective ground-fault current path

constructed, low-impedance electrically conductive path designed and intended to carry current under ground-fault conditions from the point of a ground fault on a wiring system to the electrical supply source and that facilitates the operation of the overcurrent device or ground fault detectors on high impedance grounded systems" (see figure 1-12).

Ground Fault. "An unintentional, electrically conducting connection between an ungrounded conductor of an electrical circuit and the normally non-current-carrying conductors, metallic enclosures, metallic raceways, metallic equipment, or earth."

Ground-Fault Current Path. "An electrically conductive path from the point of a ground fault on a wiring system through normally non-current-carrying conductors, equipment, or the earth to the electrical supply source.[2]

Overview of NFPA 70 Article 250

Proper grounding and bonding provide protection from electric shock hazards and facilitate operation of overcurrent devices to clear faulted circuits. Fast and

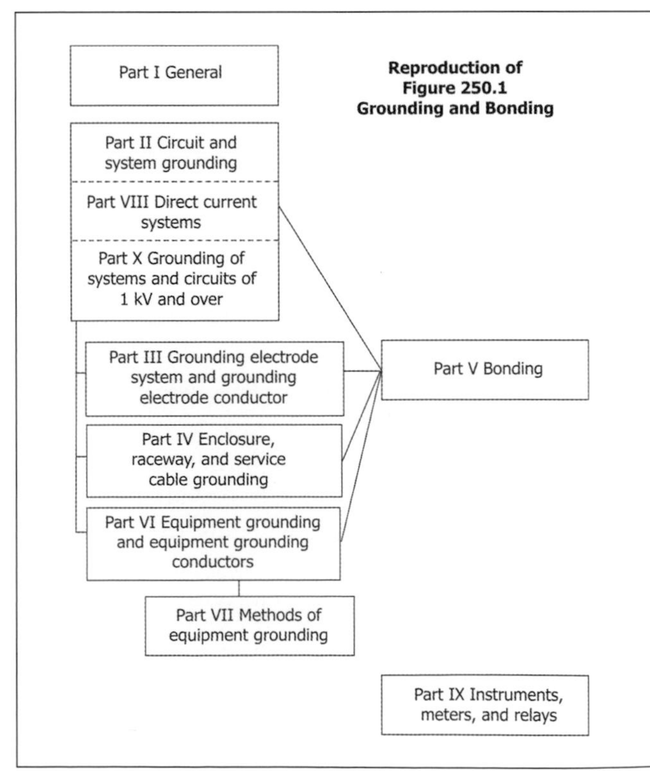

Figure 1-13. A reproduction of Figure 250.1 in the *NEC*

effective operation of overcurrent devices, covered in Article 240, depends on the proper bonding and effective fault-current paths required by Article 250. Article 250 is located in Chapter 2, Wiring and Protection, of the *NEC* and is made up of ten parts. Part I includes general requirements and some important performance requirements vital to the understanding and proper application of prescriptive requirements found in latter parts of the article.

The parts of Article 250 are very much interlocking; that is, they must be used together in many instances to properly apply the rules. Figure 250.1 is a built-in map for Article 250, which serves to enhance the usability issues in the *Code*. Figure 1-13 assists the reader in understanding the arrangement of the article at a glance.

Article 250 includes general requirements for grounding and bonding of electrical installations and six specific requirements.

"1. Systems, circuits, and equipment required, permitted, or not permitted to be grounded
2. Circuit conductor to be grounded on grounded systems
3. Location of grounding connections
4. Types and sizes of grounding and bonding conductors and electrodes
5. Methods of grounding and bonding
6. Conditions under which guards, isolation, or insulation may be substituted for grounding"[3]

Alternatives to Grounding

Item six in 250.1 indicates that three conditions—guards, isolation, or insulation—may be substituted for grounding under certain conditions. Definitions of these terms, found in Article 100, help clarify the intent of the alternatives.

Guarded. "Covered, shielded, fenced, enclosed, or otherwise protected by means of suitable covers, casings, barriers, rails, screens, mats, or platforms to remove the likelihood of approach or contact by persons or objects to a point of danger."

Isolated (as applied to location). "Not readily accessible to persons unless special means for access are used."

Insulated. "Equipment or materials that are covered with an insulating material."

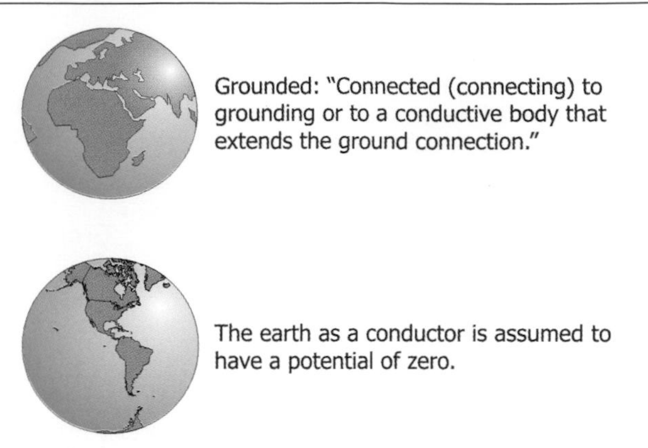

Figure 1-14. Grounded means connected to ground (earth).

The insulation must have sufficient dielectric strength to provide protection against electrically conductive parts. Power tools are an example of an insulation that is likely to become energized and present shock hazards to personnel. Double insulated is permitted to be used as a substitute for grounding.[4]

Grounding of Electrical Systems

Some features of electrical safety are so fundamental they have appeared in some form in every edition of the *National Electrical Code*. These include requirements for suitable insulation for conductors; overcurrent protection for circuits; and grounding of electrical systems and equipment for safety. The grounding of equipment and conductor enclosures, as well as the grounding of one conductor of an electric power and light system, has been practiced in some areas since the use of electricity began.

At first, there was no uniform standard for grounding. However, it was not long before it became universal to ground one conductor on all 120-volt lighting circuits. Early editions of the *Code* firmly established the practice of grounding by making it mandatory to ground all such systems where the system can be grounded so that the voltage to ground does not exceed 150 volts. These editions also recommended that alternating-current systems be grounded where the system voltage to ground did not exceed 300 volts, while at the same time stating that higher voltage systems were permitted to be ungrounded. Later, the *Code* made it mandatory to ground any system having a nominal voltage to ground of not more than 300 volts

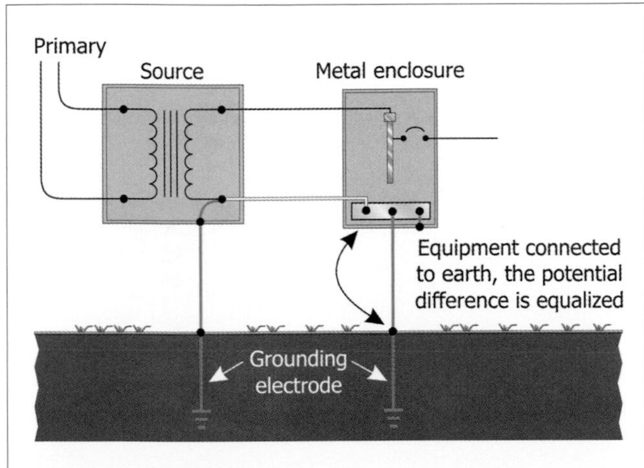

Figure 1-15. Grounding fundamentals showing proper grounding and bonding

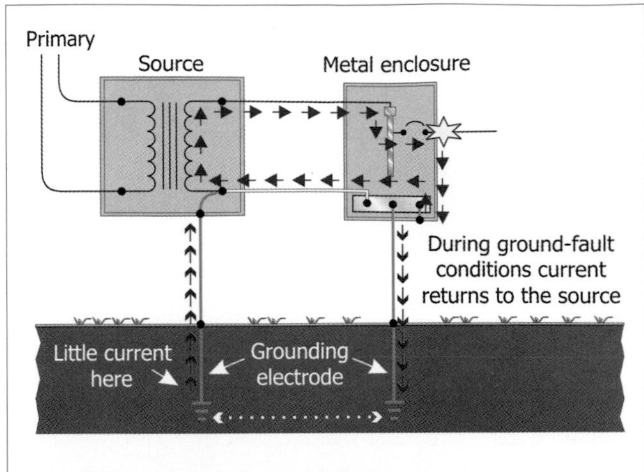

Figure 1-16. Grounding fundamentals and current paths

if the grounded service conductor (usually a neutral) was not insulated.

Grounded (Grounding)

The term *grounded* (*gounding*) is defined in Article 100 as, "Connected (connecting) to ground or to a conductive body that extends the ground connection"[5] (see figure 1-14). Conductive bodies that extend the ground connection include conduits, boxes, enclosures, equipment grounding conductors and wiring devices. These are, in fact, an extension of the ground (earth) by being electrically connected to the earth by reliable electrical and mechanical means. Metal frames of buildings that have an established connection to earth are acceptable as an electrode or extension of the earth but are not permitted to be used as an equipment grounding conductor.

The earth as a whole is properly classed as a conductor. For convenience, its electric potential is assumed to be zero. Based upon the composition of the soil or earth, the resistance of segments of the earth can vary widely from one area to another. The earth is composed of many different materials, some of which, especially when dry, are very poor conductors of electricity. Soil temperature, moisture content and chemical composition are factors that have a great influence on soil resistance. As a result of these factors, the capability of the earth to carry electrical current also varies widely (see chapter six for more information on composition of the earth and effectiveness of grounding electrodes).

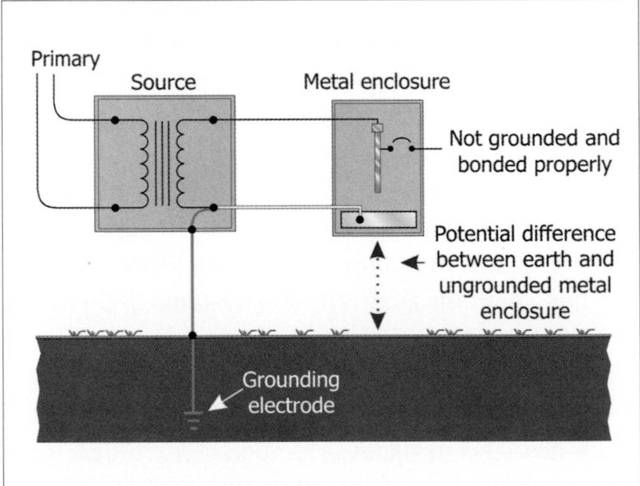

Figure 1-17. Potential hazard is present when systems and equipment are not properly grounded.

A metal object, such as a box or other equipment enclosure, that is grounded by connecting (bonding) it to the ground (earth) by means of a grounding electrode, grounding electrode conductor and/or equipment grounding conductor is thereby forced theoretically to take the same zero potential as the earth (see figures 1-15, 1-16, and 1-17). Slight differences in potential can exist due to differences in impedance of the materials or connections. Any attempt to raise or lower the potential of the grounded object results in current passing through the grounding path until the potential (voltage) of the object and the potential (voltage) of the earth (zero) are equalized. Usually, this potential above ground is caused by a line-to-ground fault. Hence, grounding is a means for ensuring that the grounded object cannot take on a potential differing enough from earth potential to be hazardous. When the equipment grounding conductor or grounding electrode conductor is broken, is inadequate in size, or has a poor connection, a hazardous, aboveground potential can be present from an abnormal condition.

Resistance and Impedance

Though a comprehensive discussion of the subject is beyond the scope of this text, a brief explanation of the terms resistance and impedance is offered.

For direct current (dc) systems and circuits, resistance (R) is properly used to describe the opposition to current. We are accustomed to using the term ohms to relate to the resistance of the circuit, such as 35 ohms. Ohm's Law can be summarized as follows: in a dc circuit, the current is directly proportional to the voltage and inversely proportional to the resistance (I = E/R). This means, as the voltage is increased, the current will increase through a fixed resistance. As the resistance is reduced, the current will increase if the voltage stays the same. A pressure (voltage) of one volt will cause one ampere of current through a resistance of one ohm.

For alternating-current (ac) systems and circuits, impedance (Z) is the proper term to describe the total opposition to current. Impedance consists of three components: inductive reactance, capacitive reactance and resistance. Impedance, rather than resistance, is used most often throughout this text since, for the most part, ac electrical systems and circuits are being considered. See chapter eleven for a detailed discussion of the importance of keeping all conductors of the circuit, including the equipment grounding conductor, close together to keep the impedance as low as possible. Also, many excellent books on electrical theory cover the subject of impedance, inductive reactance, capacitive reactance and resistance in detail.

Equipment Supplied from a Grounded System

Where the electrical system is grounded, it is critical to provide a low-impedance path with adequate capacity from all the equipment supplied by the system back to the source of the system: (1) to maintain the metal equipment enclosures as close to earth potential as possible to reduce a shock hazard; and (2) to ensure the overcurrent protection will operate in the event of a line-to-ground fault. This is required by 250.4 and is emphasized in several portions of this text.

E = Voltage R = Ohms I = Amperes

Voltage	=	*Resistance	x	**Current
120	=	60	x	2
120	=	40	x	3
120	=	20	x	6
120	=	10	x	12
120	=	5	x	24

* As the resistance or opposition to current increases, the current in the circuit decreases

** As the resistance or opposition to current decreases, the current in the circuit increases

Figure 1-18. Proper grounding and bonding facilitates the operation of overcurrent protective devices.

Electrical equipment that is left ungrounded or is poorly grounded but is supplied by a grounded system becomes a silent and often lethal source of electrical shock when ground fault occurs. Another hazard of equal significance can occur where two pieces of electrical equipment supplied from a grounded system are within reach of a person. If one piece is not grounded or is poorly grounded and becomes energized through a failure of the insulation system (ground fault), the person making contact with the two (or more) pieces of equipment becomes the circuit (path) for current to pass through as it tries to find its way back to its source. In some cases, the person will receive a mild shock. In other cases, the shock can be fatal. Even though the impedance of the human body can be relatively high, approximately 200 milliamp (mA) of current through it can be lethal. A milliamp is equal to one thousandth of an ampere (see figure 1-20).

Equipment Supplied from an Ungrounded System

Section 250.21 permits some electrical systems to be operated ungrounded. In this case, the electrical enclosures for the service, feeders, circuits and other equipment are connected to ground but the system itself is not grounded. In many parts of the country, the serving utility will provide only a grounded electrical system. From a practical standpoint, an ungrounded system exists only in theory or at the distribution transformers hanging on nonmetallic

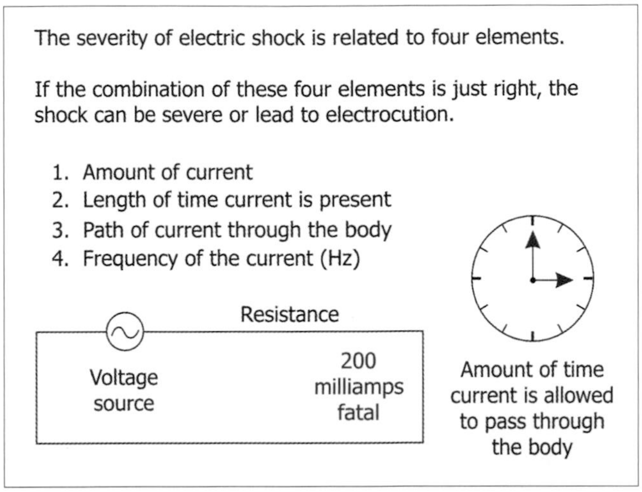

The severity of electric shock is related to four elements.

If the combination of these four elements is just right, the shock can be severe or lead to electrocution.

1. Amount of current
2. Length of time current is present
3. Path of current through the body
4. Frequency of the current (Hz)

Resistance

Voltage source

200 milliamps fatal

Amount of time current is allowed to pass through the body

Figure 1-19. Severity of electric shock is related to four elements.

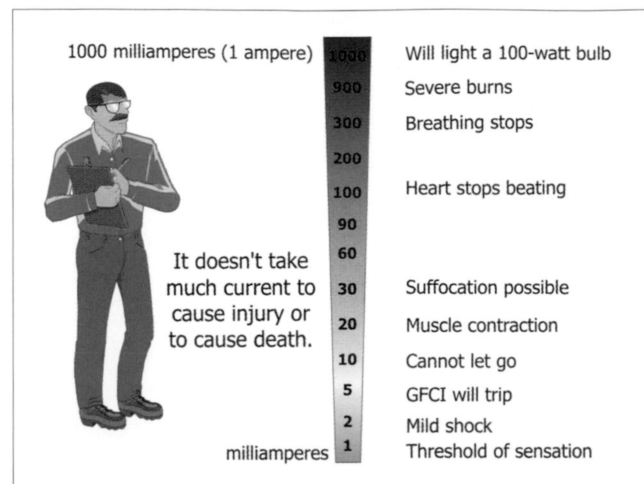

Figure 1-20. Level (in milliamperes) of current through the body

poles before the system is connected to the plant electrical system. Where the ungrounded system conductors are installed in grounded metal raceways or enclosures or connected to motors, the frame of which are grounded, the ungrounded system becomes capacitively coupled to ground (see chapter two for a more detailed discussion of this subject).

An ungrounded electrical system often becomes accidentally grounded through a line-to-ground fault. This essentially creates a poorly grounded system. The other phases of the system then rise to a voltage to ground equal to the system voltage. For example, in a 480-volt, ungrounded system, if one of the phases becomes grounded anywhere on the system, the other phases will have a voltage to ground of approximately 480 volts. Obviously, this becomes a greater shock hazard to personnel who may be servicing the installation and adds greater stress on the conductor insulation, including transformer and motor windings.

Effect of Electricity on Humans

Like the bird on the electric wire, the human body is immune to electric shock as long as it is not a part of the electric circuit. The easiest way to avoid danger from electric shock is to keep one's body from becoming a part of the electric circuit. Due to the very common use of electric tools, equipment and appliances, the risk of being exposed to electric shock is multiplied by the numbers of these items the person is exposed to.

When a person becomes a path for electricity, he or she will experience an electrical shock. The intensity and damage done to the person by the shock will be determined by the current level (the amount through the person), how long the current exists (the duration), the person's size, the pathway the current takes through the body, and the circuit frequency (see figure 1-19). Speaking in electrical terms, the person can be thought of as a resistor or impedance in the circuit. The skin of the human body can be thought of as insulation; however, its resistance is rather low and varies depending on moisture. The inside of the human body is a relatively good conductor as it is composed of primarily fluids and salt. Once electrical current has entered the body through the skin, the resistance is very low, and severe damage

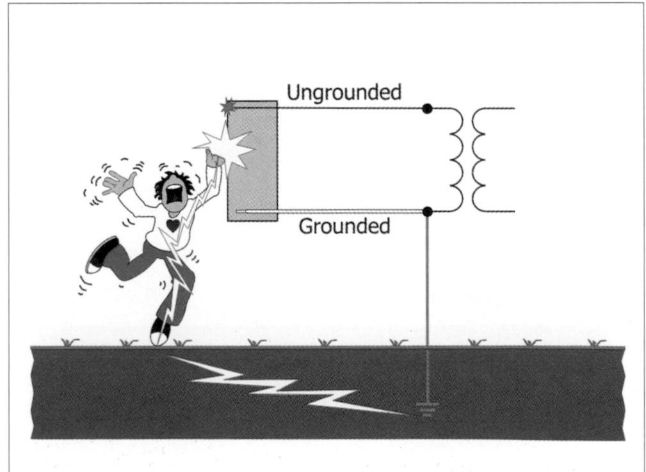

Figure 1-21. Human completing the path for current through the earth

or death is likely. The severity of the electrical shock is related to the amount of current through the body and its duration.

A person can become a pathway to ground or between conductors in one of two ways: in a series or a parallel circuit. In a series circuit, the person is the only path through which the current will attempt to return to its source. An example of this is when a person comes into contact with an ungrounded electrical appliance that is energized through a line-to-ground fault and at the same time touches another grounded appliance like an electric range or a grounded kitchen sink (see figure 1-21). In general terms, the amount of current that passes through the person's body is determined by Ohm's Law by the voltage of the circuit and the resistance offered by the person's body (see chapter fourteen for a greater discussion of current and the human body.)

In a parallel circuit, the human body and another path such as an equipment grounding conductor provide a path for current at the same time. The current will divide among the paths based on the impedance of those individual paths. The greater the impedance of the path, the less current is in that path (see figure 1-22).

The smallest risk of shock hazard occurs where the circuit protective device, usually a fuse or circuit breaker, opens the faulted circuit immediately. A factor is the time it takes for the overcurrent device to clear the fault from the circuit. During this time, the equipment subject to the ground-fault condition will have a potential above ground depending upon the voltage drop of the equipment grounding circuit.

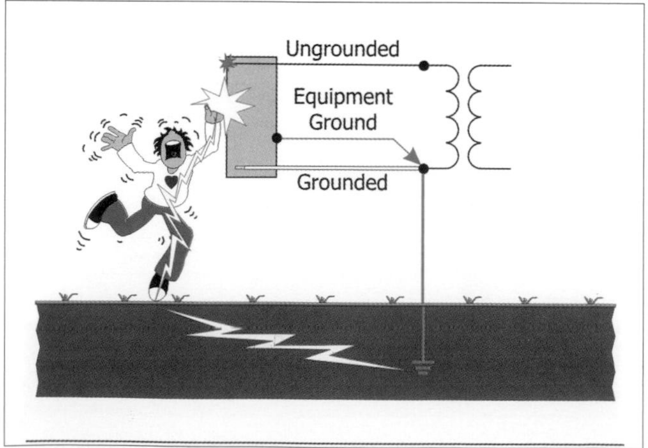

Figure 1-22. Human in parallel with equipment grounding conductor during ground fault

In some installations, the equipment may be grounded but through a high impedance. This can be the result of an equipment grounding conductor that is too small or too long for its size; the connection has become loose; or it is routed improperly. In this situation, there is usually not enough current through the equipment grounding conductor to cause the overcurrent protective device to operate and clear the faulted equipment. Part of the fault current will then be through the person in contact with the energized equipment and part through the equipment grounding conductor path. The current will divide in opposite proportions to the resistance or impedance of the paths. The greatest amount of current will be through the path providing the lowest resistance or impedance. Even though the human body can have impedance that is greater than the other path(s), it only takes a small amount of current to cause serious injury or death (see figures 1-20 and 1-22).

The primary emphasis is to provide a path for ground-fault current that is permanent and continuous, is adequately sized and of low-impedance from all electrical equipment back to the source to facilitate the operation of the overcurrent device in a reasonable time. Protection of persons from lethal levels of electric shock is the reason ground-fault circuit devices (GFCIs) have become so popular over the recent years.

Burns and Other Injuries

The most common shock-related injury is a burn. Burns suffered in electrical accidents can be of three types: electrical burns, arc burns, and thermal contact burns.

Electrical burns are the result of the current through tissues, blood vessels or bone. Tissue damage is caused by heat generated by the current through the body and is often immediately classified as a third degree burn. In many cases, the damage caused to the tissues becomes more apparent in the days or even months following the incident. Severe permanent damage can be caused to internal tissues or organs with little external indication until some time after the electric shock. Burns from electric shock are one of the most serious injuries that can be experienced and should be given immediate medical attention.

Arc or flash burns, on the other hand, are the result of high temperatures produced by an electric arc or explosion near the body. These burns can be of the more minor first-degree type, more severe second-degree, or most severe third-degree burns. They should also be attended to promptly and properly.

Finally, thermal contact burns are those normally experienced when the skin comes into contact with hot surfaces of overheated electric conductors, conduits, or other equipment. Additionally, clothing can be ignited in an electrical arc or explosion and a thermal burn will result. All three types of burns can be produced simultaneously.

The proper use of work procedures and personal protective equipment can minimize these injuries. [See NFPA 70E-2004 Standard for Electrical Safety in the Workplace for information about electrical safety requirements and safe working practice and procedures for employees in workplaces.]

Electric shock can also cause injuries of an indirect or secondary nature in which involuntary muscle reaction from the electric shock can cause bruises, bone fractures, and even death resulting from collisions or falls. In some cases, injuries caused by electric shock can contribute to delayed fatalities.

In addition to shock and burn hazards, electricity poses other dangers. For example, when an arcing type short circuit or ground fault occurs, hazards can be created from the resulting arcs. If high current is involved, these arcs can cause injury or start a fire. Extremely high-energy arcs can damage equipment, causing fragmented metal to fly in all directions, or melt steel, copper or aluminum. Even low-energy sparks can cause violent explosions in atmospheres that contain flammable gases, vapors, or combustible dusts.

Effect of Electricity on Animals

Some animals are especially sensitive to electricity. For example, past studies claimed that dairy cattle are so sensitive to electricity that a potential of as little as two volts between conductive portions of floors, walls, piping and stanchions caused behavior problems that resulted in loss of production. More recent studies claim the voltage difference to be about four volts. In other cases, severe health problems are attributed to the effects of electricity, which, if not corrected and treated can lead to death of the animal. (See chapter sixteen for methods of preventing and minimizing these problems in agricultural structures.)

Purpose of Grounding and Bonding

The general requirements for grounding and bonding are contained in 250.4. The requirements include the grounding and bonding performance requirements for grounded systems and ungrounded systems as follows:

(A) Grounded System
(1) Electrical System Grounding
(2) Grounding of Electrical Equipment
(3) Bonding of Electrical Equipment
(4) Bonding of Electrically Conductive Materials and Other Equipment
(5) Effective Ground-Fault Current Path

(B) Ungrounded Systems
(1) Grounding of Electrical Equipment
(2) Bonding of Electrical Equipment
(3) Bonding of Electrically Conductive Materials and Other Equipment
(4) Path for Fault Current

The requirements for grounded systems and ungrounded systems and the differences between them are covered in detail in chapter two of this text.

Grounding of Electrical Systems

Systems are solidly grounded to limit the voltage to ground during normal operation and to prevent excessive voltages due to lightning, line surges or unintentional contact with higher-voltage lines and to stabilize the voltage to ground during normal operation [see 250.4(A)(1)]. Several methods of grounding electrical systems are used depending on *National Electrical Code* requirements and system design and function. These methods of grounding electrical systems are covered in detail in chapter three.

Grounding of Electrical Equipment

Conductive materials enclosing electrical conductors or equipment, or that are a part of the equipment are grounded to limit the voltage to ground on these materials and bonded to facilitate the operation of overcurrent devices under ground-fault conditions [see 250.4(A)(2) and figures 1-23 and 1-24]. Where the electrical system is grounded, the equipment grounding conductor is connected to the grounded system conductor (often a neutral) at the service or the source of a separately derived system. Where

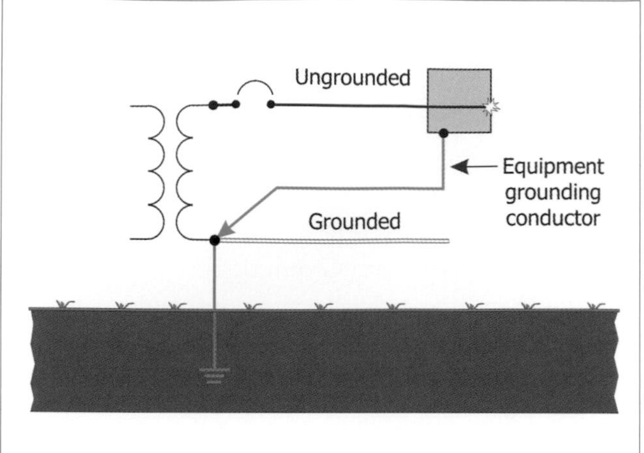

Figure 1-23. Purpose of the equipment grounding conductor

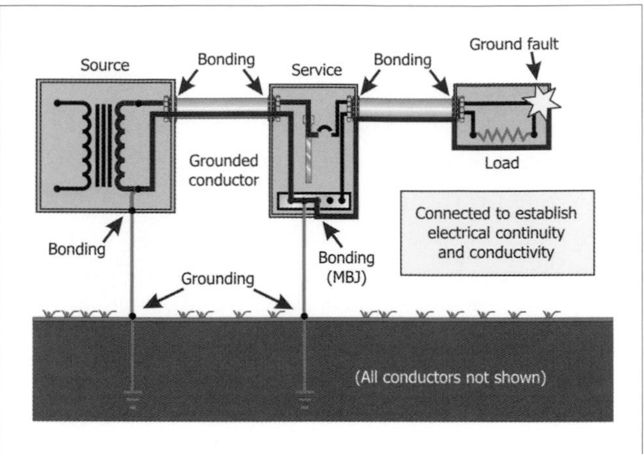

Figure 1-24. Purpose of bonding equipment and enclosures

the electrical system is not grounded, the electrical equipment is connected to earth at the service to maintain the equipment at or near earth potential and the equipment is bonded together to provide a path for fault current. This occurs where a second ground fault occurs before the first one is cleared.

Bonding of Electrically Conductive Materials and Other Equipment

Electrically conductive materials, such as metal water piping, metal gas piping and structural steel members, that are likely to become energized are bonded to provide a low-impedance path for clearing ground faults that otherwise would energize the equipment at a level above earth potential. For systems that are grounded, the equipment is bonded to the grounded system conductor (often a neutral) at the service or source of a separately derived system. Where the electrical system is not grounded, the electrical equipment is connected to earth at the service to maintain the equipment at or near earth potential and the equipment is bonded together to provide a path for fault current. Again, this occurs where a second ground fault occurs before the first one is cleared [see 250.4(A)(4)]. Chapter 2 provides more detail about ungrounded electrical systems.

Effective Ground-Fault Current Path

Section 250.4(A)(5), Effective Ground-Fault Current Path, provides requirements relative to one of the most critical elements of the grounding safety system. This section requires that the fault-current path: (1) be electrically continuous; (2) be capable of safely carrying the maximum fault current likely to be imposed on it; and (3) have sufficiently low impedance to facilitate the operation of the overcurrent devices under fault conditions.

Each of these points is important and worthy of discussion. First, a permanent, reliable and electrically continuous grounding system is vital to the overall safety of the electrical system. This includes a stable voltage reference and provides an effective path for fault current due to abnormal conditions. Intermittent connections are like an unpredictable earthquake, waiting to wreak havoc on the unsuspecting. The grounding system, including all connections, whether a wire, conduit, equipment enclosure, or other element of the path, must be electrically continuous, and have all connections made up tightly and in a workmanlike manner (see figure 1-25).

If the insulation system of other conductors of the system, like ungrounded (hot) or grounded circuit conductors, fail, it usually is obvious since sometimes such as a piece of equipment or an appliance stops functioning. This is not true of the equipment grounding path. Usually, a failure of the ground-fault path is not known until someone receives an electric shock. Also, the conductors of the equipment grounding conductor system normally only carry current in a fault situation. At this time, the equipment grounding conductor path will usually carry far more current than the ungrounded (hot) or grounded conductor (often a neutral) typically carries during normal conditions.

Second, providing adequate capacity for the fault current is also of paramount importance. The

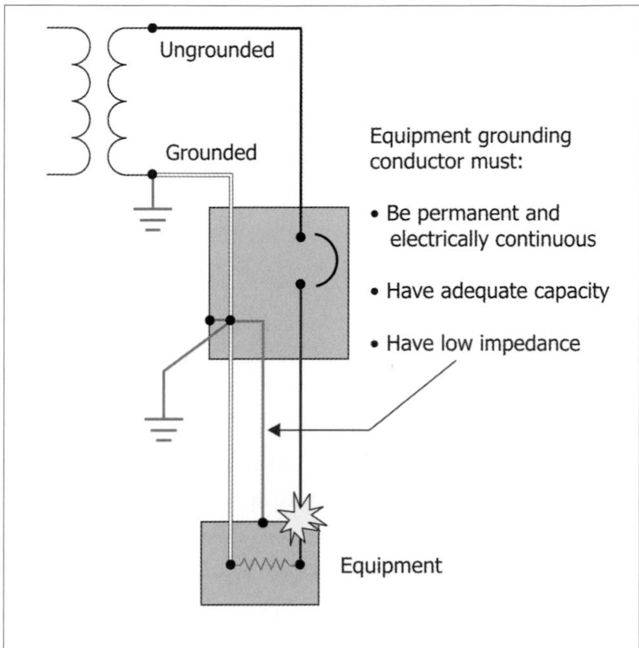
Ungrounded
Grounded
Equipment grounding
conductor must:
• Be permanent and
electrically continuous
• Have adequate capacity
• Have low impedance
Equipment

Figure 1-25. Effective ground-fault current path

minimum size of the grounded conductor, bonding conductor and equipment grounding conductor are given in several locations, many of them in Article 250. Obviously, an equipment grounding conductor or bonding jumper that burns off (fuses) due to excessive current while it is carrying out its intended purpose is of little value in the safety system. The sizes of equipment grounding conductors given in Table 250.122 are the minimum. There are cases where the withstand rating of conductors is exceeded by the available fault current and larger conductors are necessary (see chapter eleven for additional information on the subject).

Third, the ground-fault path that is effectively connected and of adequate capacity is of little value if it does not have low impedance (measured in ohms). A high-impedance path provides resistance to the circuit, thus limiting the amount of current. This allows a voltage above ground to be present on faulted equipment that can then present a shock hazard (E = I x R). The *NEC* stops short of specifying the maximum impedance acceptable for the ground-fault path, except to say that the path must have impedance that is low enough to facilitate the operation of the overcurrent protective device. Every circuit has different characteristics associated with it that contribute to the impedance of the particular circuit and, thus, the impedance value of the grounding circuit will vary, but must be kept as low as possible in all grounding circuits (see chapter eleven for additional information on the subject).

The goal of good design of the ground-fault path is to provide a permanent and adequate path of low impedance so there will be enough current in the circuit to cause a circuit breaker to trip or a fuse to open. If the opening of the breaker or fuse on the line side of the faulted circuit does not take place rapidly, thus taking the faulted equipment off the line, the grounding system will have failed to perform its critically important function. This failure can result in greater equipment damage, possibly fires or injury to personnel.

Electrical System Design

The ultimate goal of electrical system design is to prevent all faults, including ground faults, from occurring. This is accomplished by proper design, installation, operation and maintenance of electrical equipment and systems. System conductors that are ungrounded are separated from grounded conductors and grounded equipment by insulation. This insulation can be in the form of thermoplastic, thermosetting or other similar insulating material applied to wires or busbars or by separating conductors having a potential (voltage) between them. This separation is usually accomplished by mounting busbars on insulators. The air then becomes the insulation between phases and a grounded enclosure. However, it is possible to have a failure or breakdown in insulation systems in any electrical installation. Therefore, overcurrent protection and an effective ground-fault current path must be provided to safely clear line-to-ground faults that can occur. The phrase likely to become energized is included in Annex B of the *NEC Style Manual* as a standard term meaning "failure of insulation on."

Section 110.7 requires that electrical wiring installations shall be free from short circuits, ground faults, or any connections to ground other than as required or permitted in the *NEC*. Though not always practiced, it is wise to test the insulation of all installations to be certain they are free from unintended short circuits and ground faults before the wiring is energized. This can be done with some means of continuity testing, which can be a battery and a bell and leads (the origin of the term ring-out). An ohmmeter or megohm-meter (megger) can also be used. This is more than a simple continuity test. A current is passed through the conductor under test and the amount of leakage through the insulation is measured. Testing the insulation of the system before it is energized is often required in specifications prepared by architects or engineers for commercial and industrial jobs.

Also, the impedance (rather than the resistance) of the equipment ground-fault return path can be verified by testing instruments. This is especially true for branch circuits where plug-in type testers are available and greatly simplify the testing process.

Designing Electrical Systems for Safety

Electrical systems need to be designed to be certain they are adequate for the loads to be served by them. The *NEC* requires that electrical services are to be adequate for the calculated load in accordance with Article 220 [see 230.42(A)]. Methods for calculating the minimum capacity for electrical services are found in Article 220 rather than in Article 230. The calculation method for determining the minimum size of the grounded service conductor (often a neutral) is found in 220.61. Section 250.24(C) specifies the minimum sizes for grounded service conductors.

The *Code* does not require that electrical systems be designed with any additional capacity for future load expansion [see 90.1(B)]. From a practical standpoint, some additional capacity should be provided for when the system is first installed. Some designers typically plan the electrical system for at least 25 percent spare capacity.

Electrical Systems Abused

Electrical systems are intended to provide reliable service for many years. That is generally the case, provided the system is designed for the load to be carried, and the system is installed in a proper manner. Overloading the system is one of the major abuses. With the expanded use of new or more modern electrical appliances, larger electrical distribution systems are often required. It is not uncommon to find relocatable power taps, extra conductors placed under circuit breaker terminals, or splices being improperly made in attics or in crawl spaces to add extra electrical equipment to the electrical system. Often, extra fuse or circuit breaker panelboards are added to existing systems without taking proper steps to ensure that the conductors and equipment supplying the loads are adequate.

In industrial or commercial installations, additional equipment is often added to existing electrical systems. At times, this is done without consideration for whether the existing system is adequate for the additional load.

In addition, electrical systems can be exposed to overvoltage from lightning or power-line crosses, insulation failure in high-to-low voltage transformers, and to short circuits and ground faults. In some cases, electrical systems are abused until they fail. The opening of a fuse or tripping of a circuit breaker often is the first indication of a failure in the electrical system. These events should not be treated as a nuisance tripping situation.

Major Problems in Electrical Systems

The major cause of trouble in an electrical distribution system is insulation failure. The insulation can be air, such as in busways, switchboards and motor control centers, where clearance between uninsulated busbars and grounded metallic electrical equipment is maintained by air space. The clearance is maintained by insulators in the form of rubber, ceramic or thermoplastic, or insulation that is applied directly to the conductors. Insulation for wire or busbars is usually rubber, thermoplastic or thermosetting material.

Insulation failure can result in two kinds of faults: line-to-line or line-to-ground. The least likely failure is line-to-line or between any two conductors of the system, that is, from one phase conductor to another or from one phase conductor to the neutral or grounded conductor. Experience has shown that most insulation failures (as high as 80 percent or more) are line-to-ground faults or from one phase conductor to the conductor enclosure or equipment.

While the entire *National Electrical Code* and other safety electrical codes, like the Canadian Electrical *Code*, are developed and updated to provide protection against electrical fires and shocks, they are not design specifications. But an electrical design cannot be complete without using the applicable electrical code and adding specific details for the particular application. By following the rules of these codes, one will have an installation that is essentially free from hazard, but not necessarily efficient, convenient, or adequate for good service or future expansion (see 90.1).

Insulation Resistance

Previous editions of the *Code* (e.g., up to the 1965 *NEC*) contained recommended values or results for testing insulation resistance. It was found that the resistance values were incomplete and not sufficiently accurate for use in modern installations, and the recommendation was to delete them from the *Code*. However, basic knowledge of and the need for insulation-resistance testing are important.

Measurements of insulation resistance can best be made with a megohm-meter insulation tester, commonly called a megger. These instruments are available from several manufacturers and vary in cost and features. As measured with such an instrument, insulation resistance is the resistance to direct current (usually at 500 or 1000 volts for systems of 600 volts or less) through or over the surface of the insulation in electrical equipment. Energizing the conductor, wire or busbar with the potential and measuring the current that leaks through the insulation accomplish the test. By Ohm's Law, the insulation resistance is $R = E \div I$. The results are displayed in ohms or megohms, but insulation readings will be in the megohm range.

The insulation-resistance test is nondestructive, quite different from a high-voltage dielectric insulation or breakdown test, often called a hi-pot test (see photo 1-1). It is made with direct current rather than alternating current and is not a measure of dielectric strength as such. However, insulation-resistance tests assist greatly in determining when and where not to apply high voltage.

In general, insulation resistance decreases with increased size of a machine or length of cable. This occurs because there is more insulating material in contact with the conductors and the frame, ground, or sheath. The greater volume of insulating material allows more total leakage and, therefore, lowers the overall insulation-resistance reading.

Insulation resistance usually increases with higher voltage rating of apparatus because of increased thickness of insulating material. Also different types of insulation, for example, air versus thermoplastic, will have different levels of insulation resistance.

Insulation-resistance readings are not only quantitative, but are relative or comparative as well, and since they are influenced by moisture, dirt, and deterioration of the insulation, they are reliable indicators of the presence of those conditions. Cable and conductor installations present a wide variation of conditions from the point of view of the resistance of the insulation. These conditions result from the many kinds of insulating materials used, the voltage rating or insulation thickness, and the length of the circuit involved in the measurement. Furthermore, such circuits usually extend over great distances and can be subject to wide variations in temperature, which will affect the insulation-resistance values obtained. The terminals of cables and conductors will also affect the test values unless they are clean and dry, or guarded.

Records of insulation resistance should be maintained so failures of conductors or equipment can be predicted. To make valid comparisons or to indicate trends, the recorded values must be corrected to a standard temperature, typically 20°C, and differences in the relative humidity must be accounted for. Equipment or conductors can then be repaired or replaced to reduce plant downtime, which often happens at inconvenient times causing expensive loss of production.

Differentiate Between a *Short Circuit* and a *Ground Fault*

It is common practice to call all faults or failures in the electrical system *shorts* or *short circuits*. That can lead to misunderstanding, and has done so, because the terms are not used properly. A short circuit and a ground fault are different, although they both stem from insulation failure. To end this confusion, the definitions of the two terms follow:

Short circuit. A conducting connection, whether intentional or accidental, between any of the conductors of an electrical system whether it is from line-to-line or from line to the grounded conductor (see figure 1-26). The grounded conductor often is also the system or circuit neutral. The short circuit can be a solid or bolted connection, or it can be an arcing fault completing the path through a short air space.

In the case of a short circuit, the failure can be from one phase conductor to another phase conductor or from one phase conductor to the grounded conductor or neutral. For either condition, the maximum value of fault current is dependent on the available capacity the system can deliver to the point of the fault. The

Photo 1-1. Insulation-resistance (hi-potential) testing in progress Courtesy of Electro-Test, Inc.

maximum value of short-circuit current from line-to-neutral will vary depending upon the distance from the source to the fault and the impedance of the path. The available short-circuit current is further limited by the dynamic impedance of the arc, where one is established, plus the impedance of the conductors to the point of short circuit (see figure 1-26).

Ground Fault. An unintentional, electrically conducting connection between an ungrounded conductor of an electrical circuit and the normally non–current-carrying conductors, metallic enclosures, metallic raceways, metallic equipment, or earth [see *NEC* 250.2 and figure 1-27].

In the case of a ground fault, there is unintentional contact between an ungrounded phase (hot) conductor and the conductor enclosure or from the grounded phase conductor to the conductor enclosure (wire-to-conduit, wire-to-motor frame, and so forth). It is not common practice to refer to a conductor that is intentionally grounded, such as a grounded phase conductor or system grounded (neutral) conductor as being in a ground-fault condition. However, a grounded system conductor, such as a neutral conductor, that is not generally permitted to be grounded again past the service disconnecting means can be considered a ground fault. This condition is particularly important with equipment ground-fault protection systems where grounding the system grounded conductor downstream from the point of protection will desensitize the protection system.

It is often the case that a fault can easily involve both conditions: short circuit and ground fault. The fault can start as a ground fault or line-to-ground fault and escalate into a short circuit or phase-to-phase fault. A fault can also start as a short circuit and expand to a ground fault. For electrical systems of 120-volts-to-ground, the circuit protective devices will usually clear the fault or it will be extinguished when the alternating voltage passes through zero. For 277-volts-to-ground systems, destructive arcing faults can be easily sustained by the higher system voltage. Equipment ground-fault protection systems have been designed to address this problem. In some cases, the use of this equipment is required by the *NEC* such as in 230.95 and 215.10 (see chapter fifteen of this text for additional information on this subject).

NEC Requirements

Article 250 covers the subject of grounding and bonding. Grounding and bonding are practiced for the protection of electrical installations, which in turn protects the buildings or structures in which the electrical systems are installed. Persons and animals that can come into contact with the electrical system or that are in these buildings or structures are also

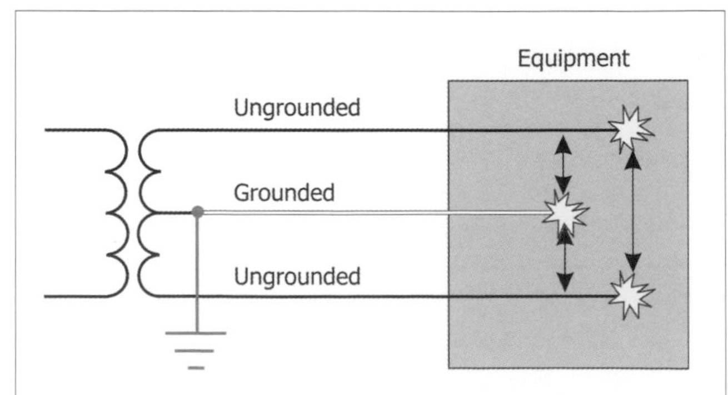

Figure 1-26. Basic diagram of a short-circuit condition

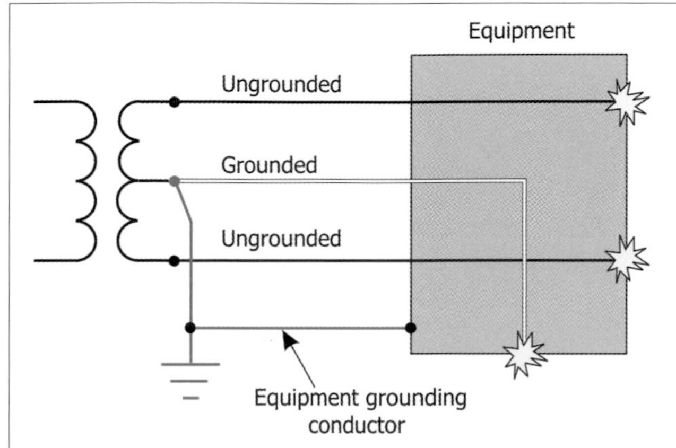

Figure 1-27. Basic diagram of a ground-fault condition

protected if the grounding system is installed and maintained properly. The *NEC* does not imply here that grounding is the only method that can be used for the protection of electrical installations, people or animals. Insulation, isolation and guarding are suitable alternatives under certain conditions.

Grounding of specific equipment is covered in several other articles. For example, health care facilities as covered in Article 517 and swimming pools in Article 680. Refer to the indices of this text, the *National Electrical Code*, or Ferm's Fast Finder Index for specific sections of the *Code* that apply to the equipment in question.

Circuit Impedance and Other Characteristics

Electrical equipment intended to interrupt current at fault levels must be designed and installed so it has an adequate rating for the nominal circuit voltage and the fault current that is available at its line terminals [see 110.9 and figure 1-28]. Failure to comply with this important requirement can result in disastrous consequences. This includes the destruction of electrical equipment, such as circuit breakers or fuses that are themselves designed to protect the system.

Equipment such as time clocks, motor controllers, lighting contactors, and so forth that are intended to interrupt current at only rated load must be suitable for the nominal circuit voltage and current that must be interrupted. These devices must also withstand the higher fault current levels until an upstream protective device, such as a fuse or circuit breaker, opens.

In addition, 110.10 requires that, "The overcurrent protective devices, the total impedance, the component short-circuit current ratings, and other characteristics of the circuit to be protected shall be selected and coordinated to permit the circuit-protective devices used to clear a fault to do so without extensive damage to the electrical components of the circuit (see figure 1-29). This fault shall be assumed to be either between two or more of the circuit conductors, or between any circuit conductor and the grounding conductor or enclosing metal raceway. Listed products applied in accordance with their listing shall be considered to meet the requirements of this section."

Listed equipment such as fuses, including the associated switch, and circuit breakers have nominal voltage and short-circuit current ratings. This is the maximum voltage and current for which the equipment is designed. In addition, other equipment such as busways may be marked similar to "Busways and associated fittings marked 'Short Circuit Current Rating(s) Maximum RMS Symmetrical Amperes _____ Volts ______' have been investigated for the rating indicated."[7] Electrical system components such as metering equipment, switchboards, panelboards, and motor control centers have short-circuit ratings that must not be exceeded.

Overcurrent protective devices such as circuit breakers and fuses should be selected and installed to ensure that the short-circuit current rating of components of the system will not be exceeded should a short circuit or ground fault occur. The overcurrent device that is selected must not only safely interrupt the fault current that is available at its line terminals, it also must limit the amount of energy that is let through to downstream equipment so as not to exceed its rating.

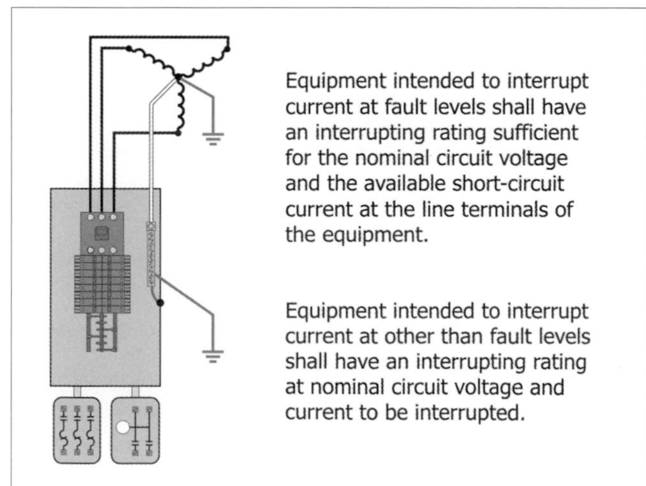

Figure 1-28. Interrupting ratings of equipment

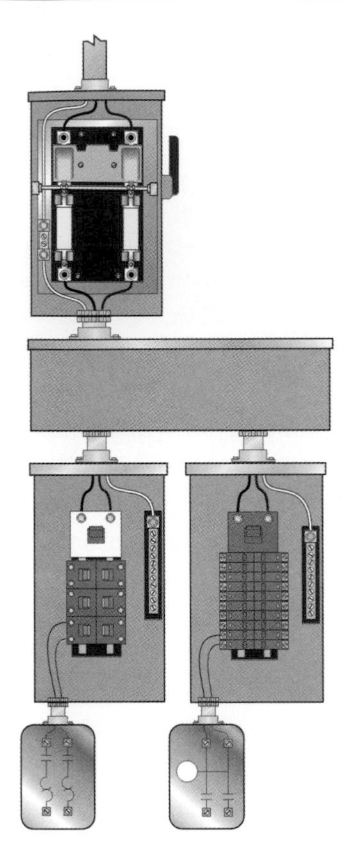
•Overcurrent protective devices
•Total impedance
•Component short-circuit current rating
•Other characteristics of circuit
Selected and coordinated to permit circuit protective devices to clear fault without extensive damage to the electrical components of the circuit
Listed products applied in accordance with their listing are considered to meet requirements of Section 110.10.
Fault assumed to be between:
•Two or more of the circuit conductors
•Any circuit conductor and the grounding conductor or enclosing metal raceway

Figure 1-29. Electrical components and equipment are required to be protected.

Electric utilities usually provide information on the short-circuit current (often referred to as the available fault current) that is available at the secondary terminals of the transformer or the service equipment. The short-circuit current must then be calculated to the point on the electrical system under consideration. Methods for calculating the available fault current at any point on the electrical system can be found in literature from many circuit breaker or fuse manufacturers. See also the Institute of Electrical and Electronic Engineers (IEEE) "Buff Book" (IEEE 242),[8] Underwriters Laboratories product safety standard and "guide card information" for the equipment under consideration[9], and Ferm's Fast Finder Index[10] where, in addition to the point-to-point method, many pages of calculations are presented in an easy-to-use table form.

Other components of the electrical system, such as wire and cable, have published withstand ratings, in addition to allowable ampacity ratings, that must be considered for safety. Withstand ratings are not marked on the wire or cable but can be determined from the manufacturer's data, IEEE Standards, or calculated as described in chapter eleven.

As can be seen upon review of this literature, a conductor can safely carry much more current than allowed in the allowable ampacity table if the time the current exists is reduced. For insulated conductors, the ampacity of the conductor can be thought of as the conductor's longtime or continuous rating and the withstand-rating as the conductor's short-time rating. This is most important when considering the proper size of equipment grounding conductors to use, as they do not have overcurrent protection on their line side and must carry fault-current levels until an overcurrent device opens (see chapter eleven for additional information on this subject).

[1] NFPA 70, *National Electrical Code* 2008, Article 100, (National Fire Protection Association, Quincy, MA, 2002), p. 70-25 and 28.
[2] NFPA 70, Article 250, p. 70-97.
[3] NFPA 70, Article 250, p. 70-97.
[4]NFPA 70, Article 100, p. 70-28.
[5] NFPA 70, Article 100, p. 70-28.

[6] NFPA 70, Article 250, p. 70-97.

[7] General Information for Electrical Equipment Directory, "Busways and Associated Fittings (CWFT)," (Underwriters Laboratories Inc., Northbrook, IL, 2007), p. 39.

[8] Available from IEEE Service Center, 445 Hoes Lane, P.O. Box 1331, Piscataway, NJ 08855-1331, (800) 678-4333.

[9] Available from Publications Stock, Underwriters Laboratories, 333 Pfingsten Road, Northbrook, IL 60062-2096, (847) 272-8800 Ext. 42612 or 42622.

[10] Available from International Association of Electrical Inspectors, P.O. Box 830848, Richardson, TX 75083-0848, (800) 786-4234, Fax (972) 235-6858.

Review Questions

The questions included here were developed using material included in this chapter. The answers can be found by reviewing the text. It is also important that students make use of *NEC*-2008, where many answers can be found.

1. A grounded electrically conductive object is forced to take the same potential as the ____.
   a. electrode
   b. metal raceway
   c. ground (the earth)
   d. system

2. Electrical systems are grounded to limit the voltage to ground during normal operation and to prevent excessive voltages because of ____, line surges or unintentional contact with higher voltage lines.
   a. low voltage
   b. overloads
   c. loose connections
   d. lightning

3. The effective ground-fault current path must be installed in a manner that creates a path that is ____.
   a. electrically continuous
   b. has the capacity to conduct safely the maximum ground-fault current likely to be imposed on it
   c. has sufficiently low impedance to limit the voltage to ground and to facilitate the operation of the circuit protective devices in the circuit
   d. all of the above

4. Insulation is considered to be air where conductors are mounted on insulators such as on poles, in busways or in equipment, or rubber or ____ material or thermosetting material.
   a. listed
   b. approved
   c. acceptable
   d. thermoplastic

5. Measurements of insulation resistance can best be made with a ____ insulation tester.
   a. voltage
   b. megohm-meter
   c. low voltage
   d. high voltage

6. Insulation resistance is the resistance to direct current (usually at ____ or ____ volts for systems of 600 volts or less) through or over the surface of the insulation in electrical equipment megohm range.
   a. 500 or 1000
   b. 600 or 1500
   c. 800 or 1800
   d. 900 or 2000

7. An insulation-resistance test is made with direct-current rather than alternating-current, and is not a measure of ____ strength as such.
   a. conductor
   b. dielectric
   c. terminal
   d. equipment

8. A conducting connection, whether intentional or accidental, between any of the conductors of an electrical system whether it be from line-to-line or to the grounded conductor describes a ____.
   a. phase fault
   b. short circuit
   c. overload
   d. line disorder

9. An unintentional electrically conducting connection between an ungrounded conductor of an electrical circuit and the normally non-current-carrying conductors, metallic enclosures, metallic raceways, metallic equipment, or earth best defines which of the following____.
   a. phase fault
   b. overload
   c. system ground
   d. ground fault

10. In a dc electrical circuit, the total opposition to current is due primarily to _____.
   a. current
   b. voltage
   c. impedance
   d. resistance

11. In an electrical circuit, current in the circuit will always attempt to return to which of the following:
   a. the earth
   b. the service
   c. the source
   d. the load

12. For current to be present there must be a complete ____.
    a. grounding conductor
    b. conduit system
    c. circuit
    d. overcurrent protective device

13. Current in an electrical circuit will take _________ to return to the source.
    a. only the path of least resistance
    b. any and all paths available
    c. a high impedance path only
    d. the path in the earth

14. The total opposition to current in an ac circuit is the_______ of the circuit.
    a. resistance
    b. impedance
    c. capacitance
    d. inductance

15. The severity of an electrical shock is dependant on the ____________.
    a. the frequency of the circuit
    b. the path through the body
    c. the amount of current through the body and how long it exists
    d. all of the above

16. The higher the impedance or resistance of a circuit, the ______ the amount of current in the circuit.
    a. lower
    b. higher
    c. same
    d. different

17. The *NEC* in recent cycles has continued to migrate away from the use of the _______ conductor for grounding of equipment downstream of the main bonding jumper at the service or bonding jumper at a separately derived system.
    a. equipment grounding
    b. grounded (neutral)
    c. bonding
    d. grounding electrode

18. Alternatives for grounding include which of the following conditions:
    a. guarded
    b. insulated
    c. isolated
    d. all of the above

19. A 480-volt, 3-phase circuit is connected to an electric heater rated at 480 volts, 3-phase/7500 watts. The current in this circuit is_______amperes.
    a. 13.6
    b. 9.03
    c. 15.7
    d. 27.9

20. The phrase *likely to become energized* generally means:
    a. near energized electrical equipment
    b. high-voltage electrical equipment
    c. equipment that is exposed to lightning
    d. failure of insulation on

21. During normal operating conditions the equipment grounding conductor should carry _______ current
    a. no
    b. return
    c. unbalanced
    d. 70% of the phase conductors

22. The equipment grounding conductor serves to _______.
    a. provide an effective ground-fault current path
    b. grounds the equipment or its enclosure by connecting to the grounding point of the system
    c. performs bonding functions
    d. all of the above

23. "Connected to establish electrical continuity and conductivity" best defines which of the following _______.
    a. grounded (grounded)
    b. bonded (bonding)
    c. adequately bonded
    d. effectively grounded

To Ground or Not To Ground

The term *electrical system* generally refers to the system as a unit of a specific voltage (potential) and often amperage (current or capacity). For example, a common system is 480Y/277 volts, 3-phase, 4-wire at some capacity such as 1600 amperes. Often, the premises are supplied by an electric utility. These systems are either over 600 volts (often referred to as *primary systems*) or 600 volts or less (often referred to as *secondary systems*). In addition, the system is referred to as *single-phase* or *three-phase*. Some two-phase, 5-wire systems still exist but are less common. Most systems are 60-hertz alternating current although some direct-current systems are in use, primarily in industrial applications.

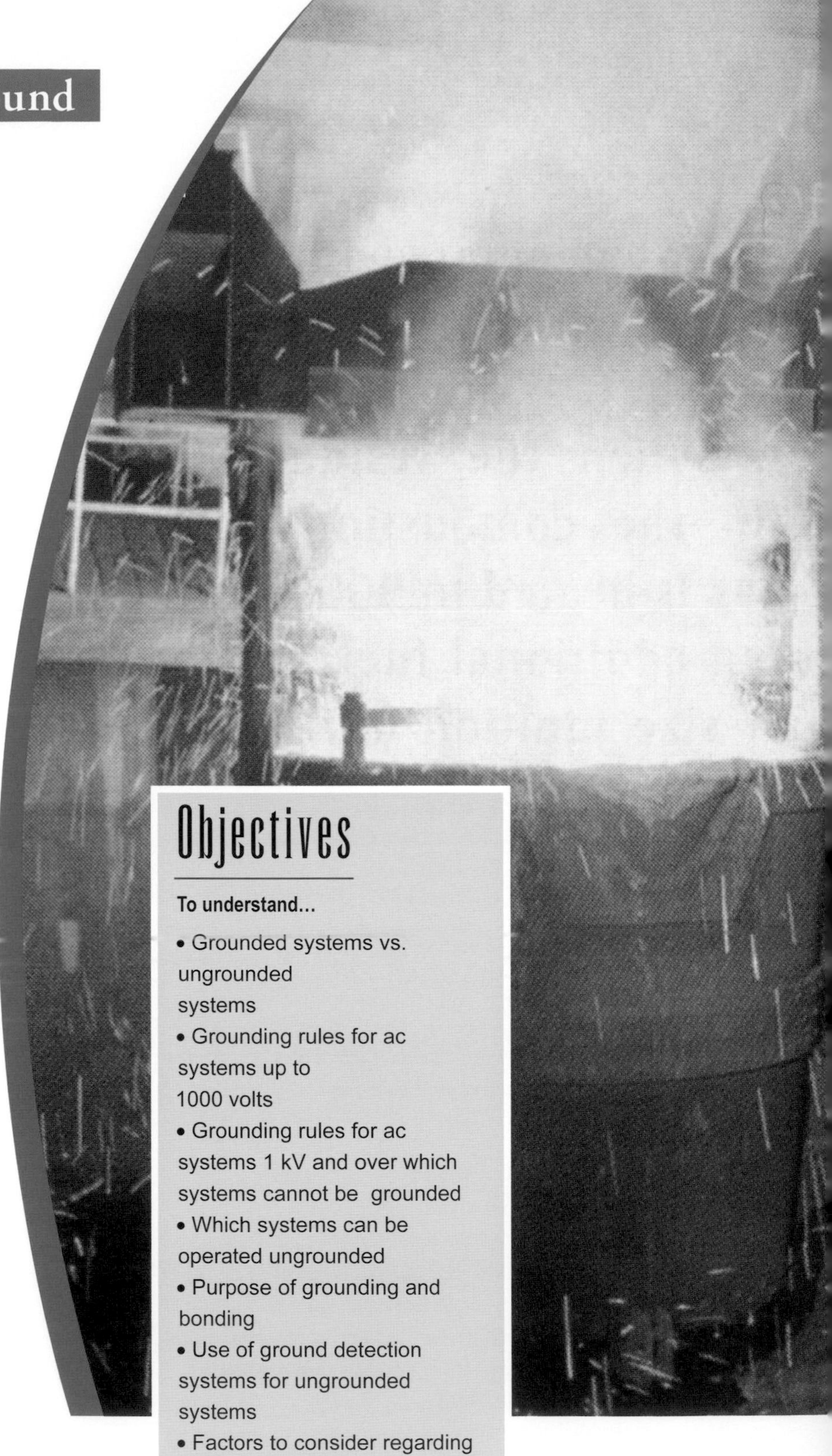

Objectives

To understand...

- Grounded systems vs. ungrounded systems
- Grounding rules for ac systems up to 1000 volts
- Grounding rules for ac systems 1 kV and over which systems cannot be grounded
- Which systems can be operated ungrounded
- Purpose of grounding and bonding
- Use of ground detection systems for ungrounded systems
- Factors to consider regarding system grounding

Chapter 2

In addition, *systems* are produced or created at various voltages and phases on-site. The most common way to produce an electrical system is through a transformer or generator. In the case of a transformer, other than an autotransformer, one electrical system ends at the primary winding(s) and another system begins at the secondary winding(s).

An example is a plant that has a service at 480 Y/277 volts and it is necessary to supply receptacle outlets at 120 volts. A single-phase, 480-volt to 240-volt transformer is installed with the winding on the secondary center-tapped to result in a 120/240-volt, single-phase system. In this example, the 480-volt system ends at the primary windings and a new 120/240-volt system begins at the secondary winding. In other cases, a 208Y/120-volt system is derived by installing the appropriate transformer(s). Similar examples can be shown for generator-supplied systems, though, for these, we consider the system supplied as a new system. A new source is developed by the transformer or generator. For the purposes of grounding, the treatment of the grounded system conductor (often a neutral) determines whether the system is considered to be separately derived for grounding purposes (see chapter twelve for more information on this subject).

Definitions

Ground Fault. "An unintentional, electrically conducting connection between an ungrounded conductor of an electrical circuit and the normally non–current-carrying conductors, metallic enclosures, metallic raceways, metallic equipment, or earth."

Ground-Fault Current Path. "An electrically conductive path from the point of a ground fault on a wiring system through normally non–current-carrying conductors, equipment, or earth to the electrical supply source."

Examples of ground-fault current paths are of any combination of equipment grounding conductors, metallic raceways, metallic sheaths of cables, electrical equipment, and any other electrically conductive material such as metal water and gas piping, steel framing members, stucco mesh, metal ducting, reinforcing steel, shields of communications cables, and the earth itself.

Effective Ground-Fault Current Path. "An intentionally constructed, low-impedance electrically conductive path designed and intended to carry current under ground-fault conditions from the point of a ground fault on a wiring system to the electrical supply source *and that facilitates the operation of the overcurrent device or ground fault detectors on high impedance grounded systems.*" [1]

Effective ground-fault current paths are created by effectively bonding together all of the electrically conductive materials that are likely to be energized by the wiring system. Effective bonding is accomplished through the use of equipment grounding conductors, bonding jumpers or bonding conductors, approved metallic raceways, connectors and couplings, approved metallic sheathed cable and cable fittings, and other approved devices. A ground-fault path is effective when it will carry the maximum ground-fault current likely to be imposed on it.

The *NEC* has been revised in recent cycles to include the above definitions of the terms now used in 250.2, 250.4 and elsewhere in Article 250. The 2005 revision added the concept of facilitating the operation of the overcurrent device or ground detector to the definition of *effective ground-fault current path.* These definitions help provide a clearer understanding of some important performance characteristics of ground faults and effective ground-fault current paths. These terms enhance the ability to properly apply the prescriptive grounding and bonding requirements in the later parts of Article 250.

Grounded Systems

Section 250.4 is divided into two main parts. Part A includes the performance requirements for grounded systems, and Part B includes the performance requirements for ungrounded systems (see figure 2-1).

"**250.4 General Requirements for Grounding and Bonding.** The following requirements identify what grounding and bonding of electrical systems are required to accomplish. The prescriptive methods contained in Article 250 shall be followed to comply with the performance requirements of this section.

"**(A) Grounded Systems.**

"**(1) Electrical System Grounding.** Electrical systems that are grounded shall be connected to earth in a manner that will limit the voltage imposed by lightning, line surges, or unintentional contact with higher-voltage lines and that will stabilize the voltage to earth during normal operation.

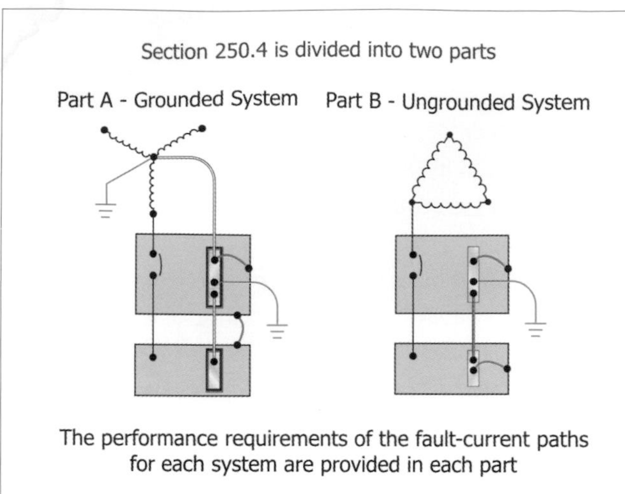

Figure 2-1. General requirements for grounding and bonding

The FPN to 250.4(A)(1) indicates that important considerations for limiting imposed voltages are to minimize excessive lengths and to avoid unnecessary bends and loops in ground and bonding conductors [250.4(A)(1) FPN]. Note that this is a recommendation and not a requirement as provided for in the fine print notes in 90.5(C).

"(2) **Grounding of Electrical Equipment.** Normally, non–current-carrying conductive materials enclosing elect-rical conductors or equipment, or forming part of such equipment, shall be connected to earth so as to limit the voltage to ground on these materials.

"(3) **Bonding of Electrical Equipment.** Normally, non–current-carrying conductive materials enclosing electrical conductors or equipment, or forming part of such equipment, shall be connected together and to electrical supply source in a manner that establishes an effective ground-fault current path.

"(4) **Bonding of Electrically Conductive Materials and Other Equipment.** Electrically conductive mater-als that are likely to become energized shall be connected together and to the electrical supply source in a manner that establishes an effective ground-fault current path.

"(5) **Effective Ground-Fault Current Path.** Elect-rical equipment and wiring and other electrically conductive material likely to become energized shall be installed in a manner that creates a low-impedance circuit facilitating the operation of the overcurrent device or ground detector for high impedance grounded systems. It shall be capable of safely carrying the maximum ground-fault current likely to be imposed on it from any point on the wiring system where a ground fault may occur to the electrical supply source. The earth shall not be considered as an effective ground-fault current path." [2]

Grounding Electrical Systems

Over the years, there have been great debates over the merits of grounding electrical systems versus installing and operating them ungrounded.

Many of the decisions about whether or not to ground electrical systems are made for us. The *National Electrical Code* requires that certain electrical systems falling within the parameters of 250.20 be grounded. Other electrical systems are permitted to be grounded, while some systems are not permitted to be grounded due to special conditions. Some systems, primarily in the industrial or agricultural sector, are operated ungrounded. Common reasons for choosing to operate an electrical system ungrounded are continuity of service for critical operations and minimizing downtime. These decisions are usually mutually agreed upon between the building owners and operators, the designers and engineering team, and others. The last sentence of 250.20 indicates that whether the system is required to be grounded, or is grounded by choice, the applicable grounding rules for grounded systems must be followed.

Systems Required To Be Grounded

In accordance with 250.20, the alternating-current systems that must be grounded are: (A) alternating-current systems of less than 50 volts (see figure 2-2), (B) alternating-current systems of 50 volts to 1000 volts (see

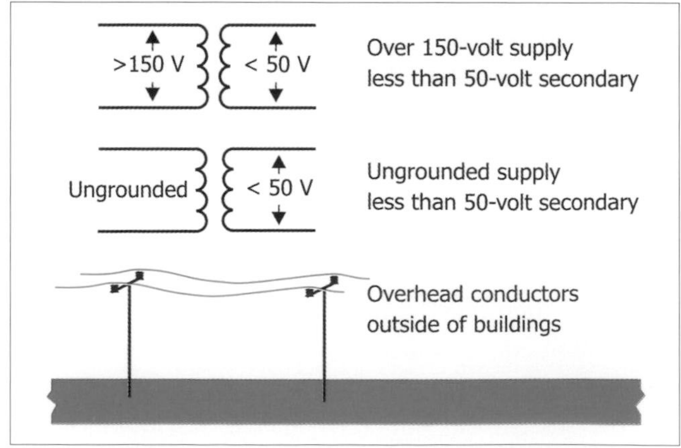

Figure 2-2. Systems less than 50 volts that must be grounded

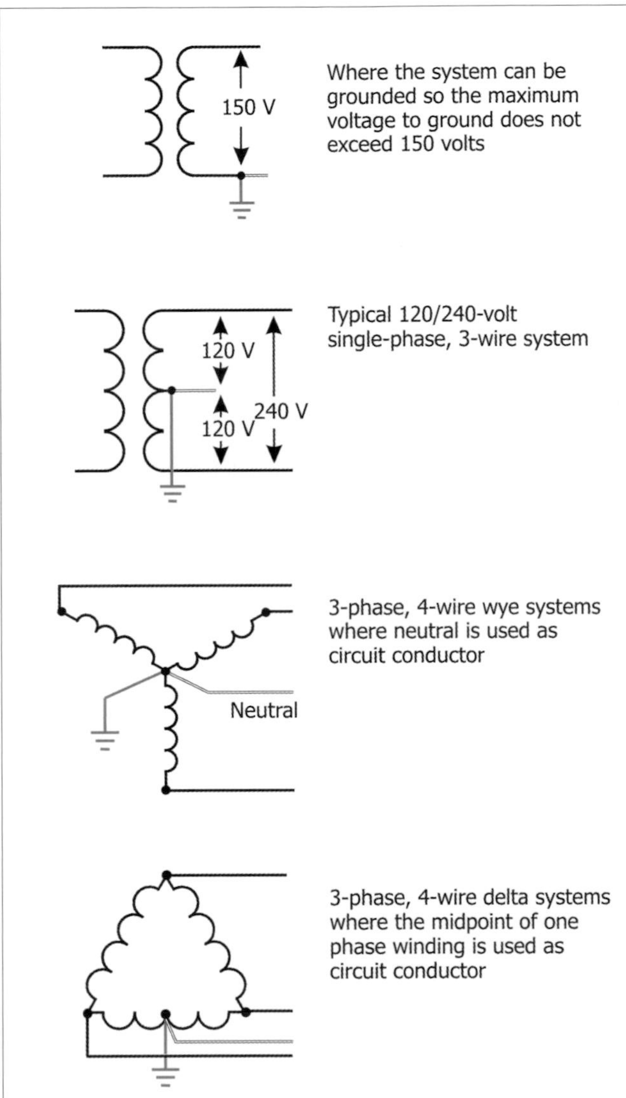

Figure 2-3. Systems rated 50 to 1000 volts that are required to be grounded

figure 2-3), (C) alternating-current systems of 1 kV and over (see figure 2-4), (D) separately derived systems (see figure 2-5), and (E) impedance grounded neutral systems. Each of these systems is discussed in the following sections.

Less than 50-volt systems

The following systems that are less than 50 volts, such as on the secondary of a transformer must be grounded, where:

1. Supplied by transformers if the supply voltage (primary) to the transformer exceeds 150 volts to ground
2. Supplied by transformers if the transformer-supply system (primary) is ungrounded
3. Installed as overhead conductors outside of buildings:

AC systems of 50 to 1000 volts

Alternating-current systems of 50 to 1000 volts that supply premises wiring and premises wiring systems are required to be grounded under any of the following conditions, where the system:

1. Can be grounded so the maximum voltage to ground on the ungrounded conductors does not exceed 150 volts. Typical systems include:
   - 120-volt, 1-phase,
   - 120/240 volt, 1-phase, 3-wire, and
   - 208Y/120 volt, 3-phase, 4-wire.
2. Is 3-phase, 4-wire wye-connected and where the neutral is used as a circuit conductor. Typical systems include:
   - 208Y/120 volt, 3-phase, 4-wire, and
   - 480Y/277 volt, 3-phase, 4-wire.
3. Is 3-phase, 4-wire delta-connected in which the midpoint of one phase winding is used as a circuit conductor. Typical systems include:
   - 120/240 volt, 3-phase, 4-wire, and
   - 240/480 volt, 3-phase, 4-wire.

Alternating-Current Systems of 1 kV and Over

Alternating-current systems supplying mobile or portable equipment shall be grounded as specified in 250.188. Where supplying other than mobile or portable equipment, such systems shall be permitted to be grounded.

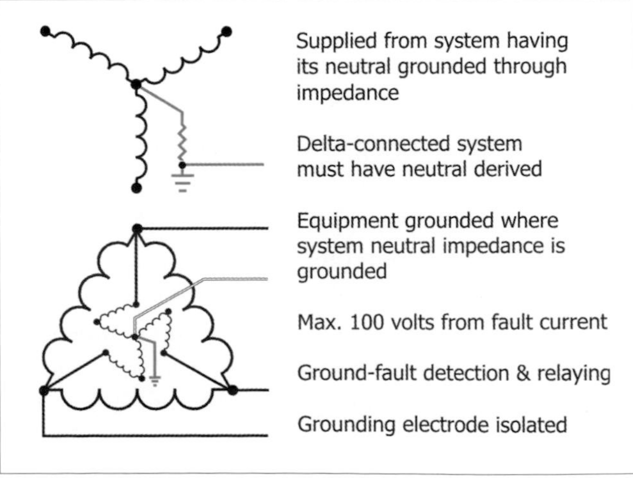

Figure 2-4. AC systems of 1 kV and over supplying portable or mobile equipment

These ac systems must be grounded if they supply portable or mobile equipment as covered in 250.188. Systems supplying portable or mobile *high-voltage* equipment, other than substations installed on a temporary basis, are required to comply with the rules in (A) through (F) below.

"**(A) Portable or Mobile Equipment.** Portable or mobile high-voltage equipment shall be supplied from a system that has its neutral conductor grounded through an impedance. Where a delta-connected high-voltage system is used to supply portable or mobile equipment, a system neutral shall be derived." This is usually accomplished by means of a *zigzag grounding transformer* (see chapter three of this text for additional information on grounding systems by using zigzag grounding transformers).

"**(B) Exposed Non–Current-Carrying Metal Parts.** Exposed non–current-carrying metal parts of portable or mobile equipment shall be connected by an equipment grounding conductor to the point at which the system neutral impedance is grounded." This point may be at the service or source of a separately derived system.

"**(C) Ground-Fault Current.** The voltage developed between the portable or mobile equipment frame and ground by the flow of maximum ground-fault current shall not exceed 100 volts." This requirement, no doubt, necessitates an engineering study to determine the voltage drop across the equipment grounding conductor.

"**(D) Ground-Fault Detection and Relaying.** Ground-fault detection and relaying must be provided to automatically de-energize any high-voltage system component that has developed a ground fault. The continuity of the equipment grounding conductor must be continuously monitored so as to de-energize automatically the high-voltage feeder circuit to the portable or mobile equipment upon loss of continuity of the equipment grounding conductor.

"**(E) Isolation.** The grounding electrode to which the portable or mobile equipment system neutral impedance is connected shall be isolated from and separated in the ground by at least 6.0 m (20 ft) from any other system or equipment grounding electrode, and there shall be no direct connection between the grounding electrodes, such as buried pipe, fence, and so forth.

"**(F) Trailing Cable and Couplers.** High-voltage trailing cable and couplers for interconnection of portable or mobile equipment must meet the requirements of Part III of Article 400 for cables and 490.55 for couplers."

This type of equipment is commonly found in mobile rock crushing plants and batch plants. Other applications are for open pit mining operations. Note that self-propelled mobile surface mining machinery and its attendant electrical trailing cable are not covered by the *Code* [see 90.2(B)(2)]. Even though exempted from the *Code*, many requirements for this equipment are incorporated by regulations enforced by the Mine Safety and Health Administration (MSHA).

Other ac systems over 1000 volts are permitted but are not required to be grounded.

Separately Derived Systems

Electrical systems derived from a battery, a solar photovoltaic system, or from a generator, transformer, or converter windings that have no direct electrical connection, including a solidly connected grounded circuit conductor, to the supply conductors originating in another system must be grounded if the system that is derived meets the the conditions in 250.20(A) or (B). Where an alternate source such as an on-site generator is provided and includes transfer equipment that introduces a switching action in the grounded conductor, the alternate source (derived system) is required to be grounded as specified in 250.30(A).

Examples of systems that are separately derived include:

- Inverters or batteries such as for uninterruptible power systems
- Solar photovoltaic systems or fuel cells
- Transformers with no direct electrical connection between the primary and secondary
- Generator systems that supply power such as for carnivals, rock crushers or batch plants where the neutral is not connected to the utility system
- Generator systems used for emergency, required standby or optional standby power that have all conductors, including a neutral, isolated from the neutral or grounded conductor of another system usually by a transfer switch
- Ac or dc systems derived from inverters or rectifiers (see figure 2-5).

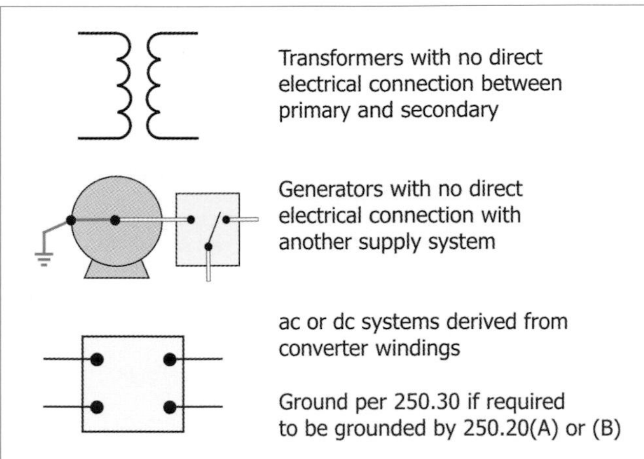

Figure 2-5. Separately derived systems that must be grounded

Figure 2-6 illustrates systems that are not separately derived and thus do not fall under the grounding requirements of 250.30. These systems include:

- A system supplied by an autotransformer since autotransformers by design have a conductor that is common to both primary and secondary
- Systems from a generator that do not have all conductors, including the grounded system conductor (often a neutral), switched by a switching mechanism in a transfer switch

The key to determining whether a generator supplied system is to be grounded as a separately derived system is often to examine electrical connections in the transfer switch (see figure 2-6).

If all the phases and the neutral or grounded conductor is switched by the transfer switch, the generator is a separately derived system that must be grounded in compliance with 250.30(A). If the neutral or grounded conductor is not switched, the system produced is not to be grounded as a separately derived system and the neutral must not be grounded at the generator (see chapter twelve of this text for thorough information on grounding and bonding requirements for separately derived systems.)

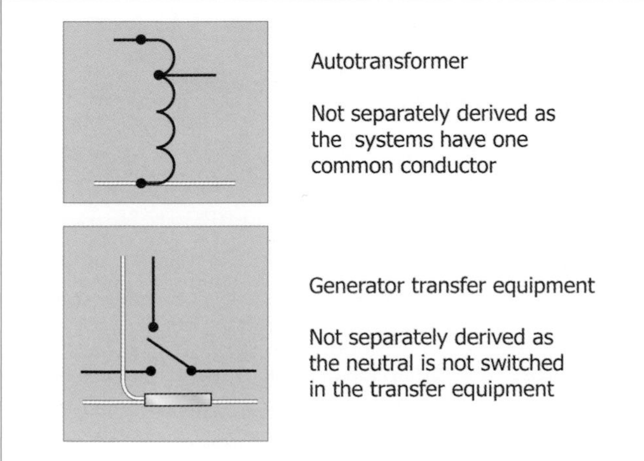

Autotransformer
Not separately derived as the systems have one common conductor
Generator transfer equipment
Not separately derived as the neutral is not switched in the transfer equipment

Figure 2-6. Not separately derived systems

High-Impedance Grounded Systems

High-impedance grounded systems are often considered in electrical designs where the facility operation cannot tolerate electrical system disruption from the first ground fault. This system has all the advantages of an ungrounded system, so far as operation of the plant or system with one phase faulted to ground is concerned, with none of the disadvantages of an ungrounded system. While the initial cost of the system is more, it can pay for itself many times over the installation cost by operational savings in more reliable and uninterrupted electrical system operation (see chapter four of this text for additional information on high-impedance grounded neutral systems.)

Ungrounded Systems

The term *ungrounded* is defined as "not connected to ground or to a conductive body that extends the ground connection."[3] Ungrounded systems are derived electrical systems that have no conductor of the system grounded, either solidly or through any resistance or impedance. Theoretically, there is no potential between any of the system conductors and ground because there is not a connection of any conductor of the system to ground. Because there is capacitance between the insulated conductors and other grounded objects, such as raceways and equipment enclosures, the system is capacitively coupled to ground. Section 250.4(B), Ungrounded Systems, includes four subsections (1) Grounding of Electrical Equipment, (2) Bonding of Electrical Equipment, (3) Bonding of Electrically Conductive Materials and Other Equipment, and (4) Path for Fault Current.

"**(B) Ungrounded Systems**

"**(1) Grounding of Electrical Equipment.** Non–current-carrying conductive materials enclosing electrical conductors or equipment, or forming part of such equipment, shall be connected to earth in a manner that will limit the voltage imposed by lightning or unintentional contact with higher-voltage lines and limit the voltage to ground on these materials.

"(2) Bonding of Electrical Equipment. Non–current-carrying conductive materials enclosing electrical conductors or equipment, or forming part of such equipment, shall be connected together and to the supply system grounded equipment in a manner that creates a low-impedance path for ground-fault current that is capable of carrying the maximum fault current likely to be imposed on it.

"(3) Bonding of Electrically Conductive Materials and Other Equipment. Electrically conductive materials that are likely to become energized shall be connected together and to the supply system grounded equipment in a manner that creates a low-impedance path for ground-fault current that is capable of carrying the maximum fault current likely to be imposed on it.

"(4) Path for Fault Current. Electrical equipment, wiring, and other electrically conductive material likely to become energized shall be installed in a manner that creates a low-impedance circuit from any point on the wiring system to the electrical supply source to facilitate the operation of overcurrent devices should a second ground fault from a different phase occur on the wiring system. The earth shall not be considered as an effective fault-current path. [4]

Circuits That Are Not To Be Grounded

Five types of circuits are not permitted to be grounded by 250.22 (see figure 2-7). They are as follows:

Circuits for electric cranes that operate over combustible fibers in Class III locations [see 250.22(1) and 503.155]. This action reduces the likelihood that sparks from faulted equipment will fall onto combustible fibers below the crane, causing a fire.

For health care facilities, those isolated power circuits in hazardous (classified) inhalation anesthetizing locations are required to be supplied by an isolation transformer or other ungrounded source [see 517.61(A)(1)]. In addition, receptacles and fixed equipment in *wet locations* of hospital patient care areas as defined in 517.2 must be protected by ground-fault circuit-interrupter devices where interruption of power to equipment under fault conditions can be tolerated. Where interruption of power under fault conditions cannot be tolerated, protection of these receptacles and fixed equipment is to be supplied from isolated, ungrounded sources by ungrounded electrical systems [see 250.22(2) and 517.20(A)].

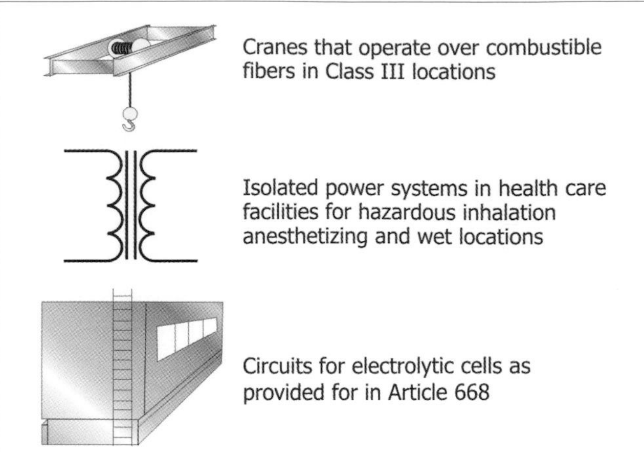

Figure 2-7. Circuits not permitted to be grounded

Circuits for electrolytic cells as provided in Article 668. Equipment located or used within the electrolytic cell line working zone or associated with the cell line dc power circuits are not required to comply with Article 250 [see 250.22(3) and 668.3(C)(3)].

Lighting systems as provided in 411.5(A). Article 411 covers lighting systems operating at 30 volts or less. The secondary circuits are required to be insulated from the branch circuit by an isolating transformer. Secondary circuits from these transformers are not permitted to be grounded.

Lighting systems for swimming pools and similar installations supplied by a listed isolating winding transformer having a grounded metal barrier between the primary and secondary windings are not permitted to be grounded [see 680.23(A)(2)].

Electrical Systems Operated Ungrounded

Electrical systems that fall outside the requirements for grounding in 250.20(A), (B), (C) or (D) are sometimes operated ungrounded. At times, the plant owner or engineer or combination will collectively choose to operate electrical systems ungrounded. These systems usually are found in industrial or agricultural applications and often are either 240-volt or 480-volt, three-phase, three-wire systems. Some higher voltage systems are also used in heavy-industrial applications. Where ungrounded

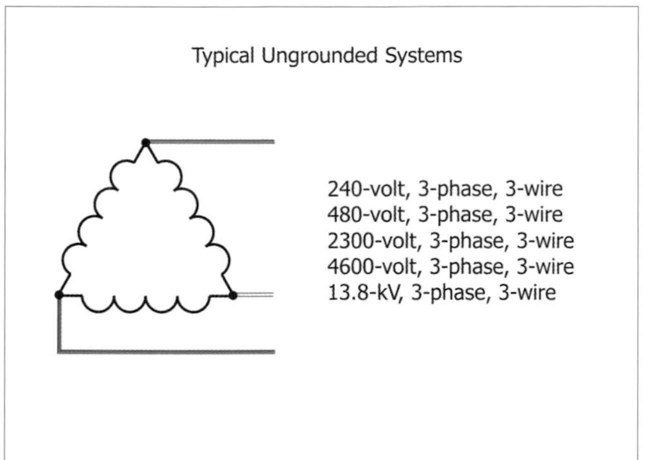

Figure 2-8. Systems permitted to be ungrounded

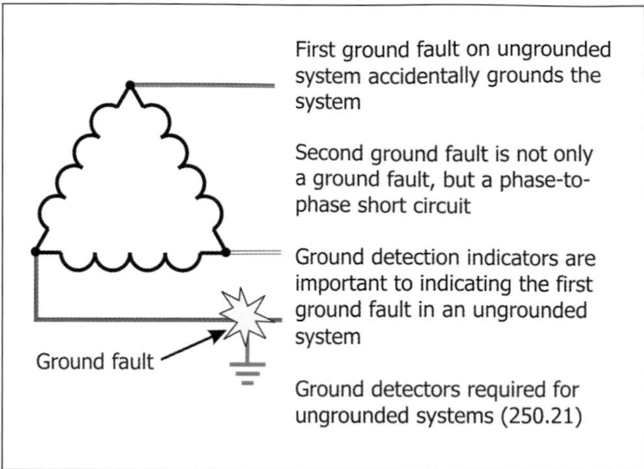

Figure 2-9. First ground fault on ungrounded systems

systems are installed, the engineering decision is often based on an effort to obtain an additional degree of service continuity while providing equal and effective means for safety of equipment by the use of ground-fault indicator equipment.

Typical systems (see figure 2-8) that are operated ungrounded include:

- 240 volt, 3-phase, 3-wire, delta-connected
- 480 volt, 3-phase, 3-wire, delta-connected
- 2300 volt, 3-phase, 3-wire, delta-connected
- 4600 volt, 3-phase, 3-wire, delta-connected
- 13,800 volt, 3-phase, 3-wire, delta-connected

Since the system is ungrounded, the occurrence of the first ground fault (not a short circuit or line-to-line fault) on the system will not cause an overcurrent protective device for the service, feeder, or branch circuit to open or operate. This fault does, however, ground the system but usually accidentally and through ineffective means (higher impedance) and in unspecified and uncontrolled locations (see figure 2-9). In essence, this system accidentally becomes a corner-grounded delta system. There will be little, if any, current when this first ground fault occurs (see chapter three of this text for additional information on equipment and enclosure grounding requirements for these ungrounded systems).

When an ungrounded system with one ground fault experiences a second ground fault on a different phase, the result is a phase-to-phase fault on the system. This will usually cause one or more overcurrent protective devices to open or operate, provided there is adequate current in this path. A major concern for this type of system happens where the first and second faults are located some distance apart (see figure 2-10). Often, these faults are from line-to-conduit or metallic enclosures, such as wireways, pull boxes, busways or motor terminal housings in different parts of the plant. Where this occurs, a relatively high-impedance path for current is often established. In some cases, it has been found that a great deal of heat along with arcing and sparking is produced along this fault path due to loose connections or inadequate bonding. Every conduit coupling and locknut connection to enclosures in the fault-current path must be tight to provide an adequate and low-impedance path and to reduce this arcing and sparking.

It is important for safety reasons and for system continuity that maintenance personnel locate and eliminate ground faults when first identified on ungrounded systems. This should be done as soon as practical and especially before the second ground fault on a different phase occurs on the system.

Ground-Detection Indicator Systems

Commercially manufactured ground-detection equipment is available. This equipment, which can be located at the service equipment or in feeder distribution panels, can be set to operate an overcurrent relay or shunt-trip circuit breaker or to operate a visual or audible signaling system to indicate a ground-fault condition.

This monitoring equipment is now required by the electrical code by the last sentence of 250.21. Successful operation of an ungrounded electrical system depends

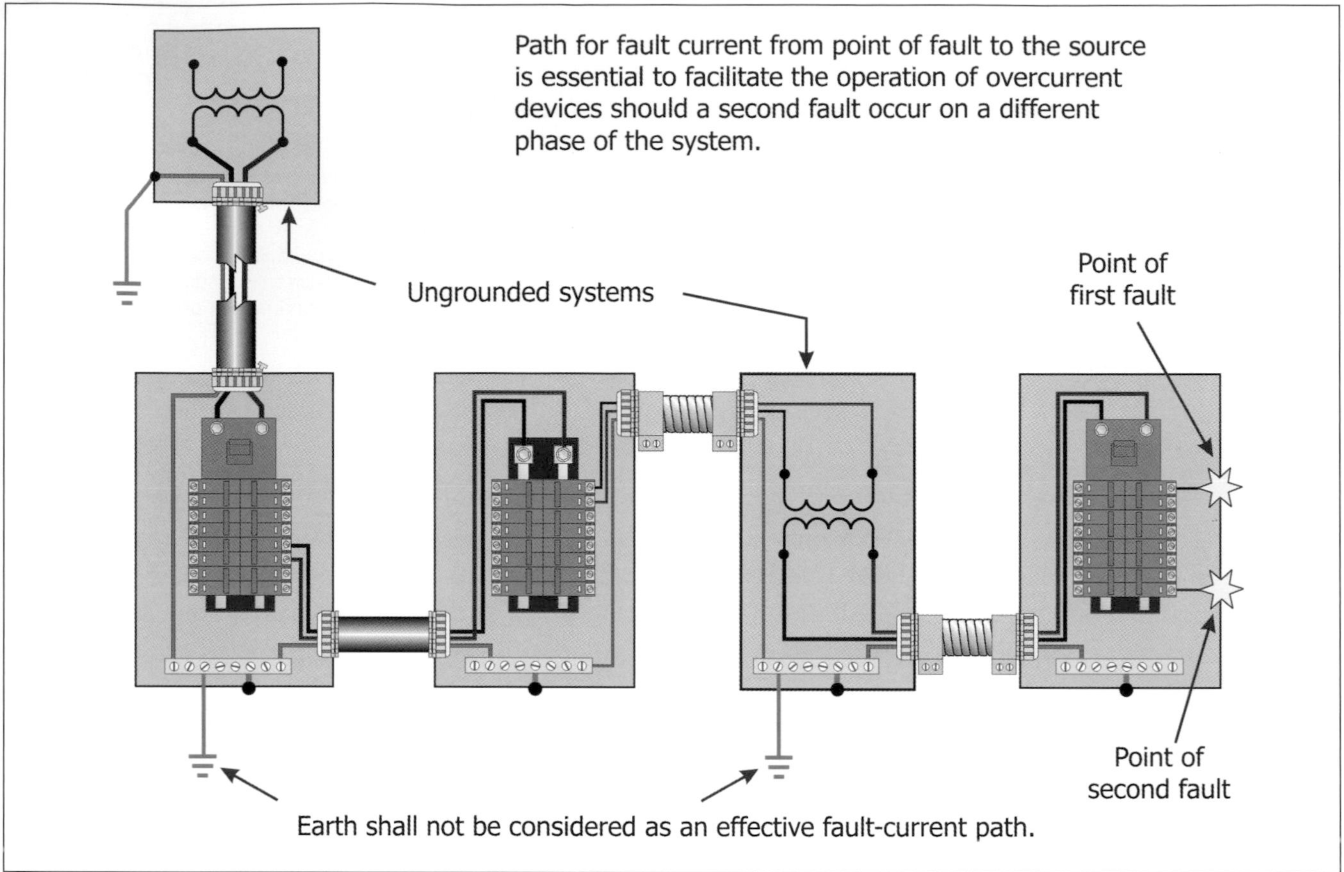

Path for fault current from point of fault to the source is essential to facilitate the operation of overcurrent devices should a second fault occur on a different phase of the system.
Ungrounded systems
Point of first fault
Point of second fault
Earth shall not be considered as an effective fault-current path.

Figure 2-10. Second ground fault on an ungrounded system

on good system maintenance and prompt elimination of the first ground fault. Sophisticated equipment is now available to identify the part of the electrical system where the fault is located while the system is energized. This significantly speeds detection and repair.

Ground Detectors

In the past, ground detector lights or neutralizer or "potentializer" plugs were installed to indicate that a ground fault had occurred on the ungrounded system. The 7 1/2-watt indicator lights are connected to the lines through 18,000-ohm resistors. A tap is made to each resistor to give 120 volts to the lamp. The lamp burns until its phase goes to ground, at which time there is no or little potential across the lamp and it stops glowing, thus identifying the faulted phase (see figure 2-11).

More modern types of ground detection indication equipment are available and offer the added benefits of no system ground connection, not even through a resistor as was the case in the older ground detection light systems. These systems are typically equipped with transformers (windings) between the indication circuit and the ungrounded conductors of the system (see figure 2-12).

Ground-fault indication is intended to alert the maintenance personnel to the problem so the ground fault can be corrected during hours when the plant is not operating. The plant can continue to operate with one-phase grounded, thus preventing costly production downtimes. In some cases, downtime in production plants can cost thousands of dollars per minute.

A wide variety of "homemade" systems have been installed over the years, some of which are downright dangerous. Some of these systems have consisted of nothing more than two lampholders with 240-volt lamps that are connected in series from phase-to-ground on 480-volt systems. Only proven and tested designs for ground detection systems should be used. Listed equipment is available for use on ungrounded system.

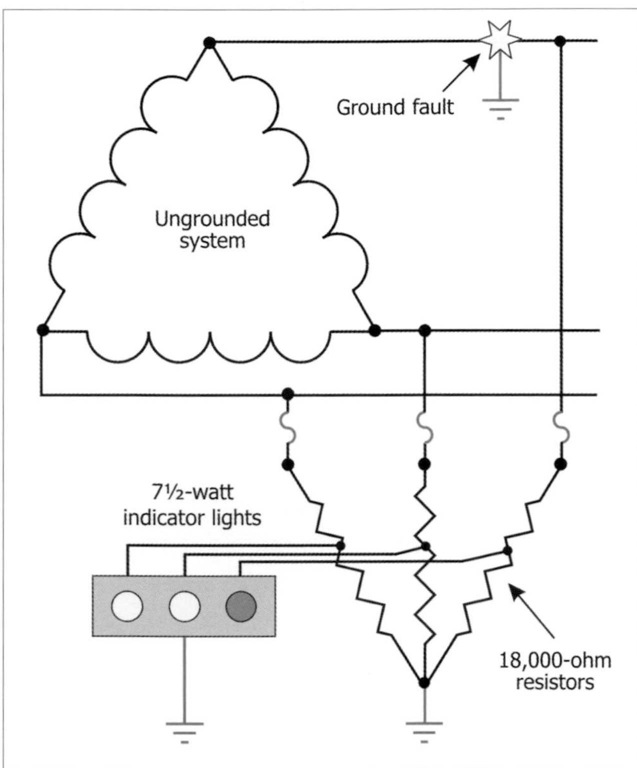

Figure 2-11. Ground detectors (lights)

Ungrounded System Problems

An ungrounded system exists only in theory, in a laboratory or at the electrical distribution transformers hanging on the pole before connection to the plant electrical system. In the real world, ungrounded systems having insulated conductors installed in metallic enclosures are grounded to varying degrees through the distributed leakage capacitance of the system (see figure 2-13). Physically, a capacitor exists whenever an insulating material separates two conductors that have a difference of potential between them.

When any conductor is installed in close proximity to grounded metal, there is a capacitance between them that is increased as the distance between the conductors is reduced. In 600-volt systems, the two greatest sources of capacitance to ground are conductors in metal conduit and windings, such as for motors and transformers. In both cases, conductors are separated from grounded metal by fairly thin insulation. The capacitance to ground is known as the *leakage capacitance*, and the current from the conductors to ground is known as the *leakage current* or *charging*

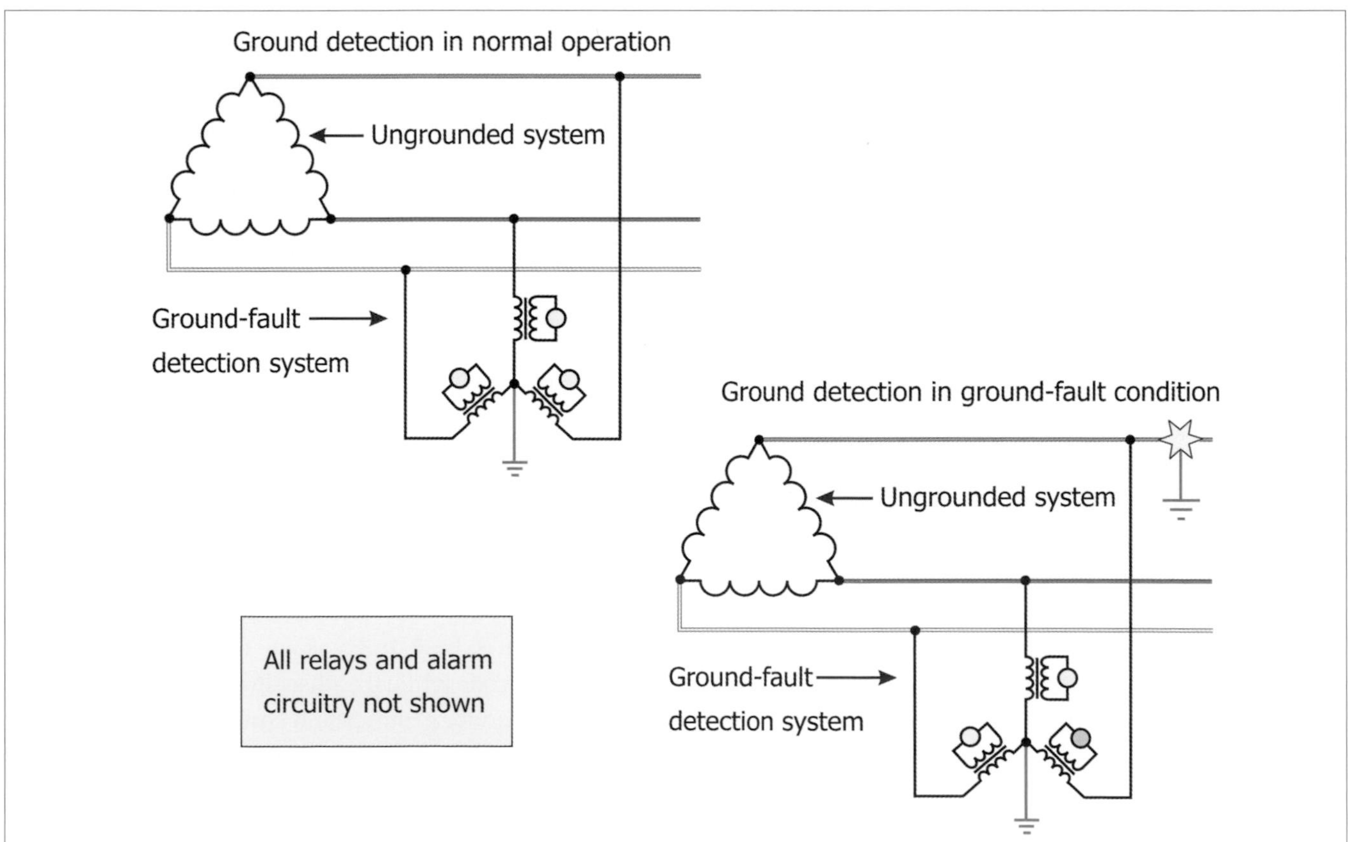

Figure 2-12. Ground detection is required on ungrounded systems (not all circuitry shown)

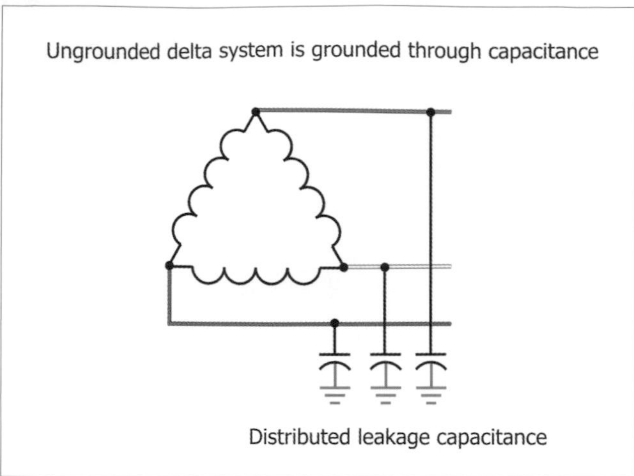

Figure 2-13. Actual ungrounded delta system indicating the presence of distributed leakage capacitance

current. This capacitance is distributed throughout the electrical system but electrically acts like it is a single, lumped capacitance.

Disadvantages of operating systems ungrounded include but are not limited to the following:

- Power system overvoltages are not controlled. In some cases, these overvoltages are passed through transformers into the premises wiring system. Some common sources of overvoltages include: lightning, switching surges and contact with a high-voltage system.
- Transient overvoltages are not controlled, which, over time, can result in insulation degradation and failure.
- System voltages above ground are not necessarily balanced or controlled.
- Destructive arcing burndowns can result if a second fault occurs before the first fault is cleared.

Ungrounded systems have the characteristic that they are subject to relatively severe transient overvoltages. Such overvoltages can be caused by external disturbances as well as internal faults in the wiring system and easily can reach a value of five to six times normal voltage. An actual case involved a 480-volt ungrounded system. Line-to-ground potentials in excess of 1200 volts were measured on a test meter. The source of the trouble was traced to an intermittent or sputtering (arcing) line-to-ground fault in a motor starting autotransformer. These faults are not uncommon on 480-volt ungrounded systems. During the two-hour period that this arcing fault existed, between 40 and 50 motor windings had failed.

Circuit-switching operations can also be responsible for the creation of transient overvoltages in ungrounded systems. These generally are of short duration and typically reach only two to three times nominal system voltage.

Experience has shown that these overvoltages that easily reach several times the system voltage can cause failure of insulation at locations on the system other than at the point of the fault and can result in future system failures. This often occurs at a system weak point such as in a motor or transformer winding.

Locating the ground fault can be troublesome. While it is easy to spot a ground fault on a one-line diagram, locating it in a plant with a complex electrical system can be much more difficult, unless sophisticated ground-fault detection equipment has been installed. The first step is to open the feeders one at a time and observe the ground detection indicator. After finding the feeder with the ground fault, branch circuits are disconnected, one at a time, until the offending circuit is located. A significant loss of plant operation time can occur during this process. This is contrasted with a grounded system where only the offending equipment is taken off the line by the circuit protective devices.

Often, overcurrent devices are set above the current level of the fault in ungrounded systems. When arcing faults occur, destructive burndowns of electrical equipment can result. The arcing fault releases a tremendous amount of energy such that conductors and metal enclosures in the vicinity are destroyed.

When the first ground fault occurs on the 480-volt ungrounded system, the other conductors of the system rise to a level of 480 volts to ground. This presents an additional risk of shock to operation and maintenance staff. This can be contrasted to a 480Y/277-volt grounded wye system where the voltage to ground does not exceed 277 volts while the phase-to-phase voltage is 480 volts, even under ground-fault conditions.

Systems Permitted but Not Required to Be Grounded

There are electrical systems of 50 volts to 1000 volts that are permitted, but not required, to be grounded. It should be noted that whether the system is required to be grounded or is grounded by choice, all grounding and bonding requirements for grounded systems must be followed (see 250.21). These alternating-current systems are:

1. "Electric systems used exclusively to supply industrial electric furnaces for melting, refining, tempering, and the like (see photo 2-1)

2. "Separately derived systems used exclusively for rectifiers that supply only adjustable-speed industrial drives

3. "Separately derived systems supplied by transformers that have a primary voltage rating less than 1000 volts, provided that all the following conditions are met:

a. The system is used exclusively for control circuits.

b. ...only qualified persons service the installation.

c. Continuity of control power is required.

4. "Other systems that are not required to be grounded in accordance with the requirements of 250.20(B)."

Ground detection is required for ungrounded ac systems addressed in 250.21(A)(1) through (4) where the system operates at not less than 120 volts and not exceeding 1000 volts [250.21(B)]. For three-phase ac systems of 480 to 1000 volts that are high-impedance grounded-neutral systems [see 250.36]. An impedance device, usually a resistor, limits the current of the first ground fault to a low value. All of the following conditions must be met before high-impedance grounded (neutral) systems are permitted:

1. Only qualified persons service the system.
2. Ground detectors are installed.
3. Line-to-neutral loads are not supplied.

Factors to Consider Regarding System Grounding

Where grounding of the electrical system is optional, the advantages and disadvantages of grounding must be carefully weighed by the plant owner or electrical designer to make the best decision.

In the long run, greater service continuity may be obtained with grounded systems rather than ungrounded ones. Faults can be isolated to the feeder or circuit affected and cleared without disrupting the entire system. This is obviously a major consideration if the equipment or circuit affected is critical to the plant operation. This has to be balanced against the ungrounded system's tolerance of the first line-to-ground fault but with possible deterioration of conductor insulation from transient overvoltages and possible serious damage caused by a second ground fault on the system.

Ungrounded systems are often utilized with industrial electric furnaces where continuity of the service is critical and an outage due to a ground fault cannot be tolerated.

Ground detectors are required on these systems to monitor the system for ground faults that can be cleared prior to a second ground fault on the system.

Photo 2-1. Industrial electric furnaces

Bolted Faults

A common myth is that ground-faults are always *bolted* or solidly connected and that there will be a great deal of fault-current, which will cause the overcurrent device to open or operate and clear the fault. Bolted faults rarely occur, while sparking, intermittent or arcing faults are quite common. The higher impedance in the arc limits the total current, so standard overcurrent devices can be ineffective.

Arcing faults produce a great deal of heat in the vicinity of the fault and can lead to destructive burndowns of electrical switchboards and motor control centers.

This is the reason the *Code* requires equipment ground-fault protection systems for 3-phase, 4-wire, wye-connected systems of certain voltage and amperage services and feeders (see chapter fifteen of this text for additional information on this subject).

Bolted faults are usually utilized in testing laboratories under controlled conditions. A bolted fault, typically a 3-phase bolted fault, is used for the purposes of determining interrupting ratings of overcurrent protective devices as well as bracing of busbars, and so forth. This is generally considered to be the worst-case condition that causes the greatest amount of fault current. For example, UL Standard 891 requires short-circuit current testing of equipment to determine a switchboard overall short-circuit current rating. For additional information, see UL 891—*Standard for Switchboards and UL Standard 67 for Panelboards.*

[1] *NEC 2008 National Electrical Code*, 250.2, (Quincy, MA, National Fire Protection Association, 2007), p. 70-97.

[2] *NEC 2008*, 250.4, p. 70-98.

[3] *NEC 2008*, Article 100, p. 70-32.

[4] *NEC 2008*, 250.4(B), p. 70-99.

Review Questions

The questions included here were developed using material included in this chapter. The answers can be found by reviewing the text. It is also important that students make use of *NEC-2008*, where many answers can be found.

1. Where operating at less than 50 volts, alternating-current systems are required to be grounded where supplied by transformers if the supply voltage to the transformer exceeds ____ volts to ground.
   a. 100
   b. 110
   c. 140
   d. 150

2. Where operating at less than 50 volts, ac systems are required to be grounded where supplied by transformers if the transformer supply system is ____.
   a. bonded
   b. ungrounded
   c. identified
   d. approved

3. Conductors installed on the outside of buildings where ac systems operate at less than 50 volts are required to be grounded when they are installed outside as ____ conductors.
   a. overhead
   b. underground
   c. optical fiber
   d. Type IGS

4. Alternating-current system grounding is required for 50 to 1000 volt systems supplying premises wiring or premises wiring systems where the system can be grounded so the maximum voltage to ground on the ungrounded conductors does not exceed ____ volts.
   a. 180
   b. 150
   c. 240
   d. 208

5. Alternating-current system grounding is required for 50 to 1000 volt systems supplying premises wiring or premises wiring systems where the system is 3-phase, 4-wire, wye connected in which the neutral is used as a ____ conductor.
   a. bonding
   b. circuit
   c. equipment grounding
   d. isolated

6. Alternating-current system grounding is required for 50 to 1000 volt systems supplying premises wiring, or premises wiring systems where the system operates at 3 phase, 4 wire, delta connected in which the midpoint of one phase is used as a ____.
   a. bonding jumper
   b. equipment grounding conductor
   c. circuit conductor
   d. switch leg

7. Alternating-current system grounding is not required for 50 to 1000 volt electric systems used exclusively to supply industrial electric furnaces for ____.
   a. melting
   b. refining
   c. tempering
   d. all of the above

8. Alternating-current system grounding is not required for 50 to 1000 volt separately derived systems used exclusively for rectifiers supplying only ____.
   a. motor control centers
   b. adjustable speed industrial drives
   c. commercial buildings
   d. Class III hazardous (classified) locations

9. Alternating-current system grounding is not required for 50 to 1000 volt separately derived systems supplied by transformers that have a primary voltage rating less than 1000 volts if which of the following conditions is or are met. ____
   a. system is used for only control circuits.
   b. only qualified persons service installation, ground detectors are installed on the control system.
   c. continuity of power is required.
   d. all of the above

10. Alternating-current system grounding is not permitted for 50 to 1000 volt systems supplying premises wiring, or premises wiring systems where ____ are permitted or required for flammable anesthetizing systems in health care facilities.
   a. grounded power systems
   b. isolated power systems
   c. impedance grounded systems
   d. medium voltage power systems

2 Soares Book on Grounding and Bonding

11. High impedance grounded neutral systems are permitted for premises wiring, or premises wiring systems where the three-phase, ac systems is rated 480 to 1000 volts if which one of the following conditions is or are met:
   a. only qualified persons will service the system.
   b. ground detectors are installed.
   c. line-to-neutral loads are not supplied.
   d. all of the above

12. Ac systems of ____ and over must be grounded if they supply mobile or portable equipment as specified in Section 250.188.
   a. 250 volts
   b. 1000 volts
   c. 480 volts
   d. 600 volts

13. Ac systems operating at over ____ volts are permitted (not required) to be grounded where they do not supply mobile or portable equipment.
   a. 1000
   b. 300
   c. 277
   d. 480

14. Equipment operating at over ____ volts is commonly found in mobile rock crushing plants and batch plants. Other applications are for open pit mining operations.
   a. 277
   b. 300
   c. 600
   d. 1000

15. A ____ premises wiring system is one that is derived from a generator, transformer or converter windings that have no direct electrical connection to the supply conductors originating in another system.
   a. identified
   b. open neutral
   c. isolated power
   d. separately derived

16. Separately derived systems must be grounded as specified in Section ____.
   a. 250.24(A)
   b. 250.36
   c. 250.30(A)
   d. 250.32(A)

17. An example of a separately derived system include(s) ____ with no direct electrical connection between the primary and secondary.
   a. phase converters
   b. transformers
   c. elevator motors
   d. Class 1 systems

18. Examples of separately derived systems include generator systems used for emergency, required standby or optional standby power that have all conductors including a neutral switched by ____.
   a. double pole
   b. single pole
   c. transfer equipment (transfer switches)
   d. backfed device

19. Examples of separately derived systems include ac or dc systems derived from ____.
   a. inverters
   b. rectifiers
   c. generators
   d. all of the above

20. Certain systems are permitted to be operated ungrounded and usually are located in industrial or agricultural applications. Typical systems that are operated ungrounded include which of the following ____.
   a. 2300-volt, three-phase, three-wire, delta connected.
   b. 4600-volt, three-phase, three wire, delta connected.
   c. 13,800-volt, three phase, three-wire, delta connected.
   d. all of the above

21. Ground detector indication systems are required to be installed to indicate that a ground fault has occurred on the ____ systems.
   a. ungrounded
   b. grounded
   c. bonded
   d. identified

22. For an ungrounded electrical system, the first phase-to-ground fault
   a. does nothing
   b. causes a fuse to blow
   c. unintentionally or accidentally grounds the system
   d. none of the above

23. For an ungrounded electrical system, a second ground fault on a different phase of the system
    a. does nothing
    b. causes an overcurrent device to operate
    c. grounds the system
    d. none of the above

24. Disadvantages of operating an ungrounded electrical system include
    a. power system voltages are not controlled
    b. transient overvoltages are not controlled
    c. destructive arcing burndowns can occur on a second ground fault
    d. all of the above

25. Systems not permitted to be grounded include which of the following:
    a. 120-volt, single-phase, three-wire system
    b. High-impedance grounded neutral systems
    c. A transformer secondary with a primary input of 460 volts
    d. A low voltage lighting system operating at 24 volts

26. Effective ground-fault current paths are created by _______all of the electrically conductive materials that are likely to be energized by the wiring system.
    a. connecting to ground
    b. connecting to the grounded conductor
    c. bonding together
    d. connecting to a grounding electrode

27. The earth shall not be considered as _______.
    a. an effective ground-fault current path
    b. being at zero potential
    c. non-conductive
    d. being round

28 Ungrounded 480 volt systems used exclusively for rectifiers that supply only adjustable-speed industrial drives are required to have_______
    a. temporary grounds
    b. ground detectors
    c. overcurrent protection at 58 percent of the full load current
    d. all of the above

29. Where a generator is the alternate source of power for a building and the transfer switch does not open the grounded (neutral) conductor, the generator shall be grounded in accordance with which of the following?
    a. 250.30(B)
    b. 250.36
    c. 250.30(A)
    d. 250.24(A)

30. A secondary circuit for a swimming pool lighting system supplied by an isolating transformer shall meet which of the following?
    a. Shall be grounded per 250.30(A)
    b. Shall be grounded per 250.36
    c. Shall not be grounded in accordance with 250.30(B)
    d. Shall not be grounded per 250.22(5)

31. A ground detection indicator is not required on which of the following systems?
    a. 480Y/277, 3-phase, 4-wire grounded wye
    b. 208Y/120, 3-phase, 4-wire grounded wye
    c. 120/240, 1-phase, 3-wire grounded
    d. All of the above

32. Ground detectors are required on all ungrounded systems, except which of the following?
    a. 480-V delta ungrounded
    b. 240-V delta ungrounded
    c. 416 kV delta ungrounded
    d. 600-V delta ungrounded

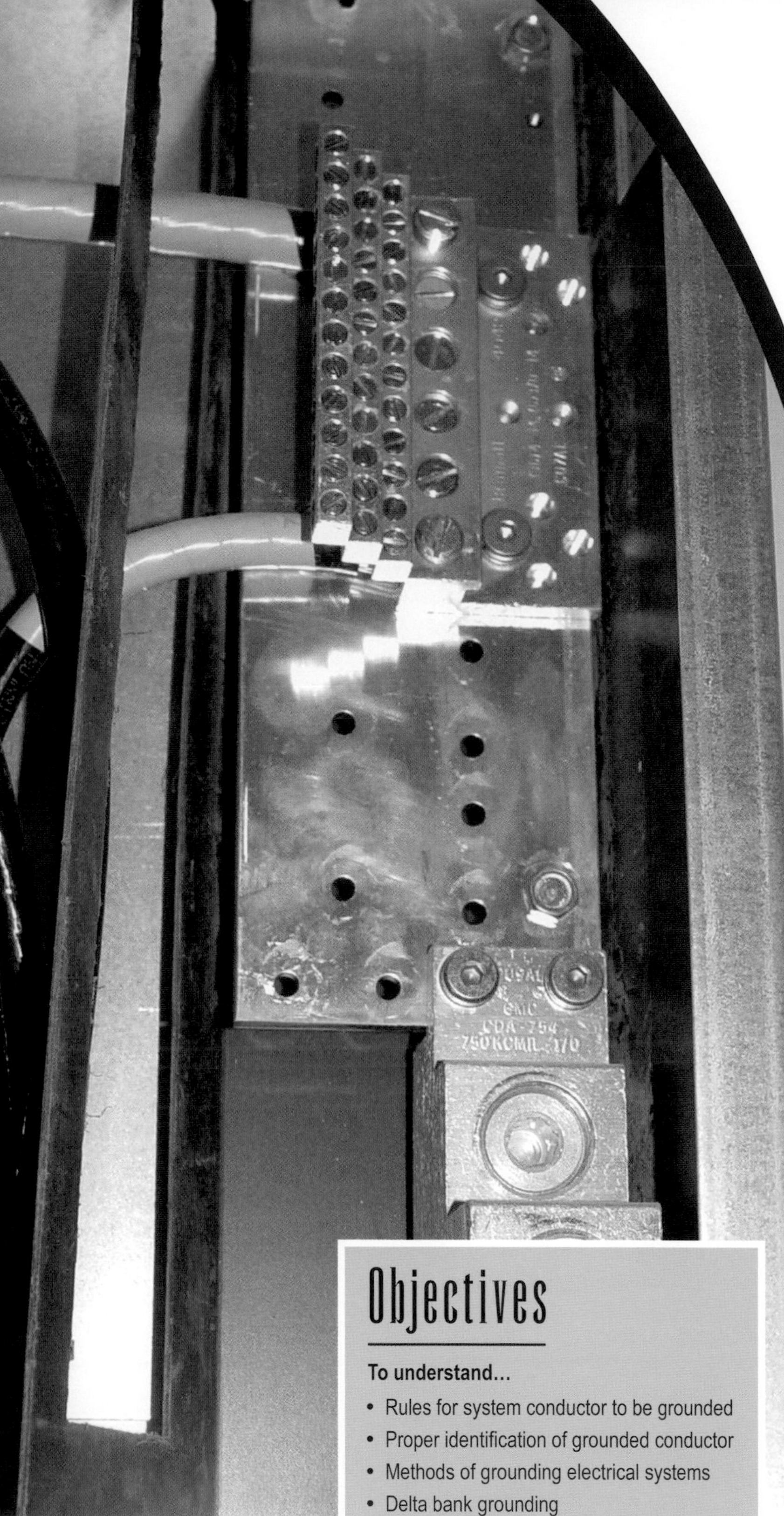

Objectives

To understand...

- Rules for system conductor to be grounded
- Proper identification of grounded conductor
- Methods of grounding electrical systems
- Delta bank grounding
- Grounding rules for ungrounded systems
- Corner-grounded delta systems

Chapter 3

Grounding Electrical Systems

Section 250.20 requires that many electrical systems be grounded. It is important to understand that one is dealing with the electrical system and not the service equipment, disconnecting means, or non-current-carrying metal parts or enclosures at this point. Electrical energy is typically delivered to the customer by the serving utility by either a grounded or ungrounded system or sometimes by both systems (see figure 3-1). Electrical utilities have tariffs and standards that dictate whether or not they will deliver a system of a certain voltage level and phase configuration as grounded or ungrounded systems. Many utilities require that all low-voltage (600 volts and under) systems be grounded. Others will supply 3-phase, 480-volt delta-connected systems ungrounded, while they insist on furnishing wye systems only in a 480Y/277-volt, grounded-wye configuration.

Large industrial plants may purchase power at medium or high voltage levels, such as 69,000-volt, 115,000-volt, or 230,000-volt levels, and own and maintain their electrical distribution systems. Transformers, capacitor banks, controls, overcurrent devices and relaying systems are then installed at customer-owned switchyards. Power is distributed to utilization points on the premises where transformers are installed as needed for the utilization voltages desired. Generally, electrical systems at the utilization level are grounded at the voltage levels and configurations required by the *NEC*.

Definitions

Ground. "The earth."

Grounded (Grounding). "Connected (connecting) to ground or to a conductive body that entends the ground connection."

Grounded Conductor. "A system or circuit conductor that is intentionally grounded."[1]

Conductor to Be Grounded

Where the electrical system is grounded, 250.26 specifies which conductor in an alternating-current system shall be grounded (see figure 3-2). For alternating-current premises wiring systems, the conductor required to be grounded is specified below:

- Single-phase, 2-wire: one conductor (either one)
- Single-phase, 3-wire: the neutral conductor
- Multiphase systems having one wire common to all phases: the common conductor

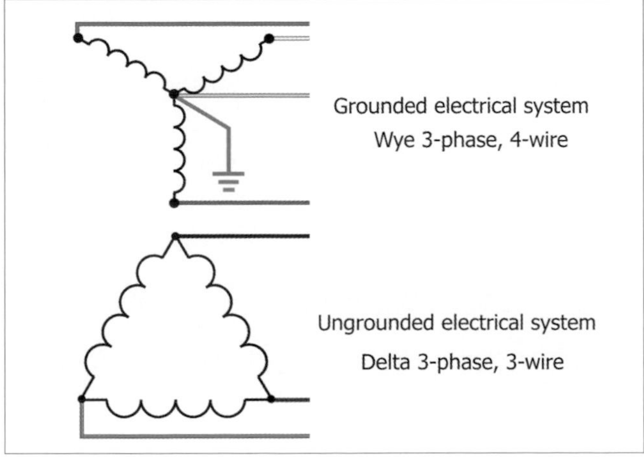

Figure 3-1. Grounded systems and ungrounded systems

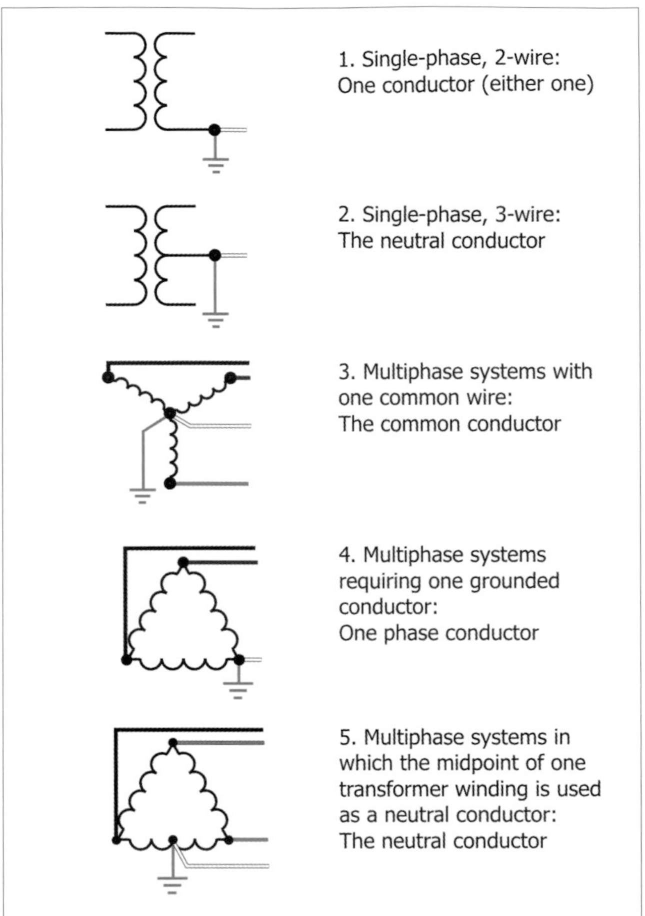

1. Single-phase, 2-wire:
One conductor (either one)
2. Single-phase, 3-wire:
The neutral conductor
3. Multiphase systems with
one common wire:
The common conductor
4. Multiphase systems
requiring one grounded
conductor:
One phase conductor
5. Multiphase systems in
which the midpoint of one
transformer winding is used
as a neutral conductor:
The neutral conductor

Figure 3-2. Conductor required to be grounded in grounded systems

- Multiphase systems requiring one grounded phase: one phase conductor.
- Multiphase systems in which one phase is used as in (2) the neutral conductor (see figure 3-2 above).

The *NEC* provides a clear differentiation between a *neutral conductor* of a system and a *neutral point* of a system. These two terms are appropriately used in each *NEC rule* where only the term neutral was used previously. The following definitions of the terms *neutral conductor* and *neutral point* are found in Article 100 (see figure 3-3).

Neutral conductor. The conductor connected to the neutral point of a system that is intended to carry current under normal conditions.

Neutral point. The common point on a wye-connection in a polyphase system or midpoint on a single-phase, three-wire system, or midpoint of a single-phase portion of a three-phase delta system, or a midpoint of a three-wire, direct-current system.[2]

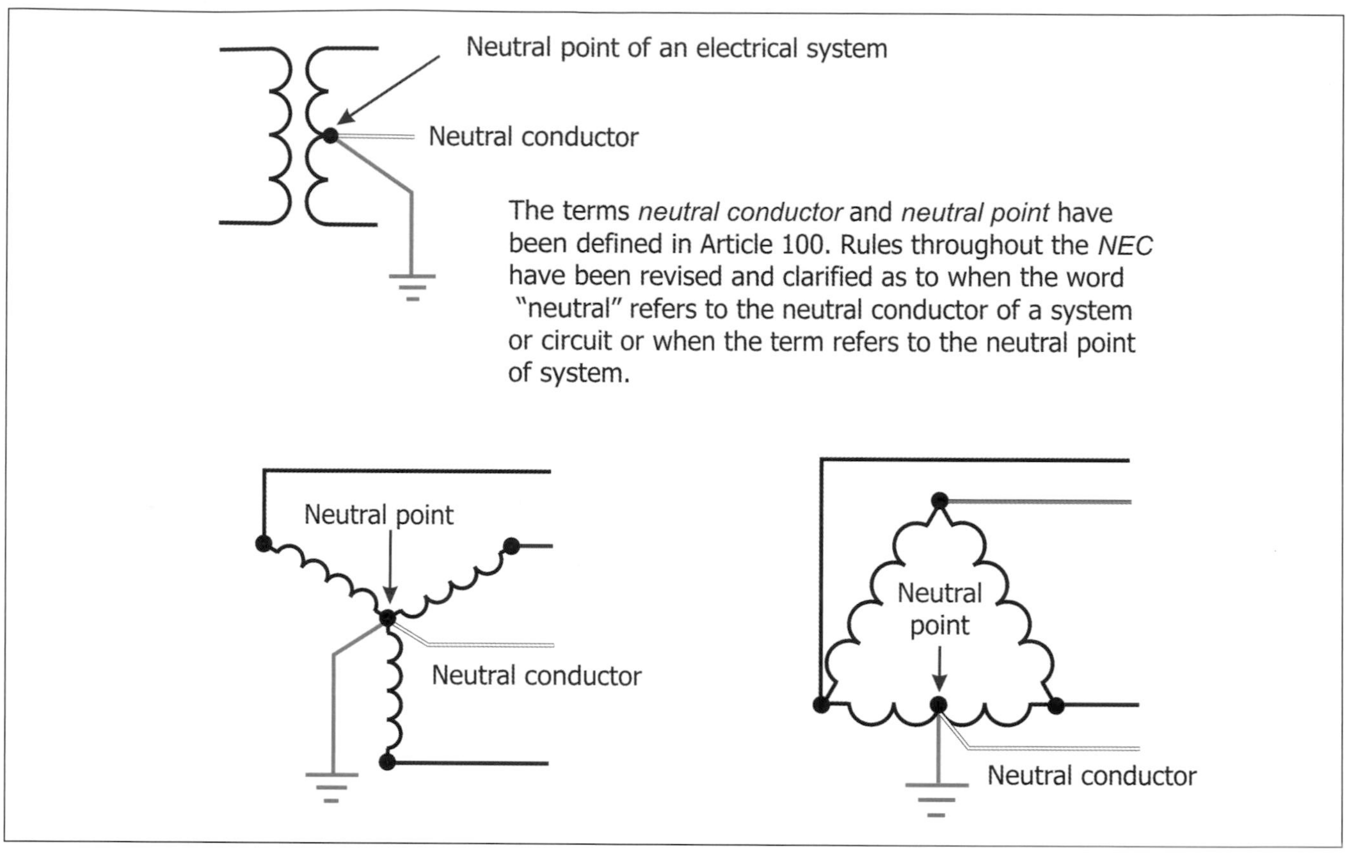
Neutral point of an electrical system
Neutral conductor
The terms *neutral conductor* and *neutral point* have been defined in Article 100. Rules throughout the *NEC* have been revised and clarified as to when the word "neutral" refers to the neutral conductor of a system or circuit or when the term refers to the neutral point of system.
Neutral point
Neutral conductor
Neutral point
Neutral conductor

Figure 3-3. Identification of grounded conductors [general requirements are in 200.6(A) and (E)]

An important aspect of these new terms is that a clear differentiation is established between neutral conductors and the point on a system where they are connected. While most neutrals of electrical systems are grounded, not all grounded conductors of systems are neutrals. For example, a ground phase conductor of a three-phase, three-wire, delta-connected system is a grounded phase conductor, not a neutral of the system (see figure 3-2.)

Identification of Grounded Conductor

Grounded conductors are required to be identified by the means specified in 200.6 (see figure 3-4). Rules are provided for identification of grounded conductors of sizes 6 AWG or smaller, conductors larger than 6 AWG, flexible cords, grounded conductors of different systems, and grounded conductors of multiconductor cables.

6 AWG or smaller

Three general means of identification of 6 AWG or smaller insulated grounded conductors are provided which include:

- a continuous white outer finish, or
- a continuous gray outer finish, or
- three continuous white stripes on other than green insulation.

This identification must be along the entire length of the conductor. It is generally not permitted to paint or phase-tape grounded conductors of 6 AWG or smaller.

Photo 3-1. Grounded neutral conductor identification

For specific conductor assemblies or applications, the following grounded conductor identification is provided regardless of size:

- For Type MI cable, the grounded conductor is permitted to be identified by distinctive marking at its termination at the time of installation.
- For single-conductor, sunlight-resistant, outdoor-rated cable for solar photovoltaic systems, the grounded conductor is permitted to be identified at terminations by distinctive white markings. This will usually be made with white adhesive vinyl marking tape.
- Fixture wire can be identified as provided in 402.8.
- For aerial cable, the grounded conductor is permitted to be identified by any of the general means identified above or by a ridge located on the exterior of the cable.

Conductors larger than 6 AWG

Conductors larger than 6 AWG are permitted to be identified either like conductors 6 AWG or smaller or, at the time of installation, by distinctive white or gray marking at each termination (see photo 3-1). This distinctive marking usually consists of adhesive vinyl tape or paint. Where so marked, the marking must encircle the conductor or insulation so it is visible from all sides.

Flexible cords

The grounded conductor within a flexible cord is permitted to be identified by one of the three methods included for conductors 6 AWG or smaller or by the methods included in 400.22. These additional methods include: colored braid, tracer in the braid, colored insulation, colored separator, tinned conductors, and surface marking.

Grounded conductors of different systems

In larger electrical installations, it is common to have more than one electrical system installed in the same enclosure, such as raceways like conduits and wireways, pull boxes and cables. For example, grounded conductors from both a 480Y/277-volt, three-phase, four-wire system and a 120/240-volt, single-phase

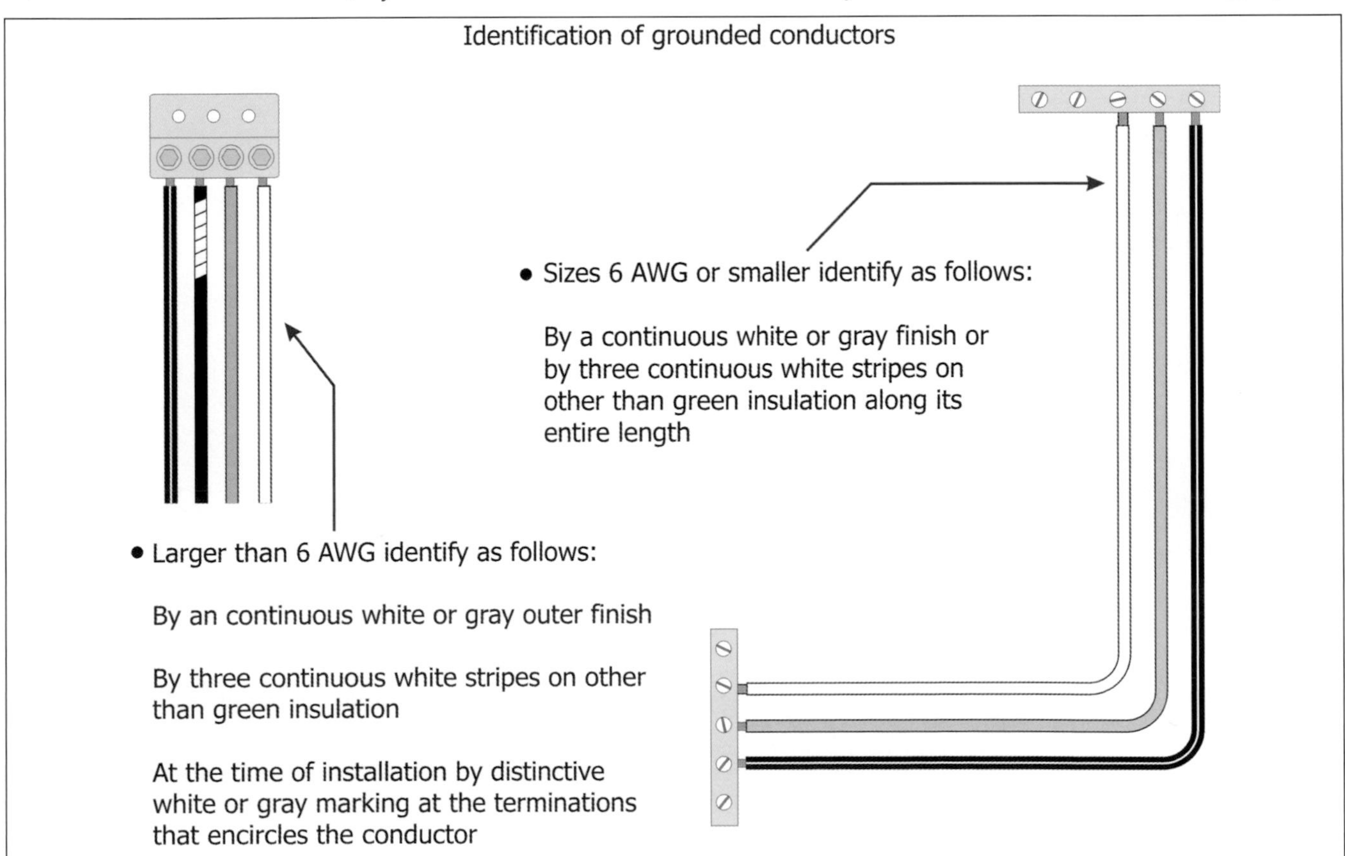

Figure 3-4. Identification of grounded conductors [general requirements are in 200.6(A) and (E)]

system may be in the same raceway or other enclosure. Where this happens, one of the grounded conductors is required to be identified in accordance with 200.6(A) or (B), as covered above, for conductors smaller or larger than 6 AWG. Each other system grounded conductor is required to be identified differently by one of the means provided for conductors smaller or larger than 6 AWG or by white or gray colored insulation with a readily distinguishable different colored stripe that is not green and runs along the insulation. Section 200.6(D) requires the means of identification of grounded conductors of different systems to be permanently posted at each branch-circuit panelboard. It is interesting that this requirement to post a means of identification applies only for each branch-circuit panelboard, even though feeders should be expected to meet the same requirement.

Grounded conductors of multiconductor cables
Insulated grounded conductors in multiconductor cables are required to be identified generally as specified in 200.6(A). Where only qualified persons will service and maintain the installation, grounded conductors of multiconductor cables are permitted to be permanently identified at terminations by a distinctive white marking or by other effective means.[3]

Use of conductors with white insulation. The previous permission to install conductors with white insulation in a conduit or other raceway and phase-tape them and use them as ungrounded conductors has been removed from the *NEC* (see 200.7 for additional requirements or restrictions on the use of conductors with white, gray, or three white stripes on other than green insulation).[4]

A fine print note in 200.6 indicates that the color gray may have been used in the past as ungrounded conductors.

Methods of Grounding Electrical Systems

A variety of methods are used to ground electrical systems. The method chosen will vary, depending upon the system voltage, code requirements, plant owner specifications or engineer's philosophy. The various methods commonly used are shown in figure 3-5 and are as follows:

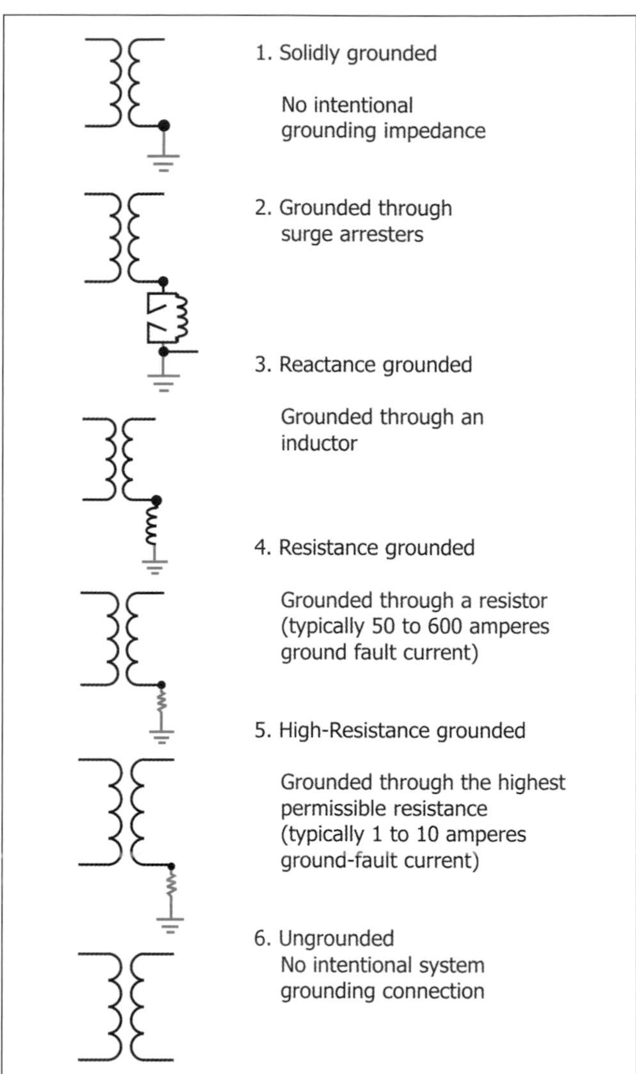
1. Solidly grounded
No intentional grounding impedance
2. Grounded through surge arresters
3. Reactance grounded
Grounded through an inductor
4. Resistance grounded
Grounded through a resistor (typically 50 to 600 amperes ground fault current)
5. High-Resistance grounded
Grounded through the highest permissible resistance (typically 1 to 10 amperes ground-fault current)
6. Ungrounded
No intentional system grounding connection

Figure 3-5. Various methods of system grounding

Solidly grounded: no intentional grounding impedance. Solidly grounded refers to the connection of the electrical system or of a separately derived system of a generator, power transformer, or grounding transformer directly to the station ground or grounding electrode system without intentionally introducing an impedance.

Surge arrester grounded: to permit the use of reduced rated (80 percent) surge arresters. Grounding of surge arresters rated near, but not less than, 80 percent of line-to-line voltage constitutes a grounded system. This will carry with it a line-to-ground circuit current of at least 60 percent of the three-phase short-circuit value.

Reactance grounded: the system is grounded through a reactor. This grounding is accomplished by a reactor (grounding transformer), a device that introduces impedance to the circuit, the principal element of which is inductive reactance.

Resistance grounded: intentional insertion of resistance into the system grounding connection. For resistance grounding, the system is grounded by connecting the system neutral to ground through a resistor. This grounding is accomplished by a resistor that introduces impedance to the circuit, the principal element of which is resistance.

High-resistance grounded: the insertion of nearly the highest permissible resistance into the grounding connection. This grounding is typically accomplished by a resistor that introduces impedance to the circuit, the principal element of which is resistance. For high-resistance grounding, the system is grounded by the system neutral through a resistor (may be a grounding transformer) that typically limits the ground-fault current to 10-amperes or less. High-resistance grounding maintains control of transient overvoltages, but might not furnish sufficient current for ground-fault relaying. The protective scheme usually associated with high-resistance grounding is detection and alarm rather than immediate *tripout* (see 250.36 and chapter four for additional information). High-resistance grounding is typically applied to 3-phase systems such as 480-volt systems and also higher voltage wye-connected systems, such as 12.47 kV systems.

Ungrounded: no intentional system grounding connection. Ungrounded systems are used in industrial plants and where desired for manufacturing processes to give an additional degree of service continuity. While a ground fault on one phase of an ungrounded system generally does not cause a service interruption, the occurrence of a second ground fault on a different phase before the first fault is cleared does result in an outage. A ground detection system is recommended, and adequate maintenance and repair will minimize interruptions.

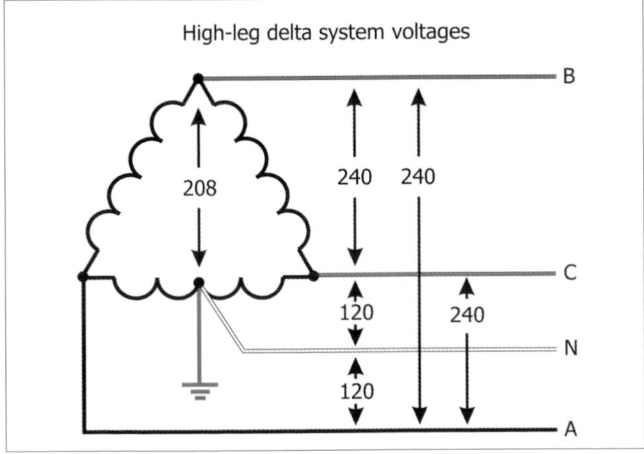

Figure 3-6. High-leg delta system voltages

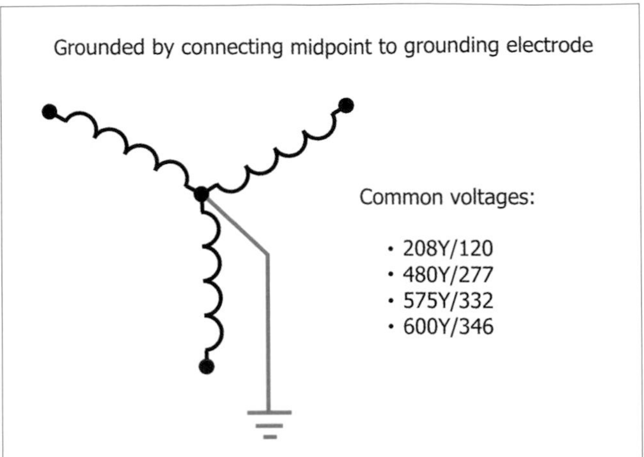

Figure 3-7. Grounding a Wye-connected system

Low-voltage systems, 600 volts and below, that are grounded are almost always solidly or high-impedance grounded; medium-voltage systems are usually either solidly or resistance grounded; and high-voltage systems above 34.5 kV are nearly always grounded through surge arresters or ungrounded.

Delta Bank Grounding

Where three transformers that have center taps are connected in a delta bank, only one transformer may have its midpoint grounded. Any attempt to use a second transformer of the delta bank to supply a second 3-wire, single-phase source would require grounding the midpoint of a second transformer. However, since there will be a difference of potential between the midpoints of the different transformers, there would be an abnormal current through both grounded neutrals. This would ultimately cause heating or a short circuit, depending on how both neutrals were connected.

With midpoint grounding of one-phase winding in a delta bank, one of the phase conductors operates at a higher voltage to ground than the other two. In practice, this *high leg* is identified as required by 110.15 and 230.56 by orange color-coding or other effective means (see figure 3-6).

The voltage of the phase conductor with the higher voltage to ground is determined by the following formula:

½ of phase-to-phase voltage x *1.732 = high-leg voltage to ground*

For example, a 240-volt, 3-phase system center-tapped to establish a neutral:

½ of 240 = 120 x 1.732 = 208 V high-leg voltage to ground

Section 408.3(F) requires switchboard and panelboard enclosures containing high-leg systems to be provided with a permanent field-applied marking as follows:[5]

CAUTION______ PHASE HAS ______VOLTS TO GROUND

Grounding Existing Ungrounded Systems

In some cases, it is desirable to ground electrical systems that originally were installed ungrounded or are permitted, but are not required, to be grounded (see chapter two for a thorough discussion of the subject). There are four methods commonly in use for grounding of ungrounded systems of 600 volts or less. Solid grounding is used in all the methods. They are as follows:

Grounding the neutral of wye-connected secondary windings of a transformer bank

As shown in figure 3-7, this is the most universal and commonly used method of grounding a system. Standard voltages are 208Y/120, 480Y/277, 575Y/332 and 600Y/346 volts. The first two voltage systems—208Y/120 and 480Y/277—are in most common use today in the United States, with the growth tending in favor of the 480Y/277-volt system owing to the better economies of that system. In general, the primary windings of the transformers serving those wye systems are delta-connected.

Grounding a delta bank with a zigzag grounding transformer

This method is best adapted to an existing 3-wire, 3-phase delta-connected distribution system that is ungrounded, and where it is desired to ground the system to obtain the advantages gained through operating the system grounded. For new systems, the use of transformers with wye-connected secondaries is advisable to obtain a grounded system. The neutral derived by the zigzag transformer can be used as a system current-carrying conductor if the zigzag grounding transformers are sized for the maximum unbalanced current (see 220.61 for the method of calculating the maximum unbalanced current on the grounded conductor).

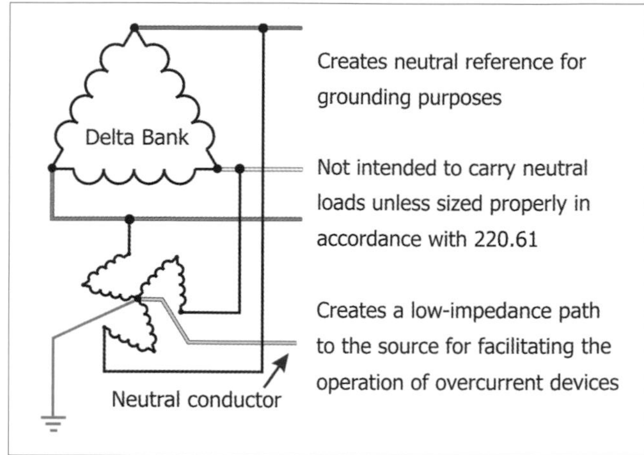

Figure 3-8. Grounding a delta system using a zigzag grounding transformer

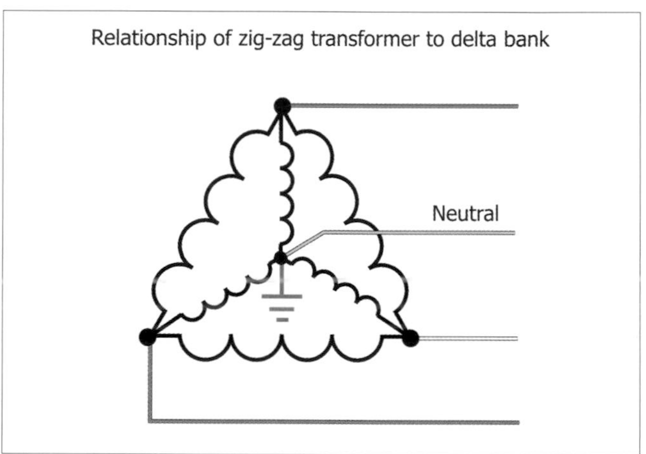

Figure 3-9. Relationship of zigzag transformer to delta bank transformer

A zigzag transformer, shown applied to a delta system in figure 3-8, obtains its name from the manner in which the windings are installed and connected. Windings for each phase are on the same core leg. All windings have the same number of turns but each pair of windings on the same core leg is wound in opposite directions. The impedance of the transformer to 3-phase currents is so high that under normal conditions there will be only a small magnetizing current. If a ground fault develops on one phase of the system, the transformer impedance to ground current is so low that there will be a high ground current. The ground current will divide into three equal parts in the three phases of the grounding transformer.

Such a 3-phase zigzag transformer has no secondary winding. This type of grounding transformer is required to carry rated current for only a short time (the duration of the ground fault). The short-time kVA rating of such a grounding transformer may thus be equal to the rating of a regular 3-phase transformer, yet be only about 10 percent of the physical size.

Electrically, a zigzag grounding transformer connection appears to superimpose a wye-connected system inside the delta-connected transformers. This wye-type connection permits the transformer to be used like a wye transformer, with phase-to-neutral loads to be utilized. In this case, it is important for the zigzag transformer to be sized for the calculated or connected load.

A wye-delta grounding transformer may also be used to provide a neutral for an existing delta-connected ungrounded system, but the use of the zigzag grounding transformer is more practical and economical.

The grounding transformer should be connected directly to the system, as shown in figure 3-9, at the power transformer secondaries. Where a grounding transformer is so connected, one grounding transformer is required for each delta-connected power transformer bank. In this manner, the switching out of any one transformer bank will not disturb the secondary system ground. The grounding transformer and the power transformer are considered a single unit, both being protected by the overcurrent protective means provided for the main power transformer.

The grounding transformer also may be connected directly to the main bus through its own overcurrent protection means. In that event, an alarm should be provided to indicate the system is operating ungrounded if the grounding transformer should be disconnected from the line.

Where a grounding transformer is used on low-voltage systems (600 volts and below), it is important that the equipment grounding conductor be connected to the neutral of the grounding transformer in such a way as to provide a low-impedance path for ground-fault currents to return back to the system. The same grounding electrode point should be used for the neutral of the grounding transformer, and the equipment grounding conductor and a common grounding conductor used for both the equipment grounding conductor and the neutral.

The same disconnecting means and overcurrent protection means for the system would be used as described under method 1.

Information about grounding autotransformers that are zig-zag or T-connected to ungrounded systems is provided in 450.5.

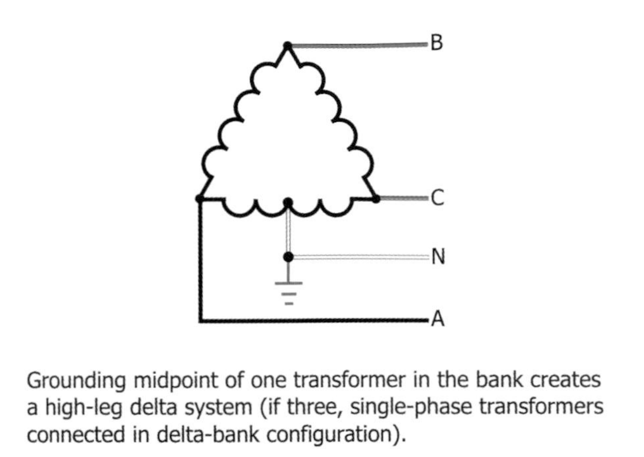

B
C
N
A
Grounding midpoint of one transformer in the bank creates a high-leg delta system (if three, single-phase transformers connected in delta-bank configuration).

Figure 3-10. Grounding a delta bank, creating a high-leg grounded delta system (closed delta shown)

Grounding the midpoint of one of the transformers in a delta-connected system

This method of grounding is commonly used, especially in smaller distribution systems, to provide a three-phase power source and a three-wire single-phase lighting source from the same transformer bank. This, then, becomes a three-phase, four-wire delta-connected system, commonly referred to as a high-leg delta system. At the same time, the advantages of a neutral grounded system are obtained. It is common for the serving utility to use three single-phase transformers connected in a delta configuration for this system. In such a system, the single-phase transformer that supplies the lighting load usually is sized larger than the other two single-phase transformers, which supply the power load only (see figure 3-10).

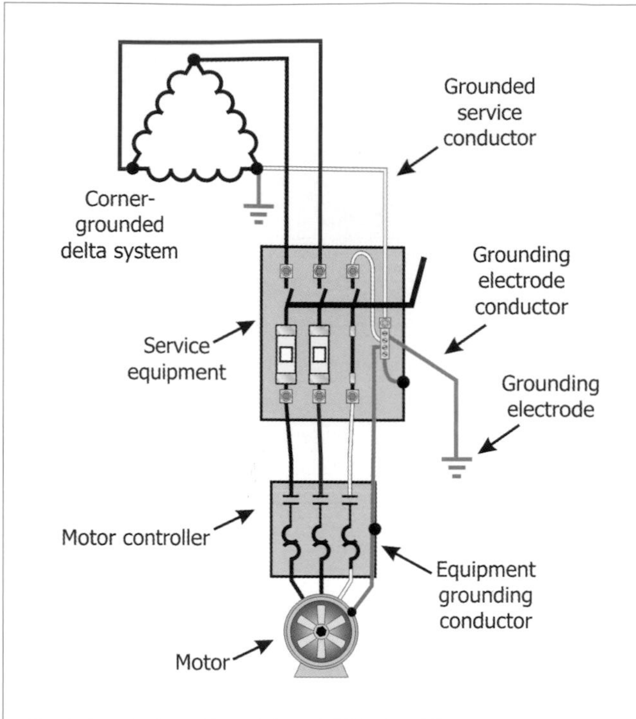

Figure 3-11. Corner-grounded system showing the use of a fused switch

Since the conductors supplying the 3-phase power will now be from a grounded source, the grounded conductor must be run to the service equipment. There, it must be connected to the equipment grounding conductor of the power system and the grounding electrode. This provides a ground-fault current return path of low impedance. Such a connection enhances the safety of the power system. If the power service and the lighting service go to the same building, which is usually the case, the grounding electrode conductor from the power service, as well as the grounding electrode conductor from the lighting service, must be connected to the same grounding electrode (see chapter four for additional information on this subject).

Grounding one corner of a delta system

In the past, most three-phase ungrounded delta distribution systems comprised three single-phase transformers that were connected delta-delta. The main reason was to be able to continue operating if one transformer failed by disconnecting the faulted transformer and operating the bank open-delta, although at reduced capacity.

When the advantages of grounding became very apparent, it first was the practice to ground one corner of the delta (see figure 3-11). The grounded leg must be positively identified throughout the system in accordance with 200.6. This grounded conductor is not a neutral conductor of the system. This system is commonly referred to as a corner-grounded delta system.

Where the grounded conductor is disconnected by the switch or circuit breaker, it is important that a bonding connection be made on the line side of the disconnect switch or circuit breaker. If this is not done, the electrical equipment will become electrically isolated from the grounded conductor when the switch or circuit breaker is in the open position. In this case, any fault that occurs would raise the voltage on the enclosure to a hazardous potential above ground equal to the system voltage.

Generally, fuses are not permitted to be installed in the grounded conductor. However, a two-pole fused switch may be used with fuses in the ungrounded phases. Also, a three-pole switch may be used with a solid neutral in the grounded phase. For services, 230.90(B) does not allow an overcurrent device, other than a circuit breaker that opens all conductors, to be inserted in the grounded service conductor. Section 240.22 generally prohibits connecting an overcurrent device in series with any conductor that is

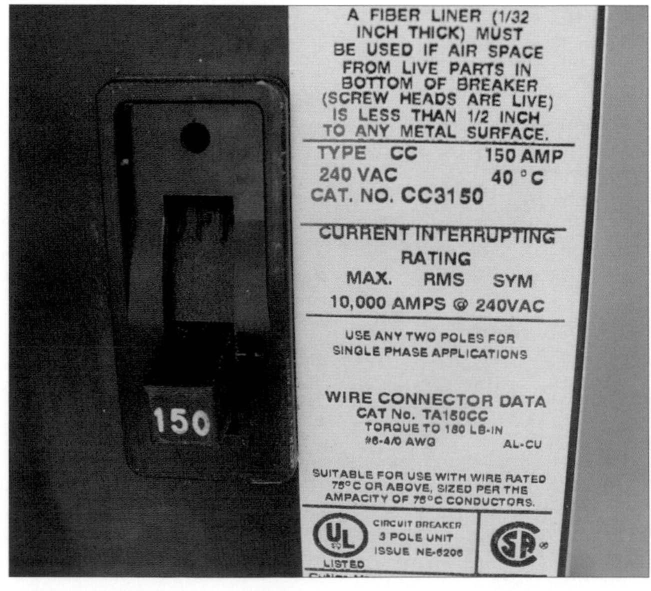

Photo 3-2. Breakers with straight voltage rating (240 V)

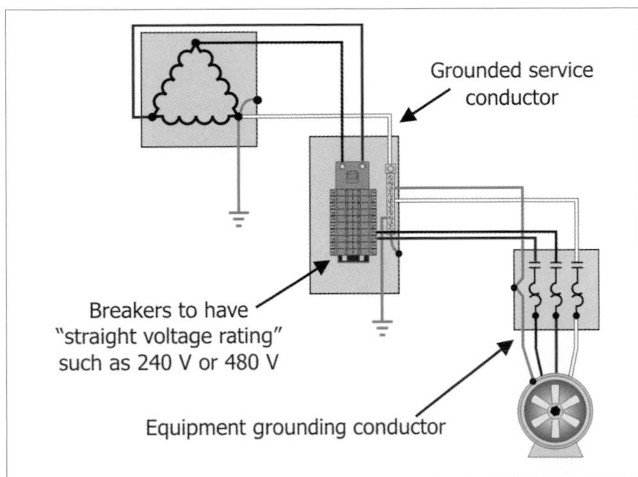

Figure 3-12. Corner-grounded system showing breakers installed in the system

intentionally grounded. Subdivision (1) permits an overcurrent device to be used that "opens all conductors of the circuit, including the grounded conductor, and is designed so that no pole can be operated independently."

This requirement means that three-pole circuit breakers can be used for this purpose while a three-pole switch with a fuse in series with each phase could not, as the switch is not the overcurrent device, the fuse is. Figure 3-11 shows an acceptable use of a fused switch for corner-grounded systems.

Section 430.36 permits a fuse to be inserted in the grounded conductor for the purpose of providing motor overload protection. However, based on the examination of *Code* requirements discussed in the previous paragraph, this application would be limited to being located downstream from the service equipment.

Keep in mind that 250.24(C) requires that the grounded system conductor be run to each service disconnecting means and be bonded to the service disconnecting means enclosure. This is usually accomplished by connecting it to a terminal bar that is mounted inside the service enclosure. Nothing permits this bonding connection to be interrupted by a switch or circuit breaker.

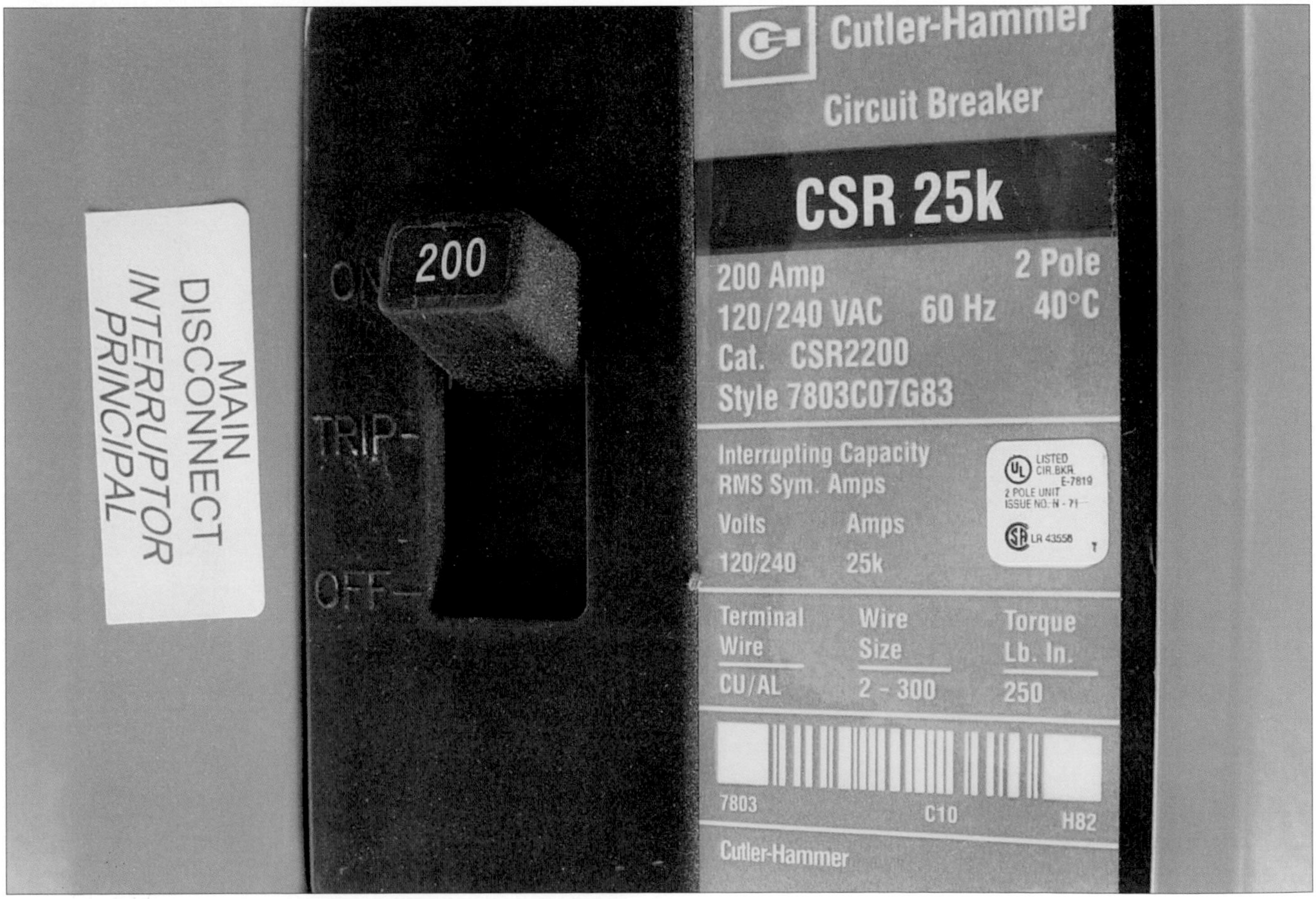

Photo 3-3. Breaker with slash voltage rating (120/240 V)

Section 200.2(B) does not permit electrical enclosures, raceways, or cable armor to be used to establish and maintain continuity of the grounded conductor. Grounded conductors (often neutral conductors) must be terminated to a grounded conductor terminal bar within equipment. (This is discussed more in depth in chapter 4).

Where it is desirable to disconnect the grounded system conductor from the feeder, it is acceptable to route the conductor from the terminal bar through the switch or circuit breaker.

For purposes of installing or grounding a corner-grounded delta system, it is helpful to think of it as a single-phase system. This is illustrated in figures 3-11 and 3-12. The system is grounded at one corner of the delta. Three conductors are taken to the service where the two ungrounded conductors connect to the disconnecting means or circuit breaker. The grounded service conductor is connected to the neutral terminal bar where it is bonded to the enclosure and connected to the grounding electrode conductor.

Where circuit breakers are used as the disconnecting means for corner-grounded systems, they must be marked with a voltage rating suitable for the system voltage and will have a *straight voltage rating* such as 240 V or 480 V (see photo 3-2). Slash-rated breakers provide two voltage rating such as 120/240 or 480Y/277 (see photo 3.3). Two-pole circuit breakers that are suitable for a corner-grounded delta system are marked "1-phase/3-phase" (see 240.85 and figure 3-12).

From the service and throughout the system, the grounded conductor must be insulated from equipment and enclosures [see 250.24(A)(5)]. An equipment grounding conductor is run with the circuit to ground equipment and enclosures, such as disconnecting means, motor controllers, and other non–current-carrying equipment that are required to be grounded.

[1] NFPA 70, *National Electrical Code* 2008, Article 100, (Quincy, MA, National Fire Protection Association, 2004), p. 70-28.

[2] NFPA 70, Article 100, p. 70-29.

[3] NFPA 70, 200.6, p. 70-45.

[4] NFPA 70, 200.7, p. 70-46.

[5] NFPA 70, 408.3(F), p. 70-266.

Review Questions

The questions included here were developed using material included in this chapter. The answers can be found by reviewing the text. It is also important that students make use of *NEC-2008*, where many answers can be found.

1. "The earth" best defines which of the following terms?
   a. grounded conductor
   b. ground
   c. grounding electrode
   d. bonded conductor

2. "Connected (connecting) to ground or to a conductive body that extends the ground connection" best defines which term?
   a. grounded (grounding)
   b. a grounded conductor
   c. being identified
   d. being bonded

3. "A system or circuit conductor that is intentionally grounded" best defines which of the following?
   a. insulated
   b. bonded
   c. a grounded conductor
   d. a grounding conductor

4. ____ AWG and smaller insulated grounded conductors are generally required to be identified by a continuous white or gray outer finish.
   a. 6
   b. 4
   c. 2
   d. 1

5. For installations that will be serviced by qualified persons only, grounded conductors of ____ cables are permitted to be marked at their terminations.
   a. Type IGS
   b. Type TFFN
   c. multiconductor
   d. single

6. Grounded conductors ____ than 6 AWG are permitted to be identified either like conductors 6 AWG and smaller, or at the time of installation, by distinctive white or gray marking at their terminations.
   a. of aluminum or smaller
   b. of copper or smaller
   c. smaller
   d. larger

7. Insulated grounded conductors 6 AWG or smaller are permitted to be identified by which of the following methods?
   a. a continuous white insulation
   b. a continuous gray insulation
   c. three white stripes on other than green insulation
   d. all of the above

8. In practice, a "high-leg" is generally required to be identified by a (an) ____ color coding or other effective means.
   a. green
   b. brown
   c. orange
   d. yellow

9. Grounding of surge arresters rated near, but not less than, ____ percent of line-to-line voltage constitutes an effectively grounded system.
   a. 70
   b. 60
   c. 50
   d. 80

10. For resistance grounding, the system is grounded by connecting the system neutral to ground through a ____.
   a. resistor
   b. terminal
   c. ground rod
   d. generator

11. The insertion of nearly the highest permissible resistance into the grounding connection results in a system that is ____.
   a. extra low resistance grounded
   b. medium resistance grounded
   c. low resistance grounded
   d. high resistance grounded

12. High-resistance grounding maintains control of transient overvoltages, but may not furnish sufficient current for ground-fault ____.
   a. detection
   b. relaying
   c. alarms
   d. conditions

13. Low-voltage systems operating at 600 volts and below are almost always solidly grounded; medium-voltage systems are usually either solidly or resistance grounded; and high-voltage systems above ____ kV are nearly always grounded through surge arresters or ungrounded.
   a. 12.7
   b. 10.4
   c. 15.3
   d. 34.5

14. A wye-delta grounding transformer may be used to provide a neutral for an existing delta-connected ungrounded system, but the use of the ____ grounding transformer is considered as being more practical and economical.
   a. converter
   b. autotransformer
   c. zigzag
   d. Scott

15. Where a grounding transformer is used on low-voltage systems operating at 600 volts and below, it is important that the equipment grounding conductor be connected to the neutral of the grounding transformer in such a way as to provide a ____ path for ground-fault current to return to the system.
   a. high-impedance
   b. low-impedance
   c. current limiting
   d. straight

16. The same grounding electrode should be used for the ____ of the grounding transformer and the equipment grounding conductor and a common grounding conductor used for both the equipment grounding conductor and the neutral.
   a. unidentified conductor
   b. equipment grounding conductor
   c. bonding jumper
   d. neutral

17. In the past, most 3-phase ungrounded delta distribution systems were comprised of three single-phase transformers that were connected ____.
   a. wye-delta
   b. delta-wye
   c. delta-delta
   d. delta

18. For purposes of installing or grounding a corner-grounded delta system, it is helpful to think of it as a ____ system, even though it is a three-phase, three-wire system.
   a. three-phase
   b. single-phase
   c. 5-wire
   d. 4-wire

19. Circuit breakers installed and used for corner-grounded delta systems must be marked
   a. with a voltage rating such as 120/240
   b. for at least 600 volt operation
   c. with a straight voltage rating such 240 V or 480 V
   d. none of the above

20. Fuses are permitted to be inserted in a grounded conductor
   a. as desired
   b. for grounded systems only
   c. for ungrounded systems only
   d. for motor overload protection

21. Where grounded conductors of different systems occupy the same enclosure they must be identified differently and ____.
   a. the means of identification is to be permanently posted at each branch-circuit panelboard
   b. have insulation that is all of the same rating
   c. be separated by permanently installed barriers
   d. be provided with an additional jacket or sleeve

22. Many utilities require that _______
   a. all grounded conductors be uninsulated
   b. metering equipment be located indoors
   c. all 600 volt and under systems be grounded
   d. all ungrounded conductors be un-insulated

23. Solidly grounded means
   a. connected to an electrode with a pressure type connector
   b. without intentionally introducing an impedance
   c. connected using irreversible or exothermic connections
   d. connected to multiple grounding electrodes

24. Continuity of the grounded conductor shall not depend on which of the following?
   a. a metallic enclosure
   b. a metal raceway
   c. a cable armor
   d. all of the above

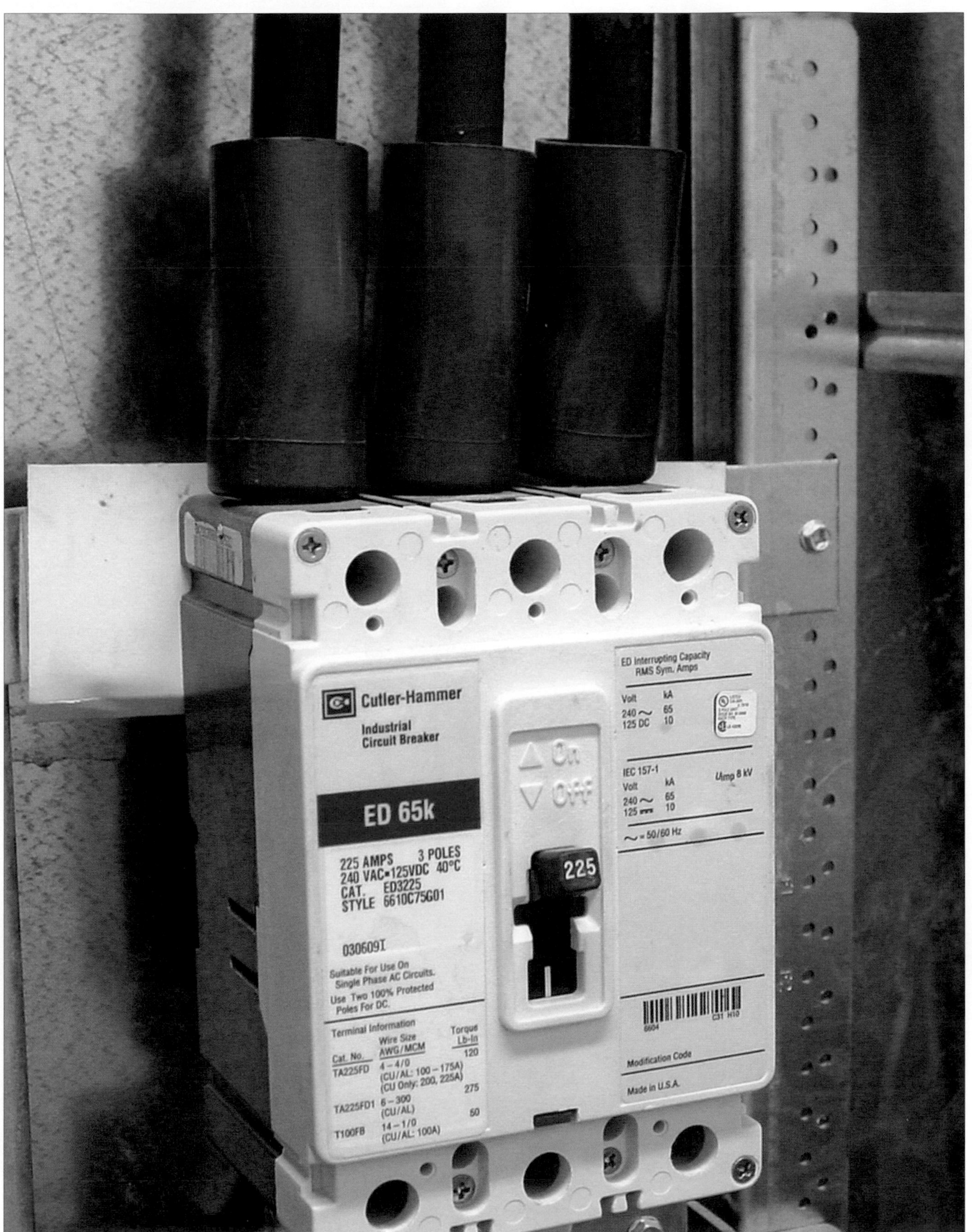
Cutler-Hammer
Industrial
Circuit Breaker
ED 65k
225 AMPS 3 POLES
240 VAC=125VDC 40°C
CAT. ED3225
STYLE 6610C75G01
030609I
Suitable For Use On
Single Phase AC Circuits.
Use Two 100% Protected
Poles For DC.
Terminal Information
225
ED Interrupting Capacity
RMS Sym. Amps
Modification Code
Made in U.S.A.

Chapter 4

Grounding Electrical Services

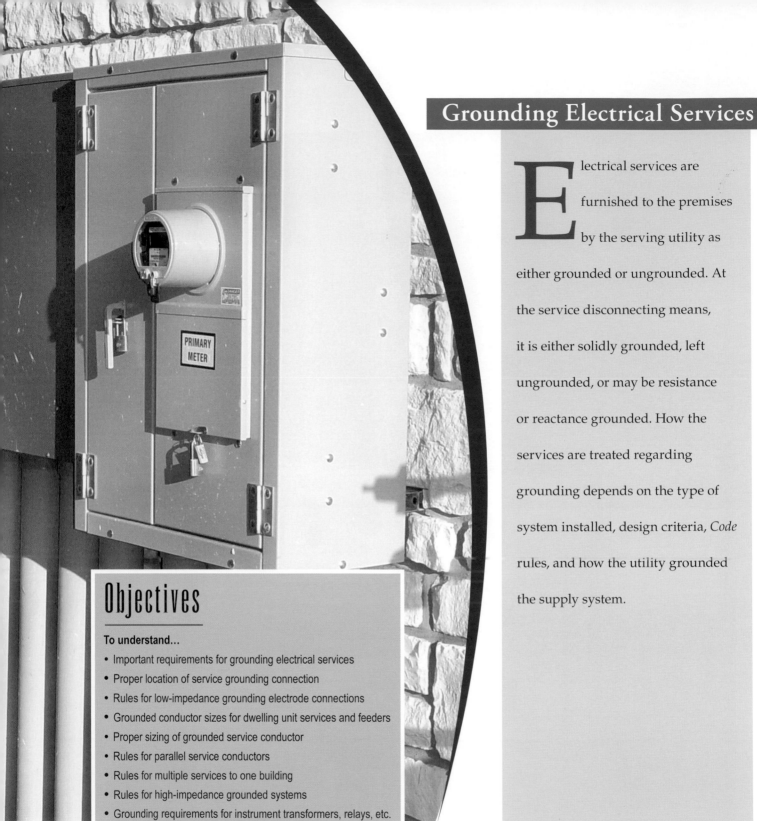

Electrical services are furnished to the premises by the serving utility as either grounded or ungrounded. At the service disconnecting means, it is either solidly grounded, left ungrounded, or may be resistance or reactance grounded. How the services are treated regarding grounding depends on the type of system installed, design criteria, *Code* rules, and how the utility grounded the supply system.

Objectives

To understand...

- Important requirements for grounding electrical services
- Proper location of service grounding connection
- Rules for low-impedance grounding electrode connections
- Grounded conductor sizes for dwelling unit services and feeders
- Proper sizing of grounded service conductor
- Rules for parallel service conductors
- Rules for multiple services to one building
- Rules for high-impedance grounded systems
- Grounding requirements for instrument transformers, relays, etc.
- Hazards of services from grounded systems without grounded conductor

Grounded Electrical Services

Important requirements for grounded services are contained in 250.24(C) (see figure 4-1). This section requires that: Where an ac system operating at less than 1000 volts is grounded at any point, the grounded conductor (usually a neutral conductor) shall be run to each service disconnecting means (see figure 4-2).

For more than one disconnecting means located in a single assembly listed for use as service equipment, an exception permits the grounded conductor to be run to the assembly (see figure 4-3). The grounded conductor must be connected to the common grounded conductor terminal or bus in the assembly enclosure. The sections of the switchboard are bolted together to form the assembly. The assembly is designed and evaluated for this purpose by the listing requirements. In addition, it is common to have an internal equipment grounding bus connected to each section. The key requirements for grounded conductors at service equipment are as follows:

- It must be bonded to each disconnecting means enclosure.
- The grounded conductor must be routed with the phase conductors.
- It must not be smaller than the required grounding electrode conductor specified in Table 250.66.
- It is not required to be larger than the largest ungrounded service-entrance phase conductor.
- For service-entrance phase conductors larger than 1100 kcmil copper or 1750 kcmil aluminum, the grounded conductor shall not be smaller than 12½ percent (0.125) of the area of the largest service-entrance phase conductor.
- Where the service-entrance phase conductors are installed in parallel, the minimum size of the grounded service conductor must be based on the equivalent area of the ungrounded parallel service conductors[1] (see figure 4-1).
- Section 250.36 provides requirements for high-impedance grounded neutral systems grounding connection requirements.

In addition to providing the return portion of the circuit for unbalanced loads, the grounded service

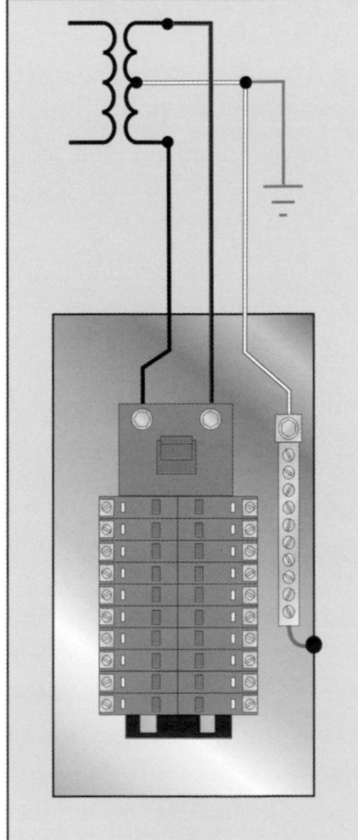

Grounded service conductor must be:

1. Run to each service disconnecting means
2. Bonded to service disconnecting means enclosure
3. Routed with phase conductors
4. Sized no smaller than grounding electrode conductor
5. Sized at least 12-½ percent of area of conductors where larger than given in Table 250.66
6. Based on equivalent area of ungrounded parallel service-entrance conductors
7. Sized in accordance with minimum requirements of 220.61

Figure 4-1. Important requirements for grounded electrical systems [250.24(C)]

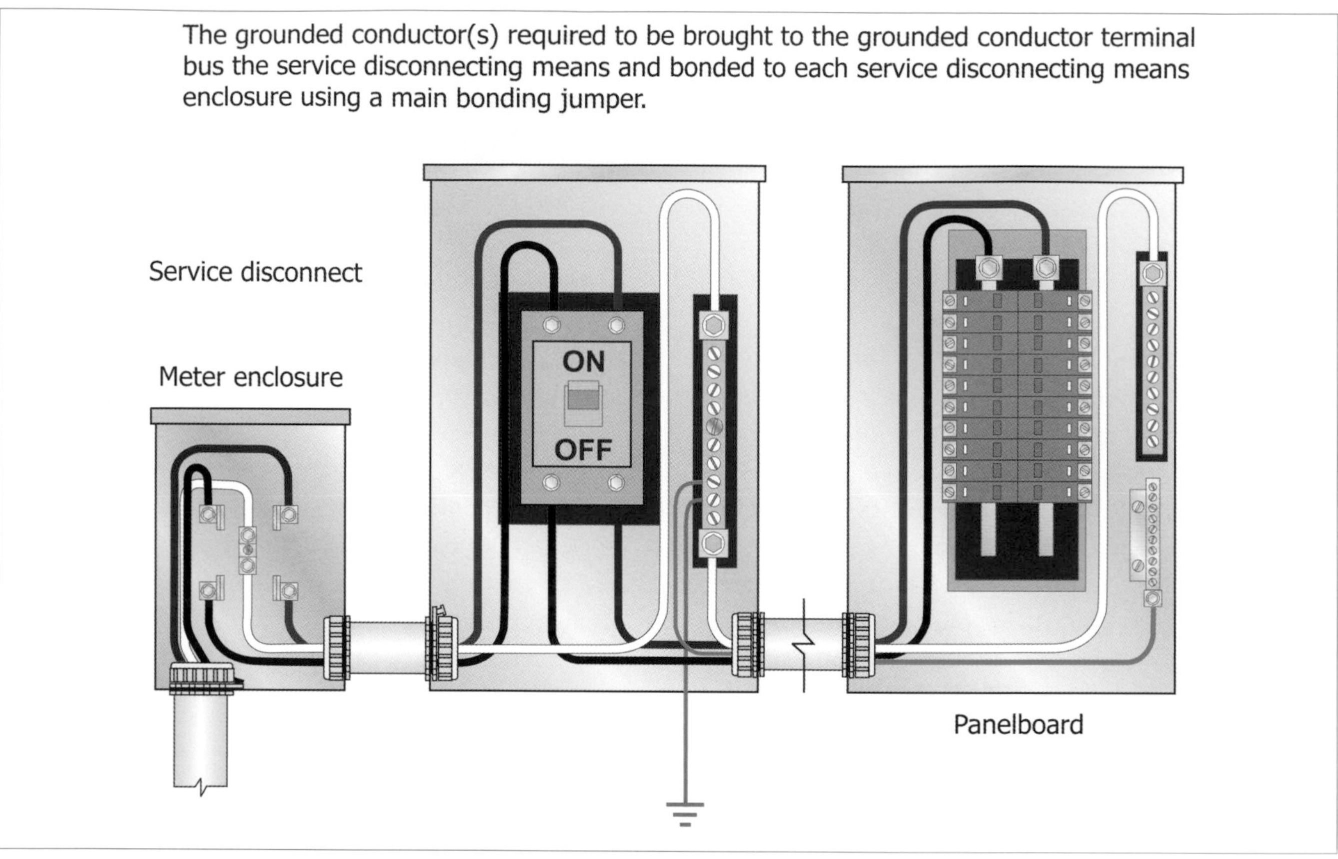

Figure 4-2. Grounded conductor is required to be run to each service disconnecting means enclosure

conductor provides the vital low-impedance path for ground-fault current to return to the source. The grounded system conductor must be run to each service disconnecting means and be bonded to the enclosure, regardless of whether or not it is needed for the service or is used to supply a load.

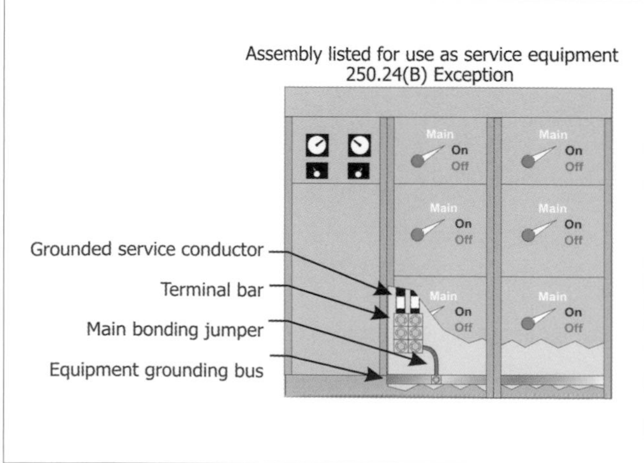

Figure 4-3. Grounded conductor to multi-section equipment that is suitable for use as service equipment (connected to the enclosure at one point)

The grounded conductor shall be run to the service disconnect, connected to the grounded (neutral) conductor terminal bar, and bonded to the enclosure through a main bonding jumper (see figure 4-2).

Figure 4-4 illustrates a 3-phase service supplied from a grounded system. The grounded conductor is not needed since only 3-phase or phase-to-phase connected loads are supplied. If the grounded conductor is not run to the service, as required in 250.24(C), a ground-fault circuit of high impedance is present as shown by the dashed lines. This is also a violation of 250.4(A)(5) and it becomes virtually impossible to clear a ground fault, thus introducing an unnecessary hazard in the system.

However, the installation complies with 250.24(C) if the grounded conductor is run to service. A low-impedance ground-fault path is provided as required by 250.4(A)(5) and the safety of the service is improved immeasurably (see the bottom portion of figure 4-4). Since the grounded conductor is there only to provide a ground-fault path, the size of the conductor to be run will depend on the size of the phase conductors in the service. This conductor must "not be smaller than the required grounding electrode conductor specified in Table 250.66." [2]

Figure 4-5 represents a 3-phase, 4-wire delta system with the midpoint of one transformer grounded. Such systems are intended to supply, economically, both a

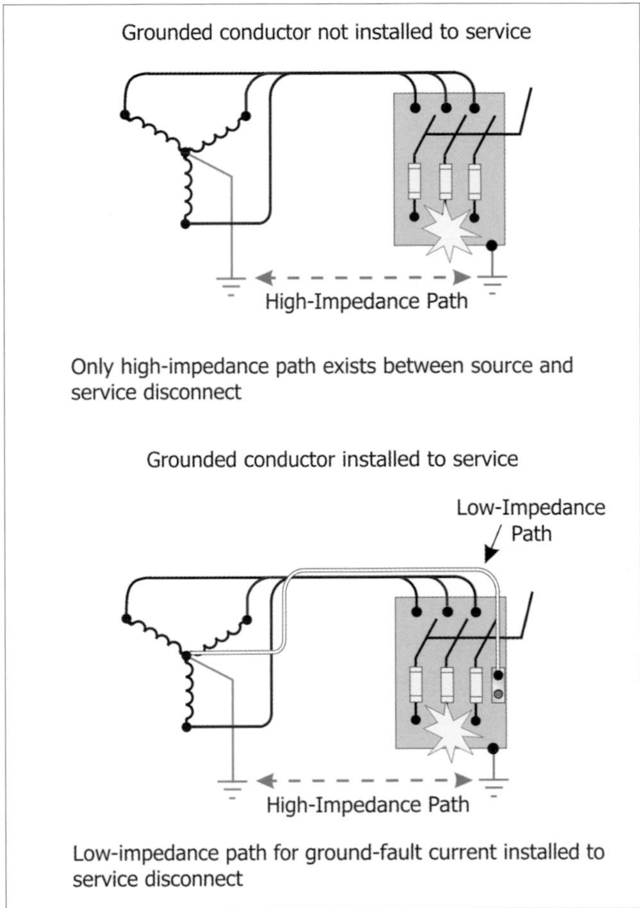

Figure 4-4. High-impedance return path as compared to the required low-impedance return path to source

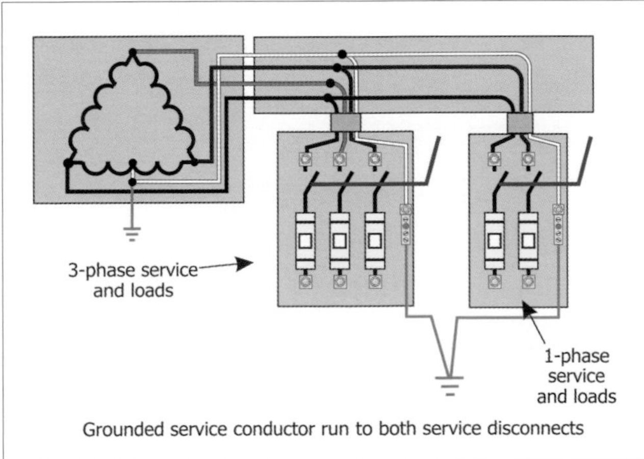

Figure 4-5. Power and lighting service disconnects for three-phase system

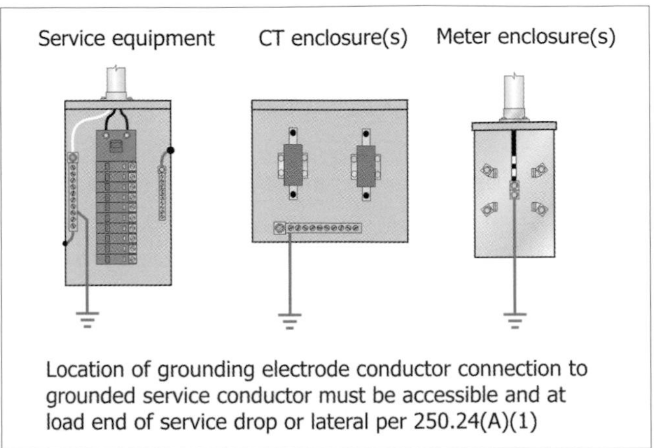

Figure 4-6. Service grounding electrode conductor connection locations are required to be accessible.

3-phase, 3-wire power service, and a single-phase, 3-wire lighting service from one transformer bank. The lighting transformer must be big enough to supply all of the lighting load plus the three-phase power only load. The other transformers are sized to carry the 3-phase, 3-wire load.

Though two separate panelboards are shown for illustration purposes, all loads could be supplied from a single 3-phase panelboard. Where this is done, the electrician needs to exercise caution in making connections for 120-volt loads. These loads must be connected to only the A and C phases as the B phase will have a voltage to ground of approximately 208 volts, which is enough to severely damage equipment designed to operate at 120 volts.

In the case of the 3-phase, 3-wire service, the neutral (or grounded service conductor) is not used for voltage or phase purposes. This is true where no line-to-neutral loads are supplied. However, it is required to run the grounded conductor of the system to the three-phase service equipment and to use the main bonding jumper to satisfactorily clear a ground fault that can develop in the service equipment or in equipment that is supplied by the service.

Location of Service Grounding Connection

Section 250.24(A) requires that a grounding electrode conductor be used to connect the grounded service conductor to a grounding electrode. The connection to the grounded service conductor must be "at an accessible point from the load end of the service drop or service lateral to and including the terminal or bus...

at the service disconnecting means" [see 250.24(A)(1)]. These locations include current-transformer enclosures, meter enclosures, pull and junction boxes, busways, auxiliary gutters and wireways as well as switchboards, panelboards or motor control centers (see figure 4-6). It is also permissible to make the service grounding connections within each service disconnecting means rather than within the wireway. Means and methods of connecting and sizing grounding electrode conductors are covered in chapter seven.

Many inspection authorities and serving utilities will not permit the grounding electrode connection to the system grounded conductor to be within current-transformer cans, meter enclosures, or other enclosures that are sealed by the utility. They interpret that connection to be no longer accessible, as utilities seal the metering equipment to prevent unauthorized access. It is important that serving utilities are consulted before beginning a project to be certain their system grounding and access policies are complied with.

The most practical and commonly accepted location for the grounding electrode conductor connection to the grounded service conductor is within the service disconnecting means or within a wireway that is not sealed at the service equipment location. The connection is usually made to the neutral bus. The grounding electrode conductor could be run up the side of a building and connected to the grounded conductor, on the load end of the service point, where it was spliced to the utilities service drop. Some believe this provides better protection from lightning transients by diverting the lightning current without having it enter the building.

Where the service is supplied by a transformer located outside the building, an additional grounding connection to an electrode must be made outside the building, usually at the transformer. This connection is usually provided by the electric utility.

Section 250.24(A)(5) prohibits grounding of the grounded circuit conductor (often a neutral) at any

Table 250.66 Grounding Electrode Conductor for Alternating-Current Systems

Size of Largest Ungrounded Service-Entrance Conductor or Equivalent Area for Parallel Conductors (AWG/kcmil)		Size of Grounding Electrode Conductor (AWG/kcmil)	
Copper	Aluminum or Copper-Clad Aluminum	Copper	Aluminum or Copper-Clad Aluminum
2 or smaller	1/0 or smaller	8	6
1 or 1/0	2/0 or 3/0	6	4
2/0 or 3/0	4/0 or 250	4	2
Over 3/0 through 350	Over 250 through 500	2	1/0
Over 350 through 600	Over 500 through 900	1/0	3/0
Over 600 through 1100	Over 900 through 1750	2/0	4/0
Over 1100	Over 1750	3/0	250

Table 250.66. Reproduction of *NEC* Table 250.66

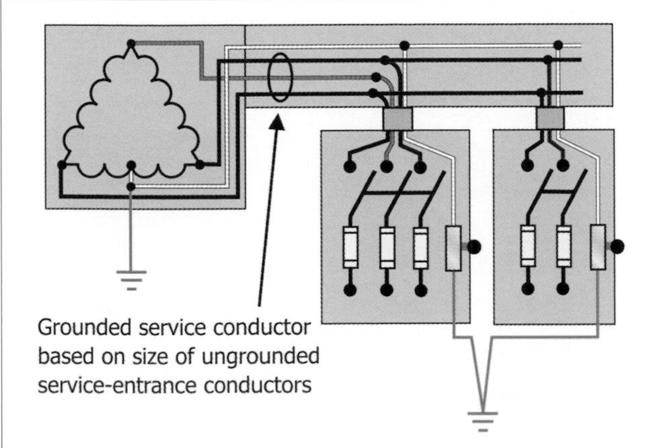
Grounded service conductor
based on size of ungrounded
service-entrance conductors

Figure 4-7. Minimum size of grounded service conductor

point beyond the service. This prevents neutral current on unintended paths such as metal piping, cable trays, cable sheaths, and so forth. Three unique and specific exceptions to this requirement are provided in the FPN that follows 250.24(A)(5), several of which will be covered in later chapters. An additional grounding connection beyond the service is permitted: (1) for separately derived systems, (2) where more than one building or structure on the same premises are supplied by a feeder or branch circuit, and (3) for existing circuits for electric ranges and dryers. The use of the words *grounded circuit conductors* is intended to cover all such conductors whether they are feeders or branch circuits.

Sizing and Routing of Grounded Service Conductor

The size of the grounded service conductor (often the neutral) varies with the load as calculated in accordance with 220.61. This becomes the minimum size grounded service conductor unless modified by 250.24(C)(1).

First, a load calculation should be performed in accordance with 220.61. Then, the resulting minimum conductor size must be compared with Table 250.66 [see table 250.66]. Section 250.24(C)(1) requires the grounded conductor to be no smaller than the grounding electrode conductor as determined in Table 250.66. The final minimum grounded conductor size must be the larger of the two.

For example, if a 400-ampere service is to be installed from a grounded system, 500 kcmil copper conductors are selected for the ungrounded service conductors. Table 250.66 shows that the grounded service conductor can be no smaller than 1/0 copper or 3/0 aluminum. This conductor must be routed with the phase conductors and be bonded to the disconnecting means enclosure.

This is the minimum size grounded service conductor permitted, and it must be installed as noted even though there is no neutral load on the system or if the calculated neutral load would permit a smaller conductor.

It is important to note that the minimum size of the grounded service conductor that must be run to the service disconnecting means is based on the size of the ungrounded (phase) service-entrance conductors and not on the rating of the circuit breaker or fuse that is installed in the service (see figure 4-7). It is helpful to remember that Table 250.66 generally applies up to the service overcurrent device, and Table 250.122 applies beyond the service overcurrent device.

Dwelling Unit Services and Feeders

Special rules or provisions for sizing service-entrance conductors and feeders are provided in 310.15(B)(6) and Table 310.15(B)(6). These rules apply only to 120/240-volt, 3-wire, single-phase dwelling services and feeders. By following the conditions of the section and table, the specified size of service-entrance or main power feeder conductors shown in the table are

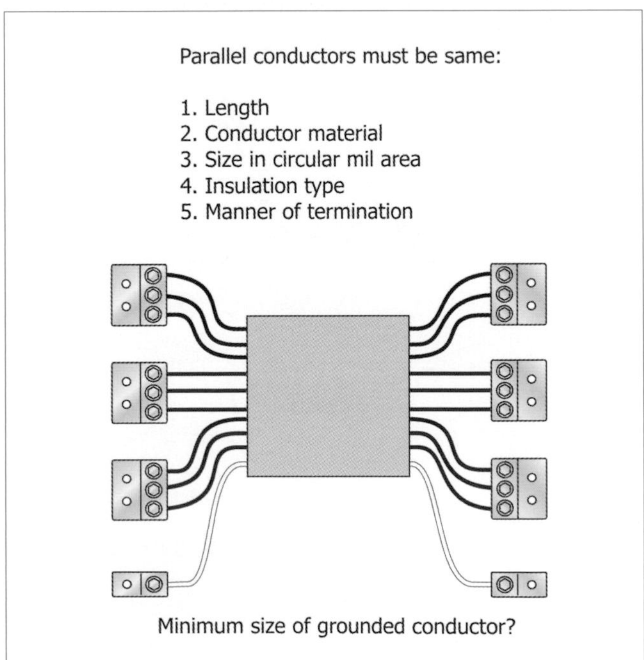
Parallel conductors must be same:
1. Length
2. Conductor material
3. Size in circular mil area
4. Insulation type
5. Manner of termination
Minimum size of grounded conductor?

Figure 4-8. Parallel service conductors (all in the same raceway or enclosure)

permitted to be used based upon the service or feeder rating.

The grounded conductor (often a neutral) is permitted to be smaller than the ungrounded (phase) conductors provided the rules of 215.2, 220.61, and 230.42 are met.

Section 215.2 provides that the feeder neutral must be adequate for the load, must be a minimum size for certain loads, and does not have to be larger than the service-entrance conductors that supply them.

Section 220.61 provides the method for calculating the feeder neutral load. The basic requirement is that the neutral conductor must carry the maximum unbalanced load from the ungrounded conductors.

Section 230.42 requires that service-entrance conductors be of sufficient size to carry the loads as calculated by Article 220. For grounded conductors in 230.42(C), the conductor cannot be smaller than required by 250.24(C).

Note that for dwelling unit services and feeders, the previous permission to size the grounded conductor not more than two sizes smaller than the ungrounded conductors given in 310.15(B)(6) has been deleted from the *Code*. As a result, the grounded system conductor now has to be sized according the above rules.

Where all service-entrance conductors are in the same raceway, the grounded conductor is to be not less than given in Table 250.66.

For conductors larger than Table 250.66, to be not less than 12.5 percent of total area of largest set of ungrounded conductor per 250.24(B)(1)

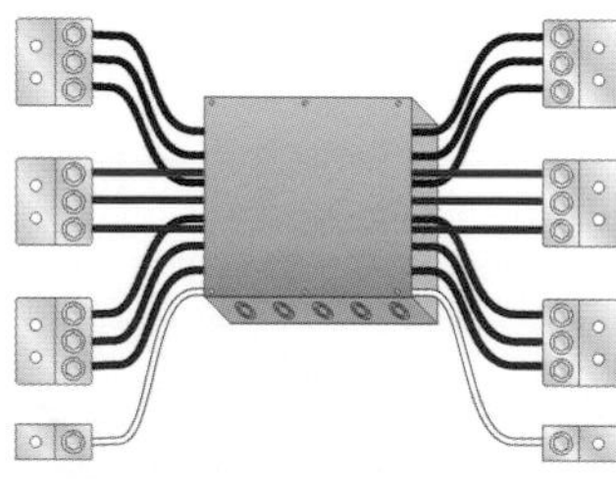

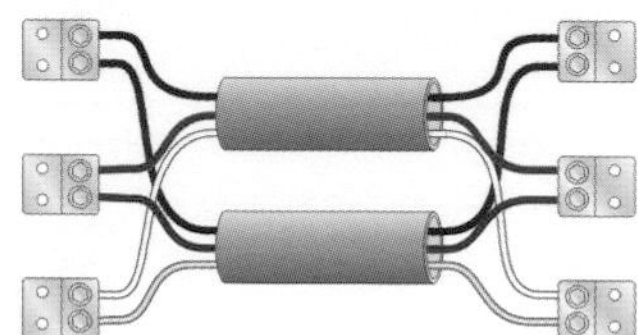

Where installed in two or more raceways, the grounded conductor size is based on the size of the ungrounded conductors in each raceway (based on Table 250.66), but not smaller than 1/0 AWG

Figure 4-9. Parallel service conductors installed in separate raceways or enclosures

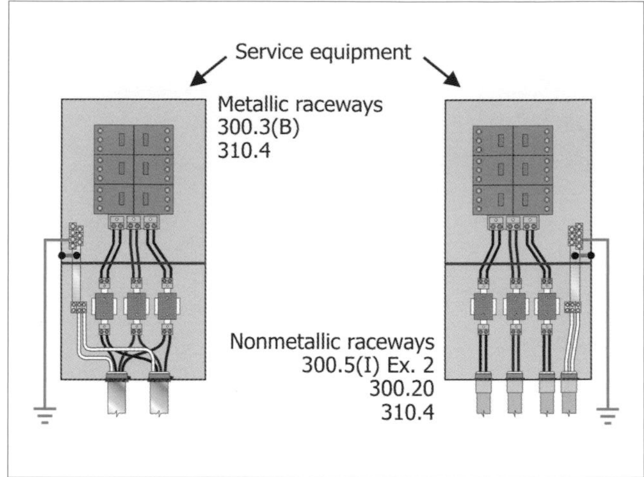

Figure 4-10. Parallel conductors installed in raceways (phases together in the same raceway and individual phases in separate raceways)

Parallel Service Conductors

Where parallel service conductors as permitted by 310.4 are installed, the minimum size grounded service conductor is determined by multiplying the circular mil area of the ungrounded conductors installed by the number of conductors installed in parallel. Installing conductors in parallel means connecting two or more conductors together at each end to form one conducting path (see figure 4-8). This is usually done for economic and practical reasons. As can be seen in Table 310.16, the ampacity of conductors does not increase in proportion to their size. Additionally terminating multiple smaller conductors is easier than terminating larger ones.

Example No. 1

Given: Three 4/0 AWG copper conductors per phase are installed in parallel. Before multiplying the conductor size (4/0), it must be converted from the American Wire Gauge (AWG) 4/0 designation to the circular mil area.

- Refer to *NEC* Chapter 9, Table 8 to determine the circular mil area of the 4/0 AWG conductors. There we find the area to be 211,600 circular mils.
- *3 x 211,600 = 634,800 circular mils*
- By referring to Table 250.66 (Over 600 through 1100), we find the minimum size grounded service

conductor is 2/0 AWG copper or 4/0 AWG aluminum.

This example assumes that all of the conductors are installed in the same raceway. This may not be practical due to the requirement that the derating or ampacity adjustment factors of Table 310.15(B)(2)(a) be applied. This obviously has the effect of requiring larger conductors to be installed than if the individual sets of service-entrance conductors were installed in separate raceways. Where conductors are installed in sheet-metal wireways, derating may be avoided if the number of conductors does not exceed 30 and the conductors do not fill more than 20 percent of the square-inch area (see 376.22 for additional information).

Where installed in separate metal raceways, 300.3(B) requires that the grounded conductor (often a neutral) must be installed in each raceway (see figure 4-9). In this case, 310.4 requires that the paralleled conductors be not smaller than 1/0 AWG, so the minimum size grounded conductor permitted in parallel is 1/0 AWG.

Example No. 2

If six 4/0 AWG copper conductors are installed in parallel, the minimum size grounded service conductor is determined as follows:

As explained above, the area of 4/0 AWG conductors (211,600 circular mils) in *NEC* Chapter 9, Table 8 is used.

- *6 x 211,600 = 1,269,600 circular mils*

Since the total conductor area exceeds the 1100 kcmil for copper conductors given in Table 250.66, the rule in 250.24(C)(1) and 250.24(C)(2) must be followed. There we find a requirement that the grounded service conductor be not smaller than 12 ½ percent (0.125) of the equivalent area for parallel conductors.

- *1,269,600 x 0.125 = 158,700 circular mils*
- By again referring to Chapter 9, Table 8, observe that the conductor that is the next size larger than 158,700 circular mils is a 3/0 AWG conductor, which has a circular mil area of 167,800.

This example also assumes that all the service-entrance conductors are installed in the same raceway, which, due to derating requirements, may not be practical.

Again, where installed in separate metal raceways, a grounded service conductor must be installed in each raceway and must not be smaller than 1/0 AWG. The grounded service conductor must also comply with 230.42(A). This section requires that the conductor be adequate to carry the load as determined by Article 220.

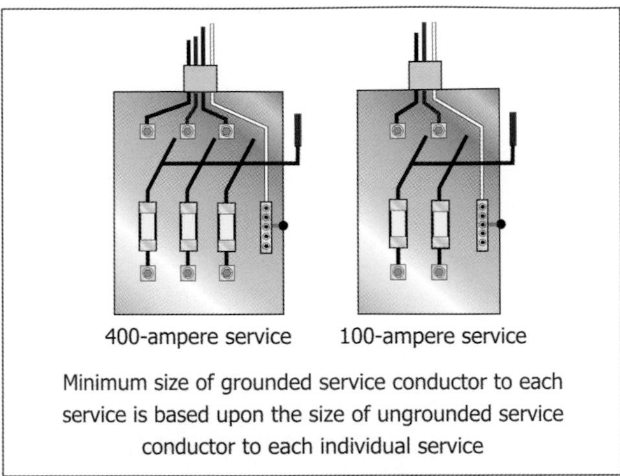

400-ampere service
100-ampere service
Minimum size of grounded service conductor to each service is based upon the size of ungrounded service conductor to each individual service

Figure 4-11. Two services to a building or structure from a grounded system (sizing grounded conductor)

Section 220.61 requires that the neutral be sized for the maximum unbalance of the load. Examples of neutral conductor load calculations are found in Annex D of the *National Electrical Code*. In addition, where the length of run of the grounded conductor from the transformer to the service equipment is long, the size of grounded conductor should be increased.

Underground Parallel Service Conductors

As illustrated in figure 4-10, for underground installations in nonmetallic raceways, it is permitted by 300.3(B)(1) Exception to install all the conductors of each phase in the same raceway. This is also permitted in 300.5(I) Exception No. 2. All the ungrounded conductors of phase A are installed in one raceway, phase B in another, C in the third and the grounded service conductors in another. This method is often chosen to allow phase conductors to readily line up with bus terminations in bottom-fed switchboards. As can be seen in the illustration, this reduces the "rat's nest" in the bottom of these enclosures caused by many conductors crossing each other for termination.

Another advantage is that it is much easier to comply with the requirement that parallel conductors be the same length. When this type of installation is

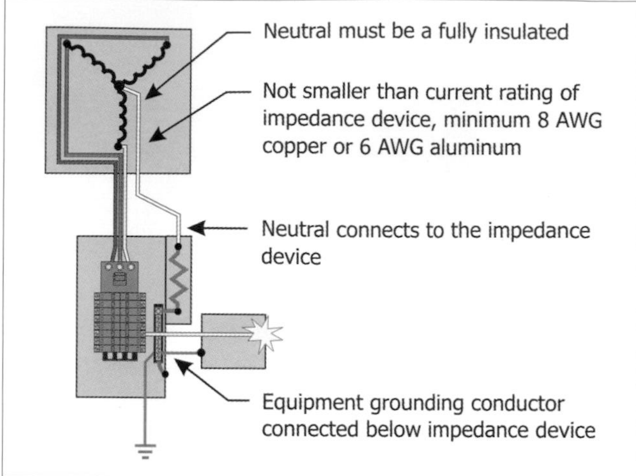
Neutral must be a fully insulated
Not smaller than current rating of impedance device, minimum 8 AWG copper or 6 AWG aluminum
Neutral connects to the impedance device
Equipment grounding conductor connected below impedance device

Figure 4-12. High-impedance grounded system fundamentals

made, care must be taken to eliminate the inductive heating of metal enclosures with magnetic properties by cutting a slot in the metal between the conduit entries or arranging for the manufacturer of the equipment to install a nonmetallic plate in the bottom of the equipment where the conduits terminate [see 300.20(A) and (B)]. Usually, a slot cut with a single hacksaw blade is adequate to provide the desired relief. Another option is to terminate these conduits above the floor in the compartment of an open-bottom, floor-standing switchboard.

See 408.5 where the conduit or raceways, including their end fittings, are not permitted to rise more than three inches above the bottom of the enclosure.

Multiple Services to One Building

Section 230.2 permits several services to one building under one of several conditions given. Each service that is supplied from a grounded system must be provided with a grounded service-entrance conductor. The size of the ungrounded service-entrance conductor for each service determines the minimum size of grounded service conductor for that service. Each service is considered individually for sizing the grounded service conductor to it.

For example, a building has a 400-ampere, 480-volt 3-phase and a 100-ampere 120/240-volt service (see figure 4-11). The service conductors in the example are aluminum, and the minimum size of grounded service conductor is determined as follows:

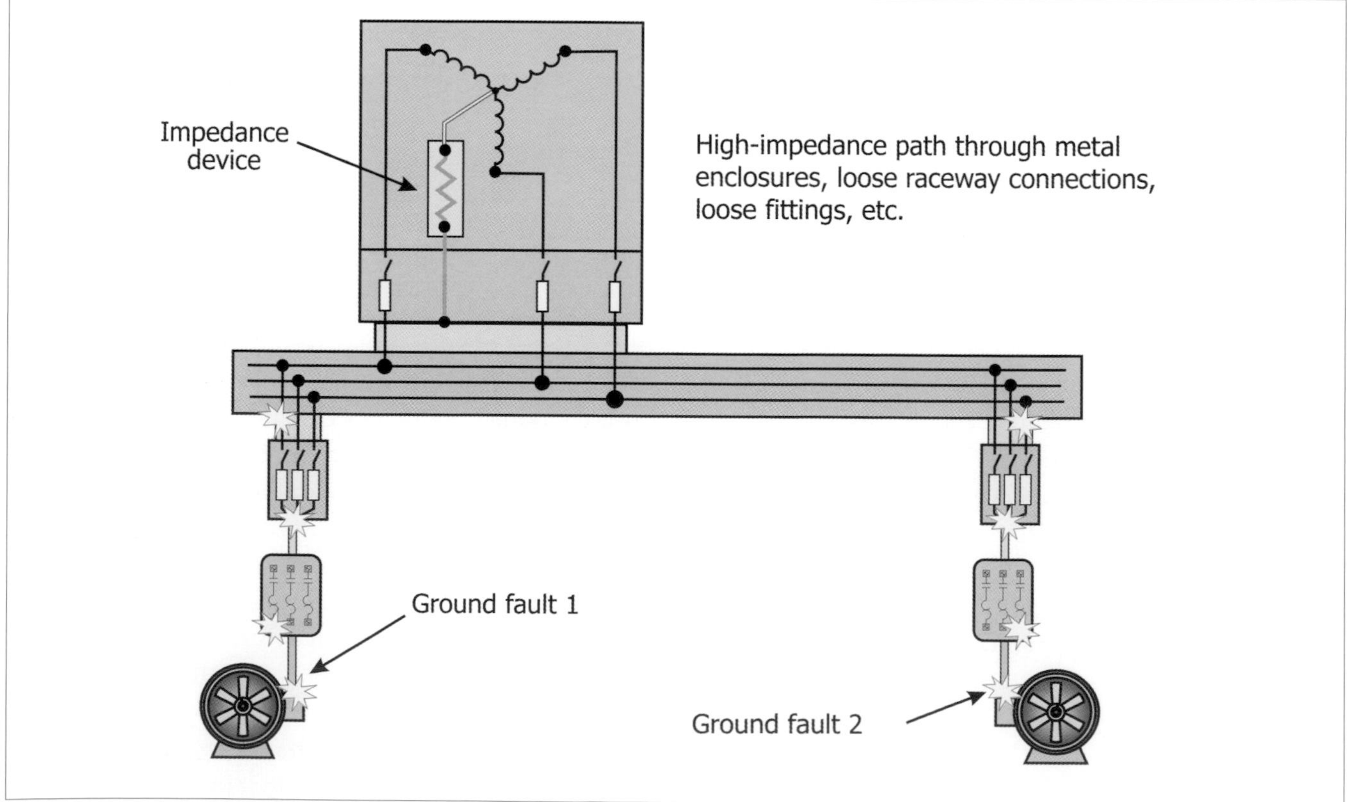
Impedance device
High-impedance path through metal enclosures, loose raceway connections, loose fittings, etc.
Ground fault 1
Ground fault 2

Figure 4-13. Faults on high-impedance grounded neutral systems

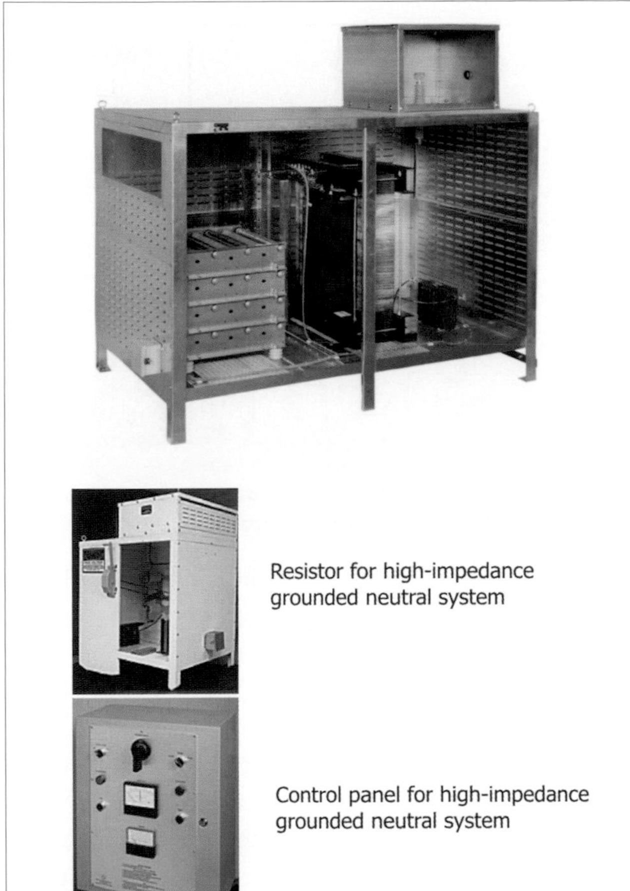

Photos 4-1 and 4-2. Equipment for high-resistance grounded neutral systems Courtesy of Post Glover

400-ampere service

• 750 kcmil THW aluminum ungrounded service conductors

• Table 250.66 = 1/0 AWG copper or 3/0 AWG aluminum grounded service conductor

100-ampere service

• 2 AWG copper ungrounded service conductors

• Table 250.66 = 8 AWG copper or 6 AWG aluminum grounded service conductor

It is emphasized that this method determines the only minimum size of grounded service conductor to comply with 250.24(C). A larger conductor may be required to carry the maximum unbalanced load on the neutral conductor as determined by 220.61

.

High-Impedance Grounded Systems

Industrial plants having a continuous industrial process often need uninterrupted electrical power and systems. It is quite common to see these plants located near a power company substation and to have more than one high-voltage feeder to improve system reliability.

Another step that is commonly taken to improve system reliability is to install high-impedance grounded neutral systems rather than solidly grounded systems. Advan-tages include improved reliability, the ability to have ground-fault relaying and fault-indicating alarms, as well as fewer problems to the system from transient overvoltages.

Three conditions must be met before the *Code* will permit high-resistance grounded neutral systems to be installed. They are as follows:

1. Qualified persons must be available to service and maintain the system.

2. Ground detectors must be installed to indicate an insulation failure.

3. Line-to-neutral loads are not served.

Specific rules are provided in 250.36 for installing these systems. The grounding impedance, usually a resistor, is installed between the transformer supplied grounded service-entrance conductor and the grounding electrode. Usually, the impedance device is sized to a value just greater than the capacitive charging current of the system. For 480-volt systems, this is usually about 10 amperes.

A fully insulated grounded service-entrance conductor must be run to the impedance device. This conductor must have an ampacity not less than the maximum current rating of the grounding impedance. The minimum size grounded service conductor cannot be smaller than 8 AWG copper or 6 AWG aluminum or copper-clad aluminum [250.36(B)]. Since the grounding impedance will limit the fault current of the first line-to-ground fault to a low value, usually 10 amperes or less, the minimum size of neutral conductor is primarily related to physical strength, not to its current-carrying capabilities (see figure 4-12).

It is not required that the grounded service conductor be routed with the ungrounded service-entrance conductors since it will carry very little current in the event of a fault [see 250.36(D)]. At the service, it is connected to the impedance device that may be located outside the service enclosure to dissipate heat. An unspliced equipment bonding jumper is installed from the load side of the impedance device to the service

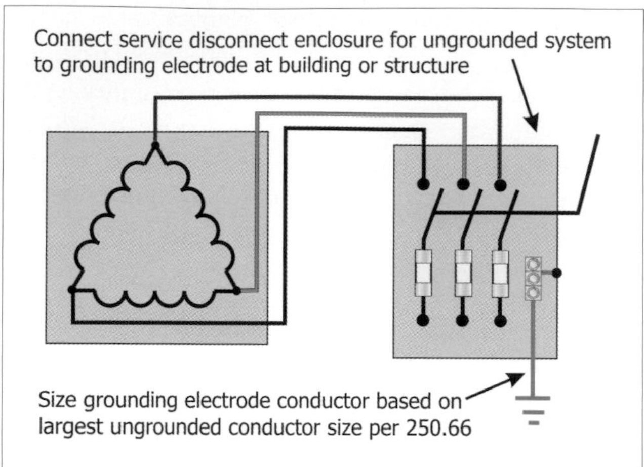
Connect service disconnect enclosure for ungrounded system
to grounding electrode at building or structure
Size grounding electrode conductor based on
largest ungrounded conductor size per 250.66

Figure 4-14. Grounding requirements where ungrounded systems are used

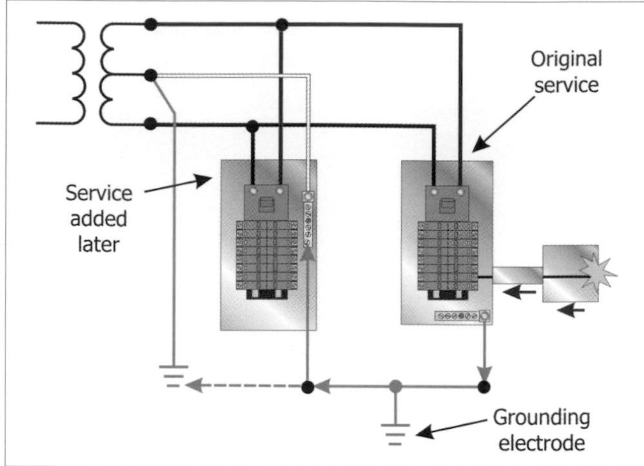
Original
service
Service
added
later
Grounding
electrode

Figure 4-15. Hazard(s) of supplying an ungrounded service from a grounded system without a grounded conductor

enclosure where a terminal bar is installed for connection of equipment grounding conductors [see 250.36(E)]. The grounding electrode conductor is permitted to be connected to any point from the grounded side of the grounding impedance to the equipment grounding connection at the service equipment or the first system disconnecting means [see 250.36(F)].

This type of system is designed to limit the fault current on the first ground fault that might occur on the system. As can be seen in figure 4-12, the impedance device is in series with the first ground fault. The electrical system will continue to function normally with the first ground fault present on the system. The ground-detection system will indicate the presence of the faulted condition by either a visual or audible signal or both. This is intended to alert qualified maintenance personnel of the ground-fault condition so corrective action can be taken, hopefully during a period the plant is not planned to be in operation (see photos 4-1 and 4-2).

One difference from ground-detection on an ungrounded system is that the faulted circuit can be identified without shutting the plant down. A second ground fault (illustrated in figure 4-13) that occurs before the first fault is cleared will be a line-to-line or phase-to-phase fault that would be cleared by the service or feeder overcurrent device, which will result in a power outage. This can, and at times has, involved two pieces of equipment in separate parts of the plant supplied by different feeders. In this situation, there can be a great deal of current in the circuit between the faults that can cause extensive damage to electrical equipment. Every metal conduit locknut and fitting connection must be wrench tight to avoid arcing.

Ungrounded Systems

Ungrounded systems are subject to relatively severe transient overvoltages that can reach several times normal voltage to ground (see chapter two for more information on this subject). Such abnormal voltages become potential hazards and often cause insulation failure and equipment breakdowns. If a system has one conductor grounded and is thus a grounded system, the value of such transient overvoltages as they develop is greatly reduced.

An ungrounded system must have its conductor and equipment enclosures connected to a grounding electrode system at the building or structure served (see figure 4-14). This keeps such enclosures as near to ground potential as possible and reduces shock hazards to a minimum. These service equipment enclosures are grounded by connecting them to a grounding electrode system per 250.24(E). Grounding electrodes and grounding electrode systems are covered extensively in chapter six.

Hazard of Services Without a Grounded Conductor Supplied from a Grounded System

Figure 4-15 illustrates the hazard of operating a service from a grounded system without installing a grounded service conductor. The original ungrounded service on the right in the figure existed before the service on the left was installed. The first and original service was supplied by an ungrounded utility system. The service and feeder shown supplying equipment were protected by large overcurrent devices. Some time later, the service on the left, which included a grounded service conductor, was installed.

At that time, the serving utility grounded the transformer bank and extended a grounded service conductor to the new service weatherhead. However, the grounded service conductor was not extended nor connected to the older, existing service disconnecting means. The older, existing service, which supplied only 3-phase equipment loads, continued to supply power to a portion of the building, though without a grounded conductor connection.

The two services were connected with properly sized grounding electrode conductors to a common grounding electrode system that included a metal water piping system.

Some time later, a ground fault occurred in the equipment being supplied from the original ungrounded service equipment. Since the system supplying the equipment was now grounded, current attempted to return to the source of power to complete the circuit. The grounding electrode conductor carried current from the ungrounded service to the grounding electrode, through the grounding electrode to the grounded service, and where it returned to the transformer bank through the grounded service conductor. In this and similar situations, the current will return to the system through all available paths and will divide among the paths based upon the impedance of the paths that are available.

Grounding electrode conductors are not designed or installed for the significant amount of current they had to carry, which caused the conductors to get extremely hot. The resulting fire burned a great deal of the industrial plant to the ground.

The action necessary to prevent the fire was to run a properly sized grounded service conductor to the existing service when the new service from a grounded supply system was installed. That conductor should have been extended to the service disconnecting means and bonded to it as required in 250.24(C). The grounded service conductor would have provided a low-impedance path back to the transformer bank and would have allowed the overcurrent devices to clear the ground fault without extensive damage. Methods of clearing ground faults and short circuits are discussed in detail in chapter eleven.

[1] NFPA 70, *National Electrical Code* 2008, 250.24(C)(1) and 250.24(C) (2), (Quincy, MA, National Fire Protection Association, 2007), p. 70-102.

[2] NFPA 70, 250.24(C)(1), p. 70-102.

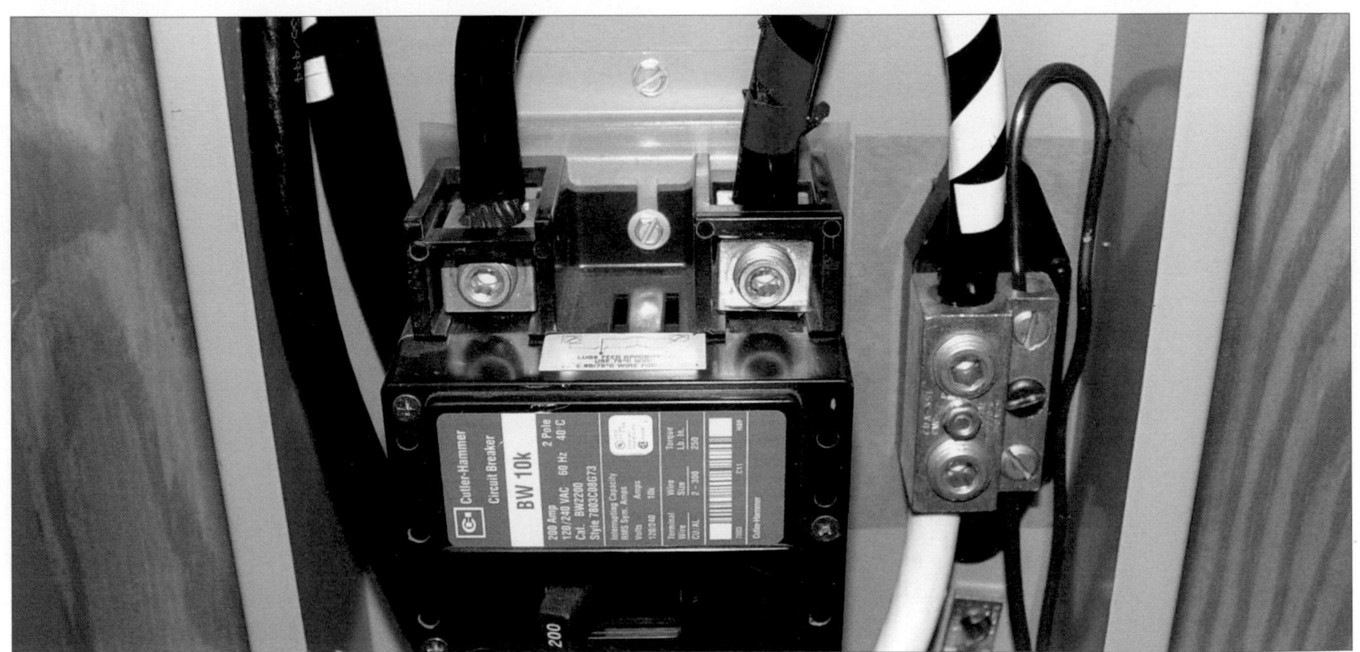

4 Review Questions

The questions included here were developed using material included in this chapter. The answers can be found by reviewing the text. It is also important that students make use of *NEC-2008*, where many answers can be found.

1. The grounded service conductor is required to be run to each ac service disconnecting means where they operate at less than ____ volts and are grounded at any point.
   a. 2000
   b. 1500
   c. 1000
   d. 1200

2. Where an ac system operating at less than ____ volts is grounded at any point, the grounded conductor is required to be bonded to each disconnecting means enclosure.
   a. 1500
   b. 1200
   c. 2000
   d. 1000

3. Where an ac system operating at less than ____ volts is grounded at any point, the grounded conductor is required to be routed with the phase conductors.
   a. 1000
   b. 1200
   c. 1300
   d. 2000

4. Where an ac system operating at less than 1000 volts is grounded at any point the grounded conductor cannot be smaller than the required ______ conductor specified in Table 250.66.
   a. service-entrance
   b. grounding electrode
   c. equipment grounding
   d. ungrounded service

5. For service-entrance phase conductors larger than 1100 kcmil copper or 1750 kcmil aluminum, the grounded conductor cannot be smaller than ____ percent of the area of the largest service-entrance phase conductor.
   a. 12 1/2
   b. 14 1/2
   c. 13 3/4
   d. 12

6. Where the service-entrance phase conductors are installed in ____, the size of the grounded service conductor is required to be based on the equivalent area of the ungrounded service conductors in parallel.
   a. a raceway
   b. a trench
   c. parallel
   d. a cable tray

7. Where more than one service disconnecting means is located in a single common assembly listed for use as service equipment, only one grounded conductor is required to be run to the assembly. It is required to ____ to the disconnecting means grounded conductor terminal or bus.
   a. be sized at 8 AWG and bonded
   b. not be bonded
   c. be connected
   d. be sized at 6 AWG and bonded

8. Some utility policies prohibit a grounding electrode conductor connection within metering equipment, and many inspection authorities prohibit that connection location because they interpret the connection to be no longer ____.
   a. accessible
   b. identified
   c. serviceable
   d. visible

9. The grounding electrode conductor connection to the grounded service conductor is usually made within the service disconnecting means and is connected to the ____ terminal or bus within the equipment.
   a. unidentified
   b. isolated ground
   c. floating neutral
   d. neutral (grounded conductor)

10. Where the transformer supplying the service is located outside the building, an additional grounding electrode conductor connection to an electrode is required to be made outside the building, usually at the ____.
   a. transformer
   b. pole
   c. meter
   d. grid

11. Where copper conductors sized at 500 kcmil are selected for the ungrounded service conductors supplying a 400 ampere service, and are installed from a grounded system,

the minimum size of the copper grounding electrode conductor generally shall not be less than ____ AWG if a metal water pipe grounding electrode is used.

a. 8
b. 6
c. 4
d. 1/0

12. Section 310.15(B)(6) requires or permits the grounded service conductor for dwelling services and the main power feeder to be:

a. sized using Table 250.122
b. used for 60-ampere feeders.
c. sized to meet the requirements of 215.2, 220.61, and 230.42.
d. Type TW conductors.

13. Industrial plants having a continuous industrial process often have a need for ____ electrical power and systems.

a. isolated
b. larger
c. two-phase, five-wire
d. uninterrupted

14. Certain conditions must be met before the *NEC* will permit high-impedance grounded neutral systems to be installed. They include ____.

a. qualified persons must be available to service and maintain the system.
b. line-to-neutral loads are not served.
c. ground detectors must be installed to indicate an insulation failure.
d. all of the above.

15. Secondary circuits of current and potential instrument transformers where the primary windings are connected to circuits of ____ volts or more to ground and, where on switchboards, are required to be grounded irrespective of voltage.

a. 300
b. 100
c. 200
d. 150

16. Where a switchboard has no live parts or wiring exposed or accessible to other than qualified persons, instrument transformer secondary circuits are not required to be grounded where the primary windings are connected to circuits of less than ____ volts.

a. 480
b. 1000
c. 277
d. 150

17. Instrument transformer cases or frames are required to be grounded where accessible to other than qualified persons. Such cases or frames of current transformers are not required to be grounded where the primaries are not over ____ volts to ground, and the current transformers are used exclusively to supply current to meters.

a. 100
b. 120
c. 130
d. 150

18. The service grounded conductor _______

a. carries unbalanced load current
b. provides an effective ground-fault path
c. must be connected (bonded) to the service disconnecting means grounding conductor terminal or bus with a main bonding jumper
d. all of the above

19. An ungrounded service must have _______connected to the metal enclosure of the service conductors.

a. the grounded conductor
b. copper conductors
c. a grounding electrode conductor
d. surge protective devices (SPDs)

20. The grounding electrode conductor is generally sized based on the _______

a. largest ungrounded service-entrance or derived phase conductor
b. overcurrent device protecting the phase conductors
c. load calculations in accordance with Article 220
d. number of service disconnecting means

Chapter 5

Service and Main Bonding Jumpers

The main bonding jumper is one of the most critical elements in the safety grounding and bonding system. This conductor is the link between the grounded service conductor, the equipment grounding conductor, and, in many cases, the grounding electrode conductor. The primary purpose of the main bonding jumper is to carry the ground-fault current from the service enclosure and from the equipment grounding system that is returning to the source during ground-fault conditions. In addition, where the grounding electrode conductor is connected directly to the grounded service conductor bus, the main bonding jumper ensures that the equipment grounding bus is at the same potential as the earth.

Objectives

To understand...

- Definitions of bonding and bonding jumpers
- Functions of the main bonding jumper
- Sizing of main and equipment bonding jumpers
- Methods for bonding at service equipment
- Use of neutral for bonding on line side of service
- Requirements for grounding and bonding of remote metering

Definitions

Bonding (Bonded): "Connect (connecting) to establish electrical continuity and conductivity."

Bonding jumper, Main: "The connection between the grounded circuit conductor and the equipment grounding conductor at the service." [1]

Main Bonding Jumper

For a grounded system, 250.24(B) requires that "an unspliced main bonding jumper be used to connect the equipment grounding conductor(s) and the service-disconnect enclosure to the grounded conductor" of the electrical system. This connection must be made within the enclosure for each service disconnect in accordance with 250.28 (see figure 5-1).

Examples of this are two or more service disconnecting means in individual enclosures that are grouped at one location (see figure 5-2). This type of installation often is made with a wireway or a short section of busway installed downstream from the metering equipment. In other cases, a wireway or short section of busway is installed ahead of metering and is supplied by a service lateral or service-entrance conductors. Sets of service-entrance conductors supply each of the service disconnecting means. Service disconnecting means are installed from the wireway or auxiliary gutter. [If there are nipples between the disconnecting means and the metal or nonmetallic trough, the *trough* meets the definition of a wireway from Article 376, Metal Wireways, or 378, Nonmetallic Wireways, rather than an auxiliary gutter covered by Article 366]. Section 250.24(C) requires that the grounded

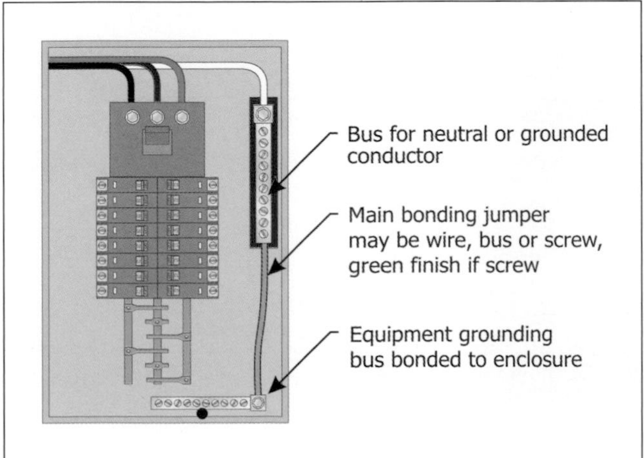

Figure 5-1. Main bonding jumper (at service by definition)

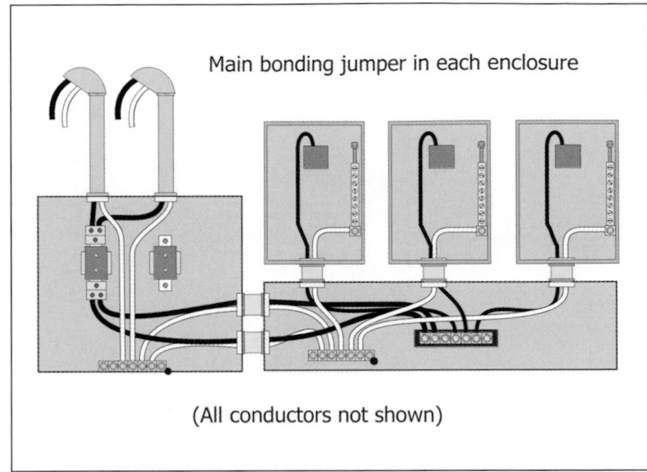

Figure 5-2. Main bonding jumpers in multiple service-disconnect enclosures

service conductor be run to each service disconnecting means and be connected to the grounded terminal bar or bus in the equipment and bonded to the service disconnecing means enclosure. The main bonding jumper is the means to accomplish this requirement.

The rules are a little different where more than one service disconnecting means is in a common enclosure. This equipment usually consists of listed switchboards, panelboards, or motor control centers. "Where more than one service disconnecting means is located in a single assembly listed for use as service equipment," 250.24(B) Exception No. 1 and 250.24(C) Exception permit the grounded service conductor(s) to be run to a common grounded conductor terminal bar or bus in the enclosure and then be bonded to the assembly enclosure (see figure 5-3). This means that only one main bonding jumper connection is required to be installed from the common grounded conductor terminal bar or bus to the assembly enclosure. The sections of the assembly are bonded together by means of an equipment grounding conductor bus and by being bolted together.

Exception No. 2 to 250.24(B) permits an alternate means for bonding of impedance grounded neutral systems (see chapter four for methods and requirements for grounding high-impedance grounded neutral systems). Also, see 250.36 and 250.186 for the specific requirements and allowances.

The main bonding jumper is permitted to consist of a wire, bus, screw, or other similar suitable conductor (see photo 5-1). It must be fabricated of copper or other

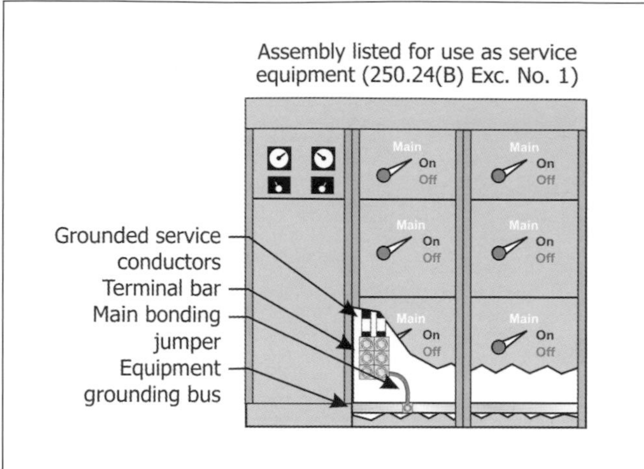

Figure 5-3. Main bonding jumper installed in listed assembly with multiple service disconnects as permitted by exception

corrosion-resistant material. Aluminum alloys are permitted where the environment is acceptable. In addition, where the main bonding jumper consists of a screw, it must have a green finish that is visible with the screw installed (see photo 5-2). This green finish assists in identifying the main bonding-jumper screw from the other screws that are on or near the neutral conductor terminal bar or bus [see 250.28(A) and (B)].

Photo 5-1. Main bonding jumper (bus type)

Functions of Main Bonding Jumper

The main bonding jumper performs three major functions:

1. Connects the grounded service conductor to the equipment grounding bus or conductor and the service enclosure.

2. Provides the low-impedance path for the return of ground-fault currents to the grounded service conductor. The main bonding jumper completes the ground-fault return circuit from the equipment through the service to the source as is illustrated in figure 5-4.

3. Connects the grounded service conductor to the grounding electrode conductor.

Under certain conditions given in 250.24(A)(4), it is permitted to connect the grounding electrode conductor to the equipment grounding terminal bar rather than the terminal or bus for the grounded service conductor. This scheme is common on larger switchboards and service equipment and is necessary for proper operation of certain types of ground-fault protection equipment (see chapter fifteen for additional in-formation on this subject).

Size of Main Bonding Jumper in Listed Enclosures

Where listed service equipment consisting of a switchboard, panelboard or motor control center is installed, the main bonding jumper that is provided with the equipment is rated for the size of conductors that would normally be used for the service. The method for sizing the main bonding jumper in listed service equipment is found in Underwriters Laboratories' safety standard for the equipment under consideration and is verified by the listing organization. Therefore, if a main bonding jumper that is a busbar, strap, conductor, or screw is furnished by the manufacturer as part of the listed equipment, it may be used without calculating its adequacy. Section 408.3(C) requires the equipment manufacturer to provide

Photo 5-2. Main bonding jumper (screw type) identified with the color green

the main bonding jumper. If equipment is listed as service equipment, it will be marked with a label that identifies it as being "Suitable for use as service equipment." This equipment has been evaluated for use as service equipment and will include appropriate provisions for connecting the grounded conductor to the equipment grounding conductor or enclosure. This connection means is usually one of the forms specified in 250.28(A) or (B). Equipment that is marked "Suitable for use Only as Service equipment" usually is manufactured with the grounded conductor bus or terminal bar electrically and mechanically connected to the enclosure.

This type of equipment generally may be installed only in the service position. There is no main bonding jumper in equipment that is suitable for use only as service equipment, but there may be provisions in larger equipment for disconnecting the grounded conductor bus from the enclosure for testing purposes only. More information on these two types of equipment can be found in UL 67 *Standard for Panelboards* and UL 891 *Standard for Switchboards.*

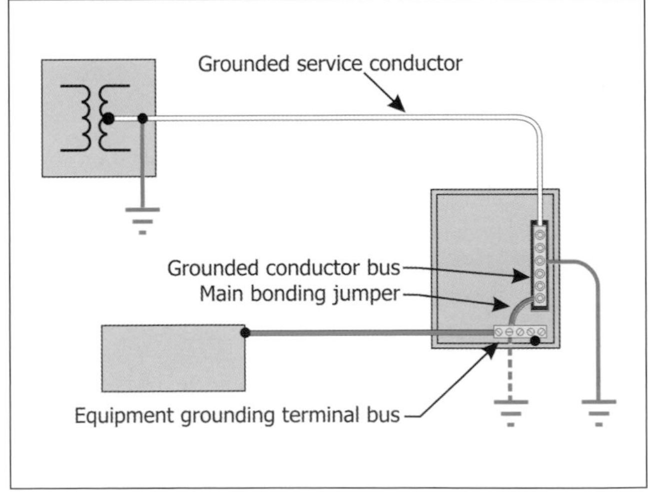

Figure 5-4. Function(s) of the main bonding jumper

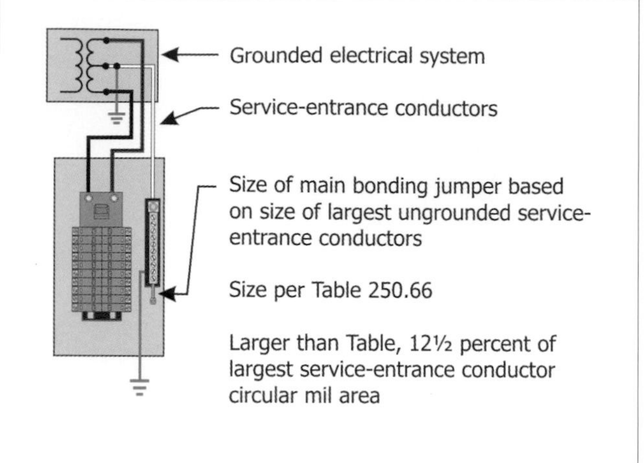

Figure 5-5. Main bonding jumper at a single disconnect

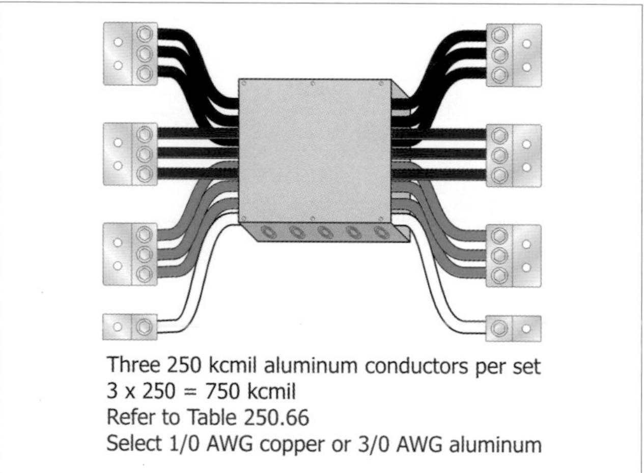

Three 250 kcmil aluminum conductors per set
3 x 250 = 750 kcmil
Refer to Table 250.66
Select 1/0 AWG copper or 3/0 AWG aluminum

Figure 5-6. Sizing main bonding jumper for parallel service conductors

Size of Main Bonding Jumper at Single Service Disconnect or Enclosure

Because the main bonding jumper must carry the full ground-fault current of the system back to the grounded service conductor (which may be a neutral), its size must relate to the rating of the service conductors which supply the service. The minimum size of the main bonding jumper is found in Table 250.66 as required by 250.28(D) (see figure 5-5). This relationship is based on the conductor's ability to carry the expected amount of fault current for the period needed for the overcurrent device to open or operate and interrupt current.

For example, where 250-kcmil aluminum service-entrance conductors are installed, the main bonding jumper is found to be 4 AWG copper or 2 AWG aluminum by reference to Table 250.66.

Where multiple service disconnect enclosures are installed as permitted in 230.71(A), and wire-type main bonding jumpers are installed in each disconnect, the minimum size for each main bonding jumper is not less than the sizes required in 250.28(D)(1) based on the largest ungrounded service conductor serving each individual enclosure [250.28(D)(2)].

The size of the main bonding jumper does not directly relate to the rating of the service overcurrent device. Do not attempt to use Table 250.122 for this purpose. Table 250.122 gives the minimum size of equipment grounding conductors for feeders and circuits on the load side of the service.

Sizing of Main Bonding Jumper for Parallel Service Conductors

Where service conductors are installed in parallel (connected together at each end to form a larger electrically conductive path), the total circular mil area of the conductors connected in parallel for one phase are added together to determine the minimum size main bonding jumper required [see figure 5-6 and 250.28(D)]. For example, where three 250-kcmil conductors are connected in parallel per phase, they are treated as a single 750-kcmil conductor. By reference to Table 250.66 the main bonding jumper, if aluminum service-entrance conductors are used, is 1/0 AWG copper or 3/0 AWG aluminum.

Where the service-entrance conductors are larger than the maximum given in Table 250.66, Section 250.28(D) requires the main bonding jumper to be not less than 12½ percent (0.125) of the area of the largest ungrounded phase conductors. This is illustrated by the following example:

Three 500-kcmil copper conductors are installed in parallel as service-entrance conductors.

3 x 500 kcmil = 1500 kcmil.

1500 x .125 = 187,500 circular mils.

Since a 187,500-circular mils conductor is not a standard size, refer to *NEC* Chapter 9, Table 8, to find the area of conductors.

The next conductor exceeding 187,500 circular mils is a 4/0 AWG conductor, which has an area of 211,600 circular mils. It is always necessary to go to the next larger size conductor since the 12½ percent size is the minimum size permitted.

Bonding of Service-Conductor Enclosures

Special rules are provided for bonding enclosures on the line side of the service disconnecting means. This is because this equipment does not have overcurrent protection, at its rating, on the line side, like feeders and branch circuits have. There must be a sufficient magnitude of fault current, during a short period of time, to cause the overcurrent device on the line side of the utility transformer to open or operate. The level of fault current and particularly the duration of the current can be much greater than in a feeder or branch circuit. This is because service conductors are usually not protected at their ampacity but only have overload protection as part of the service equipment.

The basic rule is that all metallic enclosures that contain a service conductor must be bonded together. The bonding ensures that none of the equipment enclosures can become isolated electrically and become a shock hazard should a line-to-ground fault occur. The bonding also provides a low-impedance path for fault current so the fuse or circuit breaker on the line side of the electric utility transformer will open or operate.

Sizing of Equipment Bonding Jumper on Line (Supply) Side of Service

Equipment bonding jumpers on the line side of the service and main bonding jumper must be sized to comply with Table 250.66. This bonding jumper sizing is required by 250.102(C). For example, where 250-kcmil copper conductors are installed as service-entrance conductors, Table 250.66 requires a 2 AWG copper or 1/0 AWG aluminum bonding jumper.

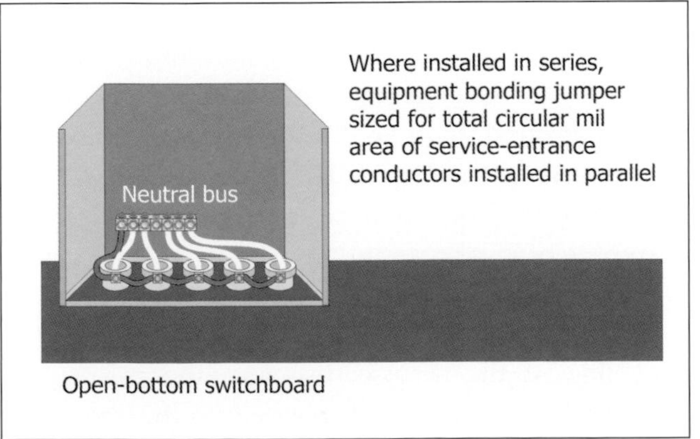

Figure 5-7. Sizing equipment bonding jumper on line side of service disconnect (single equipment bonding jumper in daisy-chain fashion)

Where the sum of the circular mil area of the service-entrance phase conductors exceeds 1100-kcmil copper or 1750-kcmil aluminum, the equipment bonding jumper must be not less than 12½ percent (0.125) of the area of the ungrounded phase conductors.

Sizing of Equipment Bonding Jumper for Parallel Conductors

Two methods are provided for bonding service raceways that are installed in parallel. The first method is to add the circular mil area of the service-entrance conductors per phase together and treat them as a single conductor. The bonding jumper size is determined from Table 250.66 and is connected to each conduit bonding bushing in a *daisy-chain fashion*. This method often results in an equipment bonding jumper that is quite large and difficult to work with.

For example, if five 250-kcmil copper conductors are installed in parallel for a phase, the equipment bonding jumper for bonding the metal raceways must not be smaller than 3/0 copper.

This is determined as follows:

5 x *250 kcmil = 1250 kcmil.*

1250 kcmil x *.125 = 156,250 circular mils.*

The next larger conductor size found in Chapter 9, Table 8, is 3/0 with an area of 167,800 circular mils.

In this case, a 3/0 copper equipment bonding jumper must be connected from the grounded service conductor or equipment grounding bus to each metal raceway in series (daisy-chain fashion from one raceway to another) (see figure 5-7).

A more practical method of performing the bonding for services supplied by multiple raceways may be to connect an individual bonding jumper between each raceway and the grounded service conductor terminal bar or equipment grounding bus (see figure 5-8). This is permitted by 250.102(C). This will usually result in a smaller equipment bonding conductor which is easier to install.

Again, using the example above and referring to Table 250.66, the minimum size equipment bonding jumper for the individual raceways containing 250-kcmil copper service-entrance conductors is 2 AWG copper or 1/0 AWG aluminum. A properly sized equipment bonding jumper is installed from the terminal bar for the grounded service conductor or

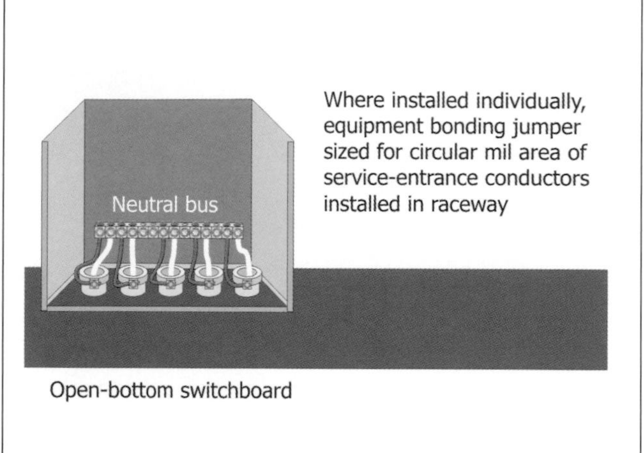

Figure 5-8. Sizing equipment bonding jumper on the line side of the service disconnect (individually from each raceway)

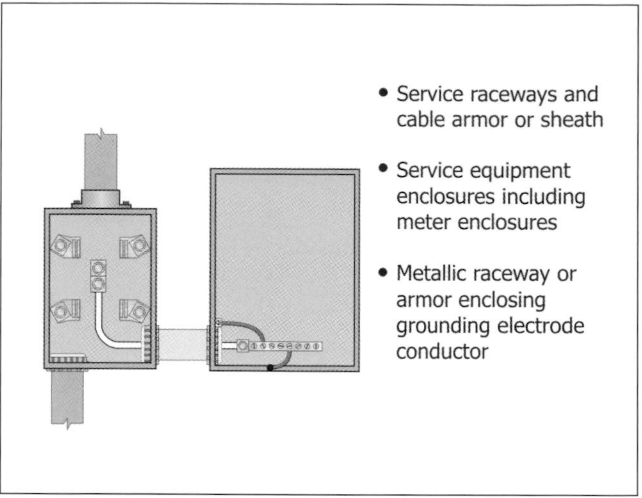

Figure 5-9. Bonding service equipment enclosures

from the equipment grounding terminal bar to each conduit individually.

Bonding Service Equipment Enclosures

The *Code* requires that electrical continuity of service equipment and enclosures that contain service conductors be established and maintained by bonding (see figure 5-9). The items required to be bonded together are stated as follows in 250.92(A):

1. The service raceways, cable-trays, cablebus framework, auxiliary gutters, or service cable armor or sheath.

2. All service equipment enclosures containing service conductors, including meter fittings, boxes or the like, interposed in the service raceway or armor.

3. Any metallic raceway or armor enclosing the grounding electrode conductor as provided in 250.64(B). (This subject is covered in detail in chapter seven.)

A provision to this requirement for bonding at service equipment is mentioned in 250.92(A)(1). It refers to 250.84, which has rules on underground service cables that are metallically connected to the underground service conduit. The *Code* points out that if a service cable contains a metal armor, and if the service cable also contains an uninsulated grounded service conductor that is in continuous electrical contact with its metallic armor, then the metal covering of the cable is considered to be adequately grounded.

Neutral Conductor for Bonding on Line Side of Service

Section 250.92(B) permits the use of the grounded service conductor (may be the neutral) for grounding and bonding equipment on the supply side of the service disconnecting means (see figure 5-10). This is also permitted by 250.142(A)(1). (Two other applications of this method of bonding are explored in later chapters of this book.) Often, connecting the grounded service conductor to equipment such as meter bases, current transformer enclosures, wireways or auxiliary gutters is a practical method of bonding these enclosures.

Usually self-contained meter sockets and meter-main combination equipment are produced with the grounded conductor terminals or bus (often a neutral) bonded directly to the enclosure. The meter enclosure is then effectively bonded when the grounded conductor is connected to these terminals within the meter enclosure. No additional bonding conductor connection to the meter enclosure is required. Current from a ground fault to the meter-main enclosure will return to the source by the grounded service conductor (may be a neutral) and will allow enough current to open or operate the overcurrent protection on the line side of the utility or other transformer.

In addition, meter enclosures installed on the load side of the service disconnecting means are permitted to be grounded (bonded) to the grounded service conductor provided that: (a) service ground-fault protection equipment is not installed; and (b) the meter enclosures are located immediately adjacent to

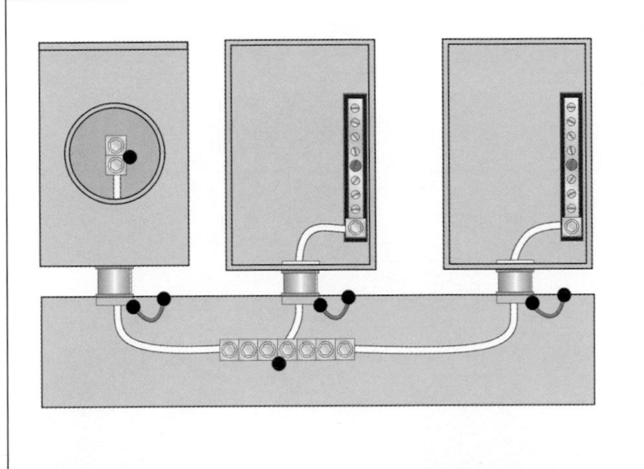

Figure 5-10. Using the grounded conductor for bonding on the supply side of the service disconnecting means

the service disconnecting means, [although there still is no distance dimension provided, it should provide more clarity for the permitted location], and (c) the size of the grounded circuit conductor is not smaller than the size specified in Table 250.122 for equipment grounding conductors.

Method of Bonding at the Service

Various methods for bonding at the service are addressed in 250.92(B) (see figures 5-11, 5-12 and photos 5-3 and 5-4). These requirements for bonding are more restrictive at services than downstream from the service disconnect. This is very important because service equipment and enclosures can be subject to heavy fault currents in the event of a line-to-ground fault.

The service conductors in these enclosures have only short-circuit protection provided by the overcurrent device on the line side of the utility transformer. Only overload protection is provided at the load end of the service conductor by the overcurrent device. This is one of the reasons the *Code* limits the length of service conductors inside a building by requiring the service disconnecting means to be nearest the point of entrance of the service conductors [see 230.70].

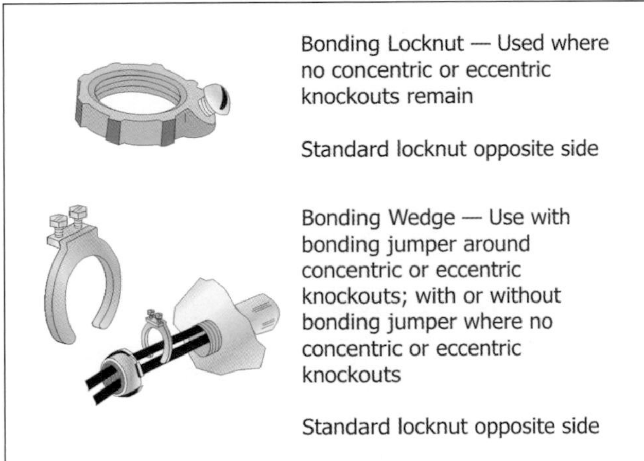

Bonding Locknut — Used where no concentric or eccentric knockouts remain
Standard locknut opposite side
Bonding Wedge — Use with bonding jumper around concentric or eccentric knockouts; with or without bonding jumper where no concentric or eccentric knockouts
Standard locknut opposite side

Figure 5-11. Methods of bonding at service equipment

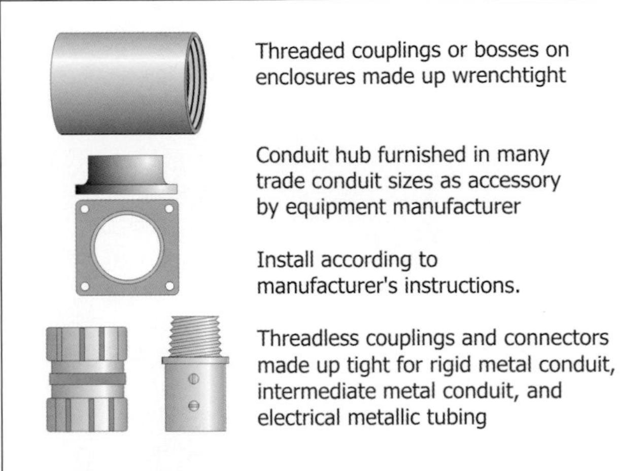

Threaded couplings or bosses on enclosures made up wrenchtight
Conduit hub furnished in many trade conduit sizes as accessory by equipment manufacturer
Install according to manufacturer's instructions.
Threadless couplings and connectors made up tight for rigid metal conduit, intermediate metal conduit, and electrical metallic tubing

Figure 5-12. Various fittings acceptable for use in bonding on the supply side of the service disconnect

Bonding of these enclosures is to be done by one or more of the following methods specified in 250.92(B):

1. Bonding to the grounded service conductor through the use of exothermic welding, listed pressure connectors such as lugs, listed clamps, or other listed means. These connections cannot depend solely upon solder.
2. Threaded couplings and threaded bosses in a rigid or intermediate metal conduit system where the joints are made up wrench-tight. Threaded bosses include hubs that are either formed as a part of the enclosure or are supplied as an accessory and installed according to the manufacturer's instructions.
3. Threadless couplings and connectors are permitted where they are made up tight for rigid and intermediate metal conduit and electrical metallic tubing and metal-clad cables.
4. Other approved devices such as bonding wedges and bonding-type locknuts and bushings.

Bonding jumpers are required to be used around concentric or eccentric knockouts that are punched or otherwise formed to impair an adequate electrical path for ground-fault current. It is important to recognize that concentric and eccentric knockouts in enclosures such as panelboards, wireways, and auxiliary gutters

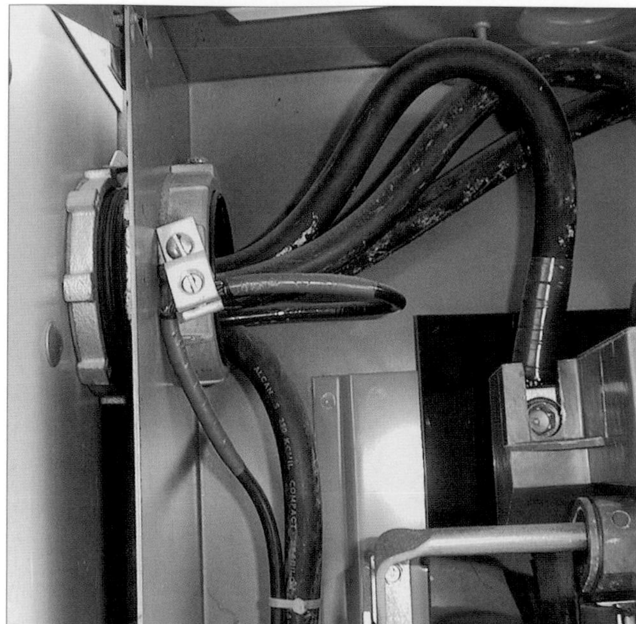

Photo 5-3. Grounding (bonding) bushing for use in bonding on the supply side of service disconnect

have not been investigated for their ability to carry fault current. Where any of these knockout rings remain at conduit connections to enclosures, they must always be bonded around to ensure an adequate fault-current path (see photos 5-3 and 5-4).

The *Code* states, "Standard locknuts or bushings shall not be the sole means for the bonding required by this section." [2]

This statement does not intend to prevent the use of standard locknuts and bushings; they just cannot be relied upon as the sole means for the bonding that is required by this section.

Standard locknuts are commonly used outside the enclosure on conduit that is bonded with a grounding bushing or bonding locknut inside the enclosure. Standard locknuts are used to make a good, reliable mechanical connection as required by 300.10. They must be tightened wrench-tight to prevent arcing due to fault current. Bonding wedges are also acceptable as a bonding means on the supply side of the service disconnecting means (see photos 5-5 and 5-6).

Different Conductor Material

Section 250.28(D) provides instructions on sizing the main bonding jumper where different conductor materials are used for the service-entrance conductors and the bonding jumper. The procedure involves assuming the phase conductors are of the same material (copper or aluminum) as the bonding jumper and that they have an equivalent ampacity to the conductors that are installed. This is illustrated as follows:

Photo 5-4. Close up of grounding (bonding) bushing
Courtesy of Thomas and Betts

- Assume aluminum ungrounded (phase) conductors and a copper bonding jumper
- Three 750-kcmil Type THW aluminum ungrounded conductors are installed.
- Determine the ampacity of the ungrounded conductors from Table 310.16:

  385 amperes x 3 = *1155 amperes.*

The smallest type THW copper conductor that has an equivalent ampacity is 600 kcmil with an ampacity of 420.

- Next, determine the total circular mil area of the copper conductors.

  3 x *600 kcmil = 1800 kcmil.*

  1800 kcmil x *.125 = 225 kcmil.*

The next standard size is 250-kcmil copper, which is the minimum size bonding jumper permitted to bond equipment at or ahead of the service equipment in this example.

Photo 5-5. Bonding locknut for bonding on the supply side of the service disconnect

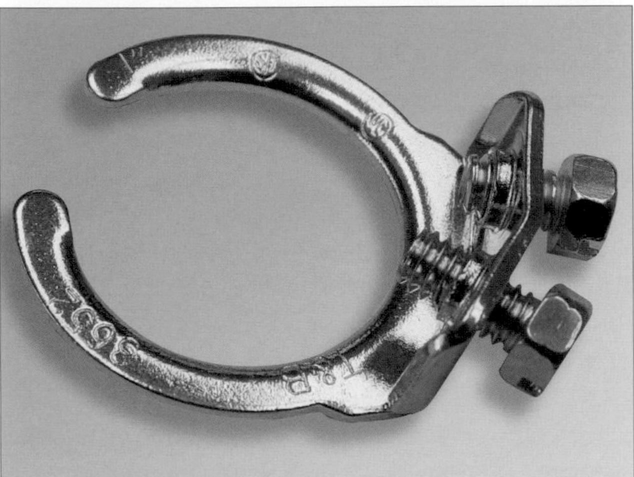

Photo 5-6. Bonding wedge suitable for bonding on the supply side of the service disconnect Courtesy of Thomas and Betts

Parallel Equipment Bonding Jumpers

Section 250.102(C) requires that where service-entrance conductors are paralleled in two or more raceways or cables and the equipment bonding jumper is routed with the raceways or cables, the equipment bonding jumper must be run in parallel (see figure 5-13).

In this case, again, the size of the bonding jumper for each raceway is based upon the size of the service-entrance conductor in the raceway by referring to Table 250.66. It is important to make the bonding jumper connections on both sides of the raceway with equipment or fittings that are suitable for that use (see photo 5-7).

Grounding and Bonding of Remote Metering

As mentioned before, 250.92(A) requires all equipment containing service conductors to be bonded together and to the grounded service conductor. This includes remote (from the service equipment) meter cabinets and meter socket enclosures.

Grounding and bonding of equipment such as meters, current transformer cabinets, raceways, and auxiliary gutters to the grounded service conductor at locations on the line side of and remote from the service disconnecting means increases safety.

This equipment should never be grounded only to a grounding electrode, such as a ground rod. Figures 5-14 and 5-15 help show why. If a ground fault occurred at this line-side equipment, and it is not bonded as required, the only means for clearing a ground fault would be through the grounding electrodes and earth. Given the relatively high impedance and low current-carrying capacity of this path through the

Photo 5-7. There are fittings for liquidtight flexible metal conduit and flexible metal conduit that provide a means for installation of externally routed equipment bonding jumpers. Courtesy of Thomas and Betts

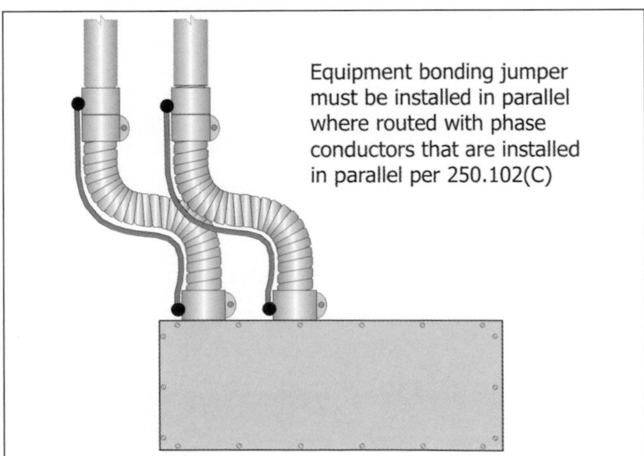

Equipment bonding jumper must be installed in parallel where routed with phase conductors that are installed in parallel per 250.102(C)

Figure 5-13. Bonding jumpers installed on the outside of the raceway or enclosure

earth and high resistance of grounding electrodes such as rods, there will be little current in this path. No overcurrent device will open or operate, leaving the equipment enclosure(s) at a dangerous voltage-to-ground potential. Any person or animal that contacts the enclosure can be shocked or electrocuted.

The voltage drop across this portion of the circuit can easily be calculated using Ohm's law (Resistance times the current gives the voltage). There are many records of livestock being electrocuted while contacting electrical equipment that was improperly grounded. Sections 250.4 and 250.54 require that the earth not be considered or used as an effective ground-fault current path.

A practical method for grounding and bonding this line-side equipment is to bond the grounded service conductor to it. As can also be seen in figures 5-14 and 5-15, a ground fault to the equipment will have a low-impedance path back to the source through the grounded service conductor. This will allow enough current in the circuit to cause the overcurrent protection on the line side of the transformer to clear the fault.

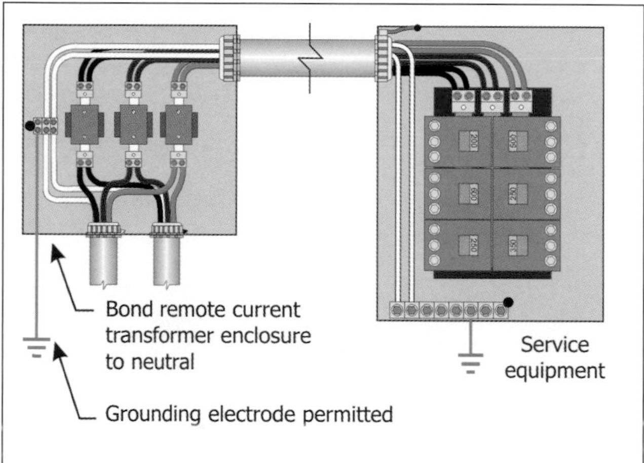
Bond remote current transformer enclosure to neutral
Service equipment
Grounding electrode permitted

Figure 5-14. Grounding and bonding of remote metering

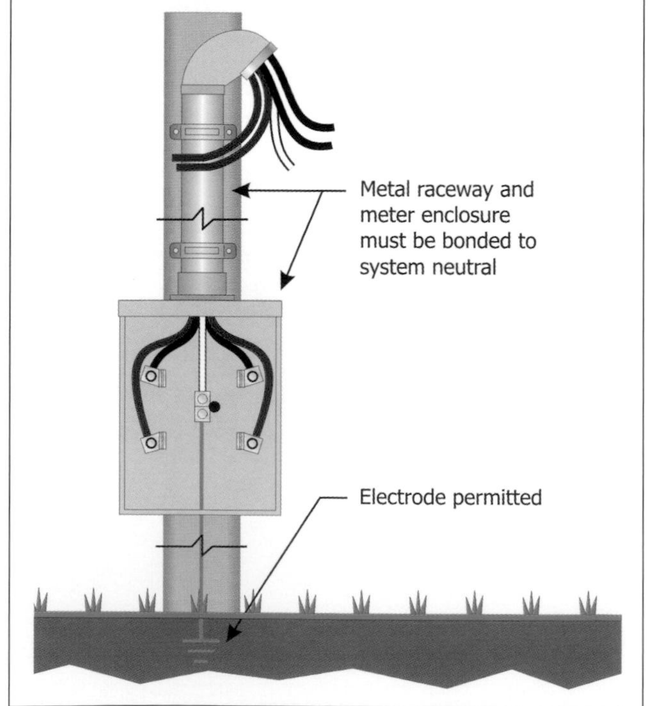
Metal raceway and meter enclosure must be bonded to system neutral
Electrode permitted

Figure 5-15. Bonding and grounding at remote meter loop

Auxiliary Grounding Electrodes

In accordance with 250.54, it is permissible to install a grounding electrode at the remote meter location shown in figures 5-14 and 5-15 as an auxillary grounding electrode. This section refers specifically to grounding electrodes being supplementary to equipment grounding conductors. Some electric utilities require a grounding electrode at meter equipment installed remote from service equipment, such as on poles. Section 230.66 makes it clear that individual meter socket enclosures are not to be considered service equipment. The same is true for metering equipment installed in remote current-transformer enclosures. As mentioned earlier, it is critically important that these meter enclosures be properly bonded as they often contain unfused or line-side service conductors.

This additional grounding electrode will attempt to keep the equipment at the earth potential in the vicinity of the meter location. Remember that the electrodes at the remote meter and those connected at the service location are bonded together by the grounded service conductor installed between the metering and service equipment.

This brings the installation into compliance with 250.50 and 250.58, which require a common grounding electrode; or where two or more electrodes are installed, they must be bonded together.

As previously stated, these grounding electrodes should never be used as the only means for grounding or bonding these enclosures or to carry fault current. An extensive discussion of this subject is found in chapter six.

Bonding of Multiple Service Disconnecting Means

Installation of multiple services as permitted by 230.2(A) through (D) and installations of services that have multiple disconnecting means can take several forms.

Additional services are permitted by 230.2 for:

1. Fire pumps, emergency, legally required, optional standby, parallel power production systems, or redundant systems for enhanced reliability.

2. By special permission, for multiple occupancy buildings where there is no available space for service equipment that is accessible to all occupants, or for a single building or structure that is large enough to make two or more services necessary.

3. Capacity requirement where the service capacity requirements exceed 2000 amperes at 600 volts or less, where load requirements of a single-phase installation are greater than the serving utility normally provides through a single service, or by special permission (related to capacity requirements).

4. Different characteristics of the services, such as different voltages, frequencies, or phases, or for different uses, such as for different rate schedules.[3]

Remember that *special permission* is defined in Article 100 as the written consent of the AHJ.

The basic rule for sizing the equipment bonding jumper for bonding these various configurations is found in 250.102(C). This section requires that the bonding jumpers on the line side of each service and the main bonding jumper be sized from Table 250.66. In addition, the size of the bonding jumper for each raceway is based on the size of service-entrance conductors in that raceway. As discussed earlier, conductors larger than given in Table 250.66 are required for larger services and are sized based on 12.5 percent of the largest phase conductor. Since different sizes of service-entrance conductors may be installed at various locations, the minimum size of the equipment bonding conductor and main bonding jumper is based on the size of the service-entrance conductors at each location.

For example, the appropriate size of bonding jumper for the installation in figure 5-16 with the assumed size of conductors is as follows: (all sizes copper)

a. 500 kcmil in service mast and nipple has a bonding jumper of 1/0 AWG

b. 1000 kcmil in wireway has a bonding jumper of 2/0 AWG

c. 300 kcmil to 300-ampere service has a bonding jumper of 2 AWG

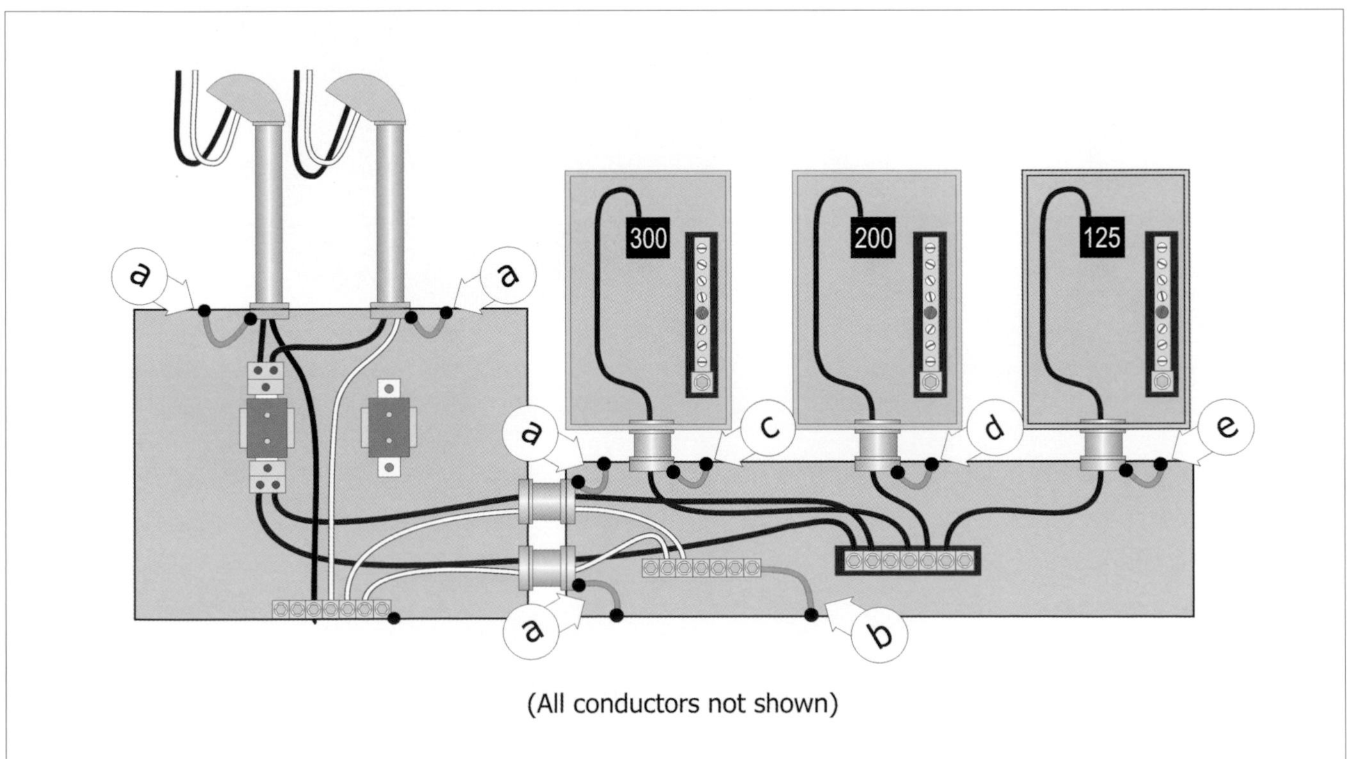

Figure 5-16. Bonding of multiple service disconnecting means

d. 3/0 AWG to 200-ampere service has a bonding jumper of 4 AWG

e. 2 AWG to 125-ampere service has a bonding jumper of 8 AWG

A practical method for bonding the current transformer enclosure and wireway (sometimes referred to as a *hot gutter*) is to connect the grounded service conductor directly to the current transformer enclosure or wireway. This may be done by bolting a multi-barrel lug directly to the wireway and connecting the neutral or grounded service conductors to the lug. Be sure to remove any nonconductive paint or other coating that might insulate the connector from the enclosure. As previously discussed, the grounded service conductor must also be extended to each service disconnecting means and be bonded to the enclosure.

[1] NFPA 70, *National Electrical Code 2008*, Article 100, (Quincy, MA, National Fire Protection Association, 2007), p. 70-25.

[2] NFPA 70, 250.92(B), p. 70-113.

[3] NFPA 70, 230.2, p. 70-76.

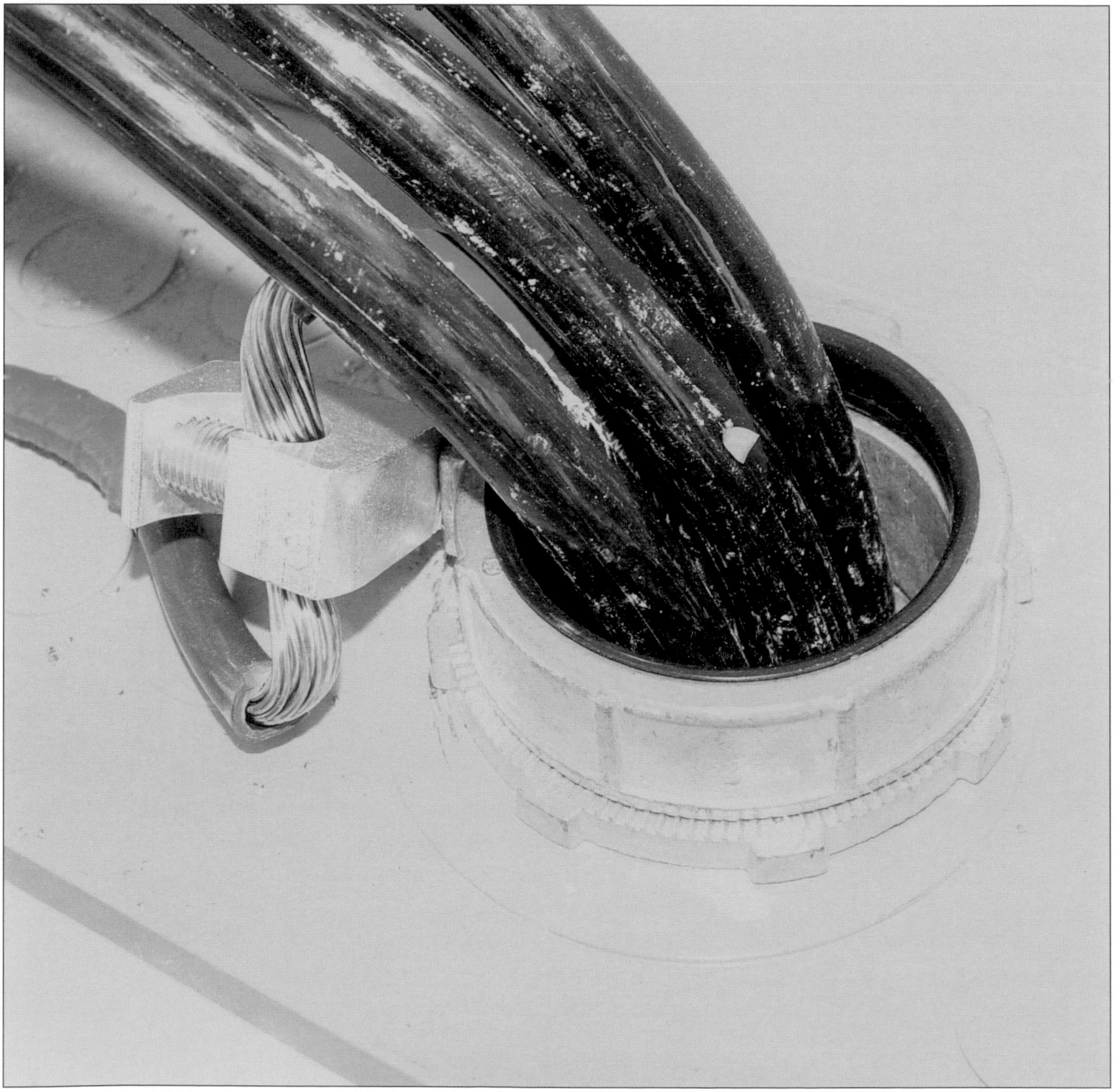

Review Questions

The questions included here were developed using material included in this chapter. The answers can be found by reviewing the text. It is also important that students make use of *NEC-2008*, where many answers may be found.

1. "Connected (connecting) to establish electrical continuity and conductivity" best defines which of the following:
   a. grounding
   b. bonded (bonding)
   c. welded
   d. grounded

2. The connection between the grounded circuit conductor and the equipment grounding conductor at the service is defined as the ____.
   a. main bonding jumper
   b. grounding electrode conductor
   c. equipment bonding jumper
   d. neutral conductor

3. The main bonding jumper is permitted to consist of a ____ or other suitable conductor.
   a. wire
   b. bus
   c. screw
   d. all of the above

4. Where the main bonding jumper consists of a ____ only, it is required to have a green finish that is visible with the ____ installed.
   a. bus
   b. screw
   c. wire
   d. jumper

5. Where ____ kcmil aluminum service-entrance conductors are installed, a wire-type main bonding jumper is required to be 4 AWG copper or 2 AWG aluminum.
   a. 1/0
   b. 2/0
   c. 3/0
   d. 250

6. Where the service-entrance conductors are larger than the maximum sizes given in Table 250.66, the main bonding jumper cannot be less than ____ percent of the area of the largest phase conductor.
   a. 9 ½
   b. 10 ½
   c. 11 ½
   d. 12 ½

7. Where service-entrance conductors are installed in parallel, and the sum of the circular mil area exceeds ____ kcmil copper or ____ kcmil aluminum, the equipment bonding jumper must be not less than 12 ½ percent of the area of the largest ungrounded phase conductor.
   a. 1100 - 1750
   b. 1000 - 1650
   c. 1050 - 1500
   d. 1075 - 1400

8. The *Code* requires that electrical continuity at service equipment and enclosures that contain service conductors be ensured by ____.
   a. grounding
   b. welding
   c. approval
   d. bonding

9. The grounded circuit conductor (may be a neutral conductor) is per-mitted to be used for grounding and bonding on the ____ side of the service disconnecting means.
   a. load
   b. supply
   c. subpanel
   d. control center

10. Grounding and bonding of equipment such as meters, current transformer cabinets and raceways to the grounded service conductor at locations on the supply side of and remote from the service disconnecting means increases ____.
   a. voltage
   b. current
   c. safety
   d. cost

11. The equipment bonding jumper for raceways containing service-entrance conductors must be sized according to:
   a. Section 250.102(C), Table 250.66
   b. Table 250.122
   c. Table 310.16.
   d. none of the above

12. Which of the following methods are NOT permitted to be the sole means for bonding enclosures for service-entrance conductors
    a. bonding bushings
    b. bonding locknuts
    c. threadless couplings
    d. standard locknuts

13. Meter enclosures are permitted to be grounded using the grounded conductor where they are_______the service disconnecting means
    a. located near
    b. located immediately adjacent to
    c. not within reach of
    d. within sight of

14. The main bonding jumper must be sized large enough to carry _______ground-fault current of the system
    a. the full
    b. approximately half
    c. 58% of the
    d. none of the

15. The main bonding jumper_______
    a. Connects the grounded service conductor to the equipment grounding bus
    b. Provides the low-impedance path for the return of ground-fault currents
    c. Connects the grounded service conductor to the grounding electrode conductor
    d. All of the above

The Grounding Electrode System

Grounding electrodes provide the essential function of connecting the electrical system and electrical equipment to the earth (see figure 6-1). The earth is considered to be at zero potential. In some cases, the grounding electrode(s) serves to ground the electrical system. In other instances, the electrode(s) is used to connect non-current-carrying metallic portions of electrical equipment to the earth. In both situations, a primary purpose of the grounding electrode(s) is to maintain the electrical equipment at the earth potential present where the grounding electrode(s) is located.

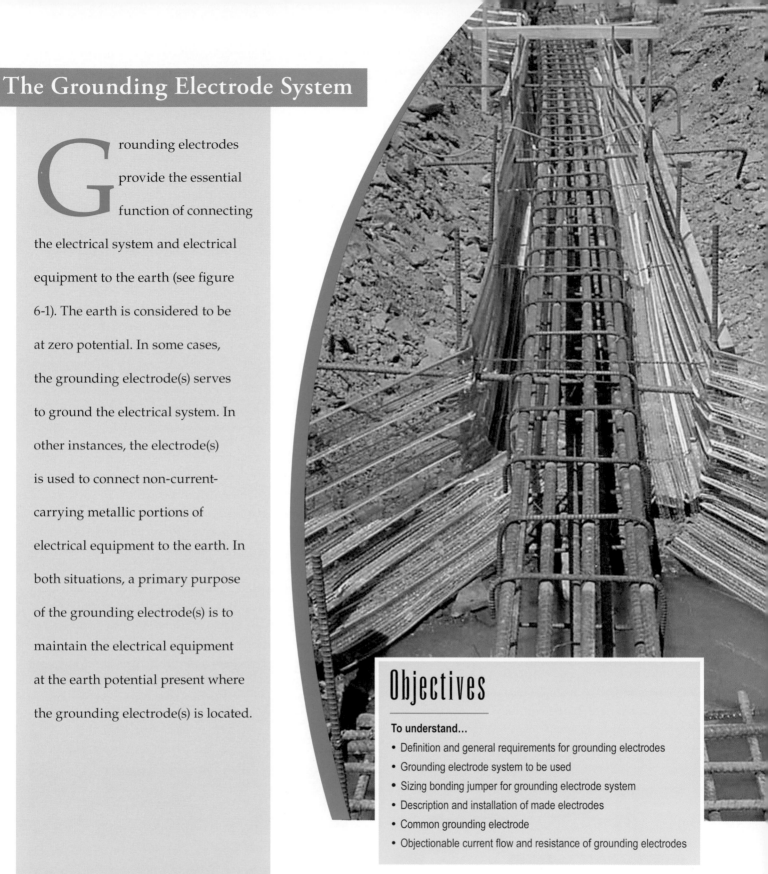

Objectives

To understand...

- Definition and general requirements for grounding electrodes
- Grounding electrode system to be used
- Sizing bonding jumper for grounding electrode system
- Description and installation of made electrodes
- Common grounding electrode
- Objectionable current flow and resistance of grounding electrodes

Chapter 6

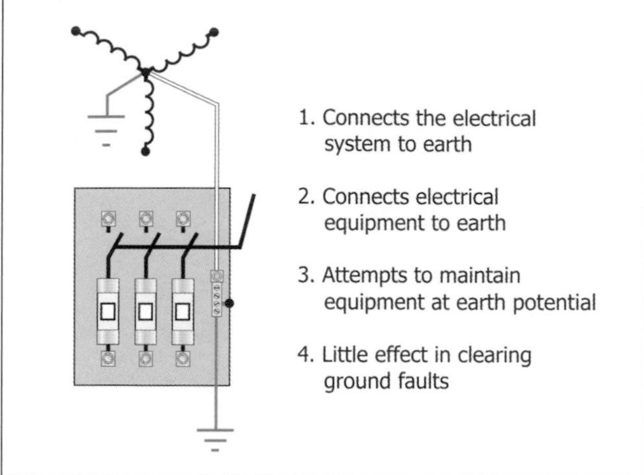

1. Connects the electrical system to earth
2. Connects electrical equipment to earth
3. Attempts to maintain equipment at earth potential
4. Little effect in clearing ground faults

Figure 6-1. Functions of the grounding electrode

Another essential function of the grounding electrode(s) is to dissipate overvoltages into the earth (see figure 6-2). These overvoltages can be caused by high-voltage conductors being accidentally connected to the lower-voltage system such as by a failure in a transformer or by an overhead conductor dropping on the lower-voltage conductor. Overvoltages can also be caused from lightning. With more equipment, even home appliances, containing microprocessors this becomes increasingly important.

Section 250.24(D) requires the equipment grounding conductors, the service-equipment enclosures, and, where the system is grounded, the grounded service conductor to be connected to the grounding electrode(s) required by Part III of Article 250. The conductor used to make this connection is the grounding electrode conductor.[1]

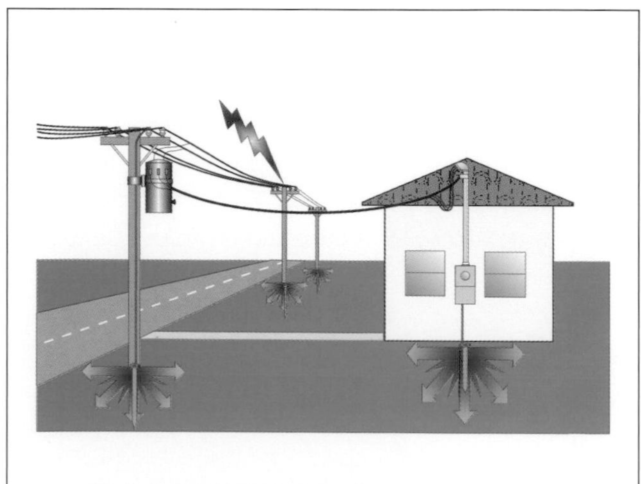

Figure 6-2. Dissipation of overvoltages

Definition

Grounding Electrode. A conducting object through which a direct connection to earth is established.[2]

To establish a true understanding of what constitutes a grounding electrode, the definition in the *Code* needs to be used cooperatively with the list of electrodes identified in 250.52(A). It can clearly be seen in this list of grounding electrodes that a grounding electrode can be a device or other conducting object such as a building footing or metal well casing that establishes and maintains a direct connection to the earth. The effectiveness of the connection is relative and is a variable item. The resistance in the connection between an electrode and the earth will vary based on soil conditions, electrode depth, type of electrode, and seasonal conditions or geographical location(s).

Grounding Electrode System

Section 250.50 requires that all grounding electrodes that are present at each building or structure served be bonded together to form the grounding electrode system (see figure 6-3). An exception to 250.50 in the 2005 *NEC* has a provision for existing buildings or structures. The reinforcing bars or rods do not have to be used as part of the grounding electrode system if the concrete would have to be disturbed to make them accessible. This applies only to an existing building or structure, not an existing foundation or footing. For new construction, coordination between the various

Photo 6-1. Grounding electrode (iron core, copper-clad rod type)

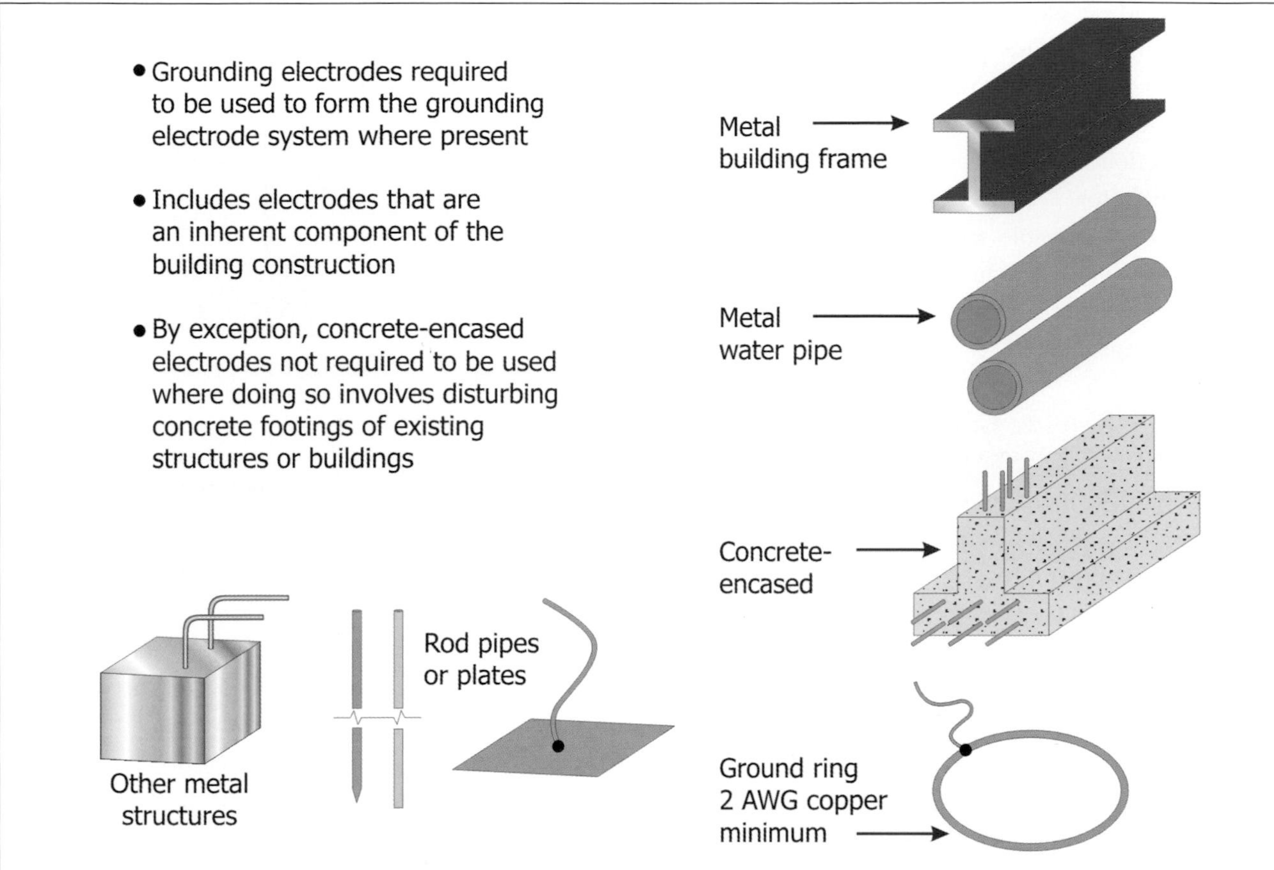
• Grounding electrodes required to be used to form the grounding electrode system where present
• Includes electrodes that are an inherent component of the building construction
• By exception, concrete-encased electrodes not required to be used where doing so involves disturbing concrete footings of existing structures or buildings
Metal building frame
Metal water pipe
Concrete-encased
Other metal structures
Rod pipes or plates
Ground ring 2 AWG copper minimum

Figure 6-3. The grounding electrode system

trades on the project may be necessary to ensure that the concrete-encased electrode becomes part of the grounding electrode system.

Section 250.52(A) contains the descriptions of what are considered acceptable electrodes. These include:

- metal underground water pipes
- metal frames of the building or structure
- concrete-encased electrodes
- ground rings
- rod and pipe electrodes
- other listed electrodes
- plate electrodes
- other local metal underground systems or structures

The general requirement is that a bonding jumper is required to be installed between the grounding electrodes to bond them together. A grounding electrode conductor is then run from one of the grounding electrodes to the grounded conductor, the equipment grounding conductor, or to both, at the service, at a separate building or structure, or at a separately derived system. The *NEC* also provides for the option of running a grounding electrode conductor to each grounding electrode individually and bonding jumpers to interconnect any of the grounding electrodes as provided in 250.64(F)(1) (2) and (3).

Where the interior metal water pipe is used as a part of the grounding electrode system or as a conductor to bond other electrodes together to create the *grounding electrode system,* 250.52(A)(1) requires that all connections to metal water piping electrodes take place within the first 5 feet from the point the water pipe enters the building. This helps minimize the possibility of having the path to the grounding electrode system interrupted if the piping was replaced with non-metallic piping. It does not require that the interior water pipe be used for the purpose of interconnecting other electrodes to form the grounding electrode system.

Any of the other electrodes, such as the metal frame of the building, concrete-encased electrode or ground ring, can be used for the purpose of interconnecting the other grounding electrodes. Where these other electrodes are used for this purpose, no restrictions are placed on where the connections are permitted to be made or how far inside the building they are permitted. Section 250.68(A) requires grounding electrode conductor connections to grounding electrodes to be accessible except for connections to a buried, driven, concrete-encased electrode, or exothermic or irreversible compression connections to fireproofed structural metal.

Description of the Required Grounding Electrodes

As previously stated, 250.52(A) contains a list of the grounding electrodes permitted to be used to form a grounding electrode system and 250.52(B) contains a list of the items that are not permitted to be used as grounding electrodes (see figure 6-4). Underground metal gas piping systems are not permitted to be used as a grounding electrode. This does not eliminate the requirement that interior metal gas piping systems be bonded (see chapter eight for additional information on bonding of metal piping systems). Aluminum also is not permitted to be used because they would corrode in many types of soil.

All of the identified grounding electrodes are required to be used if they are present on the premises at each building or structure served. The grounding electrodes are not listed in an order of preference nor is it optional to choose which ones to use. Electrodes that are required to be used where present are as follows:

Metal Underground Water Pipe

Defined in 250.52(A)(1) as "A metal underground water pipe that is in direct contact with the earth for 3.0 m (10 ft) or more (including any metal well casing that is effectively bonded to the pipe)...." There is no minimum or maximum pipe size given. Types of metal, such as steel, iron, cast iron, or stainless steel are not distinguished. Different types of water pipes such as for potable water, fire protection sprinkler systems, irrigation piping, and so forth, are also not defined. As a result, all of these metal underground water pipes are required to be used if they are present at each building or structure served.

Continuity of the grounding path of the water pipe grounding electrode or the bonding of interior piping

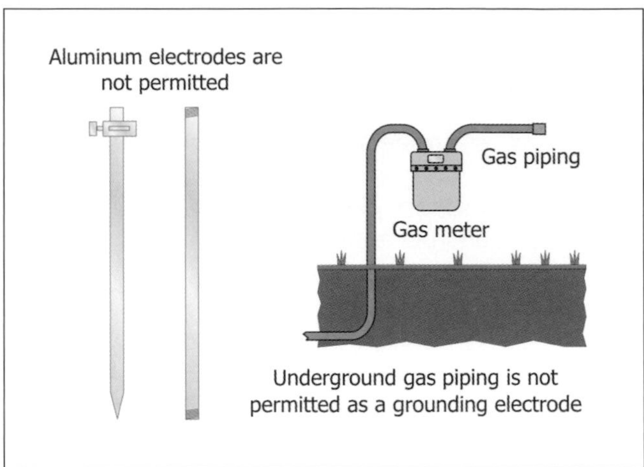

Figure 6-4. Grounding electrodes not permitted for use as provided in 250.52

systems cannot depend on water meters or on filtering devices or similar equipment. Where a water meter or filtering equipment is in this metal water piping system, a bonding jumper is required to be installed around the equipment to maintain continuity even if the water meter or filter is removed. The bonding jumper is required to be the same size as the grounding electrode conductor and long enough to allow the meter, filter, or other equipment to be removed without disconnecting the bonding jumper [see 250.68(B)].

The Metal Frame of the Building or Structure

Section 250.52(A)(2) has been revised for *NEC*-2008 to better describe if the metal frame is an acceptable grounding electrode. The concept of being effectively grounded was deleted and a list was added to clarify how to determine if there is a connection to earth. There must be 10 ft of the metal frame in direct contact with the earth, it must be bonded to an acceptable electrode, or another means must be used to establish a connection to earth. Certain backfills such as gravel or vapor barriers can render the building steel an ineffective electrode.

"250.52(A)(2) Metal Frame of a Building or Structure. The metal frame of the building or structure, that is connected to the earth by any of the following methods:

(1) 3.0 (10 ft) or more of a structural metal member in direct contact with the earth or encased in concrete that is in direct contact with the earth

(2) Connecting the structural metal frame to the

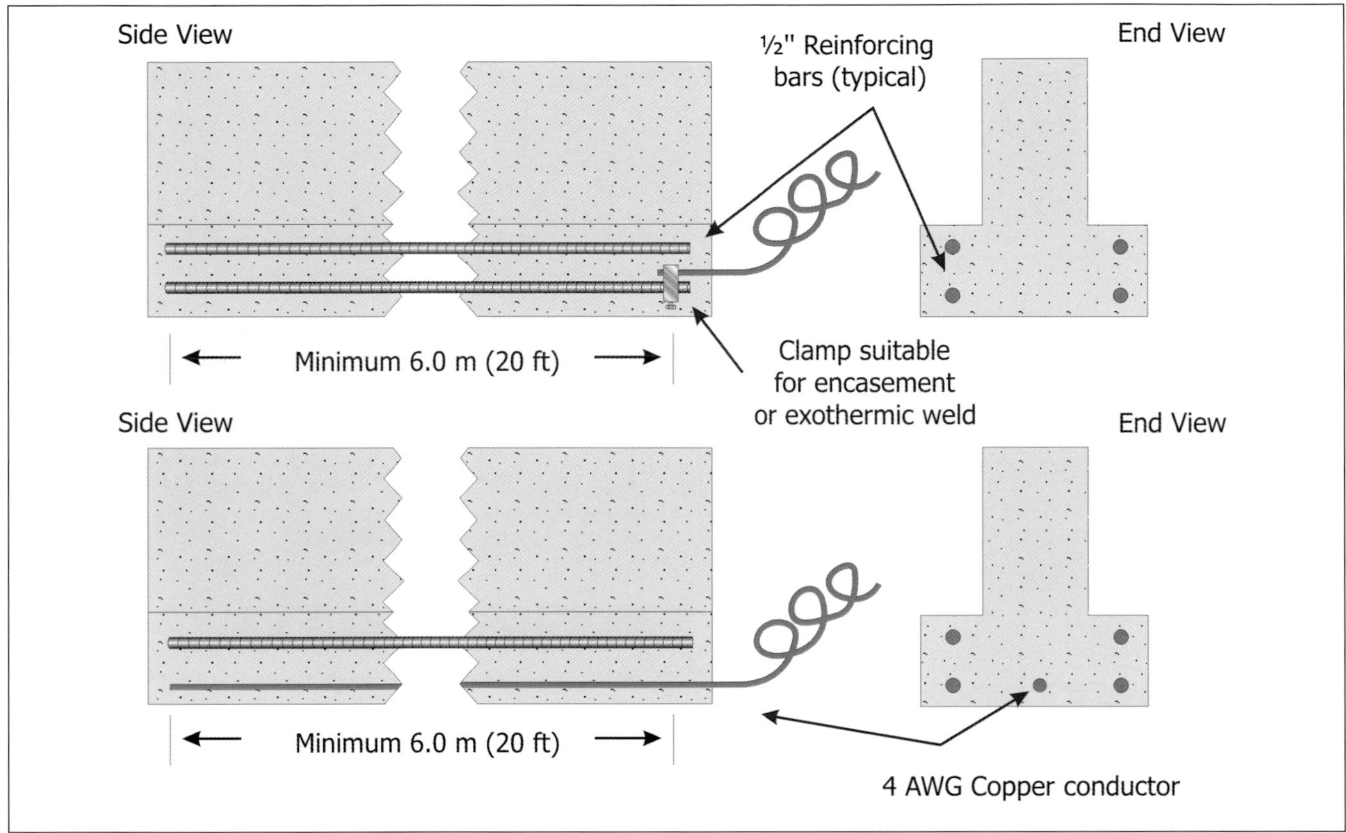

Figure 6-5. Concrete-encased electrode

reinforcing bars of a concrete-encased electrode as provided in 250.52(A)(3) or ground ring as provided in 250.52(A)(4).

(3) Bonding the structural metal frame to one or more of the grounding electrodes as defined in 250.52(A)(5) or (A)(7) that comply with 250.56, or

(4) Other approved means of establishing a connection to earth."

Concrete-Encased Electrodes

Section 250.52(A)(3) defines this grounding electrode as "encased by at least 50 mm (2 in.) of concrete, located horizontally near the bottom or vertically, and within that portion of a concrete foundation or footing that is in direct contact with the earth, consisting of at least 6.0 m (20 ft) of one or more bars of zinc galvanized or other electrically conductive coated steel reinforcing bars or rods of not less than 13 mm (1/2 in.) in diameter, or consisting of at least 6.0 m (20 ft) of bare copper conductor not smaller than 4 AWG" (see figures 6-5 and 6-6). A single 6.0 m (20 ft) length of reinforcing bar is not required. Reinforcing bars are permitted to be bonded together by the usual steel tie wires or other effective means like welding. Where subjected to events such as lightning strikes, welding might be preferred.

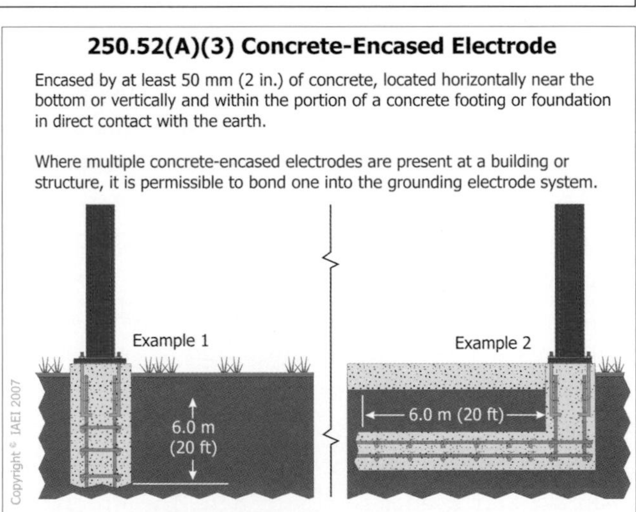

Figure 6-6. Concrete-encased electrode

Where multiple concrete-encased electrodes are present at a building or structure and are not electrically connected together, it is permissible to connect only one into the grounding electrode system, but at least one must be used to comply with 250.50.

Reinforcing rods are required to be of bare, zinc galvanized or other electrically conductive coated steel material. Obviously, insulated reinforcement rods would not perform properly as a grounding electrode. Some complaints have been made that lightning surges, which can be dissipated through this electrode, break out chunks of concrete where the surge exits the footing.

This grounding electrode is commonly referred to as the *Ufer ground* after H.G. Ufer who spent many years documenting its effectiveness. See Appendix A for additional information on the development of the concrete-encased electrode.

Several electrical inspection agencies require that a concrete-encased electrode be installed or connected to the service prior to authorizing electrical service due to its effectiveness in most any climatic and soil condition.

Ground Ring

Section 250.52(A)(4) recognizes a copper conductor, not smaller than 2 AWG and at least 6.0 m (20 ft) long, as a ground ring grounding electrode. The conductor is required to encircle the building or structure and be buried not less than 750 mm (30 in.) deep. Ground rings often are installed at telecommunication central offices, radio and cellular telephone sites. Where present on the premises served, ground rings are required to be used as one or more of the grounding electrodes making up the grounding electrode system.

Where the electrodes described in 250.52(A)(1) through (A)(7) do not exist at the building or structure served, a grounding electrode(s) is required to be installed and used (see figure 6-9). The grounding electrodes as provided in 250.52(A)(4) through (A)(8) may be either or any combination of the following types:

"(A)(4) Ground Ring. A ground ring encircling the building or structure, in direct contact with the earth, consisting of at least 6.0 m (20 ft) of bare copper conductor not smaller than 2 AWG.

"(A)(5)(a) Rod and Pipe Electrodes. Rod and pipe electrodes shall not be less than 2.44 m (8 ft) in length and shall consist of the following materials.

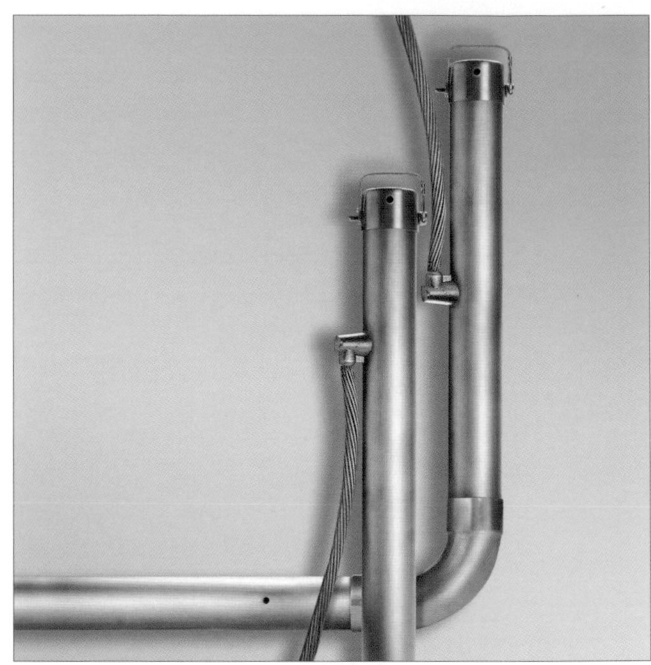

Photo 6-2. Listed grounding electrode Courtesy of Erico International

"(a) Grounding electrodes of pipe or conduit shall not be smaller than metric designator 21 (trade size ¾) and, where of iron or steel, shall have the outer surface galvanized or otherwise metal-coated for corrosion protection.

"(b) Grounding electrodes of rods of stainless or copper or zinc coated steel shall be at least 15.87 mm (5/8 in.) in diameter, unless listed and not less than 12.70 mm (1/2 in.) in diameter.

"(A)(6) Other Listed Electrodes. Other listed grounding electrodes shall be permitted." (see photo 6-2).

"(A)(7) Plate electrodes. Each plate electrode shall expose not less than 0.186 m^2 (2 ft^2) of surface to exterior soil. Electrodes of iron or steel plates shall be at least 6.4 mm (¼ in.) in thickness. Electrodes of nonferrous metal shall be at least 1.5 mm (0.06 in.) in thickness." Because a 1 sq. ft plate has two sides, it would comply with this section.

"(A)(8) Other Local Metal Underground Systems or Structures. Other local metal underground systems or structures such as piping systems, underground tanks, and underground well casings that are effectively bonded to a metal water pipe." [3] Well casings were added to the 2005 *NEC*. This clarifies that metal underground well casings are not metal underground water pipes therefore do not require a

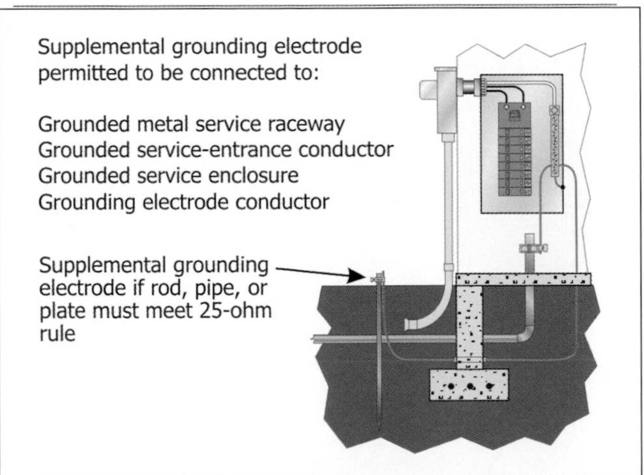

Supplemental grounding electrode
permitted to be connected to:
Grounded metal service raceway
Grounded service-entrance conductor
Grounded service enclosure
Grounding electrode conductor
Supplemental grounding electrode if rod, pipe, or plate must meet 25-ohm rule

Figure 6-7. Supplemental grounding electrode required

supplemental electrode as the metal water pipe does [see 250.53(D)(2)]. These objects are required to have the metal in direct contact with the earth. Protective coatings can render them ineffective as grounding electrodes.

Supplemental Electrode

Section 250.53(D)(2) requires that where the only grounding electrode present and connected at the building or structure served is a metal underground water pipe, it has to be supplemented by another grounding electrode. This is a required electrode versus the auxiliary electrode in 250.54 that is permitted to be connected to the equipment grounding conductor.

Electrodes suitable to supplement the metal underground water pipe include:

- the metal frame of the building,
- a concrete-encased electrode,
- ground ring,
- rod and pipe electrodes
- other listed electrodes
- plate electrodes, and
- other local metal underground systems or structures.

The electrode(s) chosen must still meet the requirements of 250.52 such as the specified lengths and burial depths, and the installation requirements of 250.53.

This supplemental grounding electrode is required because metal underground water pipes are often replaced by plastic water piping or the system continuity is interrupted by nonmetallic couplings or repairs. The effectiveness of the water pipe grounding electrode would thus be lost.

Specific locations are provided where the supplemental grounding electrode is permitted to be connected (see figure 6-7). The supplemental grounding electrode is permitted to be bonded only to the grounding electrode conductor, the grounded service-entrance conductor, the grounded nonflexible service raceway or to any grounded service enclosure. An exception to this requirement permits the bonding connection to the interior metal water piping in qualifying industrial or commercial installations to be made at any location if the entire length of interior metal water pipe that is being used as a conductor is exposed (see definition of *exposed* in Article 100). Locations where the exposed piping passes through walls or floors are required to be perpendicular to those penetrations. If all of the provisions of the exception are not met the bonding connection must be within the first 1.52 m (5 ft) of where the piping enters the building.

Often, changes, repairs, or modifications are made to the metallic water piping systems with nonmetallic pipe or fittings or dielectric unions. In this case, it is possible to inadvertently isolate portions of the grounding system from the grounding electrode conductor. This is another in several steps that has been taken over recent years to reduce the emphasis and reliance on the metal water piping system for grounding of electrical systems.

Where the supplemental grounding electrode is of the rod, pipe, or plate type, it is required to meet the 25-ohms-to-ground rule in 250.56. The supplemental grounding electrode, by itself, must have a resistance of not more than 25 ohms or a second supplemental grounding electrode is required to be used. Note that an underground metal water pipe electrode is not recognized for providing the earth connection for a metal building frame in 250.52(A)(2), (3) or (4). The metal frame should not be considered as a supplement to the water pipe electrode. The supplemental electrode covered in 250.53(D)(2) is anticipated to be the only electrode if and when the water pipe is eventually replaced with nonmetallic components.

Grounding electrodes connected together with bonding jumpers that are installed in accordance with 250.64(A), (B), and (E)
Rod and pipe electrodes
Service equipment
MAIN
ON
OFF
ON
ON
OFF
OFF
Bonding jumpers between grounding electrodes
Size bonding jumpers in accordance with 250.66
Grounding electrode conductor sized per 250.66

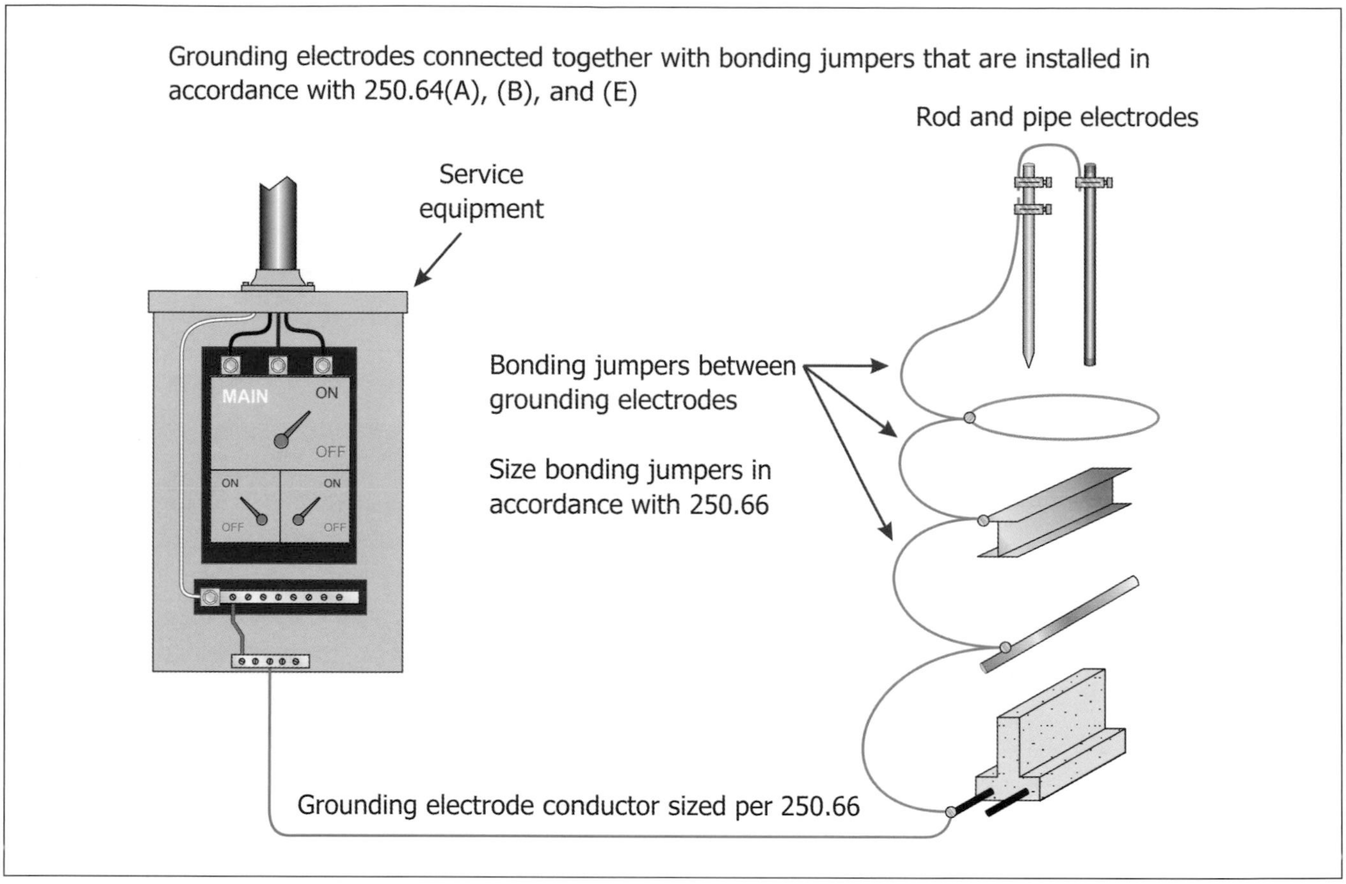

Figure 6-8. Bonding jumper(s) for the grounding electrode system

Size of Bonding Jumper for Grounding Electrode System

Section 250.53(C) requires the bonding jumper used to connect the grounding electrodes of the grounding electrode system together to be installed in accordance with the requirements of 250.64(A), (B), and (E) (see figure 6-8). The bonding jumper used to bond the grounding electrodes together to form the grounding electrode system is required to be sized in accordance with 250.66 based on the size of the ungrounded service-entrance conductor and to be connected in a manner specified in 250.70. The conductor that connects the grounding electrodes together is a bonding conductor and not a grounding electrode conductor. The bonding conductors are not required to be installed in one continuous length as grounding electrode conductors are. In addition, the exceptions for sizing the grounding electrode conductor in 250.66 apply for the sizing of the bonding jumpers.

For example, if the service-entrance conductor is 500-kcmil copper, the minimum size of bonding jumper between grounding electrodes is determined by reference to 250.66 and Table 250.66, and its installation falls under the requirements of the rules in 250.64(A), (B) and (E), which are as follows:

- To the metal underground water pipe and the metal frame of a building; 1/0 AWG copper or 250 kcmil aluminum conductor (from Table 250.66).
- To electrodes as provided in 250.52(A) such as pipes, rods, or plates; that portion of the bonding jumper that is the sole connection to the rod, pipe, or plate electrode; 6 AWG copper or 4 AWG aluminum. The term *sole connection* means that the bonding conductor is not first connected to a rod, pipe, or plate electrode and then extended to another grounding electrode [see 250.66(A)].
- To a concrete-encased electrode as in 250.52(A)(3); that portion of the bonding conductor that is the sole connection to the concrete-encased electrode; 4 AWG copper conductor [see 250.66(B)].
- To a ground ring as in 250.52(A)(4); that portion of the bonding conductor that is the sole connection

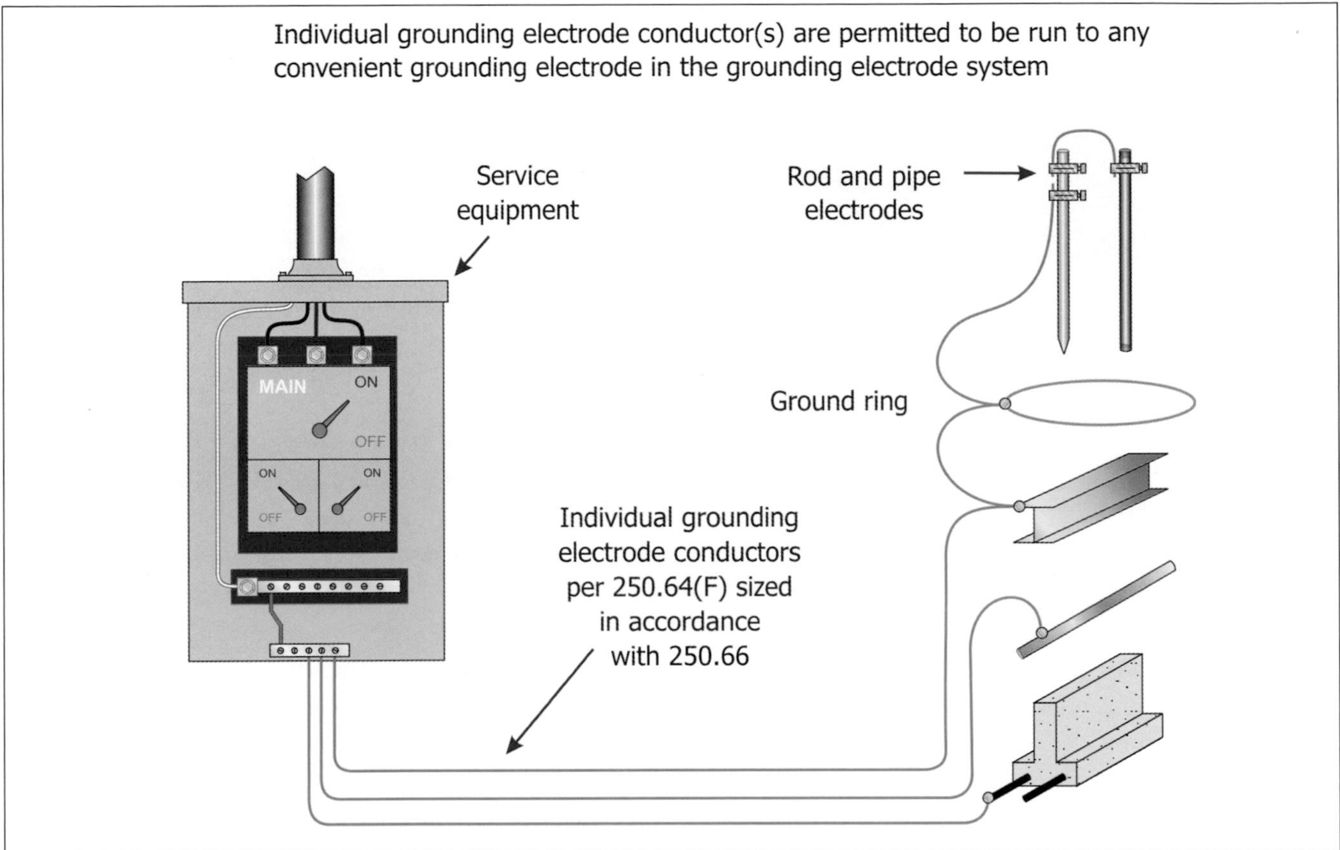
Individual grounding electrode conductor(s) are permitted to be run to any convenient grounding electrode in the grounding electrode system
Service equipment
Rod and pipe electrodes
MAIN
ON
OFF
ON
OFF
ON
OFF
Ground ring
Individual grounding electrode conductors per 250.64(F) sized in accordance with 250.66

Figure 6-9. Size of individual grounding electrode conductor

to the ground ring is not required to be larger than the ground ring conductor. The minimum size for the ground ring is 2 AWG [see 250.66(C)].

Of course, larger bonding jumpers could be used.

Section 250.64(A) does not permit bare aluminum or copper-clad aluminum conductors to be installed as a grounding electrode conductor where in direct contact with masonry or the earth or where subject to corrosive conditions. Where used outside, aluminum or copper-clad aluminum grounding conductors shall not be terminated within 450 mm (18 in.) of the earth.

No sequence or order for installing the bonding jumper or jumpers is given. However, the minimum conductor size required to the various grounding electrodes is to be observed. In addition, the point where the grounding electrode connects to the grounding electrode system is required to provide for the largest required grounding electrode conductor. For example, it would be a violation to connect a 4 AWG bonding conductor from a concrete-encased grounding electrode to a building steel grounding electrode

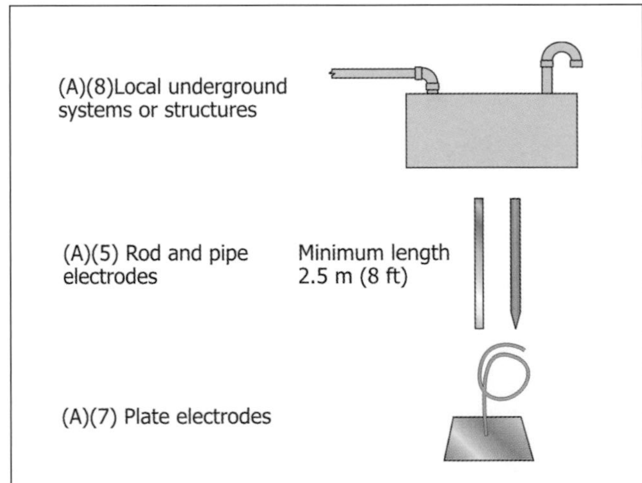
(A)(8)Local underground systems or structures
(A)(5) Rod and pipe electrodes
Minimum length 2.5 m (8 ft)
(A)(7) Plate electrodes

Figure 6-10. Installed grounding electrodes or underground structures or systems that can serve as grounding electrodes

which would require a 3/0 copper grounding electrode conductor. The installation would be acceptable if the 3/0 copper grounding electrode conductor connects to the building steel and a 4 AWG copper bonding jumper extends from the building steel to the concrete-encased electrode.

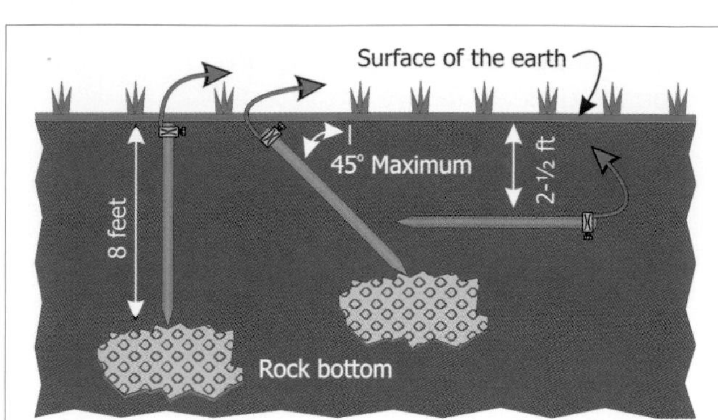

Figure 6-11. Installation requirements for rod or pipe electrodes

In addition, the unspliced grounding electrode conductor is permitted to run from the service equipment to any convenient grounding electrode. Alternately, individual grounding electrode conductors are permitted to be installed from the service equipment to one or more grounding electrodes rather than the electrodes being bonded together in a circular or *daisy-chained* manner [see 250.64(F)]. The minimum size of each grounding electrode conductor to the individual grounding electrode is based on the electrode to which each individual grounding electrode conductor is connected (see figure 6-9). A grounding electrode conductor is permitted to *supply* or *serve* any number of grounding electrodes but is sized for the largest grounding electrode conductor required. For example, a bonding jumper is permitted to be run to a concrete-encased electrode and then to the underground metal water pipe. The bonding jumper is required to be sized for the largest grounding electrode conductor required for any of the grounding electrode or electrodes served.

Installation of Grounding Electrodes

Installation requirements for grounding electrodes are covered in 250.53(A) through (H). Descriptions of the different types of grounding electrodes that are either permitted or not permitted are located in 250.52(A) and (B). Installation requirements for grounding electrodes such as depth, connections to, and electrode spacing between electrodes are found in 250.53.

Where practicable, field-installed electrodes such as rods, pipes, or plates are required to be installed below permanent moisture level. This is a key ingredient in establishing an effective electrode. They also are

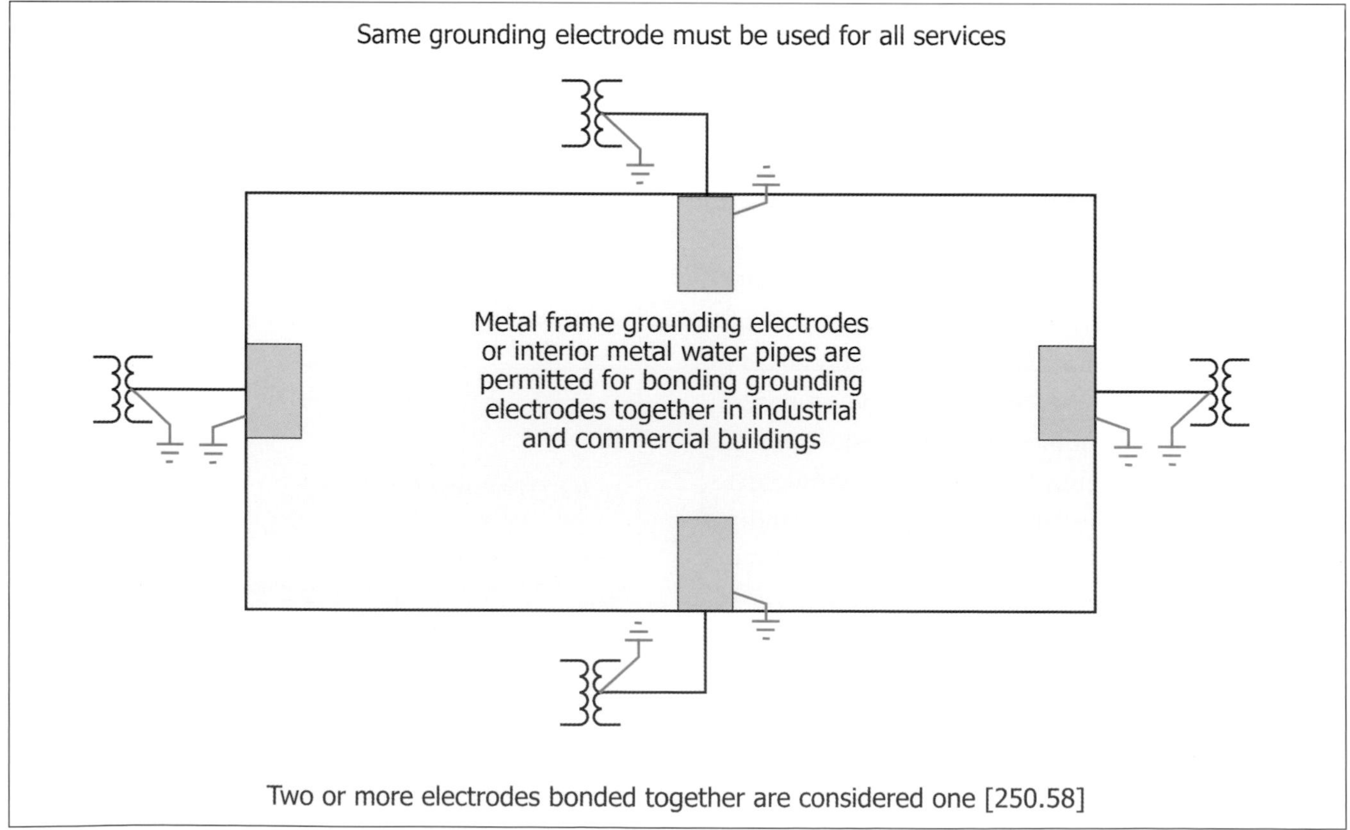

Figure 6-12. Metal building frame electrode is common to all services

required to be free from nonconductive coatings such as paint and enamel.

Rod and pipe electrodes are required to be installed so at least 8 ft. is in contact with the soil (see figure 6-11). They are required to be driven vertically unless rock bottom is encountered. If rock bottom is encountered which prevents the rod from being driven 8 feet vertically, the rod is permitted to be installed at an oblique angle of not more than 45 degrees from vertical. Where rock bottom is encountered at an angle up to 45 degrees, only then can the rod or pipe be buried in a trench that is at least 750 mm (30 in.) deep [see 250.53(H)].

The upper end of the rod is required to be flush with or below ground level unless the aboveground end of the rod and the grounding electrode attachment are protected from physical damage. This, of course, requires that a ground rod longer than 8 feet be used if any part of the rod is exposed above ground level. For an eight-foot ground rod or pipe, the ground clamp is required to be listed for direct earth burial, as the electrode is required to be driven to its full length.

Installation requirements for plate electrodes are found in 250.53(H) and require the plate to be buried not less than 2½ feet below the surface of the earth.

Section 250.10 requires that ground clamps or other fittings be approved (acceptable to the authority having jurisdiction) for general use without protection or be protected from physical damage by metal, wood or equivalent protective covering.

Common Grounding Electrode

Where more than one service supplies a building or structure, often there is more than one utility transformer or source. Multiple services or sources can have differences of potential between them, and where installed in the same building or structure, must use the same electrode. Section 250.58 requires that a common grounding electrode be used for all alternating-current system grounding in or at a building or structure. This is also required by 250.50. This section recognizes that where two or more grounding electrodes are bonded together, they are considered to be a single grounding electrode system.

Interestingly, no distance between electrodes is given beyond which the electrodes do not have to be bonded together. Buildings or structures of *large area* are permitted by 230.2(B)(2) to have more than one service. However, nothing in the *Code* defines the dimensions of a *large building or structure*. Some inspection authorities use voltage drop of major feeders for guidance in determining when a building is one of large area. Where feeder conductors would have to be increased in size unreasonably to maintain voltage regulation, one or more additional services are permitted.

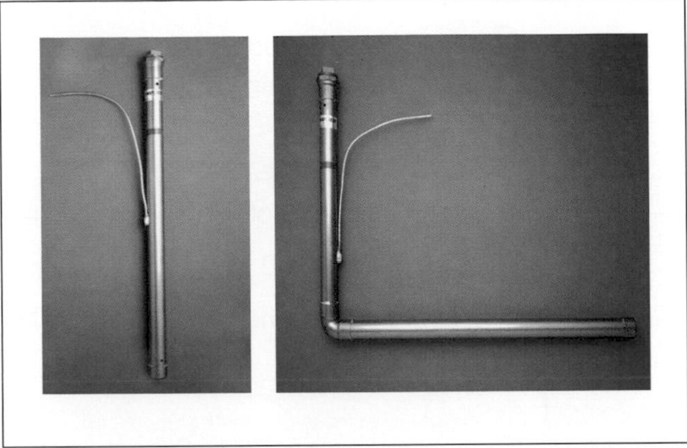

Figure 6-13. Enhanced grounding electrode Courtesy of Harger

Section 250.58 requires the grounding electrodes for the multiple services be bonded together no matter how far apart they are in the same building. This is important to keep all the equipment at the same earth potential. Section 250.53(C) requires the bonding jumper(s) used for this purpose to be sized from Table 250.66 and be installed in accordance with 250.64(A), (B) and (E) (see chapter seven for additional information on installation of grounding electrode conductors). It is permitted to use the metal frame of a building, that complies with 250.52(A)(2), or a concrete-encased grounding electrode to bond grounding electrodes from other services together.

For example, a large building is served by four services. Where each of the services is connected to the building steel for grounding electrode purposes and also has a ground rod installed at each service location, such action serves to bond the ground rods together (see figure 6-11).

Section 250.58 also requires that a common grounding electrode be used to ground conductor enclosures and equipment in or on the building and that the same grounding electrode be used to ground the system. Again this does not mean that one cannot use more than one grounding electrode. But, if more

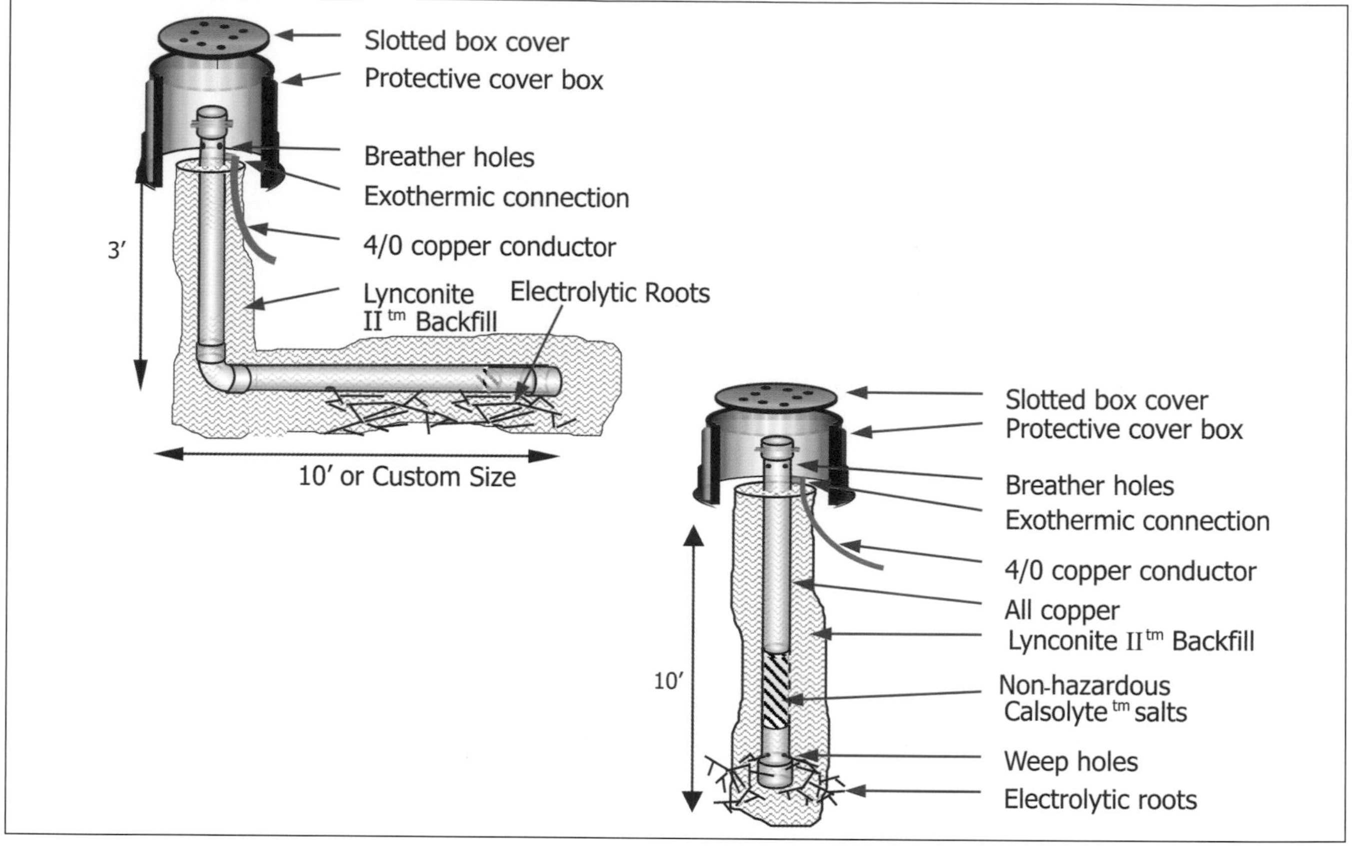

Figure 6-14. Anatomy of an enhanced grounding electrode Concept is courtesy of Lincole XIT Grounding

than one is used, then all the grounding electrodes are required to be bonded together to form a common grounding electrode. Where multiple grounding electrodes are bonded together as cited above, such multiple grounding electrodes become, in effect, a common grounding electrode system.

Enhanced Grounding Electrodes

Grounding electrodes and grounding electrode systems are covered in Part III of Article 250. These are the minimum requirements for grounding electrodes for use in grounding services, systems, and equipment. With ever-increasing installations and use of information technology equipment and sensitive electronics, there is sometimes the need to exceed the minimum requirements which are established for safety of persons and property and install electrodes or electrode systems that are extensive in nature and designed to establish and maintain a lower level of resistance in the connection to earth through the electrode. There are many listed products available to accomplish this additional grounding when desired for the electrical system. Section 250.52(A)(6) includes provisions for other listed electrodes.

Electrolytic Grounding Systems

Electrolytic grounding was invented by the XIT Rod Company in the late 1960s (see figures 6-13 and 6-14). It was specifically designed to eliminate the limitations of other typical electrodes. This active electrode consists of a hollow copper pipe filled with natural earth salts. The salts extract moisture from the air, which forms a highly conductive electrolytic solution. The solution continually weeps into the surrounding backfill material providing improved conductivity and seasonal stability. The electrode is installed in an augured hole or trench and backfilled with specially processed bentonite clay. The clay is very conductive with nearly a neutral pH that helps protect the electrode from corrosion. Due to its high conductivity, the bentonite improves the ground system performance and also provides an excellent electrical *bond* between the electrode and the surrounding earth. These active electrodes are the only

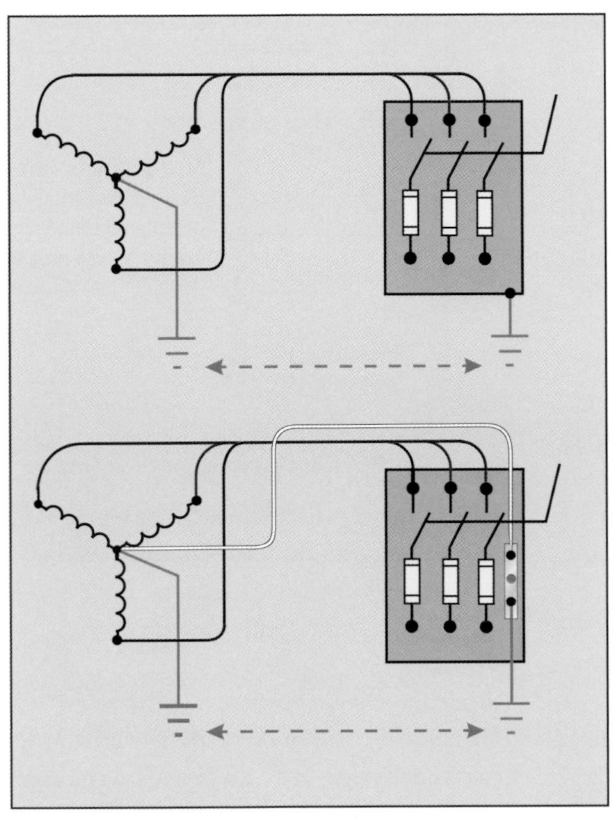

Ground fault at service attempts to return to source (transformer) through the high-impedance path through the earth

A low-impedance path for clearing ground faults is provided through the grounded service conductor

Figure 6-15. Earth is prohibited as an effective ground-fault current path (a low-impedance path is required).

ones that improve with time; all others become less effective and begin deteriorating upon installation.[4]

Enhanced Grounding Electrode Types

These enhanced electrodes can be installed in a variety of methods. The nature of these electrodes can also vary. Some are intended to be installed with no additional maintenance required. Others are installed and have to be maintained with chemicals or other effective means. The maintainable electrodes are more commonly installed under controlled conditions where there is qualified staff that will ensure proper supervision and maintenance of the electrode system. Additional information on these enhanced grounding electrodes and grounding electrode systems and their uses is covered in chapter 17.

Earth Return Prohibited

No mention is made to providing a low-resistance, low-impedance common grounding electrode for clearing ground faults. The high impedance of the earth makes it an ineffective path for the levels of current common to power systems. The earth may never be used as an equipment grounding conductor, as it is a very poor conductor. Figure 6-15 shows that the grounding electrodes provide the only return circuit through the earth. Even if the grounding electrode resistance to the earth were very low, it would have little effect on clearing a ground fault, because the reactance of the earth and the soil in the ground-fault return circuit is very high. As discussed in chapter one, the greater the resistance or impedance is, the less the amount of current. Where a parallel path exists through the earth and through the grounded service conductor, almost all of the ground-fault return current will return to the source through the grounded service conductor. A low-resistance common grounding electrode is beneficial to the electrical installation by keeping equipment and the grounded conductor at or close to earth potential. It simply is not effective in clearing line-to-ground faults. Sections 250.4(A)(5) and 250.54 make it clear that the earth is not to be depended upon to function as an effective ground-fault current path. However, grounding electrodes are permitted to be supplementary to equipment grounding conductors and connected to the equipment grounding conductor(s) as provided in 250.54.

If a ground fault should develop as shown in the upper drawing in figure 6-15 where two separate grounding electrodes are used, the fault-current will be through the service conductor then through the impedance of the ground fault, the grounding electrode conductor at the service, the grounding electrode at the service, the path through the earth to the grounding electrode at the transformer and finally through the grounding conductor at the transformer to complete the circuit to the transformer. It would be a rare case where that circuit resistance would add up to less than 12 ohms (while the impedance would be higher). Even then, the fault current would not reach a value high enough to operate a 15-ampere overcurrent device on a 120 volt-to-ground circuit (120 ÷ 12 = 10 amperes).

Considering resistance only, the circuit shown has two grounding electrodes in series. Compared to the much lower resistance parallel path of the grounded circuit conductor, a resistance ratio between the two parallel paths is about 50 times for a 100-ampere service to well over 100 times for the larger services. When impedance of the two paths is considered, the ratio will be higher. Therefore, almost all the current from a line-to-ground fault will return to the transformer over the grounded service conductor.

Under normal operating conditions there will be some unbalanced return current in the neutral back to its source. There will also be some unbalanced neutral current through the earth, but it will be a very low level compared to that through the grounded service conductor.

Any belief that the circuit to the grounding electrode can be depended on to clear a ground fault is clearly erroneous no matter how large a grounding electrode conductor is used or how good a grounding electrode is. However, when the high-impedance earth path is *short-circuited* by installing the grounded circuit conductor as shown in the lower drawing in figure 6-15, a low-impedance ground-fault return path is established as required in 250.4(A)(5). This provides or allows enough current through the equipment grounding conductor and service grounded conductor to allow the branch-circuit, feeder, or service overcurrent device to clear the fault and thus provide the safety contemplated by the *Code*.

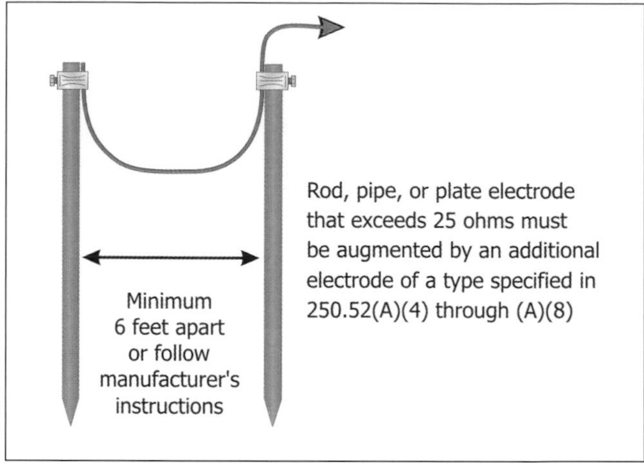
Minimum
6 feet apart
or follow
manufacturer's
instructions
Rod, pipe, or plate electrode
that exceeds 25 ohms must
be augmented by an additional
electrode of a type specified in
250.52(A)(4) through (A)(8)

Figure 6-16. Electrode spacing requirements

Resistance of Grounding Electrodes

There is no requirement in Article 250 that the grounding electrode system required by 250.50 (consisting of metal underground water pipe, metal frame of the building, concrete-encased electrode) meet any maximum resistance to ground. No doubt it is expected that the grounding electrode system will have a resistance to ground of 25 ohms or less. The *Code* specifies a maximum resistance of 25 ohms for rod, pipe, or plate electrode(s). Where the resistance of a single rod, pipe, or plate electrode exceeds 25 ohms, they are required to be supplemented by one additional electrode.

The rules change for *installed* electrodes. The *Code* states, in 250.56, that where a single rod, pipe, or plate electrode does not achieve a resistance to ground of 25 ohms or less, at least one additional electrode must be used (see figure 6-16). This means that where driven ground rods are installed, two ground rods would be the maximum required under any condition. There is no requirement that additional electrodes such as ground rods or plates be installed until the 25 ohms-to-ground resistance is obtained.

In general, metallic underground water piping systems, metallic well piping systems, metal frame buildings and similar grounding electrodes can be expected to provide a ground resistance of not over 3 ohms and in some cases as low as 1 ohm.

However, from a practical standpoint, no grounding electrode, no matter how low its resistance, can ever be depended upon to clear a ground fault on any distribution system of less than 1000 volts.

If a system is effectively grounded as pointed out under 250.4(A)(5), a path of low impedance (not through the grounding electrode) is required to be provided to facilitate the operation of the overcurrent devices or ground detectors in the circuit (see chapter eleven).

The lowest practical resistance of a grounding electrode is desirable and will better limit the voltage to ground when a ground fault occurs. It is also very important to provide a low-impedance path to clear a fault promptly, because a voltage to ground can only occur during the period of time that a fault exists. Clearing a ground fault quickly enhances safety.

Even though the grounding electrode can have a low resistance to earth, it is a high-impedance circuit and plays virtually no part in the clearing of a fault on a low-voltage distribution system. This is because there is a higher resistance path through the earth than through the grounded service conductor. In addition, the remote path through the grounding electrode and earth is a high-impedance path compared to the circuit where the grounded service conductor is installed and routed with the ungrounded (phase) conductors.

Ground Electrode System Monitoring

Grounding electrode system monitoring capability has become more and more common especially where information technology equipment and other sensitive electronic equipment are utilized. This equipment is not required by the *Code*, but often desired as an essential performance option for data centers and similar facilities. Grounding electrodes and grounding electrode systems are the cornerstones (foundation) of electrical protection of a site or facility. Grounding is an integral part of safety as well as the effectiveness of lightning protection and surge suppression systems. Grounding electrode system resistance monitoring equipment measures the grounding system performance and provides an early warning of ground system degradation or loss of integrity so remedial action can be taken. Figure 6-17 provides a conceptual creation of ground resistance monitor components.

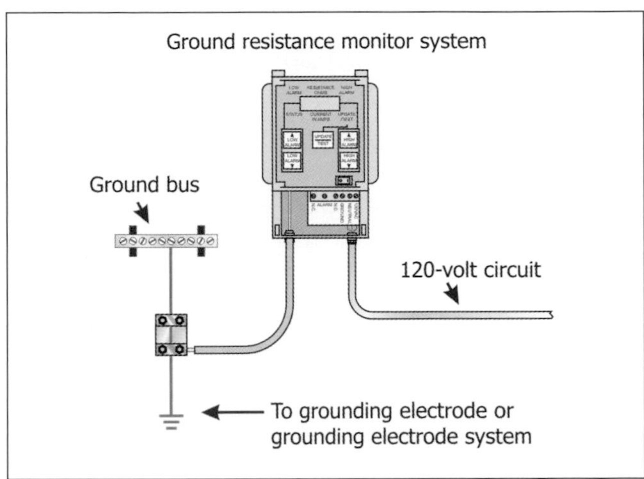

Figure 6-17. Ground resistance monitor components
Concept Courtesy of Lincole XIT

The system consists of a permanent wall-mounted meter and a sensing head (attached to the grounding system conductor). The meter features both high and low level alarms for instantaneous notification when the pre-set resistance values are exceeded.

Features of these types of systems include but are not limited to: (1) continuous monitoring of ground system resistance and current; (2) remote reading and control capability (3) local audible alarm; (4) high and low alarm values; and (5) adjustable sampling rate.

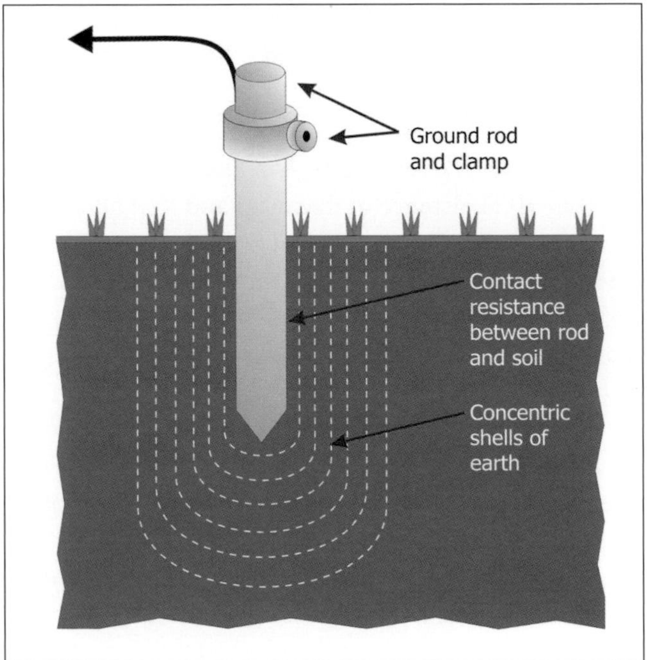

Figure 6-18. Ground rod resistance

Ground Electrodes

The term *ground* is defined as "the earth." The connection to ground (earth) is used to establish and maintain as closely as possible the potential of the earth on the circuit or equipment connected to it. A *ground* consists of a grounding conductor, a bonding connection, its grounding electrode(s), and the soil in contact with the electrode.

Grounding has several protection applications. For natural phenomena, such as lightning, grounding electrodes provide a path to earth to discharge the system of current before personnel can be injured or system components damaged. For other hazards due to faults in electric power systems using *ground returns*, effective grounding helps ensure rapid operation of the protection relays by providing low resistance fault-current paths. This provides for the removal of the hazardous voltage as quickly as possible. The grounding path should drain the hazardous voltage before personnel are injured and the power or communications system is damaged.

Ideally, to maintain a reference potential for instrument safety, protect against static electricity, and limit the system-to-frame voltage for operator safety, a ground resistance should be zero ohms. In reality, as described further in the text, this value cannot realistically be obtained.

Last, but not least, low ground resistance is essential to meet *NEC*, OSHA, and other electrical safety standards.

Figure 6-18 illustrates a grounding rod. The resistance of the electrode has the following components: (1) Resistance of the metal and that of the connection to it; (2) Contact resistance of the surrounding earth to the electrode; (3) Resistance in the surrounding earth to current; or earth resistivity, which is often the most significant factor.

More specifically:

1. Grounding electrodes are usually made of a very conductive metal (copper or copper clad) with adequate cross sections so that the overall resistance is negligible.

2. The National Institute of Standards and Technology (NIST) has demonstrated that the resistance between the electrode and the surrounding earth is negligible if the electrode is free of paint, grease or other coating, and if the earth is firmly packed.

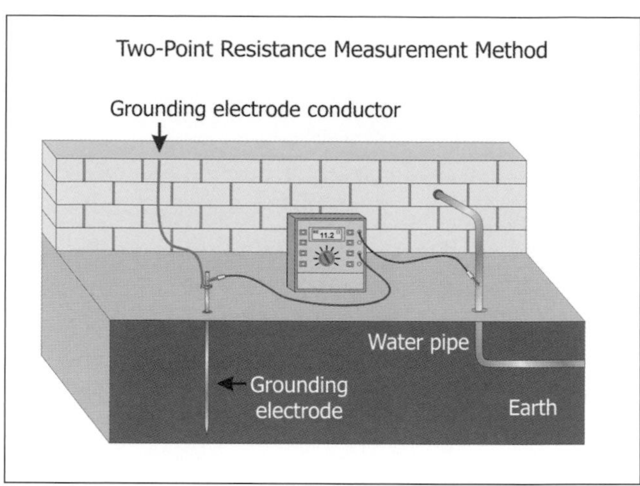

Figure 6-19. Principles of earth testing

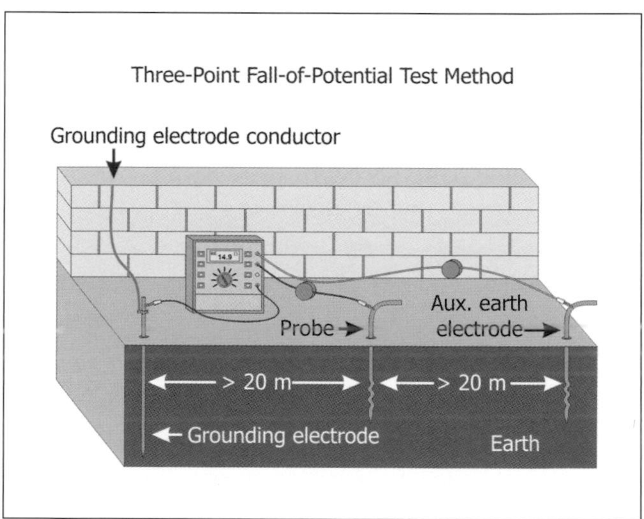

Figure 6-20. Principles of earth testing

3. The only component remaining is the resistance of the surrounding earth. The electrode can be thought of as being surrounded by concentric shells of earth or soil, all of the same thickness. The closer the shell is to the electrode, the smaller its surface; hence, the greater its resistance. The farther away the shells are from the electrode, the greater the surface of the shell; hence, the lower the resistance. Eventually, adding shells at a distance from the grounding electrode will no longer noticeably affect the overall earth resistance surrounding the electrode. The distance at which this effect occurs is referred to as the *effective resistance area* and is directly dependent on and related to the depth of the grounding electrode.

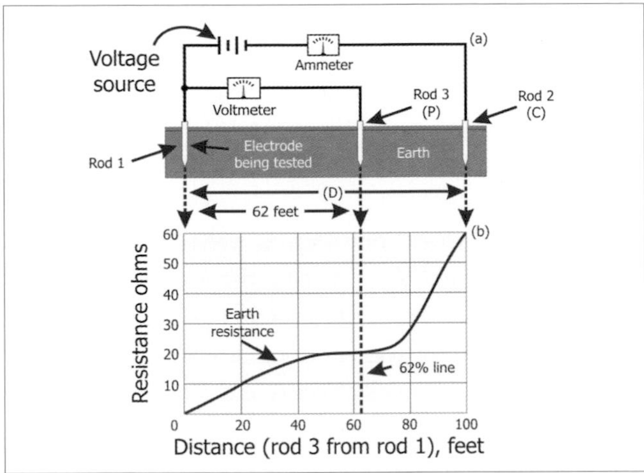

Figure 6-21. Principles of earth testing

Grounding Electrode Resistance Testing

Section 250.56 covers the resistance requirements of certain grounding electrodes and reads as follows: "A single electrode consisting of a rod, pipe, or plate that does not have a resistance to ground of 25 ohms or less shall be augmented by one additional electrode of any of the types specified by 250.52(A)(4) through (A)(8). Where multiple rod, pipe or plate electrodes are installed to meet the requirements of this section, they shall not be less than 1.8 m (6 ft) apart" (see figure 6-16).

The 25-ohm value for a single electrode is an upper limit. Much lower values are beneficial and specified in many instances.

"How low in resistance should a connection to ground be?" An arbitrary answer to this in ohms is difficult. The lower the ground resistance is, the safer the installation; and for positive protection of personnel and equipment, it is worth the effort to aim for less than one ohm. But it is generally impractical to reach such a low resistance along a distribution system or a transmission line or in small substations. In some regions, resistances of 5 ohms or less can be obtained without much trouble.

In other regions, it can be difficult to bring resistance of driven grounds below 100 ohms.

Accepted industry standards stipulate that transmission substations should be designed not to exceed one-ohm resistance. In distribution substations, the maximum recommended resistance is 5 ohms or even 1 ohm. In most cases, the buried grid system of any substation will provide the desired resistance.

In light industrial or in telecommunication central offices, 5 ohms resistance is often the accepted value. For lightning protection, the arrestors should be coupled with a maximum ground resistance of 1 ohm. These parameters can usually be met with the proper application of basic grounding theory. Circumstances can exist that will make it difficult to obtain the ground resistance required by the *NEC* or

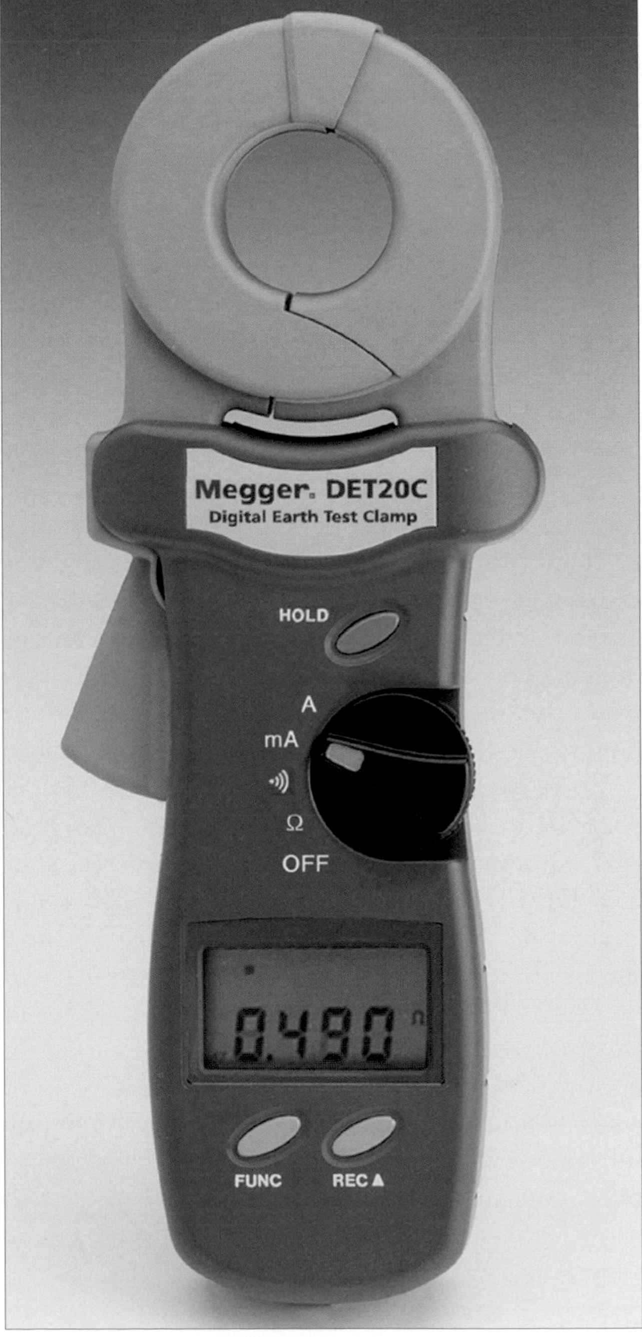

Photo 6-3. Earth/ground resistance clamp-on tester

Courtesy of Megger

Photo 6-4. Clamp-on test instrument being used to test a facility grounding electrode Courtesy of Megger

other safety standards. When these situations develop, several methods of lowering the ground resistance can be employed. These include parallel rod systems, deep driven rod systems utilizing sectional rods and chemical treatment of the soil. Additional methods, discussed in other published data, are buried plates, buried conductors (counterpoise), electrically connected building steel, and electrically connected concrete reinforced steel.

Electrically connecting to existing water and gas distribution systems was often considered to yield low ground resistance; however, recent design changes utilizing non-metallic pipes and insulating joints have made this method of obtaining a low resistance to ground questionable and in many instances unreliable.

The measurement of ground resistances can only be accomplished with specially designed test equipment. Most instruments use the fall-of-potential principle of alternating current (ac) circulating between an auxiliary electrode and the ground electrode under test; the reading will be given in ohms, and represents the resistance of the ground electrode to the surrounding earth (see figures 6-19 through 6-21). Some manufacturers of earth resistance testing instruments have recently introduced clamp-on ground resistance testers (see photos 6-3 and 6-4).

Objectionable Currents

Section 250.6 recognizes that conditions can exist which can cause objectionable current through grounding conductors, such as the grounding electrode conductor, other than temporary currents that can exist during fault conditions. We should recognize that grounding conductors are not intended to carry current under normal operating conditions. They are installed for and are intended to carry current to perform some safety function.

The *Code* does not define what is meant by *objectionable current.* Any current through a grounding electrode conductor would create a voltage drop due to the resistance of the conductor. The equipment to which the grounding electrode conductor is connected is now energized at some voltage level when compared to the earth. This energized equipment could create a shock hazard to anyone that contacted it. Anything that prevents maintaining the equipment at earth potential would be objectionable.

Section 250.6(B) permits the following corrective actions to be taken where there is objectionable current:

1. If due to multiple grounds, one or more, but not all, of such grounds may be discontinued.
2. The location of the grounding connection may be changed.
3. Interrupt the continuity of the grounding conductor or conductive path causing the objectionable current.
4. Other means satisfactory to the authority enforcing the *Code* may be taken to limit the current over the grounding conductors.

The *Code* points out that temporary currents resulting from accidental conditions, such as ground-fault currents, that occur only while the grounding conductors are performing their intended protective functions are not considered the objectionable currents covered in these sections.

Section 250.6(D) points out that currents that introduce noise or data errors in electronic equipment are not considered to be objectionable currents. Electronic data processing equipment is not permitted to be operated ungrounded or by being connected only to its own grounding electrode without being connected to an equipment grounding conductor.

Chapter seventeen includes special grounding provisions for installations, such as information technology rooms, where EMI (electro-magnetic interference) or noise in the grounding circuits or systems can cause data errors and loss of data. In these types of installations,

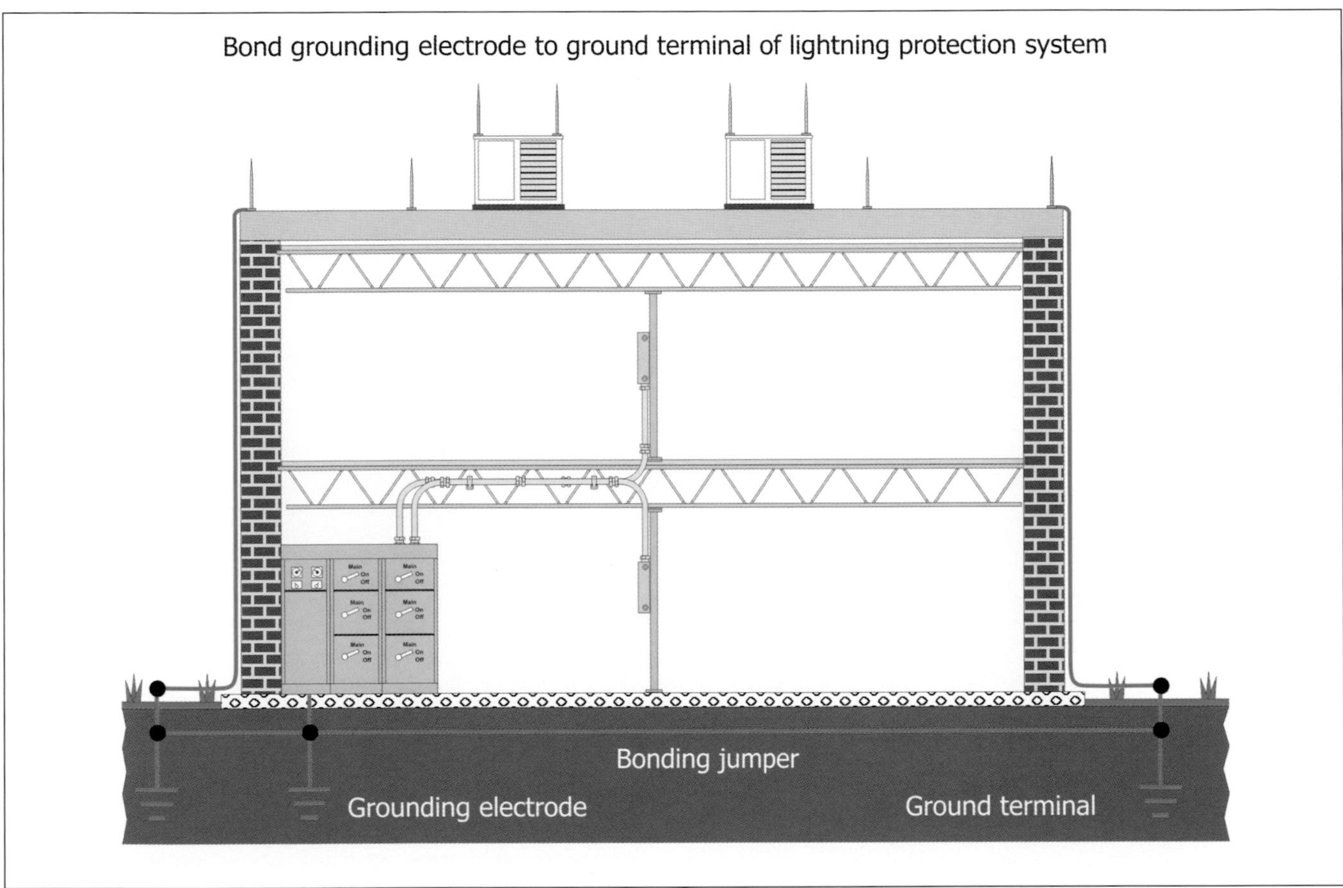

Figure 6-22. Bonding of lighting protection ground terminals to power system grounding electrode is required

the minimum grounding and bonding requirements in Article 250 may need to be expanded upon to provide better grounding (earthing) electrodes and cleaner equipment grounding conductors.

Lightning Protection System

Lightning protection systems should be installed in accordance with the NFPA-780, *Standard for the Installation of Lightning Protection Systems.*

The *Code* prohibits, in 250.60, the use of air terminal conductors and grounding terminals used for lightning protection such as driven pipes, rods, or other electrodes in place of the grounding electrodes required for a wiring system and for equipment. Note that where two grounding electrodes are installed, they are required to be bonded together. [See 250.106 for the requirement that lightning protection ground terminals (grounding electrodes of the counterpoise protection system) be bonded to the building or structure power grounding electrode system.] See figure 6-22.

The fine print notes following 250.106 provide valuable information regarding lightning protection systems. The *NEC* no longer has specific side-flash spacing requirements for separation of lightning protection system conductors from metal raceways and other metal enclosures of the building electrical system. The required spacing is typically 6 ft. through air and 3 ft. through dense construction materials such as concrete, brick, or wood. Specific requirements are given in the NFPA-780, *Standard for the Installation of Lightning Protection Systems.* See chapter 20 for information about lightning protection systems.

Conclusion

No potential exists between a conductor or equipment enclosure and earth if the system is properly and adequately grounded, except during a fault. By careful and thoughtful design, the fault clearing time can be reduced to a minimum.

While it is desirable to obtain a grounding electrode resistance as low as practical, it is also very important to provide a path of low impedance for the return of ground-fault current that will clear a ground fault when it occurs.

A hazard does not typically exist in the distribution system until there is an insulation failure to create a ground fault. The hazard only exists for the period of time it takes to clear the fault. If a ground fault clears promptly, it is unlikely that any loss of life would occur and property damage is kept to a minimum.

For maximum safety, follow a cardinal rule: only one grounding electrode system is required to be used in or on a building, with everything that should, or is required to, be grounded connected to that same grounding electrode system. If more than one grounding electrode is required, they may be used providing all the grounding electrodes used are bonded together to form a grounding electrode system which, in effect, becomes one common grounding electrode. Grounding electrodes should never be relied upon to provide the ground-fault return path for equipment. That is not their intended function in the electrical system. They cannot be relied upon to provide that effective ground-fault current path.

[1] NFPA 70, *National Electrical Code* 2008 250.24(C), (Quincy, MA, National Fire Protection Association, 2007), p. 70-102.

[2] NFPA 70, Article 100, (Quincy, MA, National Fire Protection Association, 2007), p. 70-28.

[3] NFPA 70, 250.52(A), p. 70-107 and 70-108.

Review Questions

The questions included here were developed using material included in this chapter. The answers can be found by reviewing the text. It is also important that students make use of *NEC-2008*, where many answers can be found.

1. "A conducting object through which a direct connection to earth is established" best defines which of the following:
   a. equipment grounding
   b. grounding electrode
   c. main bonding jumper
   d. earthing conductor

2. "Connected (connecting) to ground or to a conductive body that extends the ground connection" best defines which of the following:
   a. grounded (grounding)
   b. effectively grounded
   c. bonding
   d. earthing

3. "One or more steel reinforcing bars or rods not less than 13 mm (1/2 in.) in diameter, or 6.0 m (20 ft) more of bare copper conductor not smaller than 4 AWG that is encased in not less than 50 mm (2 in.) of concrete" best describes which of the following:
   a. butt ground
   b. pole ground rod
   c. feeder electrode
   d. concrete-encased electrode

4. A concrete-encased electrode must be located horizontally within or near the bottom or vertically, and within a foundation or footing in direct connection with the earth and is required to be encased by not less than ____ inches of concrete.
   a. 0
   b. 1
   c. 2
   d. 6

5. A copper conductor not smaller than 2 AWG least 6.0 m (20 ft) encircling a building or structure, and buried at not less than 750 mm (2 1/2 ft) deep defines a ____.
   a. concrete-encased electrode
   b. ground ring
   c. Ufer ground
   d. common bonding grid

6. Where an underground metal water pipe is the only grounding electrode that is present, and is connected at the building or structure served, it must be supplemented by another ____.
   a. grounding electrode
   b. main bonding jumper
   c. equipment grounding conductor
   d. circuit bonding jumper

7. An electrode that is considered as being suitable for supplementing the metal underground water pipe, includes a concrete-encased electrode, ground ring, other local metal underground systems or structures, or ____ electrodes.

a. rod
b. pipe
c. plate
d. all of the above

8. If present at each building or structure served, all grounding electrodes, including ____ electrodes, are required to be bonded together to form the grounding electrode system.

a. identified
b. approved
c. rod, pipe, or plate
d. gas pipe

9. Where the electrodes described in Section 250.52(A)(1) through (A)(7) do not exist at the premises or structure served, a grounding electrode must be installed. It may be a ____.

a. local metallic underground systems or structures, tanks, etc.
b. pipe or conduit electrodes not less than 2.44 m (8 ft) in length not smaller than metric designator 21 (3/4 in.), and if of iron or steel, must be galvanized or metal-coated for corrosion protection.
c. rod electrodes of steel or iron at least 15.87 mm (5/8 in.) diameter.
d. any of the above

10. Where two or more grounding electrodes are bonded together, they are considered to be a ____ electrode.

a. single
b. an identified
c. an approved
d. a listed

11. The *Code* recognizes that conditions may exist that may cause an objectionable current over the grounding electrode conductor, other than temporary currents that may be set up under accidental conditions. Permitted alterations include ____.

a. if due to multiple grounds, one or more, but not all, of such grounds may be abandoned
b. their location can be changed
c. continuity of the grounding conductor may be suitably interrupted
d. any of the above

12. Currents that introduce noise or data errors in electronic equipment are not considered to be ____.

a. dangerous currents
b. objectionable currents
c. harmonic currents
d. unsafe currents

13. Where a single rod, pipe, or plate electrode does not achieve a resistance to ground of ____ ohms or less, it is required to be supplemented by one additional electrode.

a. 50
b. 40
c. 30
d. 25

14. The use of continuous metallic underground water and metal well casings, as well as the metal frames of buildings, will generally provide a ground resistance not exceeding ____ ohms.

a. 3
b. 6
c. 12
d. 25

15. In general, doubling the length of a rod type grounding electrode reduces the resistance by about ____ percent.

a. 10
b. 20
c. 30
d. 40

16. The use of air terminal (lightning rod) conductors and ground terminals of lightning protection systems are ____ for the grounding electrode(s) or grounding electrode system required by the *Code* for a wiring system and for equipment.

a. permitted to be used
b. not permitted to be used
c. permitted by special permission
d. identified when labeled to be used

17. Supplementary grounding electrodes are permitted by the *Code* and are required to be connected to the __________ conductor, and the _________ shall not be used as an effective ground-fault current path.

a. grounding electrode, conduit
b. equipment grounding, earth
c. bonding, ground rods
d. grounded, equipment grounding bus

18. Underground metal well casings are _______
    a. required to be used as a grounding electrode
    b. required to be insulated from metal underground water piping
    c. required to have at least 10 feet in contact with the earth
    d. considered other local metal underground systems or structures that could qualify as a grounding electrode

19. Concrete-encased electrodes that are _______ at a building or structure served are required to be part of the grounding electrode system .
    a. present
    b. available
    c. epoxy coated
    d. visible

20. The metal frame of the building or structure must be used as an electrode_____
    a. where located below permanent moisture levels
    b. where in contact with a concrete slab
    c. where it has 10 ft in contact with the earth or is bonded to another electrode
    d. it is exposed

21. Where multiple separate concrete-encased electrodes are present at a building or structure served, which of the following applies?
    a. They are all required to be bonded into the grounding electrode system.
    b. They shall all meet the 25 ohm resistance value in 250.56
    c. It is permissible to bond only one into the grounding electrode system.
    d. They shall all be supplemented by a ground rod electrode.

Objectives

To understand...

- General requirements and definitions for grounding electrode conductors
- Functions of the grounding electrode conductor
- Grounding electrode conductor connections
- Material and protection for grounding electrode conductors

Chapter 7

Grounding Electrode Conductors

The grounding electrode conductor is used to connect the electrical system, or the equipment grounding conductor, or both, to a grounding electrode, hence the specific term, *grounding electrode conductor* (see figure 7-1). It is required to be properly sized as per Table 250.66 based on the size of the service-entrance conductors but is not required to exceed 3/0 AWG copper or 250-kcmil aluminum or copper-clad aluminum. Specific requirements are given regarding conductor material, how it is to be installed, protected from physical damage, and connected.

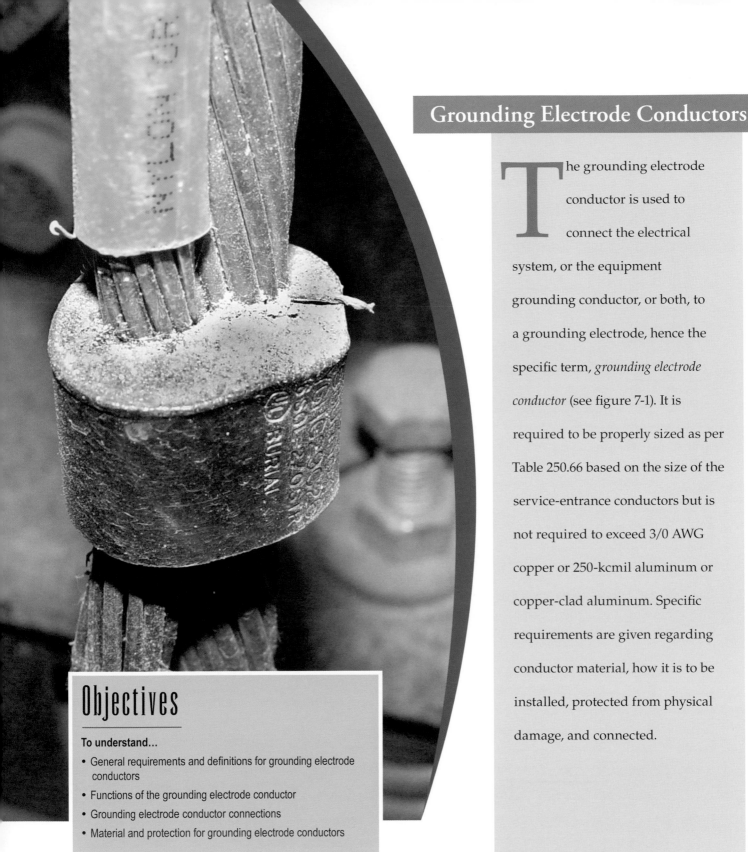

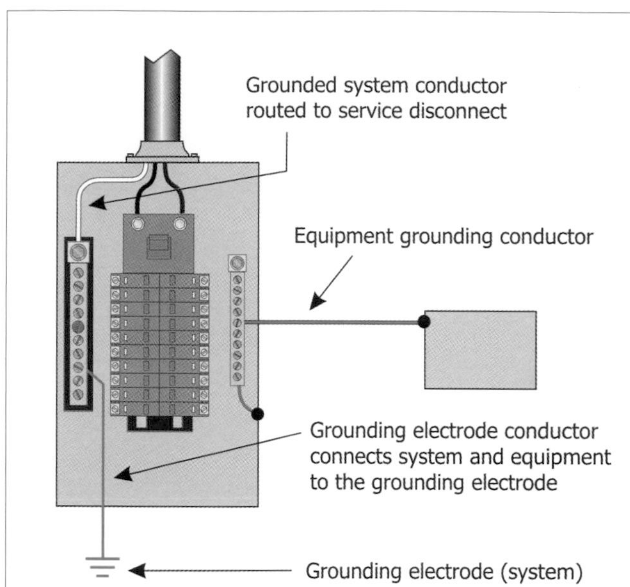

Figure 7-1. Grounding electrode conductor shown in a grounded system

Definition

Grounding Electrode Conductor. "A conductor used to connect the system grounded conductor or the equipment to a grounding electrode or to a point on the grounding electrode system. " [1]

Function of Grounding Electrode Conductor

The grounding electrode conductor is the sole connection from the grounding electrode to the grounded system conductor (may be a neutral) and the equipment grounding conductor for a grounded system; or the sole connection from the grounding electrode to the service equipment or building disconnect enclosure; or to the equipment grounding conductor for an ungrounded system [see 250.24(A) and figure 7-2].

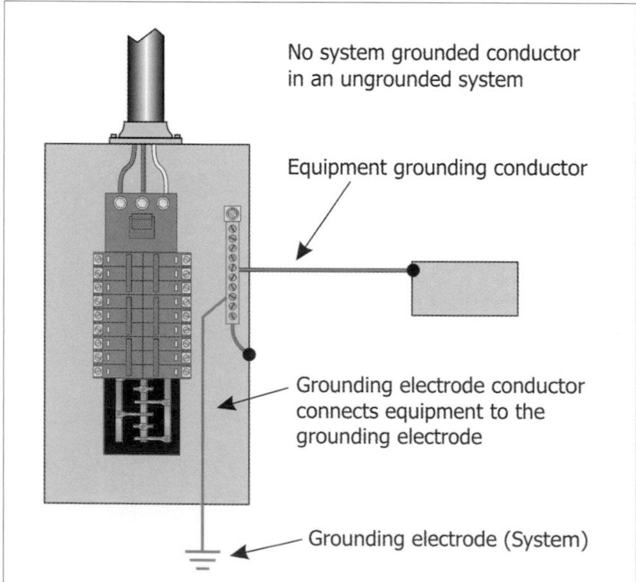

Figure 7-2. Grounding electrode conductor connected to equipment containing an ungrounded system (no system grounded conductor)

A common grounding electrode conductor is required to ground both the system grounded conductor and the equipment grounding conductor. In other words, one grounding electrode conductor cannot be used to ground the system conductor and a second grounding electrode conductor be used to ground the equipment grounding conductor even though both grounding conductors are connected to the same grounding electrode. A single grounding electrode conductor is required to be used for both the circuit and the equipment.

Maximum Current on Grounding Electrode Conductors

In all grounded systems of 600 volts or less, the maximum current in the grounding electrode conductor is limited by the impedance path through the earth. Figure 7-3 illustrates a circuit that consists of the resistance of the service conductor, the grounding electrode at the service, the grounding electrode at the transformer bank plus the resistance of the earth path between the two grounding electrodes. In addition, there is the resistance of the grounding electrode conductors themselves. In many

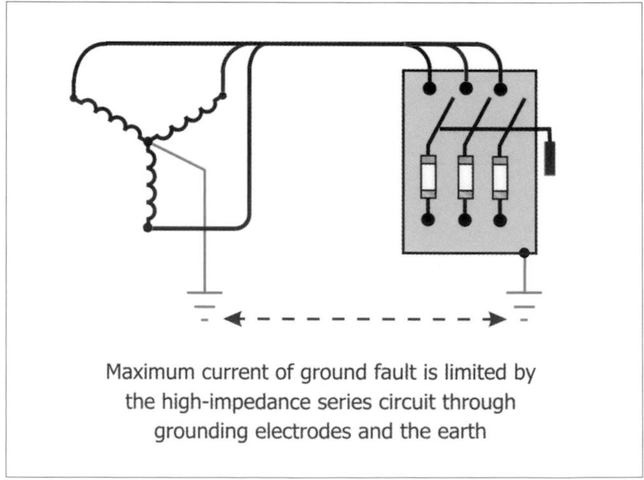

Maximum current of ground fault is limited by the high-impedance series circuit through grounding electrodes and the earth

Figure 7-3. High impedance return path through the earth to the source

installations, the length of the grounding electrode conductors is short enough that their resistance may be ignored. In addition to these resistances, there will be an inductive reactance that will vary as the spacing between the supply path and return path increases.

If we assume that the sum of these resistances is equal to 22 ohms (higher values can be common in actual practice), then, for a 120-volt-to-ground system, the maximum current through the grounding electrode conductor and grounding electrodes will be 5.5 amperes (120 volts ÷ 22 ohms = 5.5 amperes). It is obvious that this grounding connection is ineffective for facilitating overcurrent device operation. Equipment grounded only in this manner is unsafe because a ground fault would not cause an upstream overcurrent protective device to clear the fault from the equipment. The current through the high-impedance path provided by the grounding electrodes and the earth is not nearly enough to cause a 15-ampere fuse to open or a circuit breaker to operate. Voltage levels that could be fatal can be present on equipment that is grounded in this manner. This maximum current value is determined by considering resistance only. If the reactance of the circuit is included, the overall result is an impedance value much higher than the resistance value. The actual current in the grounding electrode conductor will therefore be considerably less than what is calculated based on resistance only. Ground-fault currents through the grounding electrode conductor are not likely to attain a value anywhere near the continuous rating of the conductor even under fault conditions.

As covered in chapter six in detail, the main purpose of the grounding electrode is to establish and maintain an earth reference for the system and non-current-carrying parts of the system but not to clear ground faults.

One of the reasons for requiring the grounding electrode conductors to be larger than for the normal system current they might carry is because grounding electrode conductors also carry current from other events on the system, such as lightning strikes or from accidentally being energized from high-voltage sources such as transformers or overhead lines.

Sizing the Grounding Electrode Conductor for a Single Service

The grounding electrode conductor is required to be sized in accordance with Table 250.66. That conductor is required to be a minimum size of 8 AWG and need not be larger than 3/0 AWG copper. Where aluminum or copper-clad aluminum grounding electrode conductors are installed, they are required to be not smaller than 6 AWG nor larger than 250 kcmil. The size of the grounding electrode conductor is based upon the size of the largest ungrounded service-entrance conductors or ungrounded conductors of a separately derived system. Table 250.66 is based on a conductor size relationship and not on the rating of the circuit breaker or fuse in the service equipment (see Table 250.66).

For example, where 3/0 AWG copper service-entrance conductors are installed, the minimum size grounding electrode conductor is 4 AWG copper or 2 AWG aluminum. If 750-kcmil aluminum service-entrance conductors are installed, a 1/0 AWG copper or 3/0 AWG aluminum grounding electrode conductor is required to be used.

Sizing the Grounding Electrode Conductor for Service with Parallel Conductors

Where service-entrance conductors are installed in parallel as allowed by 310.4, the circular mil area of one set of parallel conductors is added together and treated as a single conductor for purposes of sizing the grounding

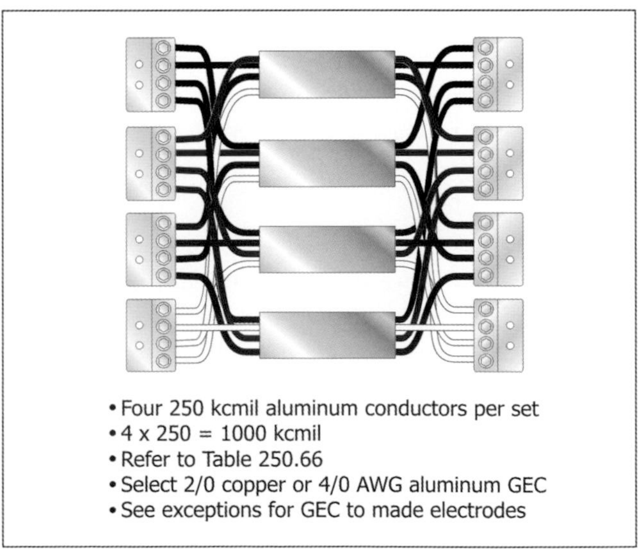

• Four 250 kcmil aluminum conductors per set
• 4 x 250 = 1000 kcmil
• Refer to Table 250.66
• Select 2/0 copper or 4/0 AWG aluminum GEC
• See exceptions for GEC to made electrodes

Figure 7-4. Grounding electrode conductor sizing where parallel service conductors are installed

Table 250.66 Grounding Electrode Conductor for Alternating-Current Systems

Size of Largest Ungrounded Service-Entrance Conductor or Equivalent Area for Parallel Conductors (AWG/kcmil)		Size of Grounding Electrode Conductor (AWG/kcmil)	
Copper	Aluminum or Copper-Clad Aluminum	Copper	Aluminum or Copper-Clad Aluminum
2 or smaller	1/0 or smaller	8	6
1 or 1/0	2/0 or 3/0	6	4
2/0 or 3/0	4/0 or 250	4	2
Over 3/0 through 350	Over 250 through 500	2	1/0
Over 350 through 600	Over 500 through 900	1/0	3/0
Over 600 through 1100	Over 900 through 1750	2/0	4/0
Over 1100	Over 1750	3/0	250

Table 250.66. Reproduction of table

electrode conductor (see figure 7-4). For example, if four 250-kcmil copper conductors are installed in parallel, they are considered to be a single 1000-kcmil conductor. By reference to Table 250.66, we find the minimum grounding electrode conductor for this set to be a 2/0 copper or 4/0 AWG aluminum conductor.

If 1/0 AWG through 4/0 AWG conductors are installed in parallel, they are required to first be converted to circular mil area before applying Table 250.66. For example, if two 2/0 AWG aluminum service-entrance conductors are installed in parallel, first refer to *NEC* Table 8 of chapter nine (reprinted in Soares chapter 21) for determining the conductor's circular mil area. There, we find the conductor to have an area of 133,100 circular mils. Then multiply the circular mil area of the conductors by the number of conductors connected in parallel.

133,100 circular mils;

133,100 cm x *2 = 266,200 cm.*

By reference to Table 250.66 we find that a 2 AWG copper or 1/0 AWG aluminum grounding electrode conductor is required.

Sizing Grounding Electrode Conductors for Multiple Enclosure Services

Services are permitted to be installed in up to six enclosures installed at one location or at separate locations. The method of sizing grounding electrode conductors is a matter of choice. One basic rule must be followed: size the grounding electrode conductor for the circular mil area of service-entrance conductor(s) at the point of connection, or as a common grounding electrode conductor for the size of the main service-entrance conductor(s) [see 250.64(D)(3)].

In figure 7-5 with assumed 500-kcmil copper service-entrance conductors, the service-entrance grounded conductor is shown grounded inside the wireway. By reference to Table 250.66, we find that the grounding electrode conductor to a water pipe

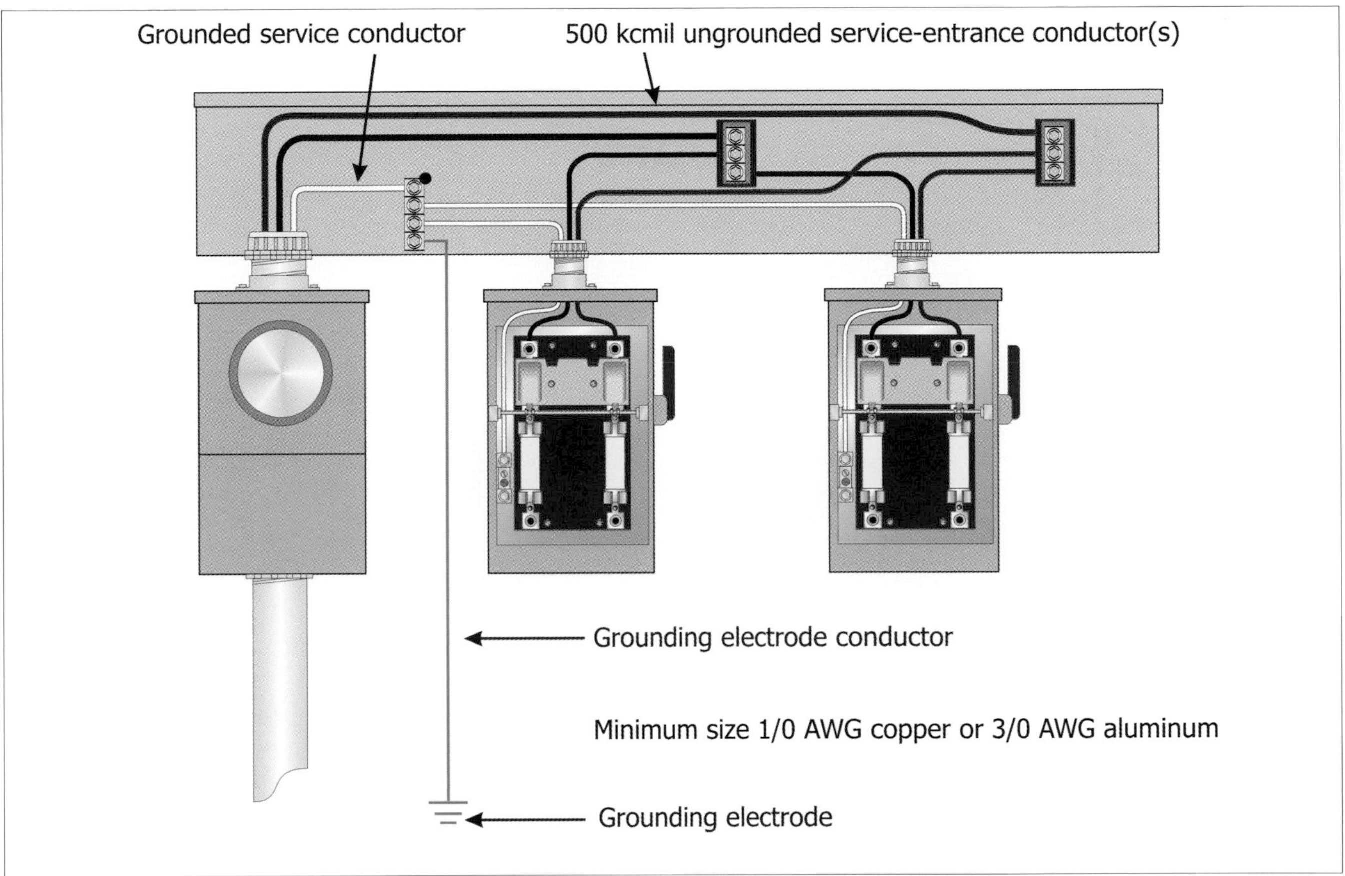

Figure 7-5. Grounding electrode conductor connection in wireway sized based on largest ungrounded service-entrance phase conductor (500 kcmil) [250.64(D)(3)]

or building steel grounding electrode is required to be 1/0 AWG copper or 3/0 AWG aluminum.

If the installer chooses to, an individual grounding electrode conductor may be installed from each service disconnecting means enclosure to the grounding electrode. In this case, the grounding electrode conductor is sized for the service-entrance conductor serving each individual enclosure as shown in figure 7-6 [250.64(D)(2)].

In addition, a single grounding electrode conductor is permitted to serve separate enclosures. The common *grounding electrode conductor* is sized based on the main service-entrance conductors. Taps that are sized based on the individual service-entrance conductors supplying each service disconnect are connected from within each service disconnecting means to the common grounding electrode. Grounding electrode conductor taps are covered in 250.64(D)(1). It is important to understand that the alternative provided in Section 250.64(D)(1) addresses two conductors. The common grounding electrode conductor that is required to be installed without a splice or joint (generally) and the grounding electrode conductor tap(s) that are permitted to be connected to the common grounding electrode conductor (see figures 7-7 and 7-8).

In figures 7-7 and 7-8, assume that 2 AWG copper conductors serve each service disconnecting means from the wireway. Shown are taps to a common grounding electrode conductor. The tap conductors are required to be connected to the common grounding electrode conductor in a manner that it remains continuous without a splice or joint. This means that the tap conductor is required to be connected to the grounding electrode conductor with a device that allows the grounding electrode conductor to remain unbroken as the connection is made. The tap conductors are required to be connected to the common grounding electrode conductor by exothermic welding or with connectors listed as grounding and bonding equipment. Acceptable methods of grounding and

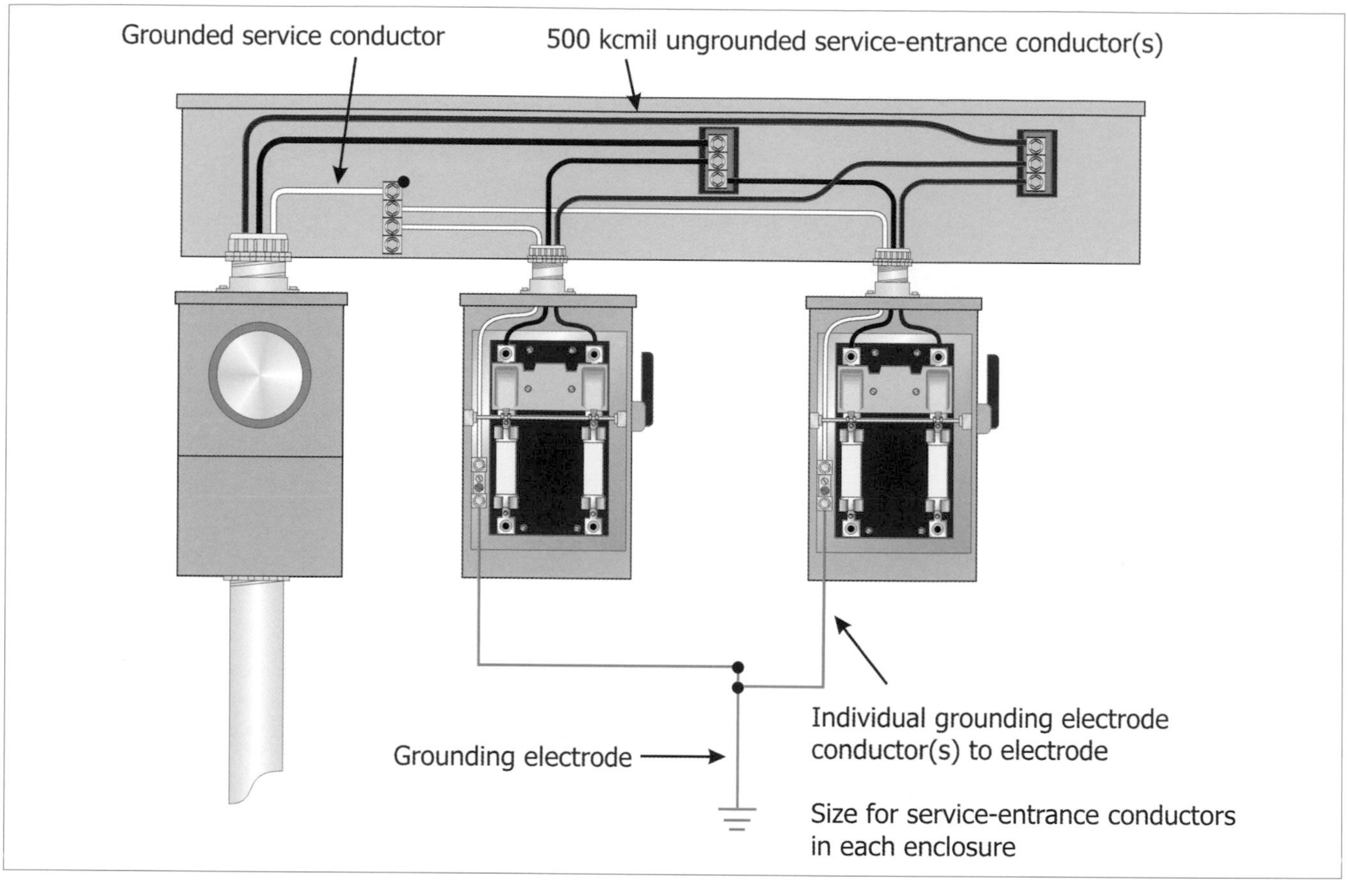

Figure 7-6. Grounding electrode conductors from individual enclosures to single grounding electrode [250.64(D)(2)]

bonding connections are provided in 250.8. The concept of tapping the grounding electrode conductor applies whether the sets of service-entrance conductors are tapped from a wireway or are installed individually as overhead or underground systems (see photo 7-1).

For figure 7-8, the size of the common grounding electrode conductor is determined as follows. Assume that 500-kcmil copper service-entrance conductors supply the service and are connected in the wireway to the service-entrance conductors that serve each enclosure.

Refer to Table 250.66. The minimum size common grounding electrode conductor is 1/0 AWG copper or 3/0 AWG aluminum. This conductor is installed from the grounding electrode to the vicinity of the wireway. The grounding electrode tap conductors from the individual enclosures are sized from Table 250.66 and are 8 AWG copper or 6 AWG aluminum. Note that there is no minimum or maximum length for these grounding electrode tap conductors. If the installer chooses to, the grounding electrode conductors from the service disconnects may be connected individually to the grounding electrode rather than being tapped to the common grounding electrode conductor [250.64(D)(2)].

Photo 7-1. Common grounding electrode conductor and taps

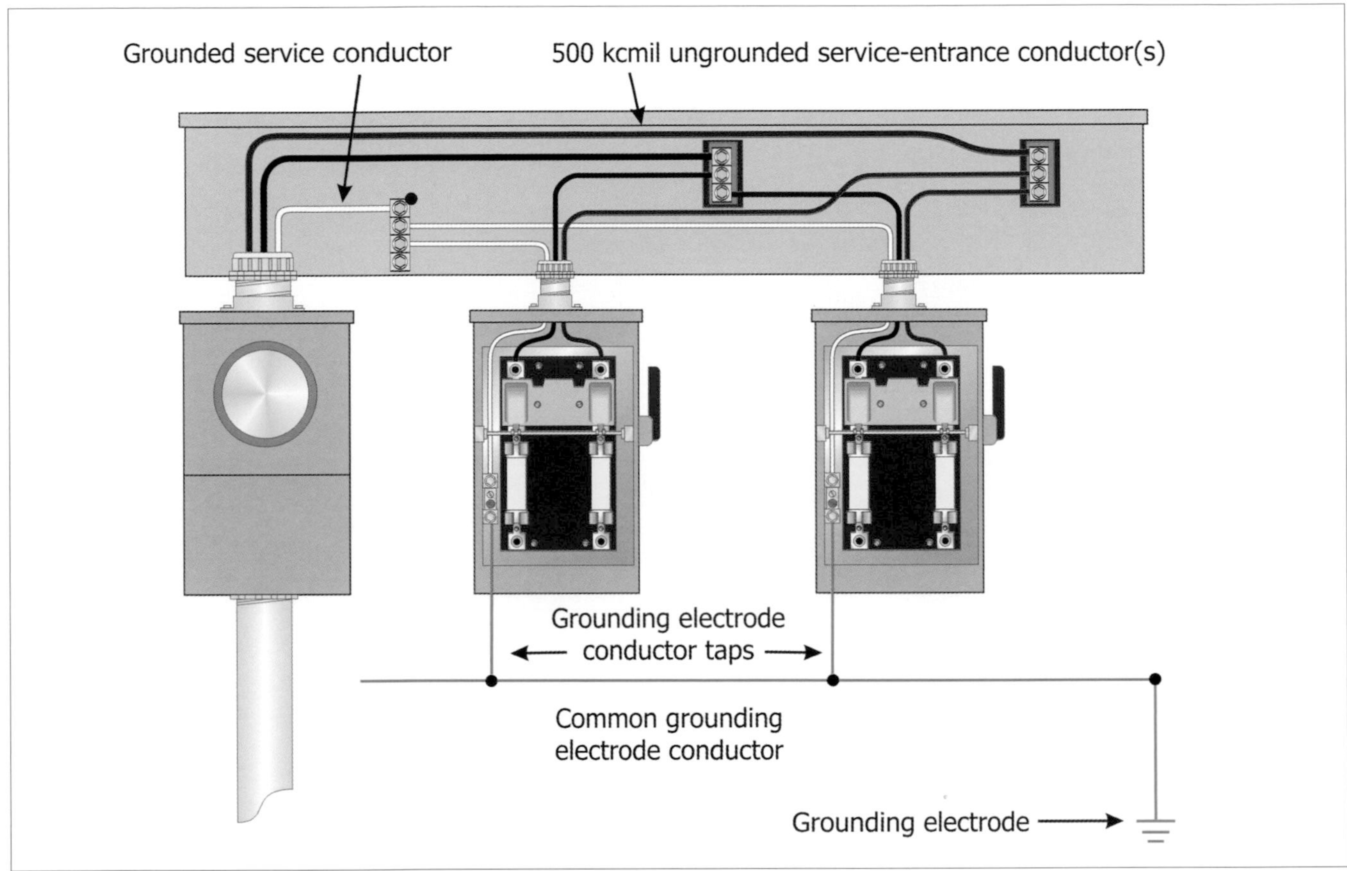

Figure 7-7. Taps to a common grounding electrode conductor [250.64(D)(1)]

Exceptions to Size of Grounding Electrode Conductor

The grounding electrode conductor is generally required to be not smaller than the values in Table 250.66 based on the size of the largest ungrounded service-entrance conductor or largest ungrounded conductor of a separately derived system. What amounts to a three-part exception to the general rule for sizing the grounding electrode conductor is provided in 250.66(A) through (C).

- Section 250.66(A) permits the grounding electrode conductor to be not larger than 6 AWG copper or 4 AWG aluminum where it is the sole connection to a rod, pipe or plate electrode.
- Section 250.66(B) provides that the grounding electrode conductor that is the sole connection to a concrete-encased grounding electrode need not be larger than 4 AWG copper. Note that aluminum wire is not permitted for this application. The use of this smaller 4 AWG conductor is based on the fact that it will never need to carry a current beyond its safe short-time rated capacity, even under ideal conditions.
- Section 250.66(C) requires that the minimum size conductor and material for a ground ring is 2 AWG copper. It also provides that, where connected to a ground ring, that portion of the grounding electrode conductor that is the sole connection to the ground ring need not be larger than the ground ring conductor. That is because the electrode resistance is the limiting factor in the circuit and increasing the grounding electrode conductor size would not serve any useful purpose. Design engineers will sometimes specify ground rings of 4/0 AWG or 250 kcmil. In this case, follow Table 250.66 to determine the minimum grounding electrode conductor size. The grounding electrode conductor is not required to be larger than 3/0 AWG copper or 250-kcmil aluminum or copper-clad aluminum.

The term *sole connection* means that the grounding electrode conductor does not serve or connect to more than one grounding electrode. For example, a grounding electrode conductor is not connected to a concrete-encased electrode and then extended to building steel or an underground metal water pipe. In

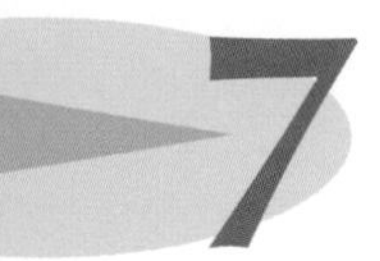

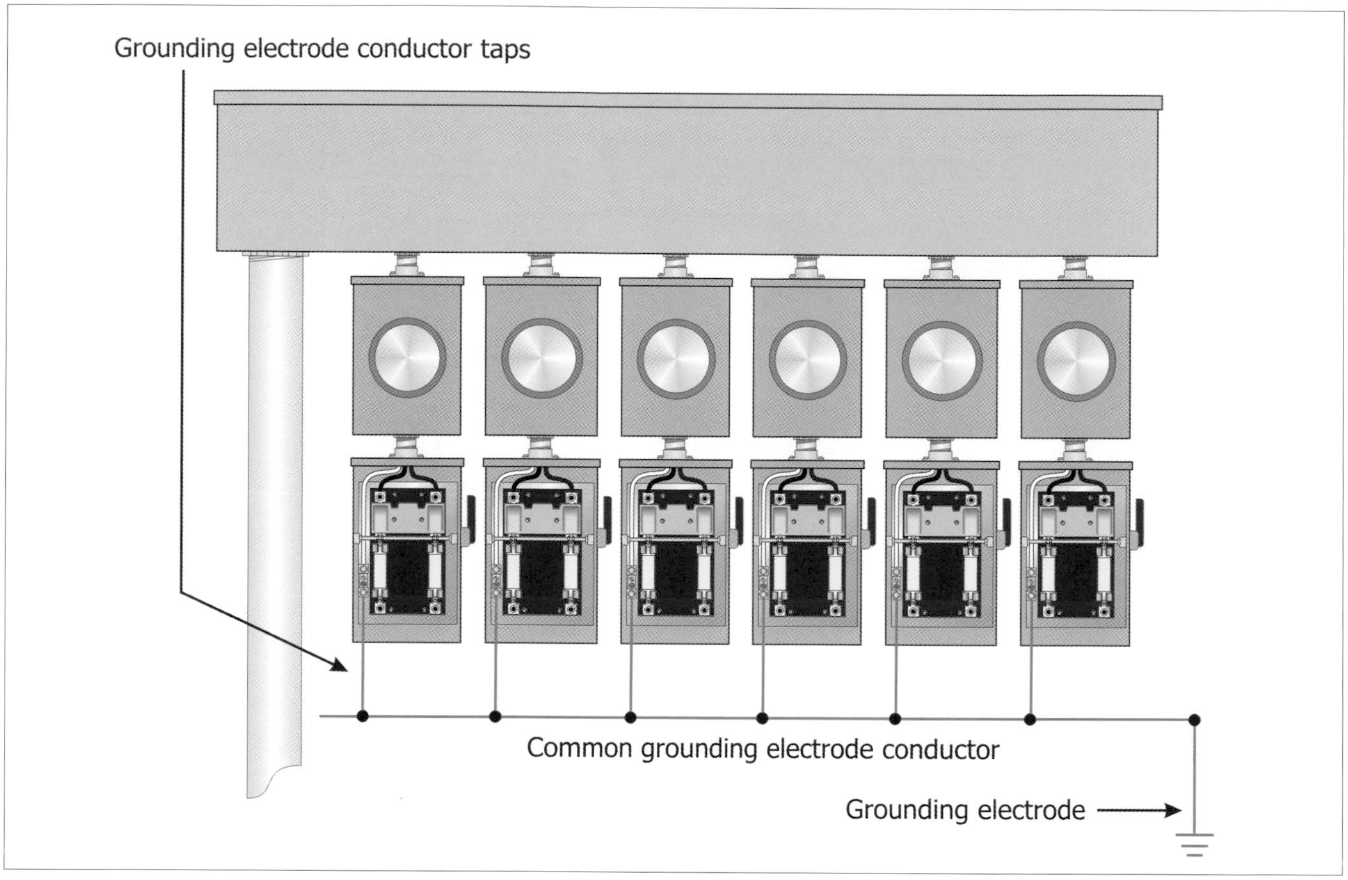

Figure 7-8. Grounding electrode conductor taps permitted to be connected to a common grounding electrode conductor [250.64(D)(1)]

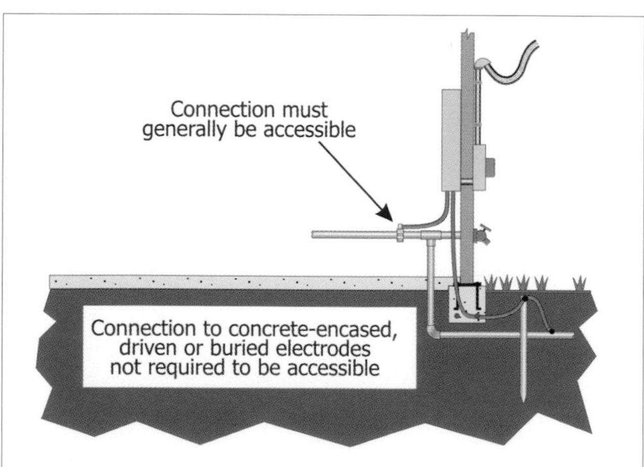

Figure 7-9. Grounding electrode conductor connections are required to be accessible (generally)

this case a grounding electrode conductor larger than 4 AWG may be required for the water pipe electrode based on the size of the service-entrance conductors.

Grounding Electrode Conductor Connections

The *Code* requires generally that the point of connection of grounding electrode conductors and bonding jumpers to grounding electrodes shall be accessible and made in a manner that will ensure a permanent and effective grounding path. An exception provides that a connection at a concrete-encased, driven, or buried grounding electrode is not required to be accessible (see figure 7-9).

Exception No. 2 to 250.68(A) provides that an exothermic or irreversible compression connection to fireproofed structural metal also does not have to be accessible [see 250.68(A) and photo 7-2]. A compression connector that is attached to the structural metal building frame by mechanical means such as nuts and bolts meets this exception.

Specific rules for connections of the grounding electrode conductor and bonding conductor to grounding electrodes are found in 250.70. The connections are required to be made by exothermic welding (see photos 7-3 through 7-8), listed lugs, listed pressure connectors, listed clamps or other listed means. The only *connection means* not required to be listed are those made by

Photo 7-2. Grounding electrode conductor connection is permitted under fireproofing materials by exception. [250.68(A) Exception No. 2]

exothermic welding, although listed exothermic weld connections are available.

Connections depending solely on solder shall not be used. Other requirements state that not more than one conductor is permitted to be connected to the grounding electrode by a single clamp or fitting, unless the clamp or fitting is listed as being suitable for connecting multiple conductors. Ground clamps are required to be listed for both the materials of the grounding electrode and the grounding electrode conductor (see figure 7-10). For example, ground clamps are required to be listed for aluminum conductors to be used for such connections. Clamps used on a pipe, rod or other electrode that is buried are required to be listed for direct earth burial (see photos 7-9, 7-11, and 7-12). Typically, these clamps are identified by the manufacturer with *direct burial* or DB or similar. Direct burial or DB means the clamp or fitting is also suitable for concrete encasement.

Sheet-metal-strap type ground clamps attached to a rigid metal base that are listed are permitted for indoor telecommunication purposes only [see

Photos 7-3 through 7-8. This series shows the steps in the exothermic welding process. Wires placed in mold Courtesy of Thomas and Betts

250.70(3)]. Also, Underwriters Laboratories' Guide Card information (KDER) states, "Strap-type ground clamps are not suitable for attachment of the grounding conductor of an interior wiring system to a grounding electrode." Use strap type ground clamps for only the

Photo 7-4. Weld disc placed in mold

Photo 7-5. Weld metallic powder added to mold

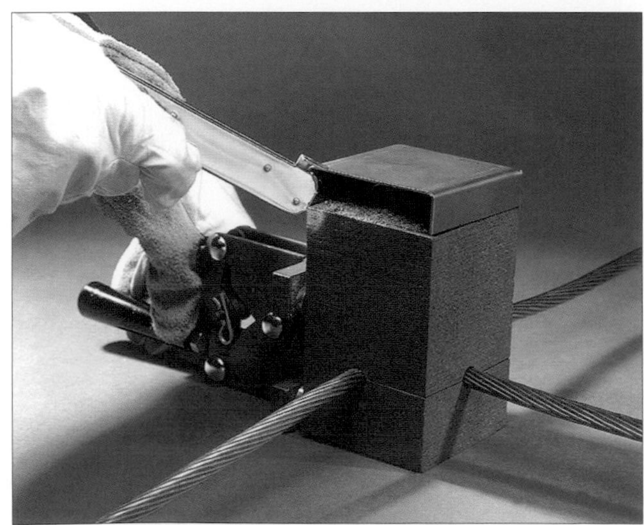
Photo 7-6. Preparing to strike an arc at mold

Photo 7-7. Ignition of metallic weld powder

Photo 7-8. Completed weld

type of conductor and grounding electrode as well as the environment for which it is listed and labeled, such as for communication circuits (see chapter eighteen for low voltage and intersystem bonding and grounding requirements).

It is required that ground clamps be protected from physical damage unless the fittings are approved for use without protection or are installed in a location where they are not likely to be damaged (see figure 7-11). Protection of the ground clamp is permitted to consist of metal, wood, or equivalent materials (see 250.10).

Clean Surfaces

Section 250.12 requires nonconductive coatings such as paint, lacquer, and enamel to be removed from threads

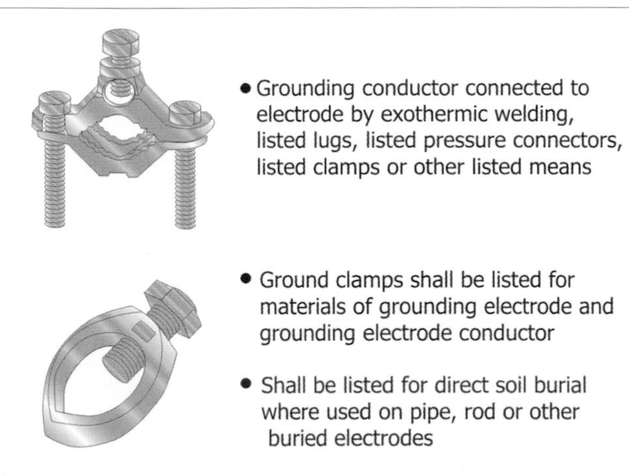

• Grounding conductor connected to electrode by exothermic welding, listed lugs, listed pressure connectors, listed clamps or other listed means
• Ground clamps shall be listed for materials of grounding electrode and grounding electrode conductor
• Shall be listed for direct soil burial where used on pipe, rod or other buried electrodes

Figure 7-10. Ground clamps listed for the application(s)

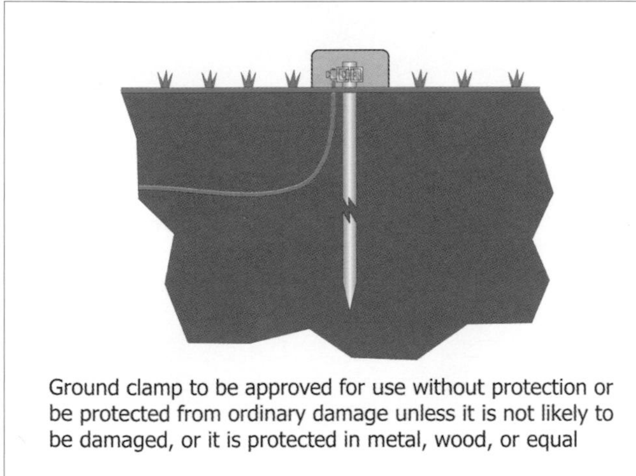

Ground clamp to be approved for use without protection or be protected from ordinary damage unless it is not likely to be damaged, or it is protected in metal, wood, or equal

Figure 7-11. Protection of grounding electrode conductor connection to the electrode [250.10]

Photo 7-9. Grounding clamp Courtesy of Thomas and Betts

Photo 7-10. Grounding clamp Courtesy of Galvan Industries

and other contact surfaces of equipment to be grounded or connected by means of fittings designed to make such removal unnecessary. (Underwriters Laboratories reported in a June 27, 1995, letter and confirmed on June 7, 1998, that no fittings listed by them incorporate this feature.) Removal of paint under connections of raceways and cable fittings is critical to make a reliable connection that can carry fault current when necessary. This will ensure a good electrical connection (see chapter eight for additional discussion on the subject). It is also necessary to consider the corrosive influence the environment can have on enclosures that have the protective coating removed.

Here, again, it is important to consider that all grounding electrode conductors and their connections form a part of a circuit that is required under certain conditions to carry current. In some cases the current is several times the full-load current rating of the conductor involved if the conductor was being used for continuous duty in an electrical circuit. Such currents are usually of short duration so the withstand rating of the conductor is not exceeded.

Grounding Electrode Conductor Material

The grounding electrode conductor is required to be of copper, aluminum, or copper-clad aluminum and may be solid, stranded, insulated, covered, or bare. The material selected shall be resistant to any corrosive condition it will be exposed to. As an option, it may be suitably protected against corrosion (see 250.62). Note that there is no color code for the grounding electrode conductor such as the

identification requirements that exist for grounded conductors or equipment grounding conductors.

The current the grounding electrode conductor will carry is limited by the resistance of the grounding electrode(s) and the earth return path. In a grounded system, the current will even be less because the grounded conductor provides a considerably lower impedance parallel path.

Grounding Electrode Conductor Installation

Where grounding of systems, equipment or both are required, grounding electrode conductors are installed and connected to the grounding electrode system. Section 250.64 provides the installation requirements for grounding electrode conductors where intalled for services, separately derived systems, or for buildings or structores supplied by a feeder(s) or branch circuit(s). The installation of grounding electrode conductors is required to comply with 250.64(A) through (F) which addresses conductor type, securing, protection, splices, multiple disconnecting means enclosures, magnetic field protection, and connections to busbars.

Specific limitations are placed on aluminum and copper-clad aluminum grounding electrode conductors. These rules apply regardless of whether the conductors are insulated or bare [see 250.64(A)].

Bare conductors are not permitted where they are in direct contact with masonry or the earth, or where subject to corrosive conditions. This rule does not prohibit installing conduit on masonry and pulling a bare aluminum grounding electrode conductor in it.

Where used outside, aluminum or copper-clad aluminum grounding electrode conductors are not permitted to be terminated within 18 in. of the earth.

Because of these restrictions on their installation, they cannot be used for connection to concrete-encased electrodes, ground rods or pipes where within 18 in. of the earth or for connection to plate electrodes. Aluminum conductors are otherwise permitted to be used as grounding electrode conductors where the clamp or connector is listed for both the conductor material and the electrode. Typical connectors or clamps would be marked *AL/CU*. The AL indicates the clamp is suitable for aluminum conductor connections.

Photo 7-11. Ground rod clamp
Courtesy of Thomas and Betts

Photo 7-12. Pipe clamp and ground rod clamp Courtesy of Greaves

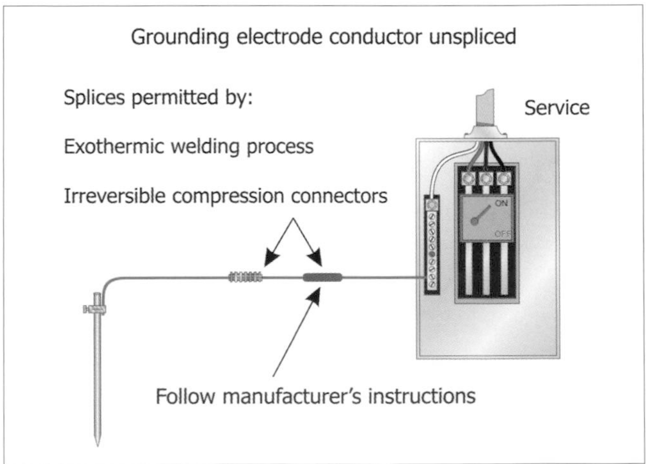

Figure 7-12. Splicing of grounding electrode conductor (generally not permitted)

Securing and Protection from Physical Damage

Section 250.64(B) requires that the grounding electrode conductor or its enclosure be securely fastened to the surface on which it is carried. Size 4 AWG or larger

Photo 7-13. Grounding electrode conductors permitted to be spliced on busbars

conductors require protection where exposed to physical damage. Size 6 AWG conductors, which are not exposed to physical damage, are required to be run along the surface of the building construction and securely fastened. Otherwise, they are required to be protected by installation in rigid or intermediate metal conduit, rigid nonmetallic (PVC) conduit, electrical metallic tubing, or cable armor. Grounding electrode conductors smaller than 6 AWG are required to be protected in one of the methods allowed for 6 AWG conductors.

Splicing Grounding Electrode Conductor

Section 250.64(C) generally requires that grounding electrode conductors are to be installed in one continuous length without a splice or joint.

Two alternatives to this requirement are as follows:

- splicing is permitted by using irreversible compression connectors, listed as grounding and bonding equipment; or by the exothermic welding process (see figure 7-12).
- busbars sections are permitted to be connected together to form a grounding electrode conductor [250.64(C)(1) and (2).

It is vital that the manufacturer's instructions be carefully followed where either of these splicing

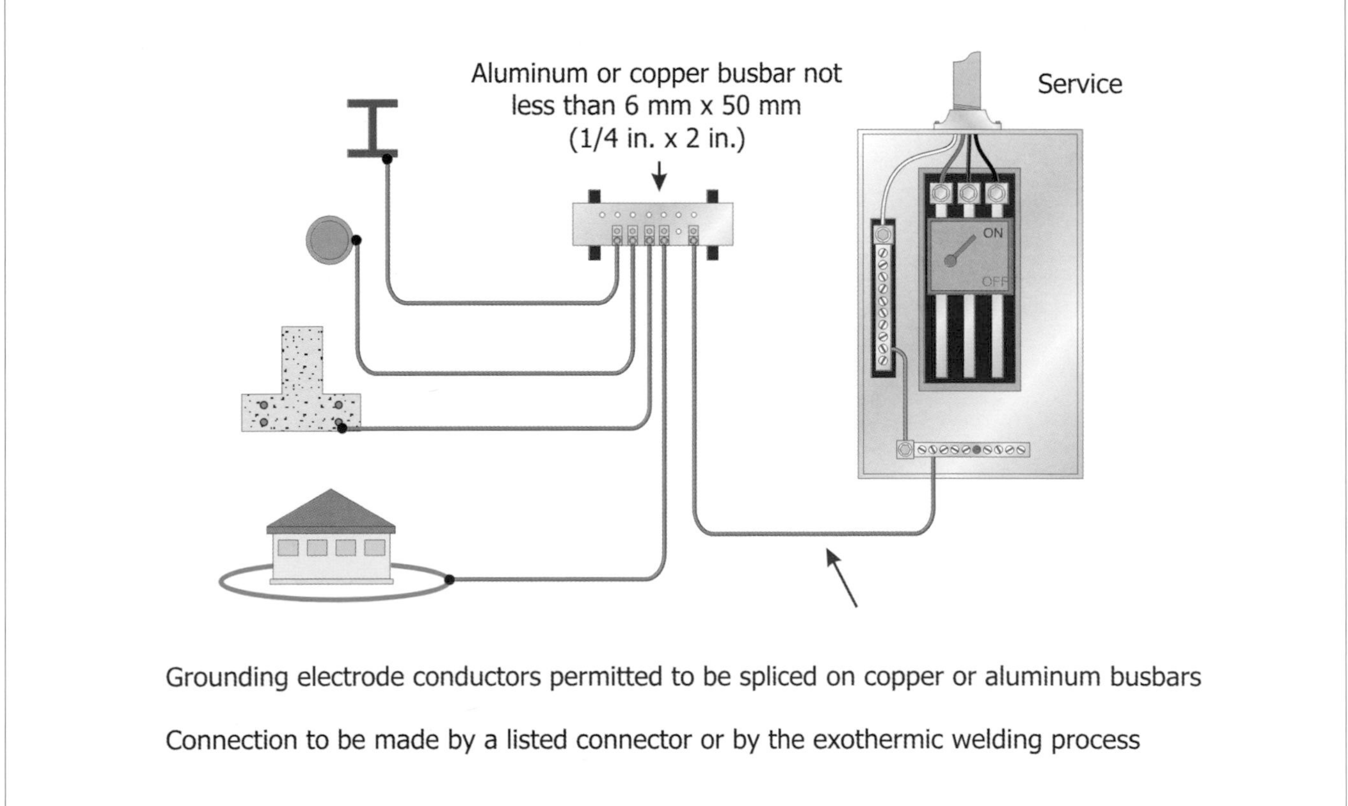
Aluminum or copper busbar not less than 6 mm x 50 mm (1/4 in. x 2 in.)
Service
ON
OFF
Grounding electrode conductors permitted to be spliced on copper or aluminum busbars
Connection to be made by a listed connector or by the exothermic welding process

Figure 7-13. Splices permitted for grounding electrode conductors on busbars

methods is chosen. Where compression-type connectors are used, the correct splicing sleeve for the conductor material and size is required to be selected. Then, the compression tool, and in some cases the proper die for the sleeve to be crimped, is required to be used. These compression-type connectors are required to be specifically listed by a qualified electrical testing laboratory for splicing grounding electrode conductors. Where exothermic welding of the grounding electrode conductor is performed, the correctly sized form or mold for the conductor to be spliced is required to be used. In addition, unless specifically permitted otherwise by the manufacturer, the conductors to be spliced by this method are required to be clean and dry. Inspect the resulting splice carefully to be certain that it has been made satisfactorily without causing damage to the conductor or otherwise affecting its integrity.

Busbars are sometimes used as grounding electrode conductors in open bottom switchboards where the main bonding jumper is at the service disconnect end of the equipment and the grounding electrode conductor connects to the grounding bus at the other end of the equipment. The bolted connections between the sections are actually splices in busbars. Busbars are often installed in electrical rooms to provide a common connection point for individual grounding electrode conductors; and bonding jumpers between grounding electrodes. Busbars often come in standard lengths such as 10 ft. and have to be joined together to achieve the required length.

Protecting Grounding Electrode Conductor from Magnetic Field

Where metal enclosures are provided for protection of the grounding electrode conductor, one has to follow some special procedures (see figures 7-13 and 7-14). This is required by 250.64(E).

- Ferrous metal enclosures must be electrically continuous from the point of attachment to cabinets or equipment to the grounding electrode. Nonferrous metal enclosures do not have the same effect on current because they are nonmagnetic.
- The grounding electrode conductor enclosure must be securely fastened to the ground clamp or fitting.

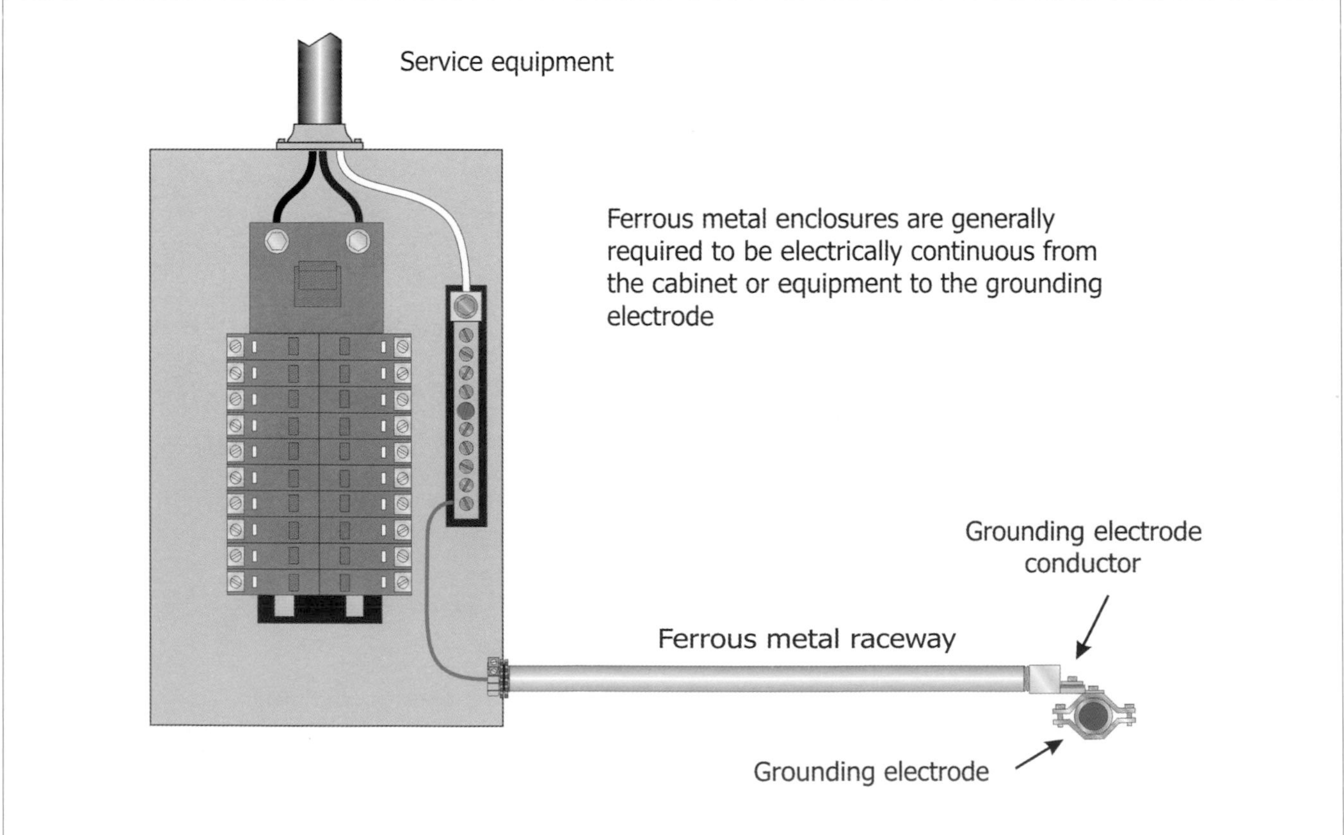

Figure 7-14. Protecting grounding electrode conductor from magnetic field as required in Section 250.64(E)

Photo 7-14. Grounding electrode conductor bonded to conduit with jumper same size as grounding electrode conductor

• Nonferrous metal enclosures are not required to be electrically continuous.

• Ferrous metal enclosures that are not physically continuous from the enclosure to the ground clamp must be made continuous by bonding both ends to the grounding electrode conductor. This bonding requirement applies to all intervening ferrous raceways, boxes and enclosures between the cabinets or equipment and the grounding electrode. The bonding jumper must be at least the same size as the grounding electrode conductor. If raceways are used as protection, they must be installed according to their respective article (see photo 7-14).

It is common practice to use an 8 AWG grounding electrode conductor protected by metallic armored cable (see photo 7-15). The need for bonding the metallic armor of such cable is required in 250.64(E). Where that bonding procedure is not followed, the impedance of the grounding electrode conductor is approximately doubled with the result that its effectiveness is markedly reduced. Such bonding requirements may be avoided by protecting the grounding electrode conductor in rigid nonmetallic conduit. Schedule 80 PVC is listed as impact and crush resistant and provides suitable protection.

Impedance of Conduit and Conductor

Table two of chapter 21 compares the continuous rating of copper grounding electrode conductors to the service conductors they are required to be used with.

Table three of chapter 21 compares the resistance and impedance of copper grounding electrode conductors where enclosed in a steel conduit for physical protection. The last two columns of that table show how the impedance of the conductor is approximately doubled where the conduit is not properly installed as required in 250.64(E). The *Code* requires that the conduit be bonded at both ends of the grounding electrode conductor to form a parallel circuit with the grounding electrode conductor. That

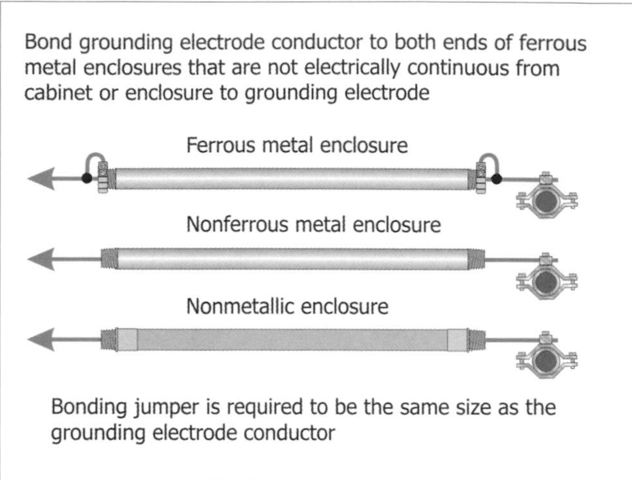
Bond grounding electrode conductor to both ends of ferrous metal enclosures that are not electrically continuous from cabinet or enclosure to grounding electrode
Ferrous metal enclosure
Nonferrous metal enclosure
Nonmetallic enclosure
Bonding jumper is required to be the same size as the grounding electrode conductor

Figure 7-15. Securing and protection from damage and magnetic field

Photo 7-15. Grounding electrode conductor in armor, bonded to clamp

important rule, if not observed, results in doubling the impedance of the grounding conductor. Of course, where the impedance of the installation is increased, the effectiveness of the grounding electrode conductor is reduced (see figure 7-16).

Where grounding electrode conductors are used on an alternating-current circuit and enclosed in ferrous metal conduit, it is necessary to compare the impedance values and not the resistance values.

The data in table three of chapter 21 shows that if an 8 AWG copper conductor is installed in ferrous metal conduit and properly bonded at each end, the impedance of the circuit is about the same as if the 8 AWG copper conductor was used alone.

For all other sizes of copper conductors installed in the proper sized conduit, the impedance of the circuit is greater where the conduit is used, as compared to using the copper conductor alone. The impedance values are from about 40 percent more for a 6 AWG copper conductor in a metric designator 21 (¾) trade size conduit to about 500 percent more for a 3/0 AWG copper conductor in

Conductor	Conduit Size	Total Amperes	Current in Conductor	Current in Conduit
6	½	100	3	97
6	½	300	5	295
2	¾	90	7	83
2	¾	350	10	340
2/0	1	150	15	135
2/0	1	590	5	585
4/0	1¼	225	15	210
4/0	1 ¼	885	15	870

The above test data confirm that, for all practical purposes, the impedance of a conductor enclosed in steel conduit, when the conduit is bonded at both ends, is approximately equal to the impedance of the conduit.

Figure 7-16. Table showing division of current between conduit and conductor (test data)

a metric designator 35 (1¼) trade size conduit, as compared to not using a metal conduit enclosure for physical protection.

Design Considerations for Grounding Electrode Conductor

The short-time rating of a copper conductor is related to the I^2t (*current* x *current* x *time*) rating of the conductor for a given temperature rise which will not damage adjacent insulated conductors or affect the continuity established by the bolted joints. For a period of five seconds the short-time rating may be taken as approximately 1 ampere for every 42.25 circular mils area. A 6 AWG copper conductor has an area of 26,240 circular mils and is thus capable of carrying about 621 amperes for five seconds safely.

Based on the safe I^2t values for the circuit comprising the various grounding electrode conductors, it can be seen that for five seconds of current, the IR drop in the different sizes of grounding electrode conductors will be approximately 37 volts per 100 ft. Using that figure as a standard, it is recommended that where a grounding electrode conductor exceeds 100 ft. in length, the conductor cross section be increased to keep the IR drop to not over 40 volts when carrying the maximum short-time current for the size conductor specified for five seconds (I^2t value). The *National Electrical Code* does not place a limit on the length of the grounding electrode conductor.

An example of selecting the proper size grounding electrode conductor for a run exceeding 100 ft. is as follows.

Given: a 1/0 AWG copper service-entrance conductor. The grounding electrode conductor specified in Table 250.66 is a 6 AWG copper and the length of the grounding electrode conductor is 150 ft.

If a 6 AWG conductor is used which has 26,240 circular mils, resulting in a short-time rating of 621 amperes, and a dc resistance of 0.0737 ohms for 150 ft. (0.491 ohms/k ft.), the voltage drop would be 621 x 0.0737 or 46 volts.

A larger grounding electrode conductor, whose resistance times the short-time rating of the 6 AWG conductor in amperes, is required to be selected and it should not exceed 40 volts.

The next larger-sized grounding conductor, a 4 AWG, has a resistance of 0.0462 ohms for 150 ft. (0.308 ohms/k ft.), so the voltage drop would be 621 x 0.0462 or 28.7 volts. That would make a 4 AWG copper grounding conductor the proper size to use for a service using a 1 AWG or 1/0 AWG copper service-entrance conductor and having a grounding electrode conductor run of 150 ft.

Direct Current Systems

Where used on a direct-current circuit, we do not destroy the value of the grounding conductor if the conduit is properly installed, that is, bonded at both ends to the grounding conductor. The resultant resistance is lower where the conduit is used for physical protection. However, that assumes a steady direct current. Special considerations are required to be given if a direct-current circuit is to be properly protected with a grounding electrode conductor against transient currents such as are produced by lightning. It is necessary to treat the selection of the grounding electrode conductor as would be done for an alternating-current circuit.

Table four of chapter 21 shows that where an aluminum conduit is used to enclose an aluminum grounding electrode conductor, the required conduit size has a lower resistance than the aluminum grounding electrode conductor in every case. In the case of aluminum wire size 6 AWG, the conduit is about one-tenth the resistance of the conductor.

For the largest aluminum grounding electrode conductor size, 250 kcmil, the conduit is about half the resistance of the conductor. Thus, where an aluminum grounding electrode conductor is protected with an aluminum conduit and properly bonded at both ends, we will always have a much lower impedance than where the conductor is not installed in an aluminum conduit. However, aluminum is subject to certain restrictions due to chemical corrosion concerns.

Since aluminum conduit is nonmagnetic and has lower resistance as well as impedance values compared to ferrous metal conduit, the use of aluminum conduit for physical protection will provide lower impedance values.

Conclusion

All of the above means that we decrease the safety of an installation (which requires a grounding electrode conductor larger than 8 AWG) if we enclose the conductor in a ferrous metal conduit. Although we provide protection from physical damage to assure the integrity of the grounding electrode conductor, another hazard is introduced by decreasing the effectiveness of the grounding electrode conductor through increased impedance in the circuit. No better case could thus be made for restricting the use of a ferrous metal conduit on a grounding electrode conductor larger than an 8 AWG.

The obvious solution, where physical protection is necessary, is to use nonmetallic or nonferrous conduits for enclosing the grounding electrode conductor. That is especially true in view of the improvement in the art of manufacturing PVC conduit, which now can be obtained in ample physical strength to meet the requirements of proper physical protection.

[1] *NEC* 2008 *National Electrical Code*, (Quincy, MA, National Fire Protection Association, 2007), p. 70-27.

Review Questions

The questions included here were developed using material included in this chapter. The answers can be found by reviewing the text. It is also important that students make use of *NEC-2008* where many answers can be found.

1. "A conductor used to connect the system grounded conductor or the equipment to a grounding electrode or to a point on the grounding electrode system" best defines which of the following:
   a. main bonding jumper
   b. grounding electrode conductor
   c. feeder bonding jumper
   d. grounded

2. A grounding electrode conductor for a service must be properly sized and is based on the size or rating of the:
   a. service breaker or fuse
   b. transformer
   c. largest ungrounded service-entrance conductor
   d. grounding electrode

3. For service conductors sized at over 1100 kcmil copper or 1750 kcmil aluminum or copper-clad aluminum, the grounding electrode conductor must generally not be smaller than a ____ copper or ____ aluminum.
   a. 2/0 AWG - 4/0 AWG
   b. 1/0 AWG - 3/0 AWG
   c. 3/0 AWG- 250 kcmil
   d. 3/0 AWG - 4/0 AWG

4. In all grounded systems of ____ volts or less, the maximum current in the grounding electrode conductor is dependent on the sum of the resistance of the grounding electrode at the service, the grounding electrode at the transformer bank, plus the resistance of the earth path between the two grounding electrodes.
   a. 600
   b. 1000
   c. 700
   d. 800

5. Where of copper, the grounding electrode conductor is required to be sized at not less than ____ AWG.
   a. 10
   b. 14
   c. 8
   d. 12

6. Where a 3/0 AWG copper service-entrance conductor is installed, the minimum size grounding electrode conductor is ____ AWG copper or ____ AWG aluminum, or copper-clad aluminum.
   a. 8 - 4
   b. 6 - 6
   c. 4 - 2
   d. 4 - 4

7. If four 250 kcmil copper service-entrance conductors are installed in parallel per ____, the total circular mil area is considered to be that of a single 1,000 kcmil conductor.
   a. service
   b. feeder
   c. cable
   d. phase

8. Generally, the minimum size grounding electrode conductor for a service that has a 1,000 kcmil ungrounded service entrance-conductor per phase is ____ AWG copper or ____ AWG aluminum, or copper-clad aluminum conductor.
   a. 1/0 - 2/0
   b. 3/0 - 4/0
   c. 1/0 - 3/0
   d. 2/0 - 4/0

9. Services are permitted to be installed in up to ____ enclosures where they are grouped at one location.
   a. one
   b. two
   c. six
   d. eight

10. A grounding electrode conductor is required to be sized at not less than ____ AWG copper or ____ AWG aluminum wire where it is the sole connection to a rod, pipe, or plate electrode.
   a. 8 - 8
   b. 6 - 4
   c. 6 - 6
   d. 4 - 8

11. A grounding electrode conductor that is the sole connection to a concrete-encased grounding electrode is not required to be larger than ____ AWG copper conductor.
   a. 6
   b. 4
   c. 8
   d. 10

12. Grounding electrode conductor connections are required to be made by ____, listed clamps, or other listed means.
   a. exothermic welding
   b. listed lugs
   c. listed pressure connectors
   d. any of the above

13. Which one of the following statements is INCORRECT?
   a. Exothermic welding connections are not required to be listed.
   b. Soldered connections are permitted for connections to a grounding electrode.
   c. Unless listed, not more than one conductor can be connected to the grounding electrode by a single clamp or fitting.
   d. Clamps used on a pipe, rod, or other buried electrode must also be listed for direct burial.

14. Where exposed to physical damage, ____ AWG or larger grounding electrode conductors require protection.
   a. 4
   b. 8
   c. 6
   d. 3

15. Size ____ AWG grounding electrode conductors must be run along the surface of the building construction and be securely fastened, or they are required to be protected by installation in rigid or intermediate metal conduit, rigid PVC conduit, electrical metallic tubing, or cable armor.
   a. 8
   b. 6
   c. 4
   d. 2

16. Bare aluminum or copper-clad aluminum conductors are not permitted to be installed where in direct contact with masonry or earth, or where subject to corrosive conditions, and are not permitted to be terminated within ____ inches of the earth.
   a. 12
   b. 18
   c. 14
   d. 16

17. An exothermic or irreversible compression connection bolted to a fire-proofed structural metal grounding electrode_______
   a. shall be identified by a green color
   b. shall be accesible
   c. shall not be required to be accessible
   d. all of the above

18._______metal enclosures for grounding electrode conductors must be electrically continuous or made electrically continuous by bonding to the grounding electrode conductor
   a. Ferrous
   b. Nonferrous
   c. Nonmetallic
   d. Aluminum

19. Grounding electrode conductors shall be permitted to be which of the following materials?
   a. copper
   b. aluminuim
   c. copper-clad aluminuim
   d. any of the above

20. A grounding electrode conductor connection to a metal water pipe electrode shall be made using which of the following:
   a. a listed bolted clamp (cast bronze or brass)
   b. a listed bolted clamp of malleable iron
   c. a pipe fitting, or pipe plug
   d. all of the above are acceptable

21. The secondary of a transformer separately derived system supplies a 400-ampere panelboard and the derived phase conductors are 600-kcmil copper (one per phase). What is the minimum size copper grounding electrode conductor to a metal water pipe electrode?
   a. 1/0
   b. 2/0
   c. 3/0
   d. 6

22. Bonding jumpers for grounding electrode systems are required to be sized according to:
   a. 250.70
   b. 260.64(A)
   c. 250.64(E)
   d. 250.66

Chapter 8

Bonding Enclosures and Equipment

Objectives

To understand...

- The purpose of bonding
- Requirements for maintaining continuity and conductivity
- Systems over 250 volts to ground
- Multiple raceway systems
- Receptacles
- Metal water piping systems
- Other metal piping systems
- Interconnected exposed structural metal framing

Bonding is an ongoing process in any electrical system from the point of service delivery to the final outlet on the system. The act of bonding metal parts or enclosures of electrical components and conductors connects them together electrically and mechanically, establishing eletrical continuity and conductivity. Essentially the desired outcome when bonding metal parts together is to make them electrically become one. Bonding has a very important function electrically for both grounded and ungrounded systems. Bonding metallic parts together puts the parts at the same potential, and through the bonding connection to the grounding electrode at the service or source of separately derived system, at the ground (earth) potential. The *NEC* defines *bonding* in Article 100 as follows: "Bonded (bonding). Connected (connecting) to establish electrical continuity and conductivity."[1]

The definition of *bonded (bonding)* is universal with how it is used in rules of the *NEC* wherever referring to an effective path for fault current or bond to establish an equipotential bonding grid or plane such as for a swimming pool or agricultural facility (see figure 8-1).

Maintaining Continuity

Section 250.96(A) requires that bonding be done around connections of metal raceways, cable trays, cable armor, cable sheath, enclosures, frames, fittings and other metal non-current-carrying parts used as equipment grounding conductors where necessary. This may be necessary to assure that these systems have electrical continuity and the current-carrying capacity to safely conduct the fault current likely to be imposed on them. Bonding of these components is required to be done regardless of whether or not an equipment grounding conductor is run within the raceway. This will ensure that the raceway will not become energized due to a line-to-enclosure fault without having the capacity and capability of clearing the fault by allowing sufficient current to operate the overcurrent protective device on the line side of the fault.

Keep in mind that the *weakest link rule* applies to the ground-fault return path. To provide adequate safety, the effective ground-fault current path is required to be (1) electrically continuous, (2) have the capacity to conduct safely any fault current likely to be imposed on it, and (3) have sufficiently low impedance to limit the voltage to ground and to facilitate the operation of the circuit-protective devices [see 250.4(A)(5)]. This ground-fault path is required to meet all three conditions from the farthest enclosure or equipment all the way back to the service equipment and ultimately to the source. This path can be through many boxes, conduits or other raceways, pull boxes, wireways, auxiliary gutters, panelboards, motor control centers and switchboards. Every connection is important. It only takes one loose locknut or broken fitting to break a link in the fault-current chain.

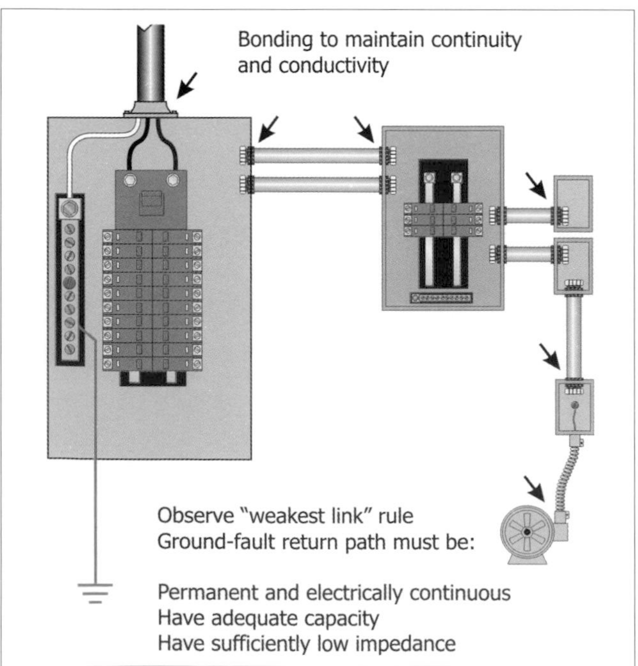

Bonding to maintain continuity
and conductivity
Observe "weakest link" rule
Ground-fault return path must be:
Permanent and electrically continuous
Have adequate capacity
Have sufficiently low impedance

Figure 8-1. Bonding to maintain continuity

Section 250.96(A) also refers to conditions where a nonconducting coating might interrupt the required continuity of the ground-fault path, and it points out that such coatings must be removed unless the fitting(s) is designed as to make such removal unnecessary. [See the section on Clean Surfaces in chapter 7 for more information.]

In some cases, the locknut can pierce painted enclosures to establish a good electrical connection. This applies to the use of heavy-type, formed-steel locknuts. General instructions are that the locknuts be tightened by hand, then be further tightened ¼ turn by means of a screwdriver and hammer. At that point, examine the connection to be sure any paint under the locknut has been adequately broken. Remove the locknut and scrape the paint off or install a bonding bushing if there is any question about the adequacy of the connection.

Testing of Conduit Fittings

The importance of removing paint from enclosures where the conduit or raceway is intended to serve as the fault-current path is further emphasized in a report on "Conduit Fitting Ground-Fault Current Withstand Capability" issued by Underwriters Laboratories on June 1, 1992. Over 300 conduit-fitting assemblies from ten different manufacturers were subjected to a current test to simulate performance under ground-fault conditions.

A sample assembly consisted of a conduit fitting secured to one end of a two-foot length of conduit and attached to a metal enclosure.

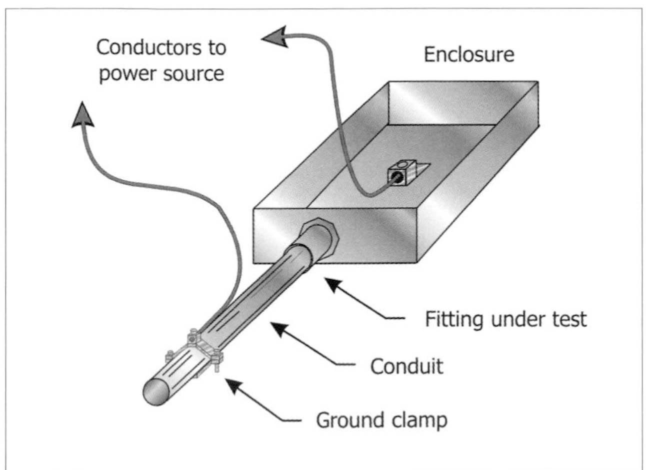

Figure 8-2. Testing of conduit fittings (sketch of actual testing assembly)

Photo 8-1. Actual enclosure used in the testing procedure

After securing the conduit fitting to the conduit properly, the conduit fitting was secured to the enclosure using the locknut provided by the manufacturer. The locknut was first hand-tightened and then further tightened ¼ turn with a hammer and standard screwdriver. The fittings were installed through holes in the enclosures that were punched rather than being installed in preexisting knockouts. A pipe clamp, wire connector, conductors and a power supply were assembled to complete the testing. Thermocouples were placed at strategic locations to record pertinent temperature data. Figure 8-2 is a drawing of the sample assembly.

Fittings for conduit in the 3/8-inch through 6-inch trade sizes were tested. The appropriate current was applied to the fittings in the test program established by Underwriters Laboratories.

This test should not be confused with a short circuit withstand test and is not intended to test the maximum short-circuit current these fittings can withstand. Due to the time and current involved, a great deal of heat is generated in the test assembly.

Seven of the more than 300 assemblies tested sustained damage. A visual examination of sample assemblies that failed showed that melting of the die-cast zinc locknuts occurred as a result of the fault current (see photo 8-1). Melting of the die-cast zinc body occurred on five sample assemblies. The painted enclosures on which the fittings were tested were also examined. The examination indicated that melting of the die-cast zinc was probably due to the inability of the locknut to penetrate through the enclosure paint and provide good electrical contact between the fitting and metal of the enclosure.

A visual examination of all the conduit fittings with die-cast zinc locknuts showed that there were three different constructions of the locknuts. The three constructions differed in that the surface of the locknut contacting the enclosure either was flat, ribbed, or serrated. The sample assemblies with die-cast zinc locknuts that did not complete the current test with acceptable results had locknuts with flat or ribbed surfaces. All fittings having die-cast zinc locknuts with serrations completed the current tests with acceptable results. It appeared as though locknuts with serrations consistently penetrated through the enclosure paint and provided better electrical contact between the fitting and the metal of the enclosure than did the locknuts with flat or ribbed surfaces.

The fittings investigated in this work were formed of die-cast zinc, steel and malleable iron. The melting point of zinc is 420°C while the melting point of steel and malleable iron is much higher, typically greater than 1400°C. Heat generated from the fault current in some sample assemblies was obviously greater than the melting point of the die-cast zinc fittings and locknuts, but not greater than the melting point of steel or malleable iron since no melting of the steel enclosure occurred. This was further evidenced by tests of two sample assemblies where the die-cast zinc body of the fittings melted, but the steel locknuts did not.

All of the conduit fittings that were constructed of steel bodies and steel locknuts completed the test with acceptable results. A visual examination of the steel locknuts indicated that the nibs on these locknuts, which in most cases were sharp and well defined from the metal forming process, provided for better penetration through the enclosure paint than the nibs on the die-cast zinc locknuts.

For most of the sample assemblies that completed the current test with acceptable results, the maximum temperatures on the fitting bodies and locknuts were about the same as or less than the temperature of the conduit. In the case of the flexible metal conduit, the temperatures on the fittings were much less than on the conduit. This would seem to indicate that if the fitting can provide good electrical contact to the enclosure metal the fitting will provide for adequate equipment grounding.

Conclusions reached by Underwriters Laboratories as a result of the testing are as follows:

"1. Over 300 conduit-fitting assemblies from ten different manufacturers were subjected to the Current Test to simulate performance under ground-fault conditions. As a result of the tests, only seven assemblies representing four different conduit fittings and three different manufacturers did not withstand the fault current without breaking or melting of the conduit-fitting assembly. All seven of these sample assemblies were compression type connectors with die-cast zinc bodies, and all but one of these assemblies utilized a die-cast zinc locknut.

"2. An examination of the seven sample assemblies that did not complete the Current Test with acceptable results showed that the failures were probably due to high resistance from the inability of the fitting locknut to penetrate through the enclosure paint and provide good electrical continuity between the fitting and the metal enclosure. Heat generated by the high-resistance arcing was sufficient to melt the zinc, but not steel or iron.

"3. Some of the sample assemblies that did not exhibit breaking or melting did show signs of arcing and welding between the locknut and the enclosure and/or the fitting and the conduit. These sample assemblies usually had higher temperatures during the Current Test, however, the temperatures were not sufficient to cause melting of the zinc or steel parts nor loss of continuity between the conduit, fitting, and enclosure.

- Threadless couplings and connectors for cables with metal sheaths
- Two locknuts, on rigid metal conduit or IMC, one inside and one outside
- Fittings with shoulders that seat firmly against the box or enclosure such as for EMT, flexible metal conduit and cable connectors with one locknut inside

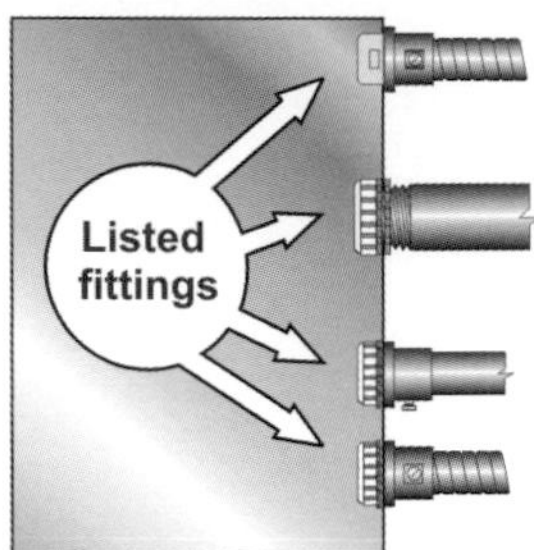

Also permitted are:

- Threaded couplings and bosses
- Threadless couplings and connectors
- Bonding jumpers
- Other approved devices such as bonding-type locknuts

Figure 8-3. Bonding methods for circuit over 250 volts to ground

"4. Most of the sample assemblies that were subjected to the Current Test attained maximum temperatures on the fitting bodies and locknuts that were about the same as or less than the temperature of the conduit. For the tests with flexible metal conduit, the temperatures of the fittings were much less than the temperatures of the flexible conduit.

"5. As a result of the tests, it was observed that if the fitting provides good electrical contact to both the enclosure and the conduit, the fitting will provide a suitable equipment ground path for fault current." [1]

Bonding for Over 250 Volts

Section 250.97 requires that for circuits having a voltage exceeding 250 volts to ground, the electrical continuity of metal raceways and metal-sheathed cables that are not used for service-conductors must also be ensured by specific methods (see figure 8-3).

Acceptable methods include any of the methods approved for bonding at service equipment found in 250.92(B)(2) through (4). Note that standard locknuts and bushings without additional bonding means are not generally permitted for bonding equipment, which has concentric or eccentric knockouts, over 250 volts to ground.

The bonding methods permitted include those for services including:

"(2) For rigid and intermediate metal conduit, connections made up wrenchtight with threaded couplings or threaded bosses on enclosures.

"(3) Threadless couplings and connectors made up tight for rigid metal and intermediate metal conduit and electrical metallic tubing.

"(4) Other listed devices like bonding type locknuts and bushings." [2]

In addition, bonding jumpers are permitted around concentric or eccentric knockouts that are punched or formed so as to impair the electrical connection to ground.

An exception to 250.97 provides that where oversized, concentric or eccentric knockouts are not encountered, or where concentric or eccentric knockouts have been tested and the box or enclosure is listed to provide a reliable bonding connection, the following methods of ensuring continuity for these connections are permitted:

- threadless couplings and connectors for cables with metal sheaths,
- for rigid and intermediate metal conduit, two locknuts, one inside and the other outside the boxes and enclosures,
- fittings that seat firmly against the box or enclosure or cabinet such as for electrical metallic tubing, flexible metal conduit and cable connectors, with one locknut inside the enclosure, or
- listed fittings.

Photo 8-2. Concentric and eccentric knockouts in boxes are acceptable for bonding.

Listed outlet boxes are specially designed and tested so knockouts perform satisfactorily for over 250-volt-to-ground applications (see photo 8-2). These boxes have only one eccentric knockout so when the solid knockout is removed, a conduit or fitting locknut makes contact with the base metal of the box to ensure good electrical and mechanical contact.

Oversized Knockouts

The installer needs to be cautious in the use of equipment that has concentric or eccentric knockouts, as their ability to carry fault current must be of concern. It is very common to find nibs of adjacent rings damaged during removal of the desired knockout. This leaves less material available for carrying fault current. The safest practice is to install bonding bushings around concentric and eccentric knockouts where there is any question about their integrity.

Concentric and eccentric knockouts in equipment such as cabinets, enclosed switches, junction and pull boxes, auxiliary gutters and wireways are not tested or certified by an electrical products testing laboratory for their current-carrying ability. The specific methods

Photo 8-3. Expansion coupling Courtesy of Thomas and Betts

Photo 8-4. Cut-away of an expansion coupling with bonding jumper Courtesy of Thomas and Betts

provided for in 250.97 must be used if those enclosures have eccentric or concentric knockouts.

In other areas, where oversized, concentric or eccentric knockouts are not present, threadless fittings which are made up tight with conduit or armored cable or the use of two locknuts, one inside and one outside of boxes and cabinets, is acceptable for bonding.

Where loosely jointed metal raceways are used and especially where there are expansion joints or telescoping sections of raceways (see photos 8-3 and 8-4), the *Code* requires that they be made electrically continuous by the use of equipment bonding jumpers or other means [see 250.98].

Reducing Washers

Reducing washers are commonly used in electrical installations where it is desirable or necessary to install conduit or fittings of a size that is smaller than the knockout available in the enclosure. These reducing washers are evaluated and listed for bonding (see UL White Book, category QCRV). Where painted or coated enclosures are encountered, one should always bond around to provide an adequate fault-return path (see photo 8-5).

Photo 8-5. Bonding around reducing washers at coated enclosures to maintain continuity and the capacity to conduct any fault current that might be imposed

Load-Side Bonding Jumper Sizes

Equipment bonding jumpers form a part of the effective ground-fault path and can carry the same fault current that the equipment grounding conductor would carry; therefore, they are required to be the same size.

The size of the bonding jumper will depend on its location and is based on the size of the nearest overcurrent device in the circuit ahead of the equipment [see 250.102(D)]. Column 1 (*left*) of Table 250.122 gives the size or setting of the overcurrent device in the circuit ahead of the equipment. Columns 2 (*middle*) and 3 (*right*) give the minimum size of the equipment grounding conductor, whether copper or aluminum or copper-clad aluminum.

Attaching Jumpers

Where bonding jumpers are used between grounding electrodes or around water meters and similar equipment, the *Code* requires that good electrical contact be maintained and that the arrangement of conductors be such that the disconnection or removal of equipment will not interfere with or interrupt the grounding and bonding continuity of the jumper [see 250.68(B)].

Bonding jumpers are required to be attached to circuits and equipment by any of the means provided in 250.8 including exothermic welding, listed pressure connectors, listed clamps or other suitable and listed means. Bonding jumper connections and equipment grounding conductor connections are required to be made using any of the following methods:

1. listed pressure connectors
2. terminal bars
3. pressure connectors listed as grounding and

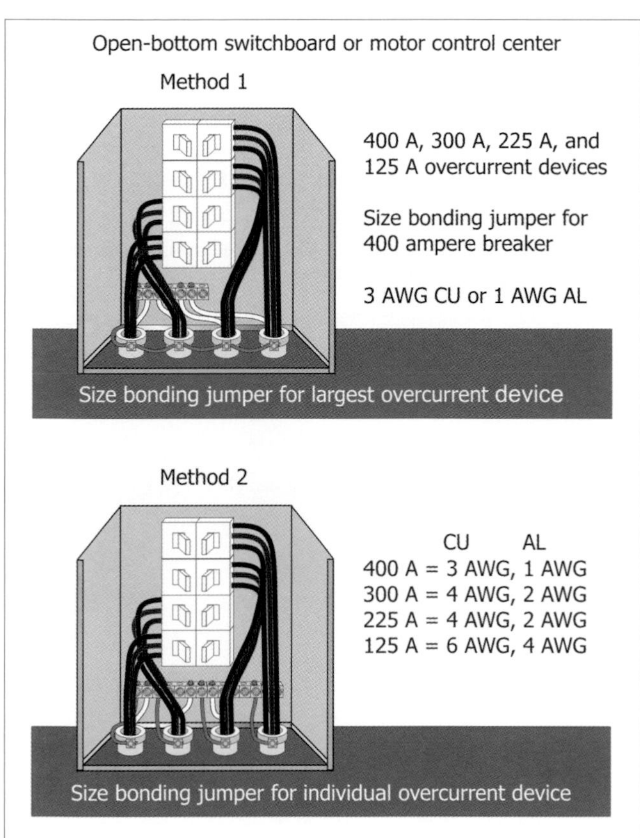

Figure 8-4. Bonding raceways to enclosures (two methods shown)

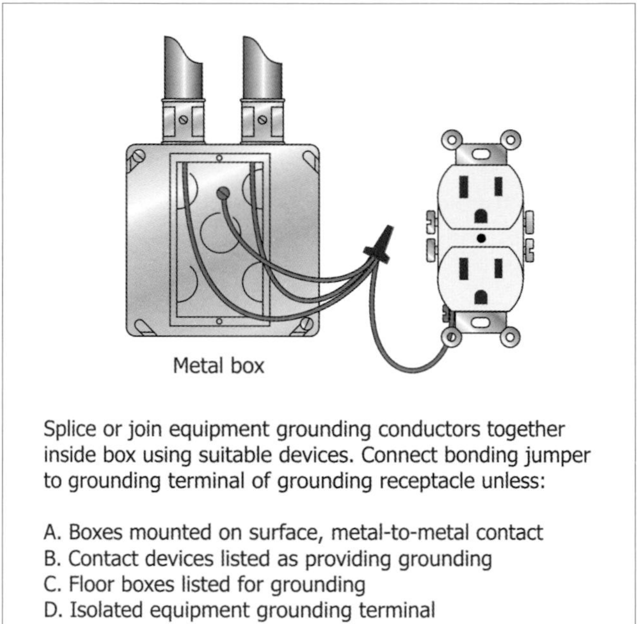

Figure 8-5. Bonding of grounding-type receptacles to boxes

bonding equipment

4. exothermic welding
5. machine screw-type fastners that engage at least two threads or are secured with a nut
6. thread-forming screws engaging not less than two threads in the enclosure
7. connections that are part of listed assemblies
8. other listed means

Connections that depend solely on solder are not acceptable [see 250.8].

Bonding Multiple Raceway Systems

Where more than one raceway enters or leaves a switchboard, pull or junction box or other equipment, it is permissible to use a single conductor to bond these raceways to the equipment. The equipment (bonding) jumper is sized for the largest overcurrent device ahead of conductors contained in any of the raceways [see 250.102(D)]. These are feeder or branch-circuit conductors and not service-entrance conductors.

For example, as shown in figure 8-4, four metallic raceways leave the bottom of an open switchboard or motor control center. The overcurrent protective devices ahead of the raceways are 400, 300, 225 and 125 amperes respectively. According to Table 250.122, the minimum size equipment bonding jumper for the raceway having conductors protected at 400 amperes is 3 AWG copper or 1 AWG aluminum. If this conductor were looped through a grounding bushing on each raceway, compliance with the *Code* would be obtained. Of course, the grounding bushings would need to be listed for both the size of conduit and conductor. In some cases, this method may require a larger conductor to bond some conduits than where individual bonding jumpers are installed.

In addition, as shown in figure 8-4, it is acceptable to install an equipment bonding jumper individually from each raceway to the equipment grounding terminals of the equipment. Each bonding conductor is sized for the overcurrent device ahead of the conductors in that raceway per Table 250.122.

Nonmetallic Boxes

Section 314.3 permits metal raceways or metal-armored cables to be used with nonmetallic boxes only where:

- internal bonding means are provided between all raceways or metal-armored cables, or
- integral bonding means with provision for attaching a grounding jumper inside the box are

provided. This type of bonding means is typically molded in the box.

The bonding jumpers are required to be sized in accordance with Table 250.122. It should be noted that the size of the bonding jumpers given in 250.122 is the minimum size. Larger bonding jumpers may be required to comply with the available fault-current requirements in 110.10.

Bonding Receptacles

An equipment bonding jumper is required to connect the grounding terminal of a grounding type receptacle to a grounded box [see 250.146 and figure 8-5]. Where more than one equipment grounding conductor enters a box, they must be spliced or joined inside the box with suitable devices. Four exceptions to the general rule requiring the bonding jumper are provided in 250.146(A) through (D) as paraphrased below.

(A) Where the box is mounted on the surface and direct metal-to-metal contact is made between the receptacle yoke and the box. The rule permits receptacles without an equipment bonding jumper only where the box is mounted on the surface. A requirement was added to the *Code* to remove at least one of the insulated washers used to retain the screws, unless the receptacle is listed as self-grounding. Cover-mounted receptacles, such as in a raised cover on 4-in. square boxes, are acceptable where the receptacle is fastened to the listed cover with screws and locking nuts or rivets and the cover is equipped with mounting holes on a flat, non-raised portion of the cover.

Photo 8-6. Self-grounding receptacles

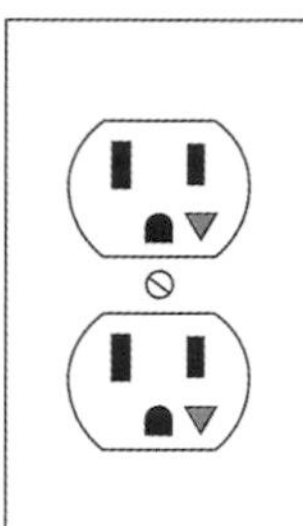

Isolated ground receptacles to be identified by orange triangle on face
To be used only with isolated grounding conductors per Section 406.2(D)(1)
Isolated ground receptacles installed in nonmetallic boxes shall be covered with a nonmetallic faceplate unless the box has a feature or accessory for grounding the faceplate

Figure 8-6. Isolated ground receptacles

(B) Contact devices or yokes designed and listed as providing bonding with the mounting screws to establish the grounding circuit between the device yoke and flush-type boxes. This is the device commonly referred to as a *self-grounding receptacle* (see photo 8-6). The device is designed and listed as maintaining good electrical contact between the yoke and box by means of a spring-type device that maintains continuity between the device and the mounting screws.

(C) Floor boxes that are listed as providing satisfactory grounding continuity

(D) A receptacle having an isolated equipment grounding terminal that is required for reduction of electrical noise (electromagnetic interference)(see figure 8-6). In this case, the equipment grounding terminal is required to be grounded by an insulated equipment grounding conductor that is run with the circuit conductors. Section 406.2(D) requires that this receptacle be identified by an orange triangle on the face of the receptacle. This equipment grounding conductor is permitted to pass through one or more panelboards, boxes, wireways, or other enclosures so as to terminate at an equipment grounding terminal of the separately derived system or service.

However, this insulated equipment grounding conductor must be connected to the equipment grounding terminal bar at the building disconnecting means or the source of a separately derived system within the same building or structure.[3]

To provide an effective ground-fault current path , the isolated, insulated equipment grounding conductor should never pass without being terminated at the separately derived system that is the source of power for the equipment being grounded. Also note that

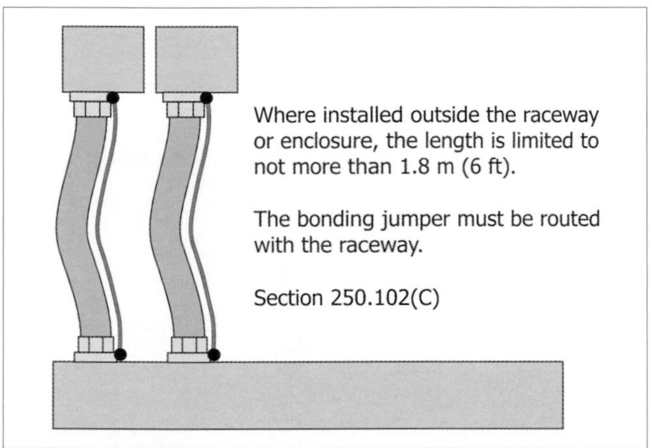

Where installed outside the raceway or enclosure, the length is limited to not more than 1.8 m (6 ft).
The bonding jumper must be routed with the raceway.
Section 250.102(C)

Figure 8-7. Installation of equipment bonding jumper on the outside of the raceway

Photo 8-7. Equipment bonding jumper outside raceway

special rules are provided in 406.2(D) for grounding receptacle covers where isolated ground receptacles are installed in a nonmetallic outlet box.

Installation of Equipment Bonding Jumper

Equipment bonding jumper is defined in Article 100 as "The connection between two or more portions of the equipment grounding conductor." This definition describes only one installation of equipment bonding jumpers on the load side of an overcurrent protective device. It should be noted that equipment bonding jumpers are also used on the supply side of an overcurrent device such as on the line side of the service disconnecting means and on the supply side of the first overcurrent device enclosure on the secondary of a separately derived system [see 250.102(C) and 250.30(A)(2)].

The equipment bonding jumper is permitted to be installed either inside or outside of a raceway or enclosure [see 250.102(E)].

Where the jumper is installed on the outside, the length is generally limited to not more than 1.8 m (6 ft). In addition, the bonding jumper is required to be routed with the raceway. This is vital to keep the impedance of the equipment bonding jumper as low as possible (see figures 8-7 and 8-8 and photo 8-7). An equipment bonding jumper longer than 1.8 m (6 ft) is permitted at outside pole locations for the purpose of bonding or grounding isolated sections of metal raceways or elbows that are installed in exposed risers of metal conduit or other metal raceway (see photo 8-7, equipment bonding jumper outside raceway).

Bonding of Piping Systems

Section 250.104 requires that metal water piping and other metal piping systems installed within or attached to buildings or structures be bonded (see figure 8-9). This requirement for bonding is not to be confused with the requirement in 250.52(A)(1) that metal underground water piping is to be used as a grounding electrode. Some requirements change depending upon whether the piping is metal water piping or other metal piping systems.

Included among the items within a building that should be adequately grounded to the one common grounding electrode system are the water piping system (hot and cold) and gas and sewer piping, any metallic air ducts installed inside or on the exterior of the building, as well as such devices as TV towers, gutters provided with a deicing system, and so forth.

The bonding jumper generally is required to be sized in accordance with Table 250.66. Also, it is required to be installed in accordance with the general rules for installing grounding conductors in 250.64(A), (B) and (E). They shall be connected in a manner specified in 250.70. The point(s) of attachment of the bonding jumper(s) to the water piping system is required to be accessible (see figure 8-10).

Metal Water Piping

This requirement for bonding applies to all metal water piping system(s) installed within or on the exterior of the building [250.104(A)(1)]. The bonding

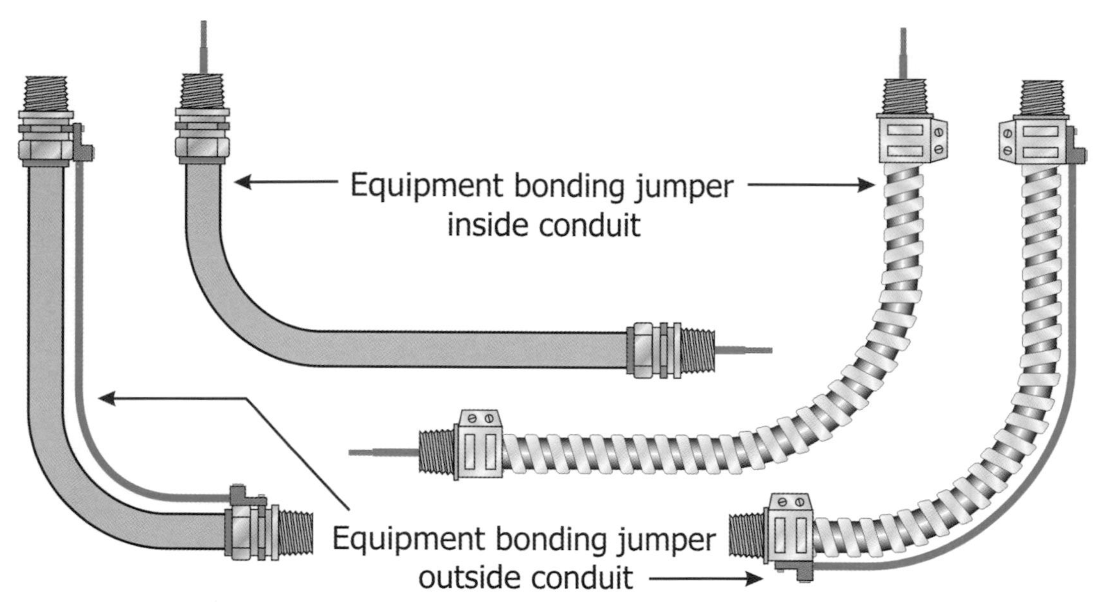

Figure 8-8. Bonding jumper installation in accordance with 250.102

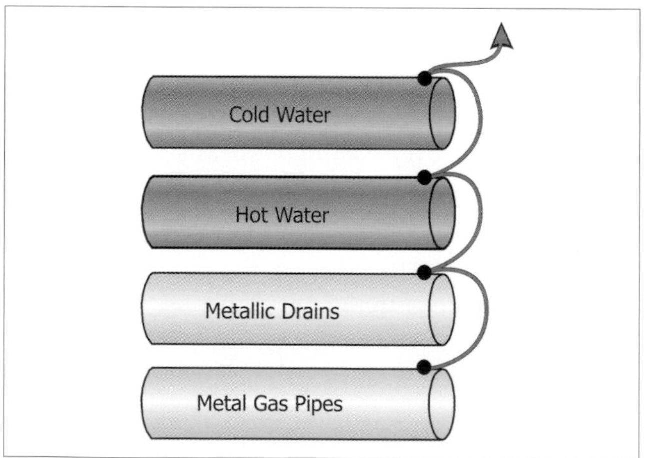

Figure 8-9. Bonding of metallic piping systems is required.

jumper in the building where the service is located is generally required to be sized in accordance with Table 250.66 and, thus, is based on the size of the service-entrance conductor and not on the rating of the service overcurrent device. In addition, the points of attachment of the bonding jumper to the metal water piping system are required to be accessible (see photo 8-8).

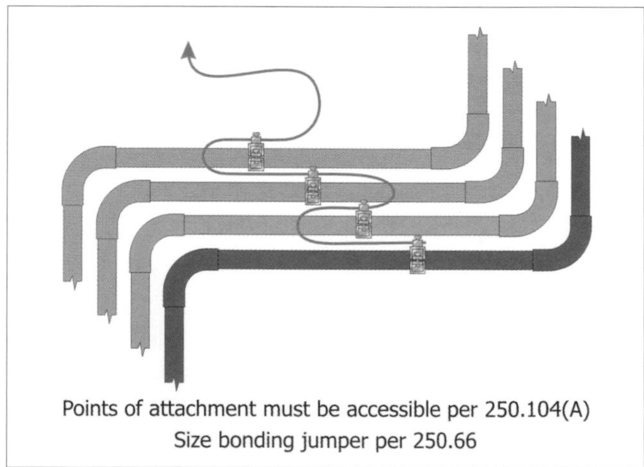

Points of attachment must be accessible per 250.104(A)
Size bonding jumper per 250.66

Figure 8-10. Bonding jumper connection to piping is generally required to be accessible.

The piping system is permitted to be bonded to the service equipment enclosure, the grounded conductor at the service, the grounding electrode conductor where large enough, or to one or more grounding electrodes.

For example, if Table 250.66 requires a 2 AWG bonding jumper to the metal water piping, it cannot be connected to a 6 AWG or 4 AWG grounding electrode conductor.

Photo 8-8. Connection to water piping is to be accessible.

Where a metallic underground water piping system exists and is connected to a metallic interior water-piping system and there is not an insulated coupling, the interior water piping system is automatically and adequately grounded (bonded) when the metallic underground water piping system is used as the grounding electrode. However, with the expanding use of nonmetallic piping and insulated couplings it becomes more important to be sure that the interior piping not only is electrically continuous, but that also it is adequately grounded by bonding it to the same grounding electrode used for the premises. That is a mandatory and essential requirement of the *Code*.

Multiple Occupancy Building

Section 250.104(A)(2) allows the metal water piping system(s) to be bonded to the panelboard or switchboard enclosure (other than service equipment) under specific conditions (see figure 8-11). The conditions are as follows:

1. The building is multiple occupancy, and
2. The metallic water piping is isolated from all other occupancies by nonmetallic water piping. In other words, the metallic piping system in each occupancy is isolated from all other piping systems of the other occupancies by nonmetallic means or individual isolation.

In this case, the bonding jumper to the water piping is sized in accordance with Table 250.122, and the ampere rating of the overcurrent device supplying the feeder to the unit or occupancy determines the minimum size of the bonding conductor [see 250.104(A)(2). The bonding jumper for the interior metal water piping runs from the equipment grounding terminal bar in the panelboard serving the unit to

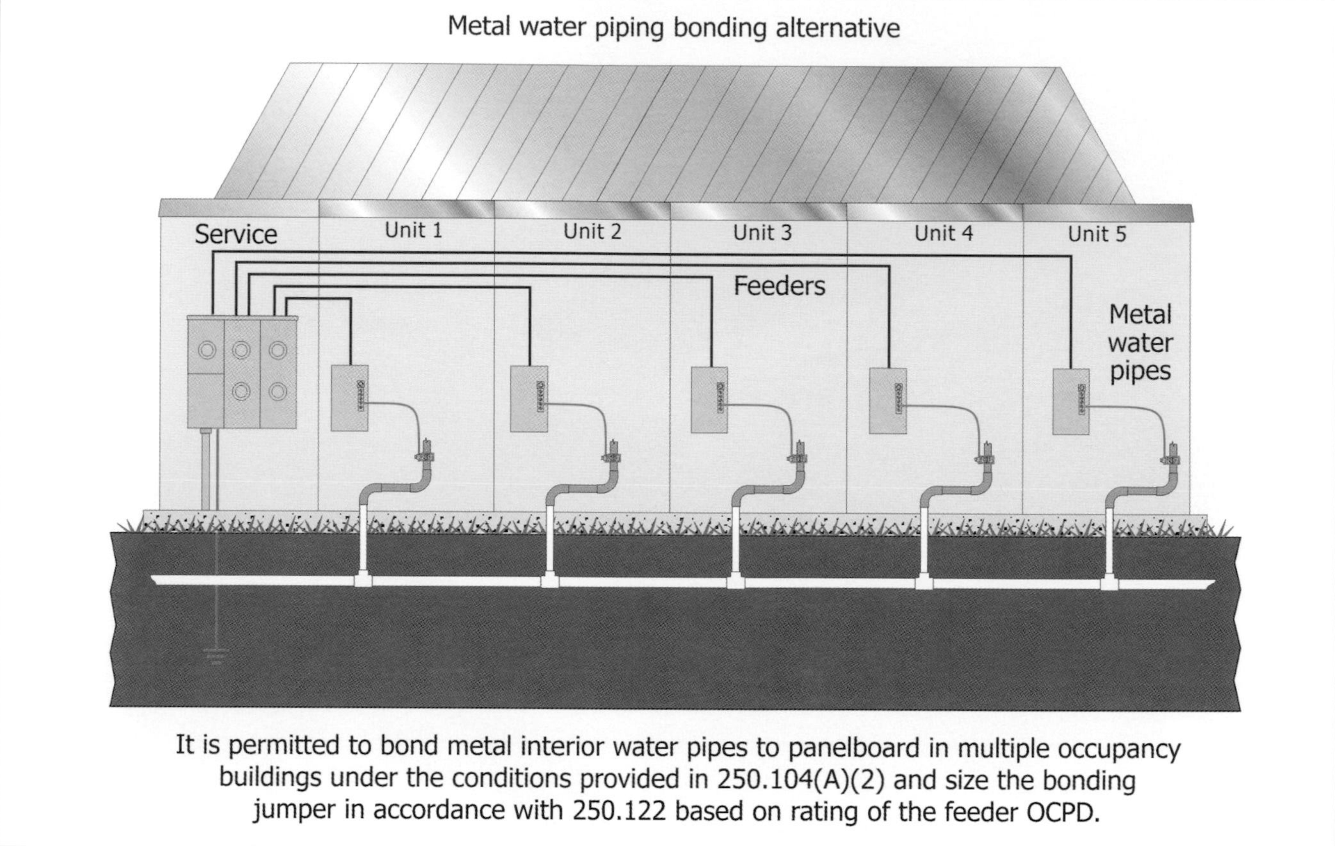

Metal water piping bonding alternative

Service
Unit 1
Unit 2
Unit 3
Unit 4
Unit 5
Feeders
Metal water pipes

It is permitted to bond metal interior water pipes to panelboard in multiple occupancy buildings under the conditions provided in 250.104(A)(2) and size the bonding jumper in accordance with 250.122 based on rating of the feeder OCPD.

Figure 8-11. Metal water piping bonding alternative for multiple occupancy buildings

the piping. In this case, the bonding jumper does not connect to the neutral terminal bar in the panelboard.

Multiple Buildings or Structures Supplied by Feeder(s) of Branch Circuit(s)

Where a building or structure is supplied by a feeder or branch circuit, bonding of metal water piping system(s) is required to be ensured by one of the following methods:

1. Bonding to the building or structure disconnecting means where it is located at the building or structure.

2. Bonding to the equipment grounding conductor that is run with the supply conductors to the building or structure. This connection would usually be made inside the building or structure disconnecting means enclosure on the equipment grounding terminal bar. Note that the equipment grounding conductor is permitted to consist of the wiring method that supplies the building or structure if recognized in 250.118.

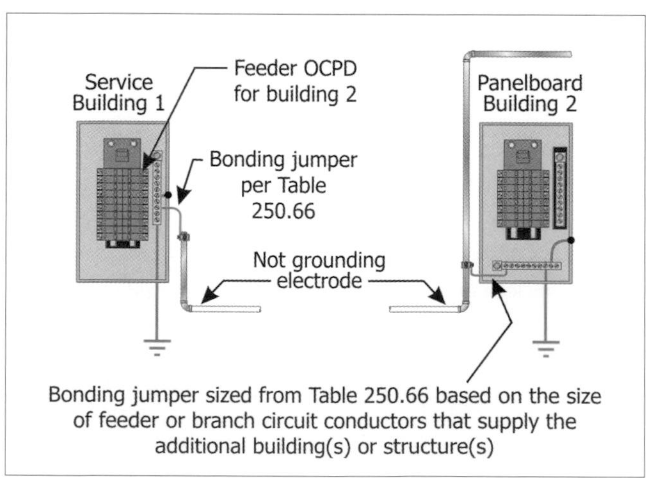

Figure 8-12. Multiple buildings or structures supplied by a feeder(s) or branch circuit(s)

3. Bonding to the one or more grounding electrodes (grounding electrode system) used.[4]

The bonding jumper to the water piping system(s) is required to be sized according to Table 250.66 based on the size of the feeder or branch-circuit conductors that supply the building or structure (see figure 8-12).

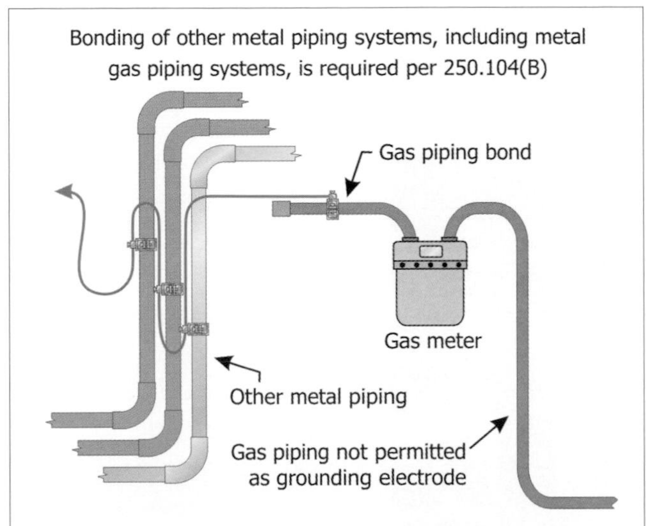

Figure 8-13. Bonding required of other metal piping systems including metal gas piping systems [250.104(B)]

Bonding Other Metal Piping

Other metal piping, installed in or attached to a building or structure, including gas piping, which is likely to become energized is required to be bonded (see figure 8-13). The piping must be bonded to the service equipment enclosure, the grounded conductor at the service, the grounding electrode conductor where of sufficient size, or to the one or more grounding electrodes used. The *Code* does not give guidance on how to determine the conditions under which metal piping is likely to become energized. Because metal piping systems are conductive, bonding all of them will provide additional safety.

Common systems that have to be bonded include interior metal: pneumatic systems; waste, drain and vent lines; and oxygen, air, and vacuum systems.

The bonding conductor is sized from Table 250.122 using the rating of the overcurrent device in the circuit ahead of the equipment.

The equipment grounding conductor for the circuit that is likely to energize the piping can be used as the bonding conductor. The point of connection of these bonding conductors to the metal piping systems is not required to be accessible as the connections to metal water piping systems are, but it is a good installation practice to locate thesse connections so they are accessible.

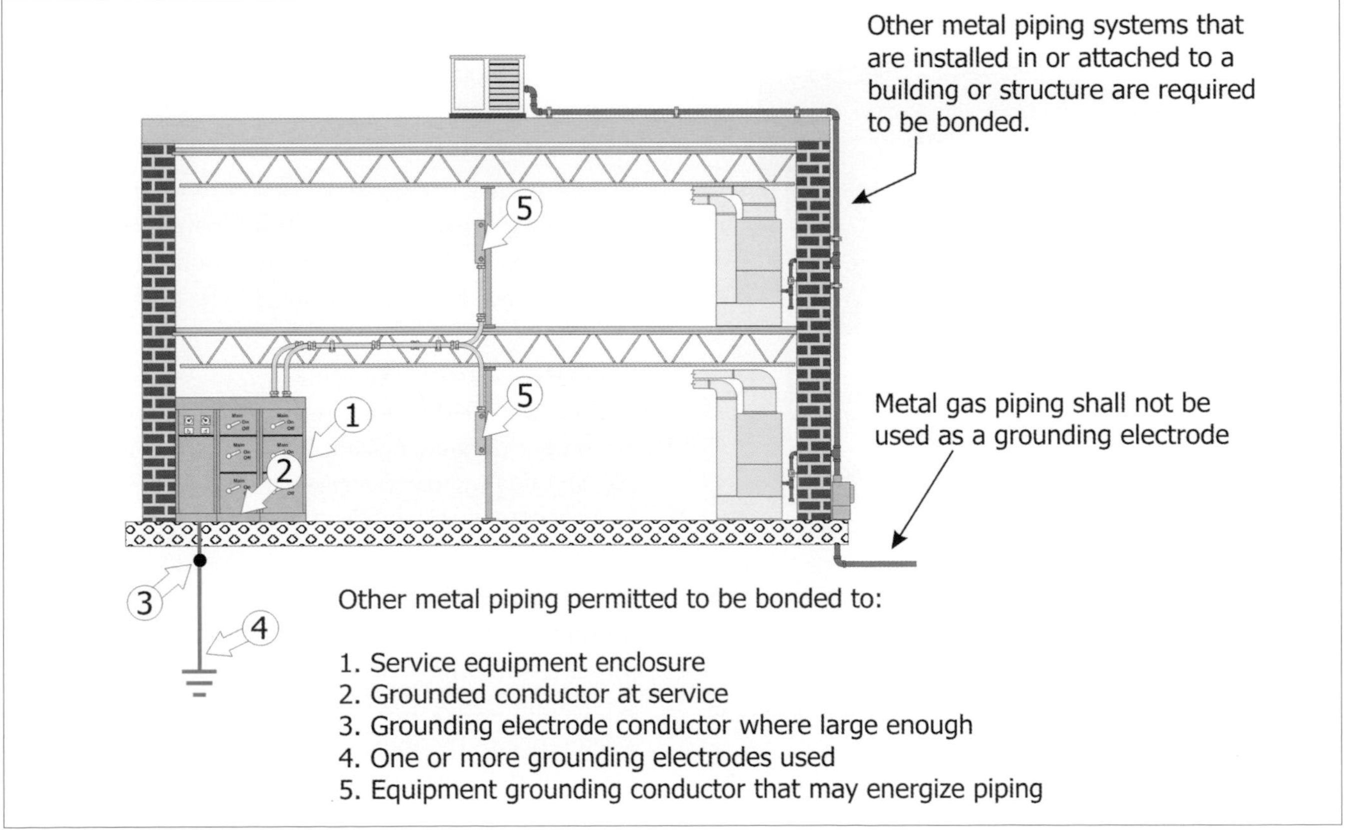

Figure 8-14. Bonding of other metal piping systems

Photo 8-9. Bonding required of exposed structural metal framing members

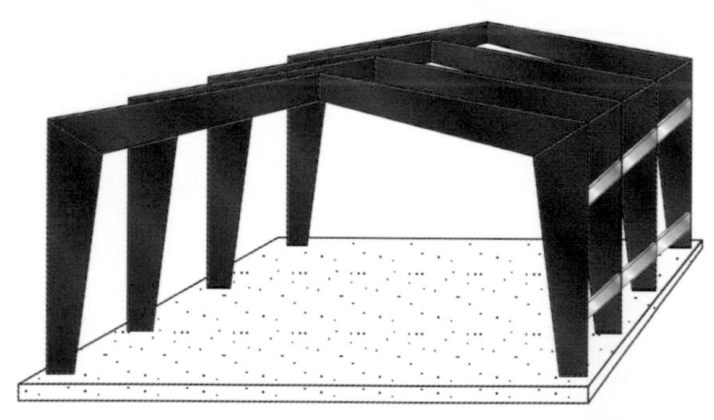

- Exposed structural metal framing that is not intentionally grounded and is likely to become energized shall be bonded
- Bond to the service equipment enclosure, the grounded conductor at service, the grounding electrode conductor where of sufficient size, or to the one or more grounding electrodes used
- Size bonding jumper per Table 250.66
- Install per Sections 250.64(A), (B), and (E)
- Attachment point(s) of bonding jumper to be accessible

Figure 8-15. Bonding structural metal framing members of buildings or structures

The bonding conductor is permitted to be connected to any of the following locations:

- Service equipment enclosure
- Grounded conductor at the service
- Grounding electrode conductor, where of ade-quate size
- One or more of the grounding electrodes used

Bonding Structural Steel

Section 250.104(C) requires exposed structural metal that is interconnected to form a steel building frame, not intentionally grounded, and is likely to become energized to be bonded (see figure 8-16 and photo 8-9). This requirement is applicable to interior or exterior structural framing members of buildings or structures. A bonding connection is required to be made to the service equipment enclosure, the grounded conductor at the service, the grounding electrode conductor where it is large enough, or to the one or more grounding electrodes used.

The bonding jumper is required to be sized in accordance with Table 250.66 and installed in accordance with the rules in 250.64(A), (B) and (E). The points of attachment of the bonding jumper to the structural steel are required to be accessible.

Separately Derived Systems

Section 250.104(D) addresses the requirements for separately derived systems (see figure 8-16). Section 250.30(A)(6) provides the correlation between 250.104(D) and 250.30(A)(6) since 250.30(A) covers ground requirements for separately derived systems.

Bonding a building's service to metal piping or the metal building frame does not provide a reference for the separately derived system. Bonding the separately derived system is necessary to establish a reference to the metal water piping and structural metal in the area served by the separately derived system. The area served can be determined by any equipment or outlets that are supplied from the separately derived system. Bonding also provides a fault-current path in the event the metal water piping or structural metal becomes energized due to an insulation failure. Where a common grounding electrode conductor is used it also must be bonded to the metal water piping and structural metal in the area. Exceptions allow bonding jumpers between metal water piping and structural metal with one bonding jumper to the separately derived system. Section 250.104(D) is as follows:

"250.104 Bonding of Piping Systems and Exposed Structural Metal.

"(D) Separately Derived Systems. Metal water piping systems and structural metal that is interconnected to form a building frame shall be bonded to separately derived systems in accordance with the following:

"(1) Metal Water Piping System(s) The grounded conductor of each separately derived system shall be bonded to the nearest available point of the metal water piping system(s) in the area served by each separately derived system. This connection shall be made at the same point on the separately derived system where the grounding electrode conductor is connected. Each bonding jumper shall be sized in accordance with Table 250.66 based on the largest ungrounded conductor of the separately derived system.

"Exception No 1: A separate bonding jumper to the metal water piping system shall not be required where the metal water piping system is used as the grounding electrode for the separately derived system and the water piping system is in the area served.

"Exception No 2: A separate water piping bonding jumper shall not be required where the structural metal frame of a building or structure is used as the grounding electrode for a separately derived system and is bonded to the metallic water piping in the area served by the separately derived system.

"(2) Structural Metal. Where exposed structural metal that is interconnected to form the building frame exists in the area served by the separately derived system, it shall be bonded to the grounded conductor of each separately derived system. This connection shall be made at the same point on the separately derived system where the grounding

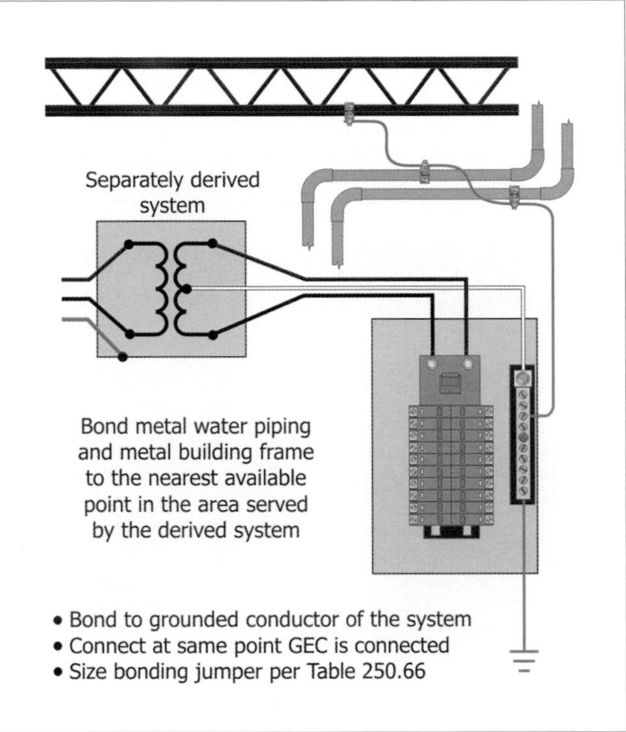

Figure 8-16. Bonding of metal piping systems and structural metal framing members to separately derived system

electrode conductor is connected. Each bonding jumper shall be sized in accordance with Table 250.66 based on the largest ungrounded conductor of the separately derived system.

"Exception No 1: A separate bonding jumper to the building metal shall not be required where the metal frame of a building or structure is used as the grounding electrode for the separately derived system.

"Exception No 2: A separate bonding jumper to a structural metal building frame shall not be required where the metallic water piping is used as the grounding electrode for a separately derived system and is bonded to structural metal frame of a building or structure in the area served by the separately derived system.

"(3) Common Grounding Electrode Conductor. Where a common grounding electrode conductor is installed for multiple separately derived systems as permitted by Section 250.30(A)(3), and exposed structural metal that is interconnected to form the building frame or metal piping exists in the area served by the separately derived system, the metal piping and the structural metal member shall be bonded to the common grounding electrode conductor.

"Exception: A separate bonding jumper from each derived system to metal water piping and to structural metal members shall not be required where the metal water piping and the structural metal members in the area served by the separately derived system are bonded to the common grounding electrode conductor."

[1] "Conduit Fitting Ground-Fault Current Withstand Capability," Underwriters Laboratories, June 1, 1992.

[2] NFPA 70, *National Electrical Code* 2008, 250.92(B)(2) through (B)(4), (National Fire Protection Association, Quincy, MA, 2007), p. 70-110.

[3] NFPA 70, 250.146(A) through (D), p. 70-119.

[4] NFPA 70, 250.104(A)(3), p. 70-111.

[5] NFPA 70, 250.104(B), p. 70-112.

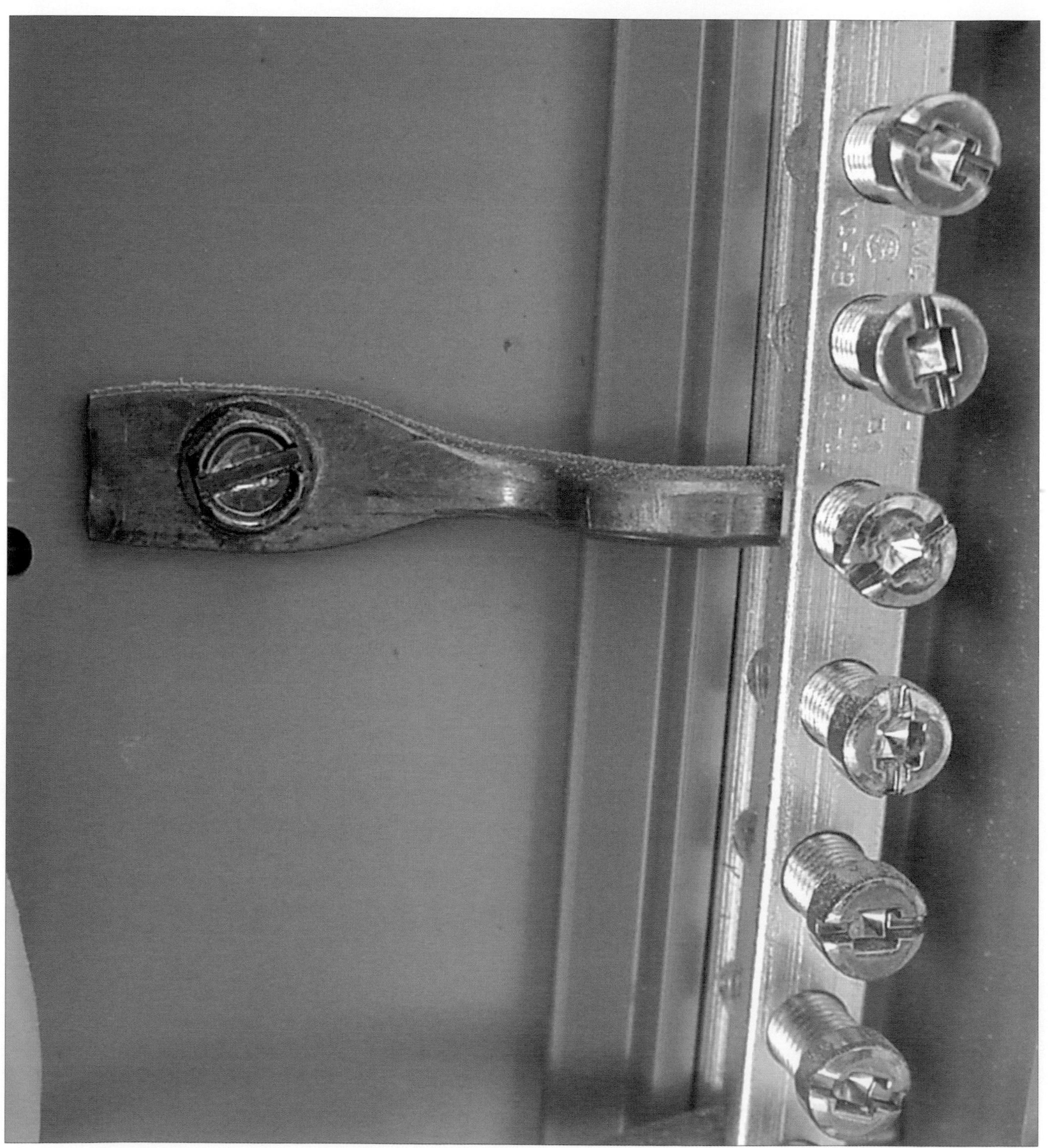

Review Questions

The questions included here were developed using material included in this chapter. The answers can be found by reviewing the text. It is also important that students make use of *NEC-2008*, where many answers can be found.

1. For circuits having a voltage exceeding ____ volts to ground, the electrical continuity of metal raceways or metal-sheathed cables, that are not service-entrance cable, must also be assured.
   a. 125
   b. 150
   c. 100
   d. 250

2. Bonding jumpers are required to be attached to circuits and equipment by means of ____, or other listed means.
   a. exothermic welding
   b. listed pressure connectors
   c. listed clamps
   d. any of the above

3. Metal raceways or metal-jacketed cables are permitted to be used with nonmetallic enclosures only where ____.
   a. internal bonding means are provided between all raceways
   b. integral bonding means with a provision for attaching an equipment bonding jumper inside the box is provided
   c. internal bonding means are provided between all cables
   d. any of the above

4. Without exception, where circuit conductors are spliced within a box, or terminated to equipment or devices within or supported by the box, equipment grounding conductors associated with those circuit conductors must be spliced or joined with____ devices.
   a. labeled
   b. listed
   c. suitable
   d. bonding

5. An equipment grounding conductor is permitted to pass through one or more panelboards, boxes, wireways, or other enclosures within the same ____ to terminate at an equipment grounding terminal of the applicable separately derived system or service.
   a. equipment
   b. building or structure
   c. enclosure
   d. cabinet

6. Where the equipment bonding jumper is installed on the outside of a raceway, generally the length is limited to not more than ____ feet. In addition, the equipment bonding jumper must be routed with the raceway.
   a. 7
   b. 8
   c. 6
   d. 10

7. The metal water piping system is permitted to be bonded to the service equipment enclosure, the grounded conductor at the service, the grounding electrode conductor where large enough, or to one or more grounding____.
   a. clamps
   b. devices
   c. fittings
   d. electrodes

8. Other common systems and metal piping that must be bonded include metal ____.
   a. gas piping and pneumatic systems
   b. waste, drain and vent lines
   c. oxygen, air and vacuum systems
   d. all of the above

9. The equipment grounding conductor for the circuit that may energize any other metal piping systems can be used as the ____ means for other metal piping systems.
   a. equipment bonding
   b. grounded
   c. bonding
   d. identified

10. Expansion joints or telescoping sections of metal raceways must be made electrically continuous by the use of an __________ or other means.
   a. steel strap
   b. equipment bonding jumper
   c. welding cable
   d. equipment grounding conductor

11. The locknut/bushing and double-locknut types of installations are not acceptable for bonding in ____.

a. hazardous (classified) locations
b. commercial locations
c. industrial location
d. computer rooms

12. If four metal raceways have their conductors protected by overcurrent protective devices sized at 400, 300, 225 and 125 amperes and leave the bottom of an open switchboard or motor control center, the minimum size of a single equipment bonding jumper to be used to bond them together is ____.
a. 4 AWG copper or 4 AWG aluminum
b. 6 AWG copper-clad aluminum
c. 3 AWG copper or 1 AWG aluminum
d. 4 AWG copper or 3 AWG aluminum

13. The points of attachment of the bonding jumper to the metal water piping system are required to be ____.
a. acceptable
b. marked
c. accessible
d. soldered

14. Concentric and eccentric knockouts in enclosures such as wireways and panelboards:
a. are not suitable for grounding and bonding in circuits over and under 250 volts.
b. are tested by a qualified electrical contractors for their current-carrying ability.
c. are not tested by a qualified electrical testing laboratory for their current-carrying ability.
d. are not capable of carrying fault current.

15. Bonding connections to metal water piping systems:
a. are permitted to be made with solder.
b. must be accessible.
c. are permitted to be connected to the neutral in a subpanel.
d. are not required.

16. Structural metal must be bonded where:
a. it is exposed.
b. it is interconnected to form a metal building frame.
c. it is not intentionally grounded.
d. all of the above.

17. ______ metallic parts together puts the parts at the same potential
a. Bonding
b. Grounding
c. The grounding electrode
d. Earthing

18. _______the insulated washers must be removed from receptacles that do not have a contact yoke or device that is listed as being self-grounding
a. Both of
b. None of
c. At least one of
d. All of

19. The grounded conductor of a separately derived system shall_______
a. not be bonded to the exposed structural metal that is interconnected to form the building frame
b. be bonded to the exposed structural metal that is interconnected to form the building frame
c. always be connected to an isolated ground rod
d. be identified by green insulation or green markings

20. Where an equipment bonding jumper is used to connect a grounding type receptacle to a grounded metal box, it shall be sized in accordance with which of the following:
a. Table 250.66
b. Table 250.122
c. The same size as the circuit conductors
d. One size smaller than the circuit conductors

21. A listed exposed work cover shall be permitted as the grounding and bonding means for a receptacle under which of the following conditions?
a. The device is attached to the cover with either rivets or thread-locking or screw-locking means.
b. The raised cover provides a flat non-raised portion to contact the grounded metal box.
c. The receptacle is a self-grounding type.
d. Both a and b.

Equipment Grounding Conductors

Equipment grounding conductors are intended to connect equipment to ground (earth) and to provide an effective ground-fault current path to facilitate overcurrent device operation in ground-fault conditions. This is accomplished by connecting suitably sized conductors from the system or equipment grounding point to the equipment supplied. Equipment grounding conductors also prevent an objectionable potential above ground on conductors and equipment enclosures.

Objectives

To understand...

- General requirements for equipment grounding conductors on grounded and ungrounded systems
- Sizing requirements for equipment grounding conductors
- Rules applied to multiple raceways or cables
- Rules for flexible cords
- Use of building steel that is properly grounded by an equipment grounding conductor
- Grounding of equipment by the grounded circuit conductor

Chapter 9

The equipment grounding conductor or path must also:

1. be electrically continuous;
2. have ample capacity to conduct safely any currents likely to be imposed on it; and
3. be of the lowest practical impedance [see 250.4(A)(5)].

The equipment grounding conductor or path is required to extend to the grounding electrode, in a low-impedance path; and if the system is grounded, it is required to be connected through a low-impedance path to the grounded service conductor (often a neutral conductor of a system or service).

Definition

Grounding conductor, equipment (EGC). "The conductive path installed to connect the normally non-current-carrying metal parts of equipment together and to the system grounded conductor, or to the grounding electrode conductor, or both" [1] (see photo 9-1).

Clearly, it can be seen in this definition that the equipment grounding conductor (EGC) performs both grounding and bonding functions and this is reinforced by FPN No. 1 following this definition. Acceptable equipment grounding conductors are listed in Section 250.118.

Grounded Systems

The equipment grounding conductor or path is required to extend from the furthermost point on the circuit to the service equipment where it is connected to the grounded conductor and grounding electrode on a grounded system, or connected to the disconnecting means enclosure and the grounding electrode conductor for an ungrounded system. This connection is made through the main bonding jumper.

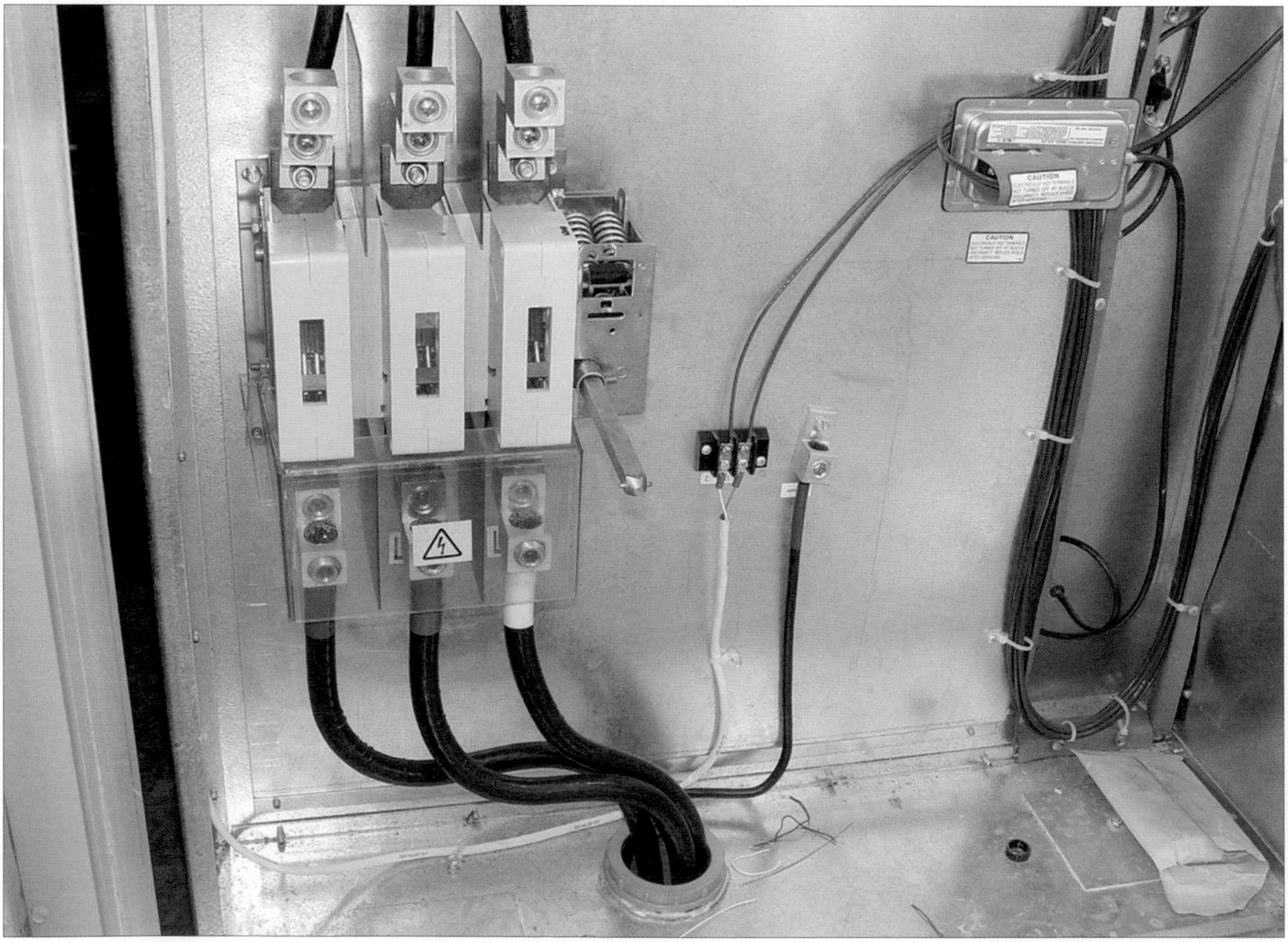

Photo 9-1. Equipment grounding conductor connected to equipment enclosure

Photo 9-2. Equipment grounding conductor is installed to ground and bond electrical equipment. Exhaust fan motor is shown and equipment grounding conductor is installed in flexible metal conduit.

Often, this equipment grounding conductor or path is the conductor enclosure (conduit, cable jacket, etc.).

For enclosed panelboards typically installed at dwelling services, the grounded and equipment grounding conductors connect to the same terminal bar (see 408.40). Where this type of equipment is installed, the main bonding jumper connects the enclosure to the grounded service conductor and to a separate equipment grounding conductor terminal bar where one is installed.

Should insulation failure occur anywhere on a phase conductor and a ground fault develop between the energized conductor and the conductor enclosure, a ground-fault circuit will be established (see figure 9-1).

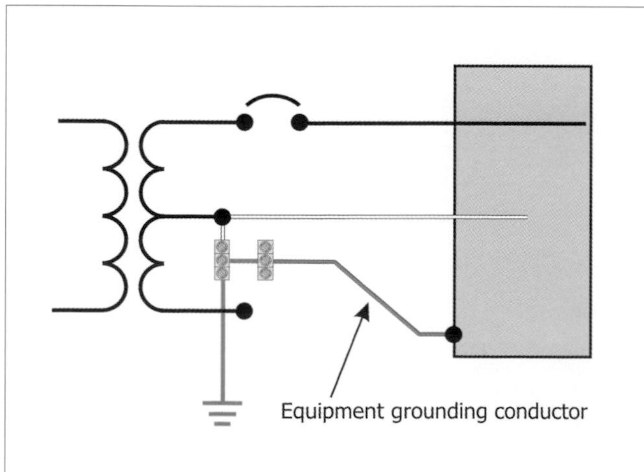

Figure 9-1. Functions of equipment grounding conductor in grounded systems

The ground-fault circuit will therefore be from the source, through the supply conductors, through the overcurrent devices, to the point of fault on the phase conductor, usually through an arc, through the equipment grounding conductor, through the main bonding jumper, to the grounded conductor (may be a neutral) and back to the source.

If this circuit is complete, of adequate capacity and low impedance, the equipment and persons who can contact it are protected because the overcurrent device will open the faulted circuit. A break in this equipment grounding circuit or other grounding system failure will expose persons to possibly lethal shocks if a ground fault from a source having sufficient potential (usually more than 50 volts) energizes the enclosure and the person provides the path for current.

Ungrounded Systems

In an ungrounded system, the equipment grounding conductor or path is required to be permanent and continuous to the grounding point to keep all equipment and conductor enclosures at or near ground potential (see figure 9-2). It also provides a fault-current path in the event a second fault occurs. The equipment grounding conductor must be the same size as called for in a grounded system if we are to have maximum safety. The minimum sizes of equipment grounding conductors given in Table 250.122 apply equally to an ungrounded system and a grounded system.

In general, it may be said that any conductor or equipment enclosure, be it conduit, electrical metallic

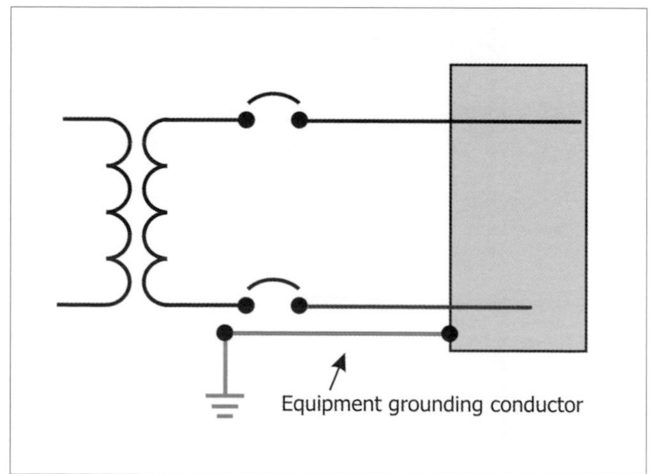

Figure 9-2. Functions of equipment grounding conductor in ungrounded systems

tubing, raceway or busway enclosure, provides a satisfactory equipment grounding conductor for an ungrounded system if all joints are made electrically continuous. It may be necessary to use equipment bonding jumpers at certain points. Such equipment bonding jumpers are also required to be sized per Table 250.122.

Equipment Grounding Conductor Material

The *Code* specifies in 250.118 the conductors that are permitted to be used for equipment grounding conductors. They are as follows:

1. A conductor of copper, aluminum, or copper-clad aluminum. The conductor shall be solid or stranded; insulated, covered, or bare; and in the form of a wire or busbar of any shape. [Some sections of the *Code* specifically require an equipment grounding conductor to be insulated or solid.]
2. Rigid metal conduit
3. Intermediate metal conduit
4. Electrical metallic tubing
5. Listed flexible metal conduit that is not listed for grounding, meeting all the following conditions:

   a. The conduit is terminated in listed fittings

   b. The circuit conductors contained in the conduit are protected by overcurrent devices rated at 20

Photo 9-3. Flexibility necessary after installation (for aiming heater)

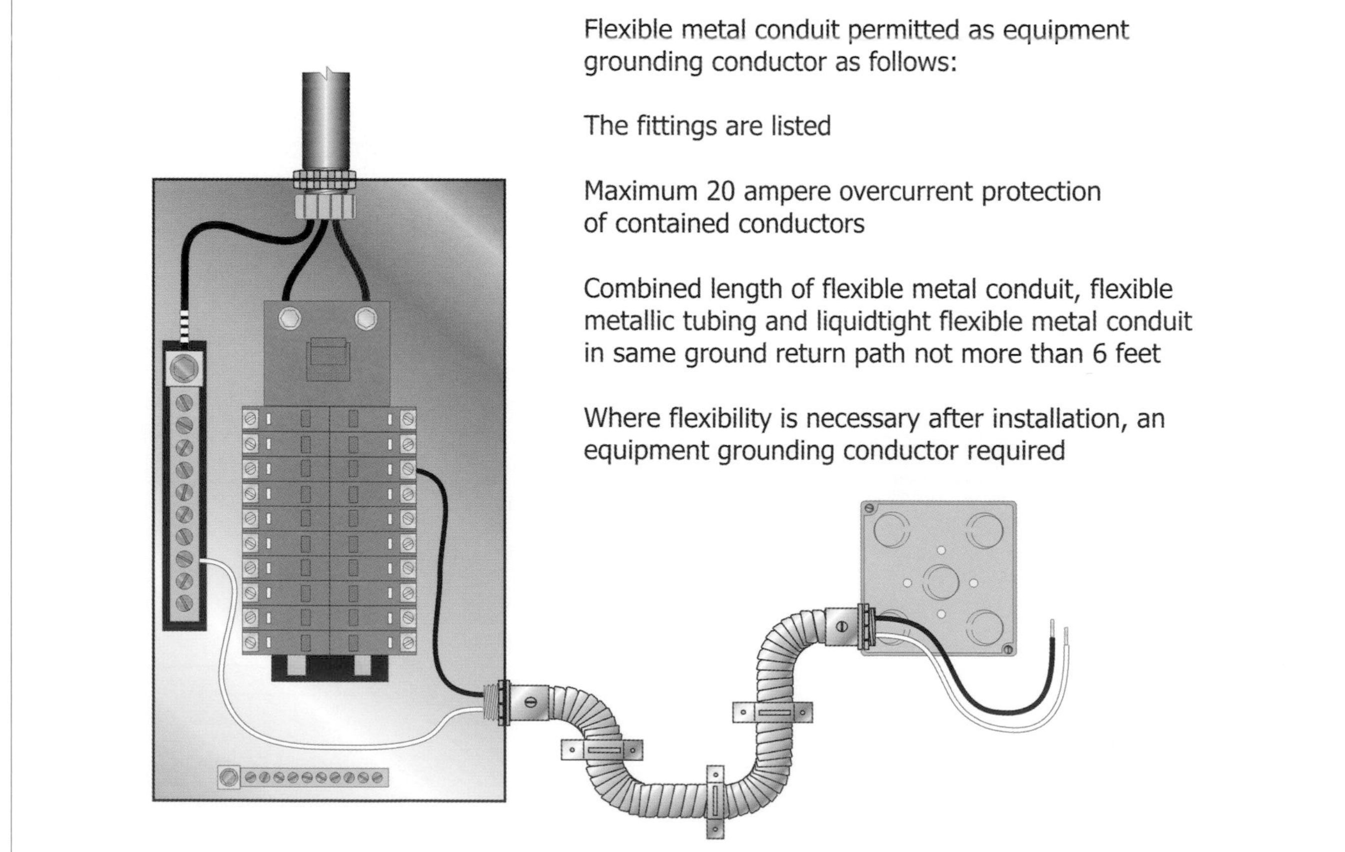
Flexible metal conduit permitted as equipment grounding conductor as follows:
The fittings are listed
Maximum 20 ampere overcurrent protection of contained conductors
Combined length of flexible metal conduit, flexible metallic tubing and liquidtight flexible metal conduit in same ground return path not more than 6 feet
Where flexibility is necessary after installation, an equipment grounding conductor required

Figure 9-3. Flexible metal conduit as equipment grounding conductor [250.118(5)]

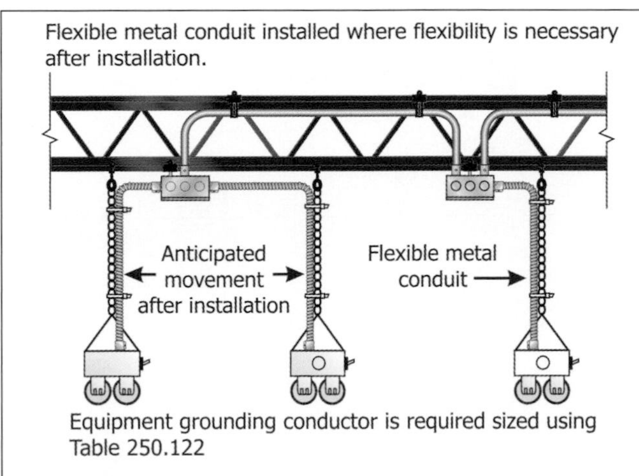

Figure 9-4. Flexibility is needed after installation so equipment grounding conductor is required

amperes or less.

c. The combined length of flexible metal conduit and flexible metallic tubing and liquidtight flexible metal conduit in the same ground return path does not exceed 1.8 m (6 ft).

d. Where used to connect equipment where flexibility is necessary after installation, an equipment grounding conductor shall be installed[2] (see figures 9-3 and 9-4).

Flexible metal conduit is commonly available as a listed product but has not been listed for grounding by a nationally recognized electrical testing laboratory. The conduit has been recognized for several years for use as an equipment grounding conductor under the limitations indicated above. Note, however, the flexible metal conduit must be terminated in fittings that are listed for grounding.

6. Listed liquidtight flexible metal conduit meeting all the following conditions:

a. The conduit is terminated in listed fittings.

b. For metric designators 12 through 16 (trade sizes 3/8 through ½), the circuit conductors contained in the conduit are protected by overcurrent devices rated at 20 amperes or less.

c. For metric designators 21 through 35 (trade sizes

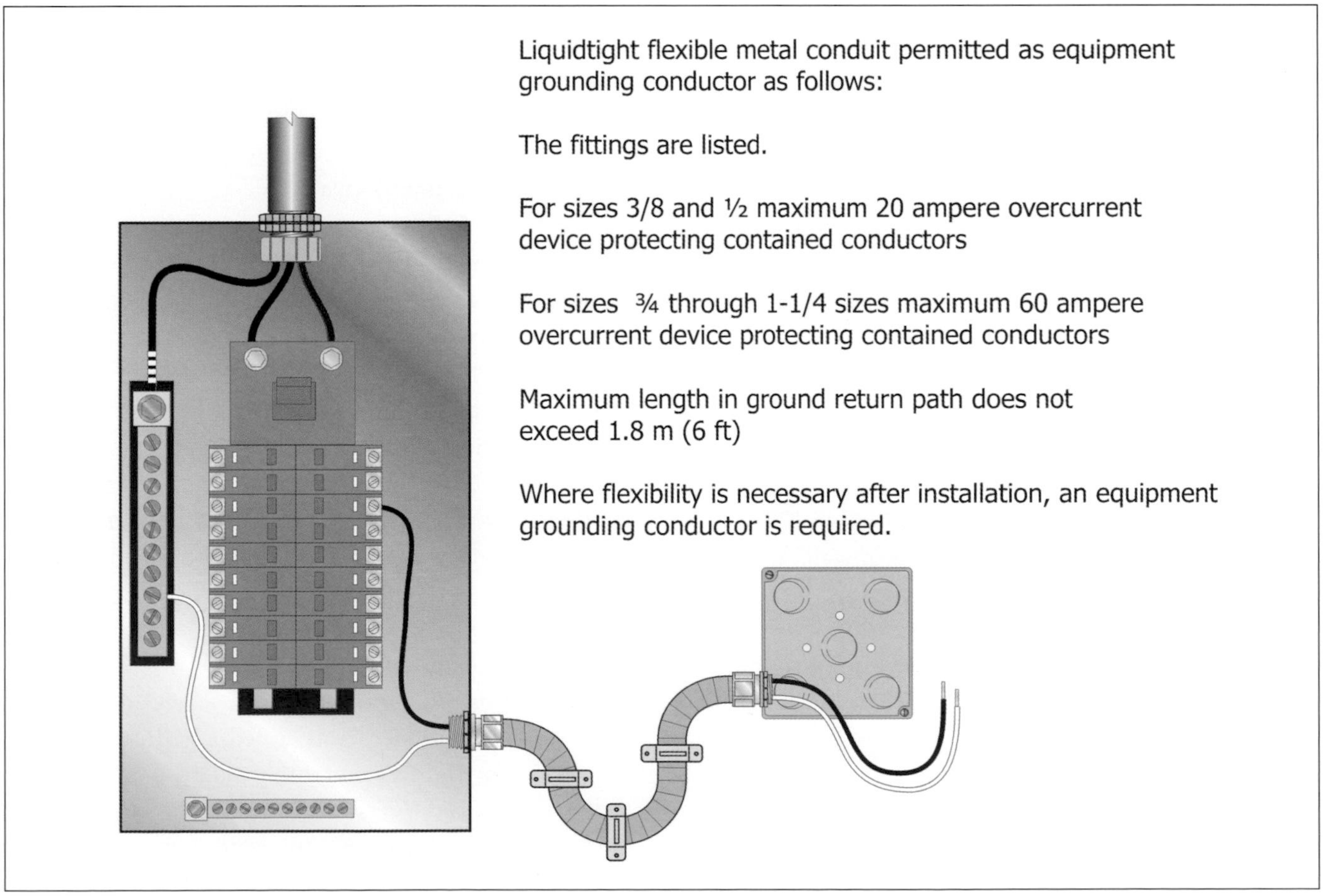

Liquidtight flexible metal conduit permitted as equipment grounding conductor as follows:

The fittings are listed.

For sizes 3/8 and ½ maximum 20 ampere overcurrent device protecting contained conductors

For sizes ¾ through 1-1/4 sizes maximum 60 ampere overcurrent device protecting contained conductors

Maximum length in ground return path does not exceed 1.8 m (6 ft)

Where flexibility is necessary after installation, an equipment grounding conductor is required.

Figure 9-5. Liquidtight flexible metal conduit as equipment grounding conductor

¾ through 1¼), the circuit conductors contained in the conduit are protected by overcurrent devices rated not more than 60 amperes and there is no flexible metal conduit, flexible metallic tubing, or liquidtight flexible metal conduit in trade sizes metric designators 12 through 16 (trade sizes 3/8 or ½) in the grounding path.

d. The combined length of flexible metal conduit and flexible metallic tubing and liquidtight flexible metal conduit in the same ground return path does not exceed 1.8 m (6 ft).

e. Where used to connect equipment where flexibility is necessary after installation, an equipment grounding conductor shall be installed.

7. Flexible metallic tubing where the tubing is terminated in listed fittings and meeting all the following conditions:

a. The circuit conductors contained in the tubing are protected by overcurrent devices rated at 20 amperes or less.

b. The combined length of flexible metal conduit and flexible metallic tubing and liquidtight flexible metal conduit in the same ground return path does not exceed 1.8 m (6 ft) (see figure 9-5).

8. Armor of Type AC cable as provided in 320.108.

9. The copper sheath of mineral-insulated, metal-sheathed cable.

10. Type MC cable where listed and identified for grounding in accordance with the following:

a. The combined metallic sheath and grounding conductor of interlocked metal tape-type MC cable.

b. The metallic sheath or the combined metallic sheath and grounding conductors of the smooth or corrugated tube type MC cable.

11. Cable trays as permitted in 392.3(C) and 392.7.

12. Cablebus framework as permitted in 370.3.

13. Other electrically continuous metal raceways and auxiliary gutters listed for grounding.[3]

Included are auxiliary gutters (not specifically a raceway as defined in Article 100 but esssentiallly the same equipment as wireways), wireways with associated fittings, busway enclosures, and in some cases an additional ground bus, surface metal raceways, and pull and junction boxes that are installed in the ground-fault path.

It should be noted here that these requirements or provisions in 250.118 are general in nature. Many

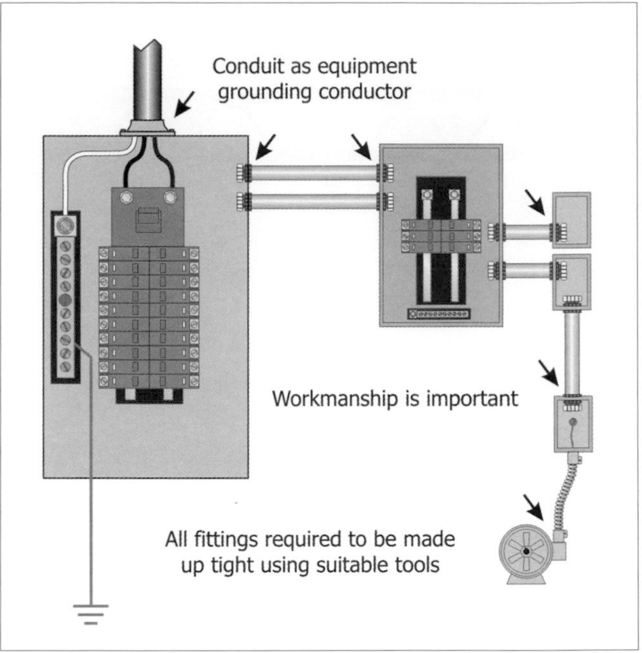

Figure 9-6. Equipment grounding conductor path back to grounding point at applicable service or separately derived system

sections of the *Code* contain specific requirements that must be complied with. A few examples follow. Section 501.30 does not recognize the standard double locknut-type conduit connections for Class I hazardous (classified) locations.

Section 517.13(B) requires an insulated equipment grounding conductor installed in a metal raceway or flexible cable assembly that qualifies as an equipment grounding conductor in patient care areas of health care facilities. Section 550.33(A) generally requires the equipment grounding conductor for the feeder to a mobile home to be insulated. Several sections of Article 680 require an insulated equipment grounding conductor. It is always best to carefully examine the specific requirements for the equipment grounding conductor for the type of installation being made.

Conductor Enclosures

Conduit runs of rigid or intermediate metal that are properly threaded and in which the couplings are made up tightly, preferably using a joint sealer that will not reduce continuity, can be expected to perform satisfactorily as an equipment grounding conductor for runs of limited length. Listed electrically, corrosion-resistant compounds aid in assuring an effective equipment grounding path because they act as lubricants and permit the joint to be screwed up tighter. Under poor conditions, the conduit impedance with couplings should not show an increase of over 50 percent when compared with a straight run of conduit. The use of this higher impedance value would provide a factor of safety. In the case cited, there is no economic justification for using an additional equipment grounding conductor.

The *Code* further requires that where conduit is used as an equipment grounding conductor, all joints and fittings shall be made up tight using suitable tools [see 250.120(A) and figure 9-6]. This calls attention to the fact that conduit, where used as an equipment grounding means, is a current-carrying conductor under fault conditions and is required to be made electrically continuous by having joints made up tight.

Usually, large and often parallel conduits are installed from the utility transformer to the service equipment. Then, smaller and smaller conduits are installed for feeders and branch circuits. For instance, at one point the equipment grounding path may be three 4-inch conduits in parallel; at another point, two 4-inch conduits in parallel while down the line it may be only one 1¼-inch conduit, all being connected together to form a permanent and continuous path. As the circuit changes from large overcurrent protective means to smaller ones,

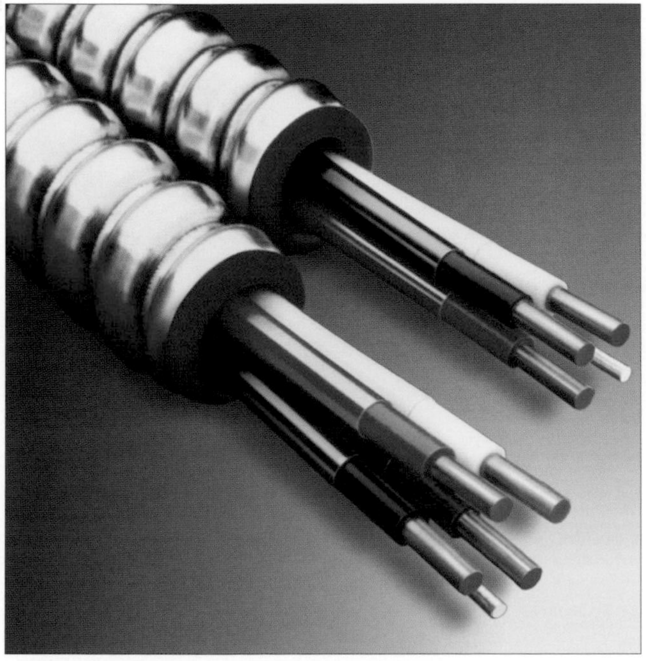

Photo 9-4. Standard AC cable where armor is the only equipment grounding conductor path [250.118(8)]. Courtesy of AFC Cable Systems.

Photo 9-5. Installation of AC cable that is acceptable for use in *patient care areas* because it provides two equipment grounding conductor paths (sheath and conductor)

the conductivity of the equipment grounding path becomes lower. The conduit or tubing at the end of the circuit may be no larger than ½-inch electrical metallic tubing or 3/8-inch flexible metal conduit.

The *NEC* does not dictate any particular size of conduit or tubing to serve as the equipment grounding conductor for an upstream overcurrent device, other than as mentioned in the previous section. It is expected that a metallic raceway that is sized properly for the conductor fill will provide an adequate equipment ground-fault return path. The fine print note to Section 250.120(A) provides an important reference to the UL Guide Information (FHIT) for equipment grounding conductors (wire types) that are part of an electrical circuit protective system or fire-rated cable listed to maintain circuit integrity for a duration of time under fire conditions.

Cables as Equipment Grounding Conductors

Several cables used as wiring methods are suitable for use as an equipment grounding conductor or contain an equipment grounding conductor. These include:

Type AC Cable

Armored cable (Article 320) is manufactured with conductors in sizes from 14 AWG through 1 AWG

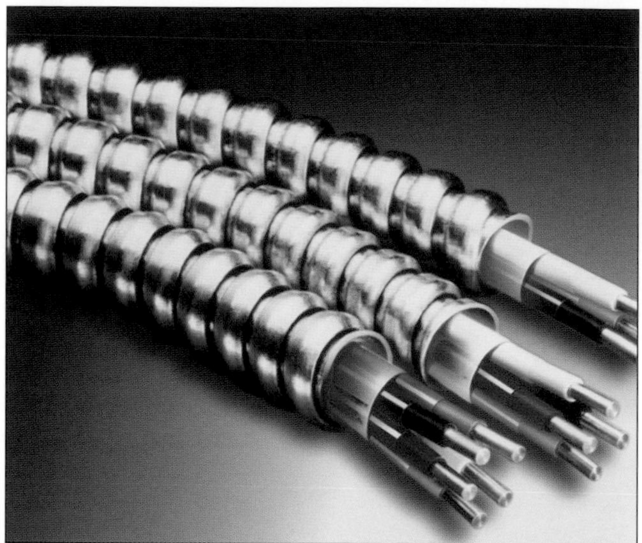

Photo 9-6. Metal Clad cable (interlocking metal tape-type shown). Courtesy of AFC Cable Systems.

copper and from 12 AWG through 1 AWG aluminum (see photo 9-4). Type AC cable is required to have an armor of flexible metal tape.

The insulated conductors are required to be in accordance with 320.104. Cables of the AC type are required to have an internal bonding strip of copper or aluminum in intimate contact with the armor for its entire length. It is suitable as an equipment grounding conductor in accordance with 250.118(8). Additionally, AC cable conductors are required to have an overall moisture-resistant and fire-retardant fibrous (paper) covering. Another type of AC cable construction includes an insulated equipment grounding conductor and is acceptable for use as the branch circuits serving patient care areas as provided in 517.13 and for use in branch circuits for isolated grounding receptacles as permitted in 250.146(D) and 408.40 Exception (see photo 9-5).

Type MC Cable

Type MC cable is covered in Article 330 (see photo 9-6). Type MC cable is produced in three configurations: spiral interlocking metal tape, corrugated metal tube, and a smooth metal tube.

1. The spiral-interlocking-metal-tape Type MC cable must always have an equipment grounding conductor which may be insulated or bare. The jacket itself is not suitable as an equipment grounding conductor. The principal equipment grounding conductor may be divided (sectioned) into more than one conductor, often to facilitate spacing in the cable construction for larger sizes. Additional equipment grounding conductors have green insulation and either a yellow stripe or other identification.

2. The sheath of the smooth or corrugated tube

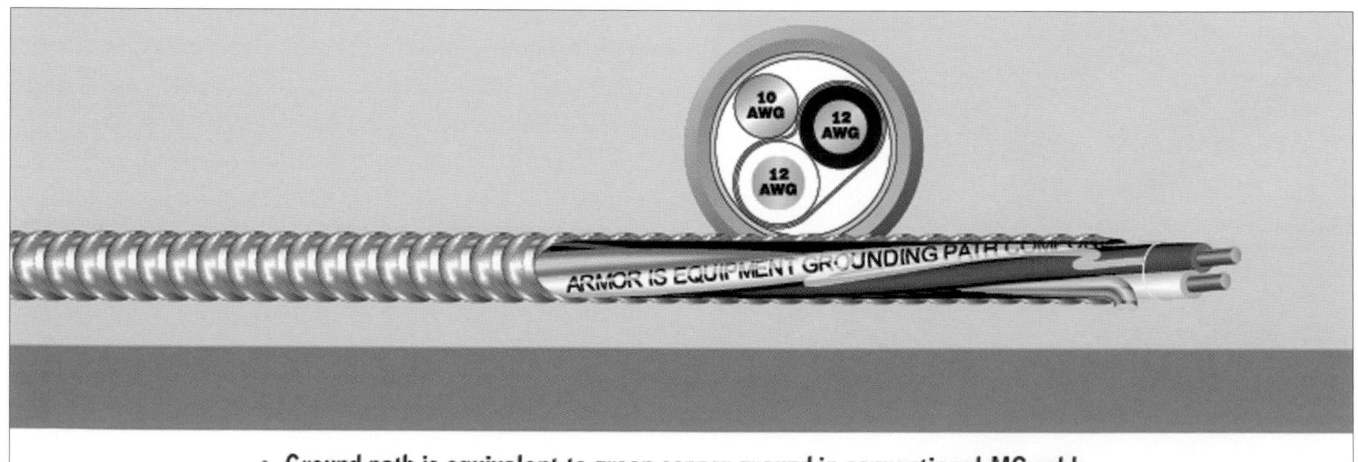

10 AWG
12 AWG
12 AWG
• Ground path is equivalent to green copper ground in conventional MC cable
• Armor and AL ground are in direct contact throughout entire cable length
• Ground increased based on size of circuit conductor
• Fittings bond the cable to the box
• Less stress on electrical connections

Photo 9-7. MC cable that provides an equipment grounding conductor (wire-type) in the assembly, and the sheath is suitable as an equipment grounding conductor. Courtesy of Southwire Company

Type MC cable or a combination of the sheath and a supplemental bare or green insulated conductor is suitable for the required equipment grounding conductor. The principal equipment grounding conductor may be divided (sectioned) into more than one conductor, often to facilitate spacing in the cable construction for larger sizes. Additional equipment grounding conductors have green insulation and either a yellow stripe or other identification.

A specific type of metal-clad cable is manufactured that includes an equipment grounding conductor in the cable and has an armor that is also recognized as an equipment grounding conductor.

This MC cable is listed under UL standard 1569. As with any cable assembly, installation in accordance with the manufacturer's instructions is required to meet the requirements of *NEC* 110.3(B). This type of MC cable that provides two equipment grounding conductor paths lends itself as suitable for use when installing isolated grounding circuits for sensitive electronic equipment as well as branch circuits serving patient care areas in health care facilities (see photo 9-7).

Nonmetallic-Sheathed Cable

NM cable is covered in Article 334. This cable is permitted to be produced in three styles: Type NM, Type NMC and Type NMS. The power conductors are permitted to be in sizes 14 AWG through 2 AWG copper and 12 AWG through 2 AWG aluminum. Type NMS is permitted to contain signaling conductors and typically contains an equipment grounding conductor sized in compliance with Table 250.122.

Service-Entrance Cable

Service entrance cable (Type SE) is covered by Article 338. Type SE cable is produced in a variety of configurations. The type most commonly used for internal wiring is Type SE style U and Type SE style R. Specific rules for Type SE cables are contained in 338.10.

Type SE cables are permitted in interior wiring systems where all of the circuit conductors of the cable are of the rubber-covered or thermoplastic type.

Type SE cables without individual insulation on the grounded circuit conductor are not to be used as a branch circuit or as a feeder within a building, except a cable that has a final nonmetallic outer covering and is supplied by alternating current at not over 150 volts to ground is permitted as a feeder to supply only other buildings on the same premises. Type SE cables are permitted for use where the fully insulated conductors are used for circuit wiring and the uninsulated conductor is used for equipment grounding purposes.

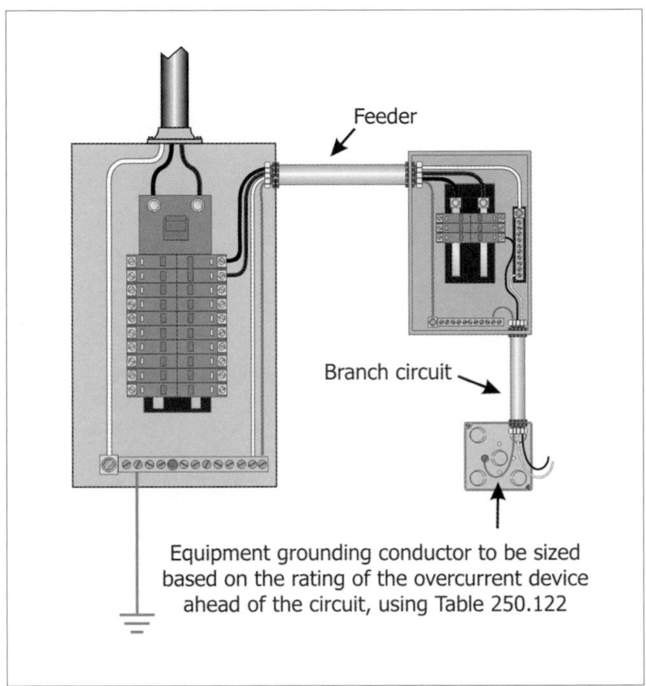
Feeder
Branch circuit
Equipment grounding conductor to be sized based on the rating of the overcurrent device ahead of the circuit, using Table 250.122

Figure 9-7. Minimum size of equipment grounding conductor

Underground Feeder and Branch-Circuit Cable

Underground feeder and branch-circuit cable is covered by Article 340. Type UF cable is permitted to be produced in sizes 14 AWG copper or 12 AWG aluminum through 4/0 AWG. Multiconductor cables are permitted to be installed in accordance with Article 340. In addition to the insulated conductors, the cable is permitted to have an insulated or bare conductor for equipment grounding purposes only. As such, it is required to comply with Table 250.122.

Size of Equipment Grounding Conductor

The entire equipment grounding conductor or path of any raceway system will be as shown in figure 9-7. Starting at the service, we have a large overcurrent protective device that is in series with other, and usually smaller, feeder or branch overcurrent protection devices. The ungrounded (phase or hot) conductor usually

Table 250.122 Minimum Size Equipment Grounding Conductors for Grounding Raceway and Equipment (in part)

Rating or Setting of Automatic Overcurrent Device in Circuit Ahead of Equipment, Conduit, etc., Not Exceeding (Amperes)	Size (AWG or kcmil)	
	Copper	Aluminum or Copper-Clad Aluminum
15	14	12
20	12	10
30	10	8
40	10	8
60	10	8
100	8	6
200	6	4
300	4	2
400	3	1

Note: Where necessary to comply with 250.4(A)(5) or 250.4(B)(4), the equipment grounding conductor shall be sized larger than given in this table.

Reproduction of Table 250.122

decreases in size as it progresses through smaller and smaller overcurrent devices.

Section 250.122(A) provides the general rules for sizing the equipment grounding conductor. It refers to Table 250.122 for determining the minimum size of conductor that is permitted to be used as an equipment grounding conductor. The size is based on the ampere rating of the overcurrent protective device ahead of the conductor. [Table 250.122 is reprinted as table 6 in chapter twenty.]

For example, if the overcurrent protection ahead of the circuit or feeder is 225 amperes, the minimum size equipment grounding conductor is found as follows:

In Table 250.122, follow the first column, which gives the rating of the overcurrent device, down to find the rating that equals or exceeds 225 amperes. Since 225 amperes is not found, go to the next larger size, which is 300 amperes. Follow that line across to find the minimum size copper wire to be 4 AWG and for aluminum, a 2 AWG minimum size conductor.

Follow a similar process to determine the minimum size conductor for any installation. In addition, the note below Table 250.122 requires that, "Where necessary to comply with 250.4(A)(5) or 250.4(B)(4), the equipment grounding conductor shall be sized larger than given in this table." [4] Notes that are part of tables in the *NEC* are mandatory. A comprehensive analysis of the withstand rating of these equipment grounding conductors can be found in chapter eleven.

Specific requirements are provided for: Equipment grounding conductors that are increased in size for any reason, as provided in 250.122(B); for multiple circuits in 250.122(C); for motor circuits in 250.122(D); for flexible cord and fixture wire in 250.122(E); and for conductors in parallel in 250.122(F).

Increasing the Size of Equipment Grounding Conductor

Section 250.122(B) requires that, "Where conductors are increased in size," (for example to compensate for voltage drop or for any other reason), "equipment grounding conductors, where installed, shall be

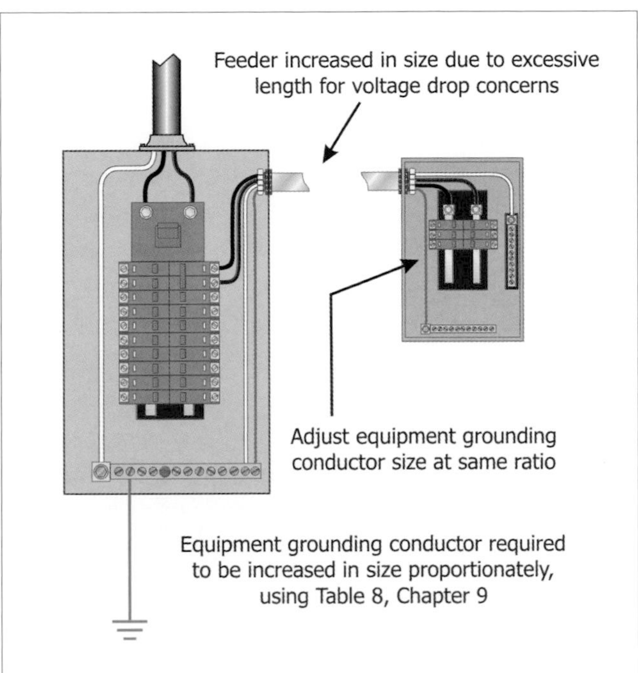

Figure 9-8. Increasing the size of the equipment grounding conductor for long circuits or feeders

increased in size proportionately according to circular mil area of the ungrounded conductors." [5] This means that where a feeder or branch-circuit conductor is increased in size, the equipment grounding conductor, where run, is required to be increased at not less than the same ratio the feeder or circuit conductors are increased. For example, a 200-ampere feeder is to be installed. It is determined that the voltage drop would be excessive. A 250-kcmil conductor is selected for the feeder rather than installing the 3/0 copper conductor as is permitted by Table 310.16. Table 250.122 requires a 6 AWG equipment grounding conductor for the 200-ampere overcurrent device.

Determine the minimum size equipment grounding conductor required for the feeder by the following formula: (Use Table 8 of *NEC* chapter 9 to determine the area in circular mils where the conductor size is given by a non-circular mil designation).

Selected Feeder Conductor Area ÷ Required Feeder Conductor Area = Ratio.

Table 250.122 Equipment Grounding Conductor X Ratio = Required EGC

250,000 kcmil ÷ 167, 800 kcmil = 1.49.

26240 (Circular mil area of 6 AWG) x 1.49 = 39098 circular mils

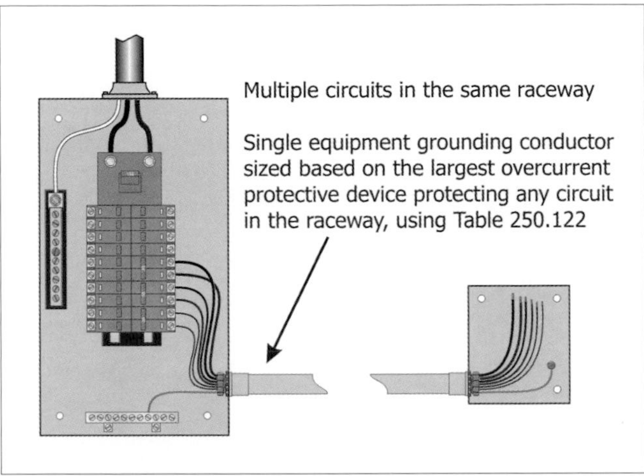

Multiple circuits in the same raceway
Single equipment grounding conductor sized based on the largest overcurrent protective device protecting any circuit in the raceway, using Table 250.122

Figure 9-9. Sizing equipment grounding conductor where multiple circuits are installed in the same raceway

Next larger size = 4 AWG copper required equipment grounding conductor

Equipment Grounding Conductors for Multiple Circuits

The *Code* permits a single equipment grounding conductor to serve several circuits that are in the same raceway or cable. To use this concept, the equipment grounding conductor is required to be sized for the rating of the largest overcurrent device of the group.

For example, a conduit contains multiple branch-circuit conductors that have overcurrent protection rated: 20-amperes, 30-amperes, 50-amperes and 60-amperes. A single 10 AWG equipment grounding conductor is permitted to serve all the branch circuits in the raceway. The minimum size is determined from Table 250.122 based on the rating of the 60-ampere overcurrent device (see figure 9-9).

Equipment Grounding Conductors for Motor Circuits

The general rule for sizing the equipment grounding conductor for motor circuits is contained in 250.122 (see figure 9-10). Determine the minimum size conductor from Table 250.122 based on the rating of the overcurrent protective device. In some cases, this could result in an equipment grounding conductor that is the same size as the branch-circuit conductors. This is illustrated as follows: a 30-hp, 460-volt motor is being installed. From Table 430.250, the full-load amperes of the motor is 40 amperes. The minimum size branch-circuit conductors can be determined from Table 310.16 by calculating 40

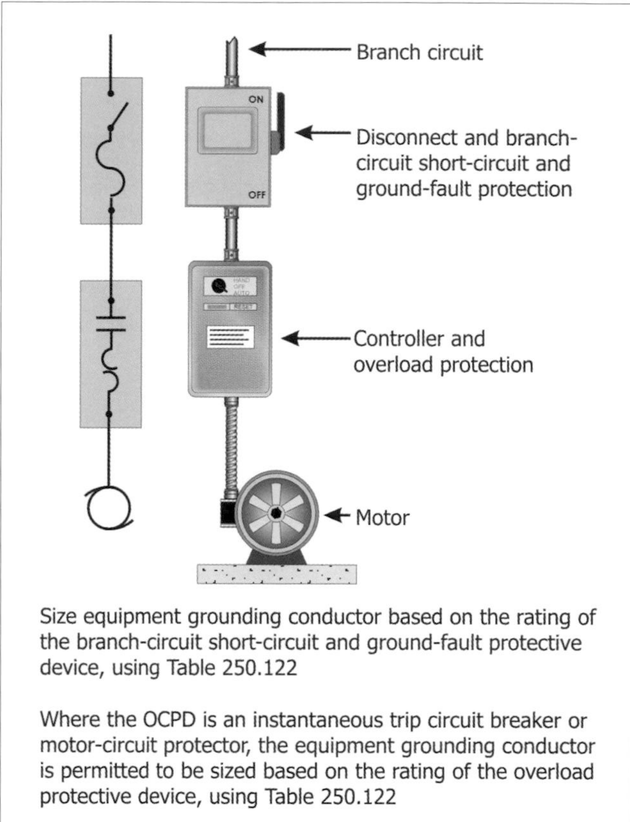

Figure 9-10. Sizing equipment grounding conductors for motor circuits

amperes x 1.25 = 50 amperes, which is 8 AWG copper conductors (75°C insulation and terminations). Maximum rating of overcurrent device of a circuit breaker type is 250 percent of the motor full-load amperes = 40 A x 250% = 100 amperes (Table 430.52), unless one of the exceptions to 430.52(C) applies. From Table 250.122, the minimum size of equipment grounding conductor based on a 100-ampere overcurrent device is 8 AWG copper, which is the same size as the branch-circuit conductors. Note that 250.122(A) provides that the size of the equipment grounding conductor is not required to be larger than the branch-circuit conductors.

If the overcurrent device for the motor consists of an instantaneous-trip circuit breaker (rather than a more standard inverse-time circuit breaker) or a motor short-circuit protector, the equipment grounding conductor size is permitted to be based on the rating of the motor overload protective device but not less than Table 250.122. Note that the instantaneous-trip circuit breaker is permitted to be used only if it is a part of a listed combination motor controller having coordinated motor overload protection.

Using the above example, the instantaneous-trip circuit breaker that serves as the branch-circuit, short-circuit and ground-fault protective device is permitted to be up to 1300 percent of the motor full-load current (up to 1700 percent for Design B energy efficient motors). The minimum size of branch-circuit conductors is determined as 40 amperes x 1.25 = 50 amperes. The minimum conductor from Table 310.16 is an 8 AWG copper conductor with 75°C insulation and terminations. The maximum rating of an overcurrent device of an instantaneous circuit breaker type is 1300 percent of the motor full-load amperes = 40 x 13 = 520 amperes, unless one of the exceptions following 430.52(C)(3) applies.

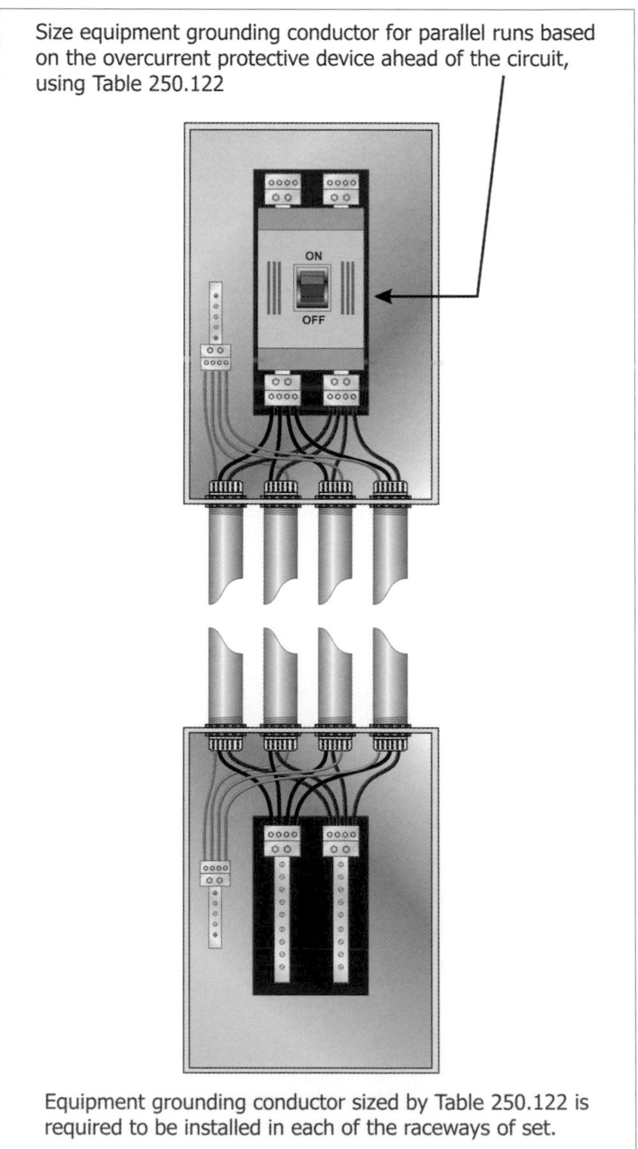

Figure 9-11. Equipment grounding conductors for feeders installed as parallel conductors per 310.4

Photo 9-8. Equipment grounding conductors in each raceway of parallel feeder

Size equipment grounding conductor for parallel runs based on the overcurrent protective device ahead of the circuit, using Table 250.122

ON

OFF

Cables installed in parallel

Equipment grounding conductor sized by Table 250.122 is required to be installed in each of the cables of the set.

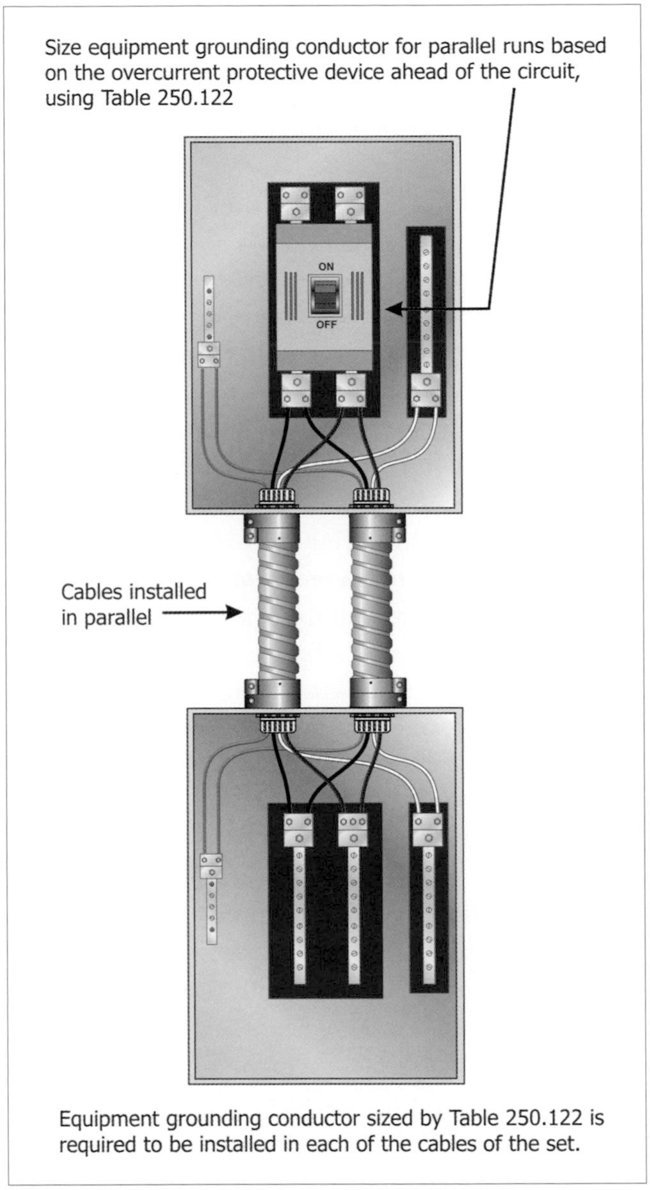

Figure 9-12. Sizing equipment grounding conductors in parallel circuits

However, assume the motor FLA on the nameplate is 38.2 amperes and the running overload protection for the motor is set at 115 percent [see 430.32(A)]. The equipment grounding conductor is permitted to be based on 43.93 amperes (38.2 x 1.15). From Table 250.122, the minimum size equipment grounding conductor based on a 50-ampere overcurrent device is 10 AWG copper.

Equipment Grounding Conductors for Flexible Cord and Fixture Wire

The use of an equipment grounding conductor in a supply cord is permitted providing the cord is used as specified in 400.7. The method of grounding non-current-carrying metal parts of portable equipment may be by means of the equipment grounding conductors in the flexible cord supplying such equipment. The proper type attachment plug is required to be used to terminate the conductors, and the attachment plug must have provision to make contact with a grounding terminal in the receptacle.

For the grounding of portable or pendant equipment, the conductors of which are protected by fuses or circuit breakers rated or set at not exceeding 20 amperes, 240.5 permits the use of an 18 AWG copper wire as an equipment grounding circuit conductor, provided the 18 AWG grounding conductor is a part of a listed flexible cord assembly [see 250.122(E)].

Equipment Grounding Conductors in Parallel

Special rules apply where more than one raceway or cable is installed with parallel conductors and an equipment grounding conductor is installed in the raceway. (Parallel conductors consist of two or more conductors that comply with 310.4 and are connected together at each end to form a single conducting path.) In this case, 250.122(F) requires that an equipment grounding conductor be installed in each raceway or cable. Generally, each equipment grounding conductor is required to be sized in compliance with the ampere rating of the overcurrent device protecting the conductors in the raceway or cable (see photo 9-8).

Section 310.4 permits equipment grounding conductors, smaller than 1/0 AWG, to be sized in compliance with Table 250.122. However, all other requirements for installing conductors in parallel must be met. These rules require that each set: (1) be the same length; (2) be of the same conductor material (all copper or all aluminum); (3) be the same size in circular mil area; (4) have the same insulation type; (5) be terminated in the same manner; and (6) the raceways or cables must have the same physical properties. However, the sets of conductors are not required to be identical (see figure 9-11).

One reason for this requirement for installing equipment grounding conductors in parallel is shown in figure 9-10. In the event of a line-to-ground fault in the equipment supplied by the circuit, the fault current should divide equally between the equipment grounding conductors. However, if a line-to-ground fault occurs in the raceway or cable, current will be fed to the fault from both directions. The equipment grounding conductor will thus be called upon to carry the entire amount of fault current until the overcurrent protective device ahead of the fault opens.

Equipment Grounding Conductors in Cables in Parallel

In some cases, where cables are installed in parallel, special constructions will be required to comply with *Code* rules on the equipment grounding conductor in the cable. Listed cables are generally produced with the equipment grounding conductors sized in compliance with a construction standard that complies with Table 250.122. For example, a copper cable construction suitable for a 300-ampere overcurrent device will have a 4-AWG copper equipment grounding conductor placed within the cable by the manufacturer. If two of these cables are installed in parallel and connected to a 600-ampere overcurrent protective device, a 1-AWG copper equipment grounding conductor would be required in each cable to comply with Table 250.122. These "special" cables can be ordered from the manufacturer although conditions such as minimum length requirements may apply. In addition, a significant amount of time may be required to produce these special cables (see figure 9-12).

Auxiliary Grounding Electrode

Engineers often specify that ground rods be installed to ground metal lighting standards or poles and at

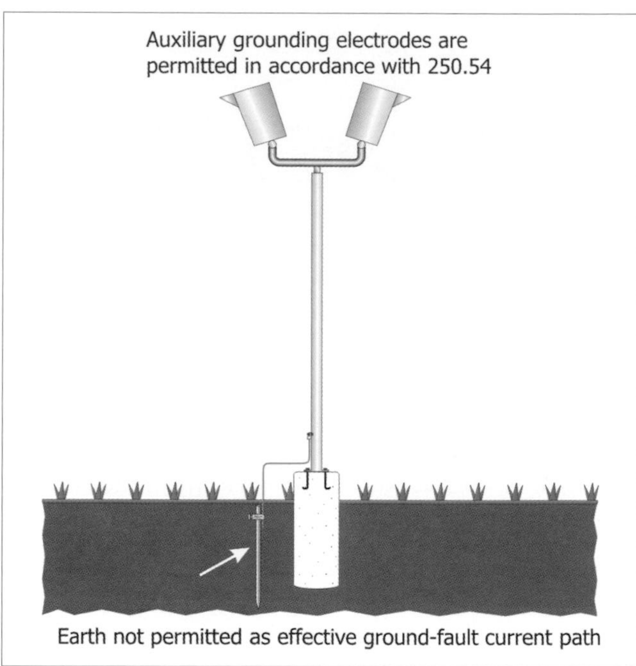

Figure 9-13. Auxiliary grounding electrodes are permitted (light pole is a common example)

metal poles for electric signs (see figure 9-13). Some manufacturers of computer-controlled machine tools specify that a ground rod be used to ground their equipment (see figure 9-14).

These rods are permitted to be used but are required to be considered auxiliary grounding electrodes. They can supplement the equipment grounding conductor that is run with the branch circuit but cannot be the only means of grounding this or similar equipment (see 250.54). To use these ground rods as the only means of grounding would constitute an earth return, which is unsafe and prohibited by *Code*. The concept of the earth being used for a circuit conductor should never be considered. The *NEC* strictly prohibits this in multiple sections. The earth is a poor conductor.

Equipment Grounding Conductor with Circuit Conductors

A very important requirement for installing equipment grounding conductors is contained in 250.134(B). This requirement is that the equipment grounding conductor is generally required to be installed in the same raceway, cable or cord, or otherwise be run with the circuit conductors. This requirement is repeated in 300.3(B) where, in addition to the requirement for raceways, equipment grounding conductors are

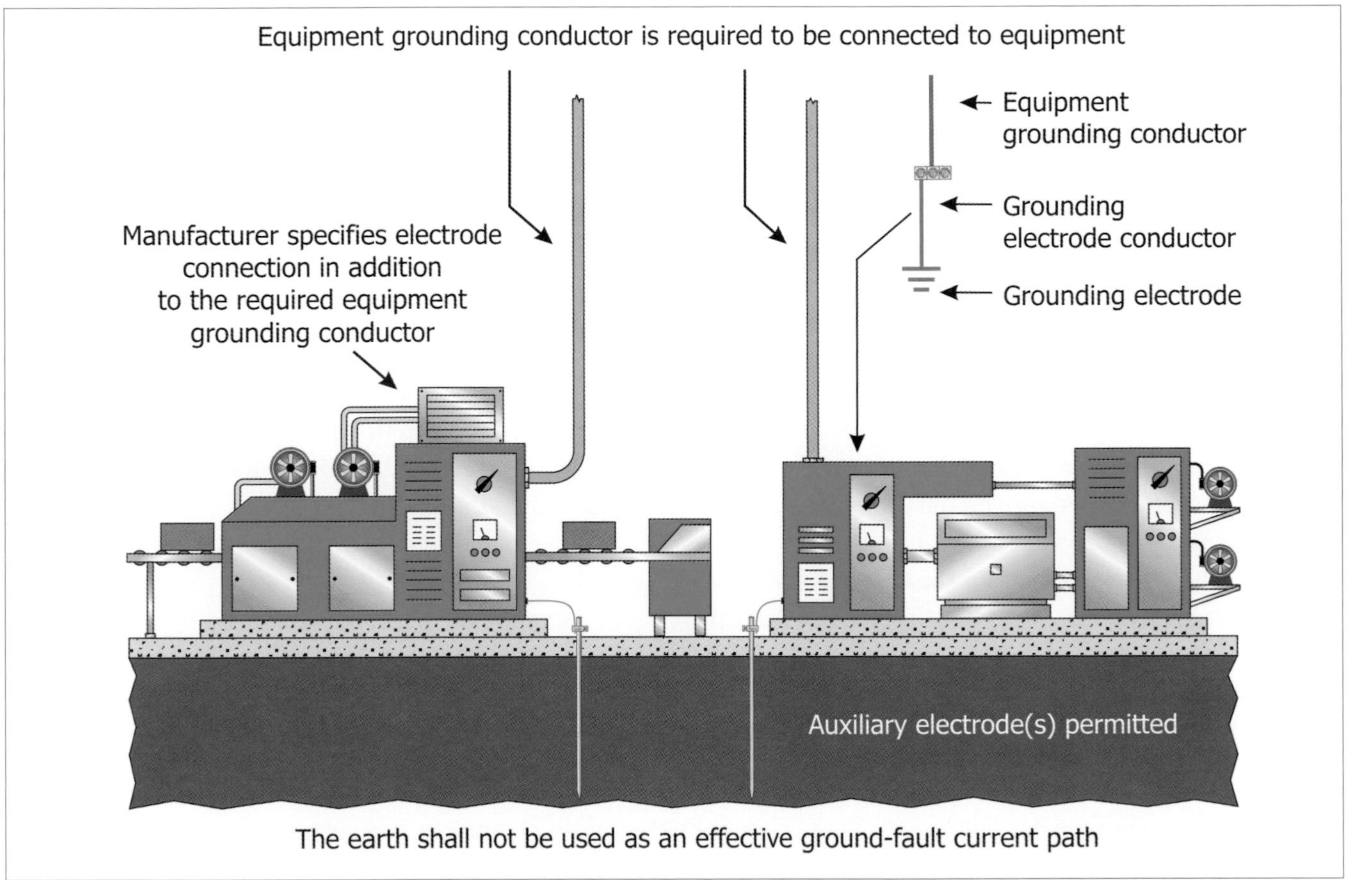

Figure 9-14. Auxiliary grounding electrodes permitted but must meet the rule of the *Code* where installed. The earth shall not be used as an effective ground-fault current path

required to be contained in the same trench with other circuit conductors. This requirement is critical for the installation of alternating-current systems.

It has been proven that separating the equipment grounding conductor from the circuit conductors greatly increases the impedance of the circuit. Separation of these conductors will increase the inductive reactance of an ac circuit, which in turn increases grounding circuit conductor impedance values. The impedance of the equipment grounding conductor of a circuit should be kept as low as practicable.

This excessive separation can render an adequately sized equipment grounding conductor ineffective in carrying enough current to operate the circuit protective device and clear the faulted equipment (see figure 9-15). In this case, providing a properly sized equipment grounding conductor, but installing it improperly, results in an ineffective and possibly unsafe installation (see chapter eleven for additional information on this subject).

Nonmetallic Raceway

Where the wiring method or means is nonmetallic, it is necessary to run an equipment grounding conductor along with the circuit conductors. Do not separate them at any point in the circuit by any metallic material regardless of whether the metallic material is magnetic or not. It is true that if the material is nonmagnetic, the increase in impedance of the circuit will not be as great as if the material was magnetic. In any case, such separation is to be avoided.

Use of Building Steel for Grounding

Section 250.136(A) permits a metal rack or structure to ground electric equipment that is secured to it and in electrical contact, provided the support means is grounded by an equipment grounding conductor as specified by 250.134. However, the structural metal frame of a building is not permitted to serve as an equipment grounding conductor to ground equipment.

That is due to the uncertain path that ground-fault current must take in an effort to clear a fault (see figure 9-15).

This section emphasizes the requirement in 250.134(B) and 300.3(B) that the equipment grounding conductor must be in the same raceway, cable or cord, or otherwise be run with the circuit conductors. Again, this is so the grounding circuit impedance will be as low as possible to allow adequate ground-fault current so the circuit protective device will clear the fault. Separating the equipment grounding conductors from the circuit conductors increases the inductive reactance of the circuit in ground-fault conditions and thus increases the impedance on the grounding circuit, which is required to be kept at a minimum.

In the same manner, and on the same basis, metal car frames, supported by metal hoisting cables attached to or running over sheaves or drums of elevator machines, are considered grounded when the machine is grounded as required by the *Code* [see 250.136(B)].

Grounding for Direct-Current Circuits

All of the previous text applies to the grounding of alternating-current systems where reactance of the circuit plays a large part in the impedance of the ground-return path. In the case of direct-current circuits the concern is with ohmic resistance only. Owing to that fact, the current-carrying capacity of the grounding conductor for a direct-current supply system is required to be equal to that of the largest conductor of the system. However, if the grounded circuit conductor is a neutral derived from a balancer winding or a balancer set which has overcurrent protection as required under 445.12(D), then the grounding conductor size shall be not less than the size of the neutral.

The requirement for overcurrent protective devices in 445.12(D) states that the two-wire direct-current generators used in conjunction with balancer sets shall be equipped with overcurrent protective devices that will disconnect the 3-wire system in the case of excessive unbalancing of voltages or currents.

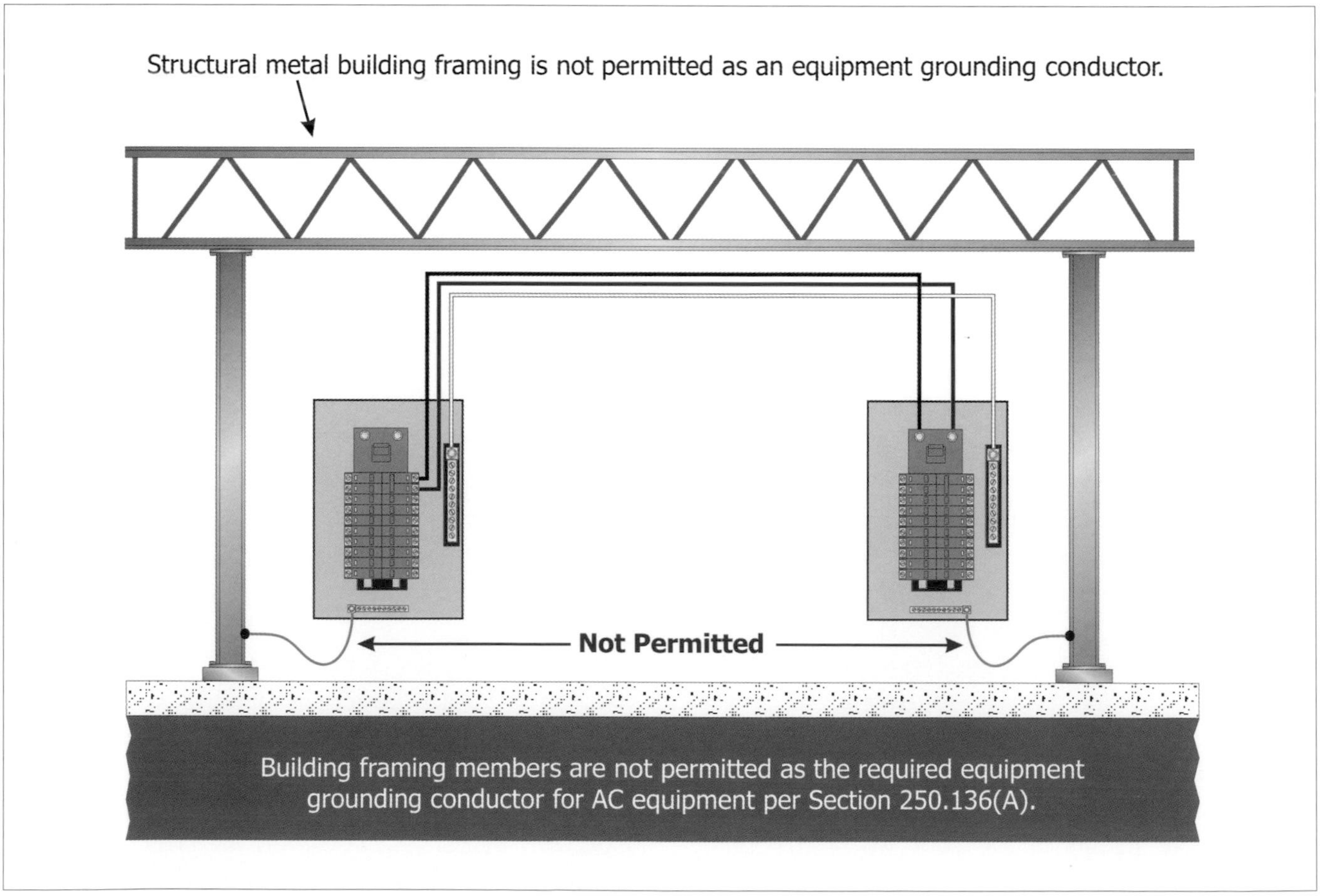

Figure 9-15. Structural metal building framing is not permitted as an equipment grounding conductor

Grounding Conductors for Direct-Current Circuits

For direct-current circuits, the size of the grounding conductor is specified in 250.166. The size can be larger than would be required for the same size alternating-current circuit. That is because resistance is the only factor in determining current in a direct-current circuit. Therefore, fault-currents are larger in a dc circuit. In an alternating-current circuit, impedance becomes the important factor for not only resistance, but reactance also must be taken into account.

Long Term Reliability of Metal Raceways

In the above discussions, it is assumed that a conductor enclosure (conduit or other raceway) has been properly installed with good tight joints that will provide a permanent and continuous electrical circuit when it is first installed. However, time and corrosion will affect the continuity of the conduit (see photo 9-9).

The safety of an electrical system will therefore depend on how long we can expect the equipment grounding conductor of the conduit to remain permanent and continuous. The answer will vary depending on the type of metal raceway, the environment it is installed in, and the quality of the installation (see table 9-1).

For design purposes, two categories can be created:

1. Where little corrosion will exist and where it can be reasonably expected that the equipment grounding conductor, in the form of a metal raceway, will remain permanent and continuous for a period of fifty years or more.

2. Where corrosion in varying degrees will exist and where the permanency of the equipment grounding conductor provided by the metal raceway can be questioned.

Photo 9-9. Metal raceway that has been severely damaged due to corrosion

In Concrete:	Required	Optional
Rigid Steel		X
Intermediate Steel		X
Aluminum Rigid	X	
Steel EMT	Below grade may be needed	On or above grade
Aluminum EMT	X	
In Soil:		
Rigid Steel		X
Intermediate Steel		X
Aluminum Rigid	X	
Steel EMT	Generally Required	
Aluminum EMT	X	

Table 9-1. Corrosion protection for metal raceway is required.

Most commercial and residential buildings are in the first category. That being the case, conductor enclosures, which are approved for the purpose can be used as part of the equipment grounding conductor (with the use of bonding jumpers where required).

Some industrial and most areas of petrochemical plants are in the second category where a metallic equipment grounding conductor, sized per Table 250.122, is usually specified to be run in parallel with and within the conductor enclosure so as to ensure continuity if the conduit circuit is broken owing to eventual corrosion (see table 9-1).

Some electrical design engineers and local elect-rical inspection agencies require that an equipment grounding conductor be installed in each metal conduit and tubing

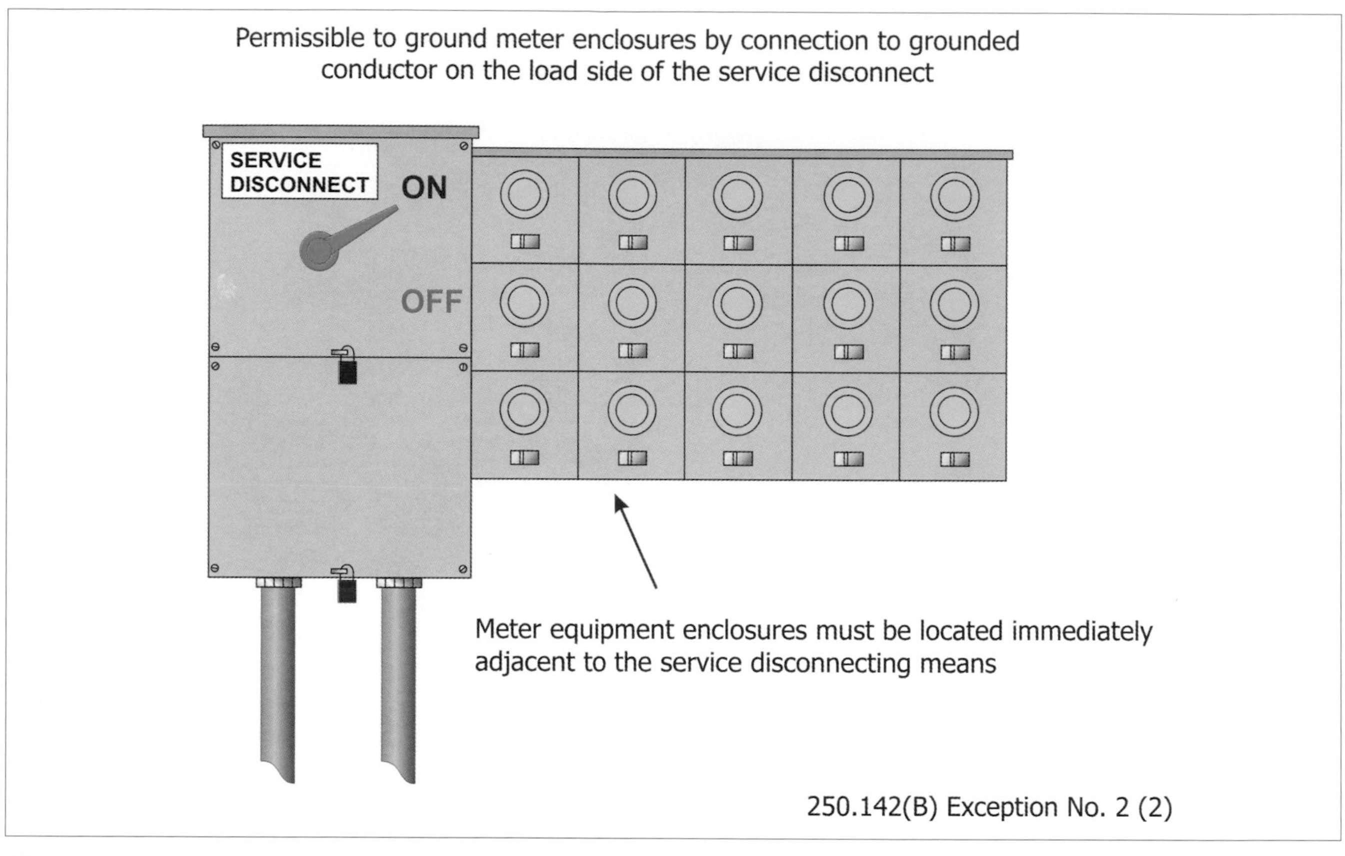
Permissible to ground meter enclosures by connection to grounded conductor on the load side of the service disconnect
SERVICE DISCONNECT
ON
OFF
Meter equipment enclosures must be located immediately adjacent to the service disconnecting means
250.142(B) Exception No. 2 (2)

Figure 9-16. Grounded conductor permitted for grounding meters on load side of service disconnect

to help ensure the reliability of the equipment grounding conductor path.

Metal Conduit Underground

Care must be taken when installing metallic conduit and electrical metallic tubing in the earth, in concrete on or below grade, or where exposed to moisture (344.10 and 300.6). The Underwriters Laboratories' 2007 guide card information for *Rigid Ferrous Metal Conduit* (DYIX) and *Intermediate Ferrous Metal Conduit* (DYBY) contains the following information regarding corrosion protection:

"Galvanized rigid (and intermediate) steel conduit installed in concrete does not require supplementary corrosion protection. Galvanized rigid (and intermediate) steel conduit installed in contact with soil does not generally require supplementary corrosion protection.

"In the absence of specific local experience, soils producing severe corrosive effects are generally characterized by low resistivity (less than 2000 ohm-centimeters).

"Wherever ferrous metal conduit runs directly from concrete encasement to soil burial, severe corrosive effects are likely to occur on the metal in contact with the soil.

"Conduit that is provided with a metallic or nonmetallic coating, or a combination of both, has been evaluated for resistance to atmospheric corrosion. Nonmetallic outer coatings that are part of the required resistance to corrosion have been additionally evaluated for resistance to the effects of sunlight.

"Rigid metal conduit with or without a nonmetallic coating has not been evaluated for severely corrosive conditions." [6]

In addition, experience has shown that steel conduit fails rapidly where exposed to corrosive environments found at some seacoast marinas, boatyards and plants as well as at some chemical plants. Experience has also shown that metal conduit systems are particularly vulnerable to failure from corrosion where they pass from concrete that is on or below grade to exposure to an atmosphere containing corrosive elements, particularly in combination with atmospheres containing oxygen.

For electrical metallic tubing [see 358.10(B)], the following instructions are given in the UL 2007 guide card (FJMX):

"Galvanized steel electrical metallic tubing installed in concrete on grade or above generally requires no supplementary corrosion protection. Galvanized steel electrical metallic tubing in concrete slab below grade level may require supplementary corrosion protection.

"In general, galvanized steel electrical metallic tubing in contact with soil requires supplementary corrosion protection. Where galvanized steel electrical metallic tubing without supplementary corrosion protection extends directly from concrete encasement to soil burial, severe corrosive effects are likely to occur on the metal in contact with the soil.

"Aluminum electrical metallic tubing used in concrete or in contact with soil requires supplementary corrosion protection. Supplementary nonmetallic coatings presently used have not been investigated for resistance to corrosion." [7]

As a result, the authority having jurisdiction is required to make a decision regarding the suitability of these raceways for these applications. This, of course, affects the reliability of the raceway serving as an equipment grounding conductor. Several reports have been made where electrical metallic tubing installed to provide an equipment grounding means has failed due to corrosion.

To maintain the integrity of the equipment grounding means, some inspection agencies require that a copper equipment grounding conductor be installed in parallel with the electrical metallic tubing.

In addition, the authority having jurisdiction must make a decision regarding the suitability of supplementary nonmetallic coatings intended for resistance to corrosion.

Grounding of Equipment by Using the Grounded Circuit Conductor

The *Code* does not generally permit the grounded circuit conductor (often a neutral) to be grounded again on the load side of the service disconnecting means [see 250.24(A)(5) and 250.142(B)]. Four exceptions to this rule exist:

1. Grounding separately derived systems.
2. Grounding the grounded circuit conductor at a remote building or structure (existing installations only).
3. For existing branch circuits only, grounding the frame of an electric range or electric clothes dryer.
4. Grounding meter enclosures that are located immediately adjacent to the service disconnecting means (see figures 9-16 and 9-17 and photo 9-10).

Where the electrical system produced by a separately derived system meets the conditions of 250.20(A) or (B), the system is required to be grounded according to 250.30(A). A system that falls within the parameters that require it to be grounded is required to have a grounding electrode conductor connected to the grounded conductor of the separately derived system (see chapter twelve for additional information on this subject).

Under the specific conditions given in 250.32(B) Exception, a grounded circuit conductor is permitted to be grounded again at a separate building or structure. This is allowed only for existing installations. Feeders and branch circuits that supply separate buildings or structures in accordance with Part II of Article 225 are now required to include an equipment grounding conductor. Where so installed, the grounded conductor serves as both a grounded conductor and an equipment grounding conductor between the

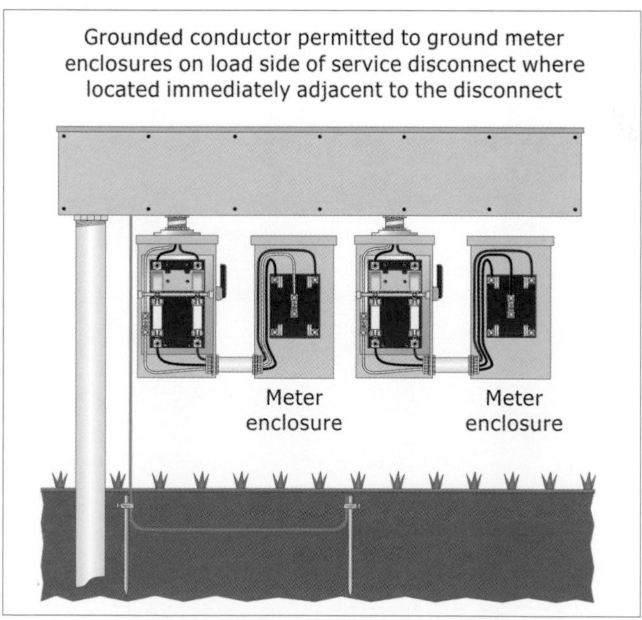

Figure 9-17. Grounded conductor is permitted for grounding where the meter equipment enclosure is located on the load side of the service disconnect, but immediately adjacent to the disconnect.

buildings or structures [see chapter thirteen for additional information on this subject].

Section 250.142(B) covers rules on the use of the grounded circuit conductor for grounding equipment on the load side of the service equipment. As stated previously, such practice is generally prohibited. Four exceptions to the general rule are provided.

Exception No. 1: The frames of ranges, wall-mounted ovens, counter-mounted cooking units, and clothes dryers under the conditions permitted for existing installations by 250.140 shall be permitted to be grounded by a grounded circuit conductor. [See chapter ten for additional information on this subject.]

Exception No. 2: It shall be permissible to ground meter enclosures by connection to the grounded circuit conductor on the load side of the service disconnect if:

(a) No service ground-fault protection is installed. [This condition is important, as grounding the grounded circuit conductor downstream from the service will desensitize the equipment ground-fault protection system.]

(b) All meter enclosures are located immediately adjacent to the service disconnecting means.

(c) The size of the grounded circuit conductor is not smaller than the size specified in Table 250.122 for equipment grounding conductors.

Exception No. 3: Direct-current systems shall be permitted to be grounded on the load side of the disconnecting means or overcurrent device in accordance with 250.164. [Rules are different depending on whether the direct-current supply is from an off-premises or on-premises source.]

Exception No. 4: Electrode-type boilers operating at over 600 volts shall be grounded as required in 490.72(E)(1) and 490.74. [8]

Underground Metal-Sheathed Cable System

The *Code* recognizes that if an underground service originates from a continuous underground metal sheath cable system and the sheath or armor is metallically connected to the underground system, or if an underground service conduit is used which contains a metal-sheathed cable which is itself bonded

Photo 9-10. Meter equipment enclosures installed immediately adjacent to disconnecting means

to the underground system, then the service run need not be grounded at the building and may be insulated from the interior conduit or piping (see 250.84).

The *Code* here recognizes that since the service armor is already bonded to a continuous underground metal sheath cable system, a path for fault-currents in the service cable is provided so there is no necessity to bond back to the service equipment in order to provide a return path to the neutral point of the supply system.

[1] NFPA 70, *National Electrical Code* 2008, Article 100, (National Fire Protection Association, Quincy, MA, 2007), p. 70-27.

[2] NFPA 70, 250.118, p. 70.114.

[3] NFPA 70, 250.118, p. 70-114.

[4] NFPA 70, Table 250.122, p. 70-117.

[5] NFPA 70, 250.122(B), p. 70-116.

[6] *General Information for Electrical Equipment 2007*, (Underwriters Laboratories, Northbrook, IL, 2001), p. 101.

[7] *General Information for Electrical Equipment 2007*, (Underwriters Laboratories, Northbrook, IL, 2001), p. 124.

[8] NFPA 70, 250.142(B), p. 70-118.

9 Review Questions

The questions included here were developed using material included in this chapter. The answers can be found by reviewing the text. It is also important that students make use of *NEC-2008*, where many answers can be found.

1. "The conductive path installed to connect normally non-current-carrying metal parts of equipment together and to the system grounded conductor of the grounding electrode conductor, or both" best defines which of the following:
   a. equipment grounding conductor
   b. main bonding jumper
   c. grounding systems conductor
   d. circuit bonding jumper

2. An equipment grounding conductor is intended to prevent an objectionable voltage above ground on conductor and equipment enclosures and to provide a low-impedance path for fault-currents. This path must also ____.
   a. be electrically continuous
   b. have ample capacity to conduct safely any currents likely to be imposed on it
   c. be of the lowest practical impedance
   d. all of the above

3. The equipment grounding conductor or effective ground-fault current path must extend from the ____ point on the circuit to the service equipment where it is connected to the grounded conductor.
   a. closest
   b. service
   c. furthermost
   d. bonding

4. Where the overcurrent protection ahead of a branch circuit or feeder is sized at 225 amperes, the minimum size of the equipment grounding conductor is to be a ____ copper conductor.
   a. 4 AWG
   b. 6 AWG
   c. 8 AWG
   d. 10 AWG

5. Which of the following is recognized as a conductor or raceway for use as an equipment grounding conductor?
   a. a conductor of copper or other corrosion-resistant material such as aluminum
   b. rigid or intermediate metal conduit
   c. electrical metallic tubing
   d. All of the above

6. Listed flexible metal conduit and listed flexible metallic tubing are permitted to be used for equipment grounding purposes. Which of the following statements is NOT true?
   a. The total combined ground return in the same path cannot exceed 20 m (6 ft)
   b. They must be terminated with listed fittings
   c. They cannot be used on a circuit exceeding 15 amperes
   d. They cannot be used on a circuit exceeding 20 amperes

7. Listed liquidtight flexible metal conduit in sizes metric designator 21 through 35 (3/4-in. through 1 1/4-in.) trade size is to be used as an equipment grounding conductor. Which of the following statements is NOT true? ____
   a. The total length of the combined ground return in the same path cannot exceed 6 feet.
   b. Listed fittings must be used
   c. The circuit is permitted to be protected by a 100 ampere or less overcurrent device.
   d. The circuit is permitted to be protected by a 60 ampere or less overcurrent device.

8. Where rigid metal conduit is used as an equipment grounding conductor, all joints and fittings are required to be ____.
   a. approved
   b. tested
   c. made up tight using suitable tools
   d. sealed

9. For the grounding of portable or pendant equipment and where protected by fuses or circuit breakers rated or set at not over ____ amperes, the *Code* permits the use of a 18 AWG copper wire as an equipment grounding conductor provided it is a part of a listed flexible cord assembly.
   a. 25
   b. 20
   c. 30
   d. 35

10. Under what conditions may the structural metal frame of a building serve as an equipment grounding conductor to ground equipment? ____
   a. Where it is effectively grounded

b. Never
c. Where approved
d. By special permission

11. Where equipment grounding conductors are installed in parallel in separate nonmetallic raceways, which of the following statements is true?
   a. A full-size equipment grounding conductor is required in only one of the conduits.
   b. A smaller equipment grounding conductor than required by Table 250.122 is permitted if the total area is not less than given in the table.
   c. A full size equipment grounding conductor is required in each of the conduits.
   d. Various size copper and aluminum conductors can be used together so long as they are not smaller than given in Table 250.122.

12. Equipment grounding conductors in parallel listed cables are permitted to be smaller than given in Table 250.122 if
   a. they are protected by an equipment ground fault protection device that is listed for the purpose of protecting the equipment grounding conductor
   b. the total area of the conductors is not less than the area required divided by the number for conductors
   c. they do not leave the building or structure they originate in
   d. none of the above

13. Metal equipment supplied from ungrounded systems
   a. must be isolated from the supply source
   b. is not required to be grounded by connection to an equipment grounding conductor
   c. is required to be grounded by connection to an equipment grounding conductor
   d. is not permitted to be grounded by an equipment grounding conductor

14. Flexible metal conduit that is listed for grounding
   a. is permitted to be used without restriction
   b. is suitable for grounding if the circuit conductors are protected at not over 20 amperes
   c. is suitable for grounding if the circuit conductors are protected at not over 60 amperes
   d. is not available

15. Flexible metal conduit and liquidtight flexible metal conduit that is used where flexibility is necessary after installation_______
   a. must have an equipment grounding conductor installed
   b. is suitable for grounding if the circuit conductors are protected at not over 20 amperes
   c. is suitable for grounding if the circuit conductors are protected at not over 60 amperes
   d. is not permitted

16. Auxiliary grounding electrodes are permitted to connect to equipment grounding conductors as long as:
   a. the earth is not used as an effective ground-fault current path
   b. a green insulated conductor is used
   c. the resistance to ground does not exceed 25 ohms
   d. the electrode is not less than 3.0 m (10 ft) in length

17. The minimum size equipment grounding conductor for a 2500-ampere feeder shall not be less than _____.
   a. 350 kcmil aluminum
   b. 400 kcmil copper
   c. 350 kcmil copper
   d. 2 AWG copper in each raceway

18. The equipment grounding conductor for a switch-leg of a 20-ampere lighting circuit shall not be smaller than
   a. 14 AWG copper
   b. 12 AWG aluminum
   c. 10 AWG copper
   d. 12 AWG copper

19. Where the ungrounded conductors of a 1000-foot long feeder are increased in size, the equipment grounding conductors of the feeder shall
   a. be increased proportionately
   b. be permitted to be sized per Table 250.122
   c. must be sized per Table 250.66
   d. permitted to be reduced in size

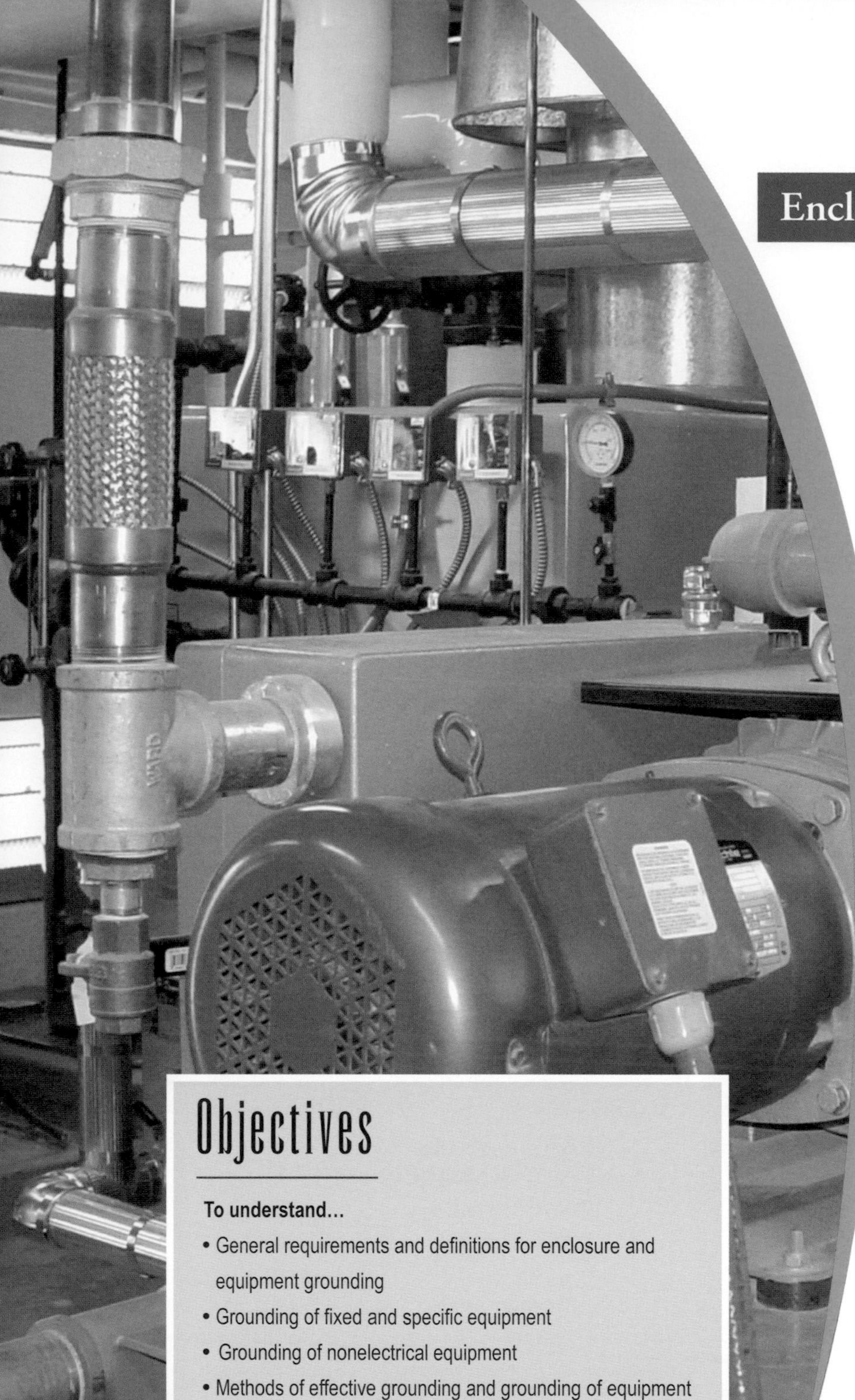

Chapter 10

Enclosure and Equipment Grounding

Objectives

To understand...

- General requirements and definitions for enclosure and equipment grounding
- Grounding of fixed and specific equipment
- Grounding of nonelectrical equipment
- Methods of effective grounding and grounding of equipment
- Grounding of metal enclosures and panelboards
- Installation of grounding-type receptacles
  Installation of isolated grounding-type receptacle circuits
- Grounding of cord- and plug-connected equipment
- Equipment spacing from air terminals

Both enclosures for service conductor and other conductor enclosures, where of metal, are required to be grounded (see 250.80 and 250.86). This does not mean that simply connecting equipment to a grounding electrode is acceptable or permitted. The installation must comply with the requirements of 250.4(A)(5) where the concept of the effective path for fault current is carefully outlined. It is important to realize that wherever the *Code* states, *shall be grounded*, it means effectively grounded as spelled out in 250.4. Note that nothing in the *Code* permits equipment that is supplied by a grounded system to be grounded to only a grounding electrode. An effective ground-fault path always includes providing a low-impedance path consisting of an equipment grounding conductor that has adequate capacity to conduct the maximum fault current it is likely to carry. It also must be electrically continuous.

While the grounding methods are different, electrical equipment associated with both grounded and ungrounded systems must be effectively grounded. Where grounding is not effectively accomplished, the situation, while bad in an ungrounded system, becomes worse in a grounded system.

Definitions

Bonding jumper: "A reliable conductor to ensure the required electrical conductivity between metal parts required to be electrically connected." [1]

Bonding jumper, equipment: "The connection between two or more portions of the equipment grounding conductor." [2]

Grounded (grounding): "Connected (connecting) to ground or to a conductive body that extends the ground connection." [3]

Grounding Conductor, Equipment (EGC): "The conductive path installed to connect normally non-current-carrying metal parts of equipment together and to the system grounded conductor or to the grounding electrode conductor, or both."[4]

Equipment Grounding Conductor

It is important to recall that an equipment grounding conductor is required to be used for grounding equipment. The equipment grounding conductor performs bonding functions and serves as an effective ground-fault current path to facilitate overcurrent device operation. The equipment grounding conductor is permitted to consist of any of the conductors or wiring methods identified in 250.118.

Installing a grounding electrode to ground equipment without having an equipment grounding conductor connected to the grounded service conductor is unsafe and not permitted (see 250.4 and 250.54).

By referring to figure 10-1, it can be seen that the grounding as shown literally meets the wording of the *Code* in that the enclosures are *grounded* which means "connected to ground …." But the wiring does not comply with 250.4(A)(5) because an effective ground-fault current return path has not been provided. In addition, the grounding shown in this figure violates 250.4(A)(5) and 250.54 as an earth return grounding circuit is indicated.

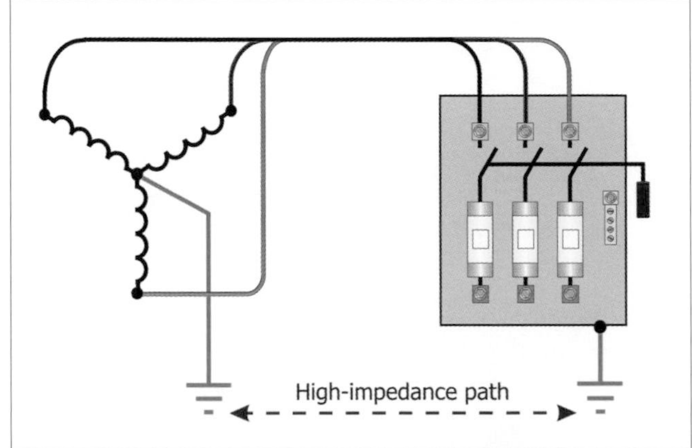

Figure 10-1. Grounded improperly (only a high-impedance return path through the earth to source)

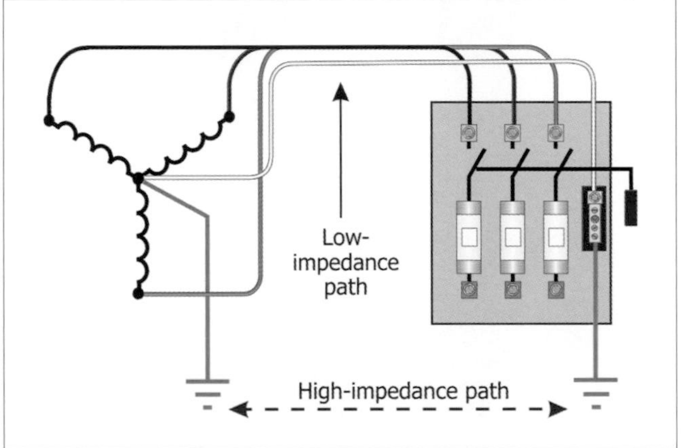

Figure 10-2. Grounded properly (in addition to the path through the earth, there is a low-impedance, effective ground-fault current path)

The wiring in figure 10-2 complies literally with 250.80 and 250.86 and also complies with 250.4(A)(5) as a grounded system conductor is installed.

It is obvious that only a high-impedance fault-current return path is indicated in figure 10-1, while in figure 10-2 there is a path having sufficiently low impedance to limit the voltage to ground and to facilitate the operation of the circuit protective devices in the circuit.

In figure 10-2 there are two paths for current to return to the source. The primary and low-impedance path is over the grounded system conductor (often a neutral conductor); while a second, high-impedance path in parallel with the grounded service conductor is through the grounding electrodes and the earth.

Service Raceways and Enclosures

We have dealt in earlier chapters with the requirement for bonding service raceways and equipment. By

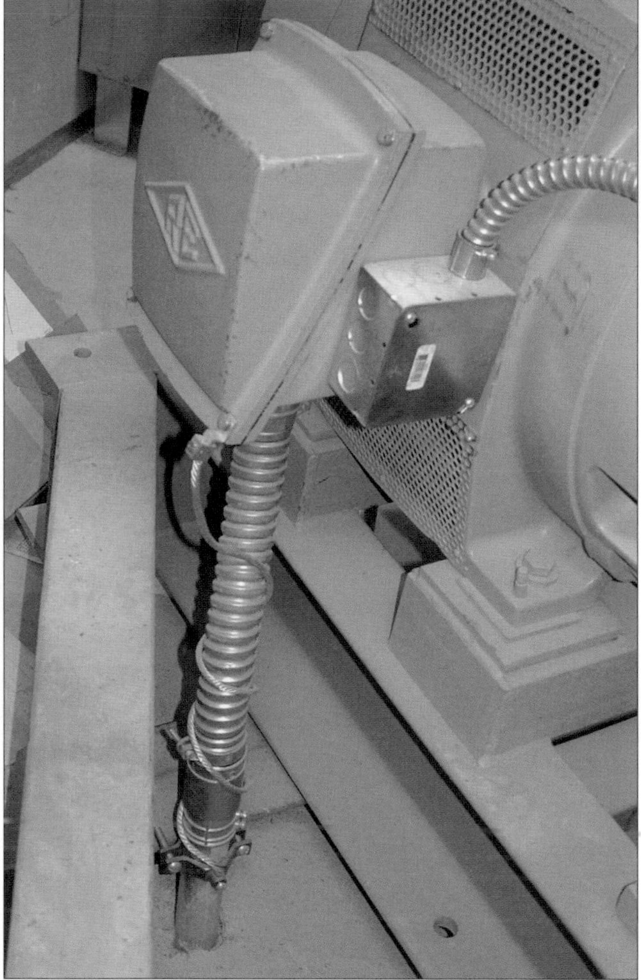

Photo 10-1. Equipment bonding jumper installed outside of flexible metal conduit (less than 6 feet)

connecting (bonding) the service equipment to the grounded service and grounding electrode, we have complied with the requirements of 250.80.

An exception to 250.80 exempts a metal elbow from the requirement that it be grounded where it is installed in an underground nonmetallic raceway(s) and is isolated from possible contact by a minimum cover of 18 inches to any part of the elbow. These metal elbows are often referred to as *pulling elbows* and are commonly installed in duct banks or other underground runs of nonmetallic raceways because they are more durable than PVC elbows during the cable-pulling process.

A similar exception regarding metal elbows used in underground runs of nonmetallic raceways (Exception No. 3) for other than service raceways has been added to 250.86. This exception exempts a metal elbow from the requirement that it be grounded where it is installed in nonmetallic raceway and is isolated from possible contact by a minimum cover of 450 mm (18 in.) to any part of the elbow or is encased in a minimum 50 mm (2 in.) of concrete, such as in decks above finished grade where these isolated metal elbows might be installed. There are additional requirements for bonding of isolated sections of metallic raceways or enclosures installed at pole locations. Section 250.102(E) permits an equipment bonding jumper to be installed either inside or outside the raceway or enclosure. Where installed on the outside of the raceway or enclosure, it is required to be routed with the raceway and not exceed 6 feet in length (see photo 10-1). The exception to this rule allows a bonding jumper to exceed the length of 1.8 m (6 ft) at pole locations for the purposes of bonding isolated portions of metallic raceways or enclosures or elbows in a run of nonmetallic raceways.[5]

Exceptions from Grounding Requirements

For other than service conductor enclosures, three exceptions are provided from the requirement that metallic conductor enclosures be connected to the equipment grounding conductor (see 250.86).

Exception No. 1 covers "metal enclosures and raceways for conductors added to existing installations of open wire, knob-and-tube wiring, and nonmetallic-sheathed cable." Conditions that must be met are as follows:

1. An equipment grounding conductor is not provided by the wiring method.
2. The metal enclosure or raceway must be less than 7.5 m (25 ft) long.

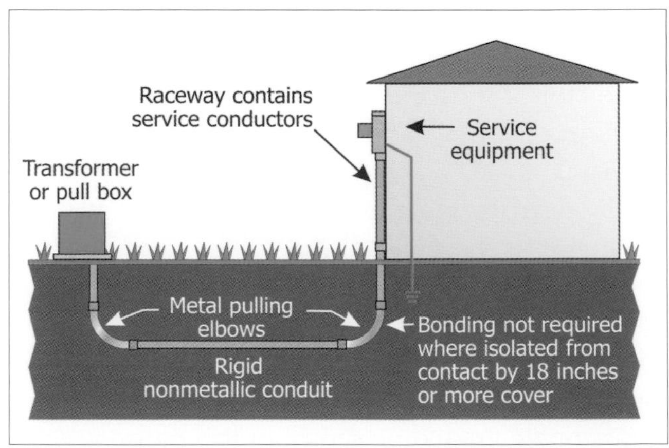

Figure 10-3. Metal elbows installed in underground nonmetallic raceway

3. The metal enclosure or raceway must be free of probable contact with ground or a grounded object.

4. The metal enclosure or raceway is guarded against contact by persons.

Exception No. 2 exempts short sections of metal enclosures from the requirement to be connected to an equipment grounding conductor of the circuit where used to protect cable assemblies from physical damage. No explanation is given for the meaning of *short sections* of metal enclosures. Since the standard length is 10 feet, short sections are often considered to be less than 10 feet long but in some cases a longer section may be acceptable.

Exception No. 3 permits a metal elbow to not be connected to the equipment grounding conductor of the circuit where it is installed in an underground run of nonmetallic raceway and is isolated from possible contact by a minimum cover of 450 mm (18 in.) to any part of the elbow or is encased in not less than 50 mm (2 in.) of concrete. These provisions are similar to that for services as provided in 250.80 Exception but include the isolated metal elbows that may be installed in runs of nonmetallic conduit such as in deck slabs (see figure 10-3).[6]

Grounding of Fixed Equipment

It is mandatory that non-current-carrying metal parts of fixed equipment that are likely to become energized be connected to the equipment grounding conductor of the circuit under the six conditions cited in 250.110. Here, again, the term *shall be grounded* must not be interpreted literally to mean "connect to a grounding electrode" but must be interpreted in the light of all of Article 250 where providing an effective fault-current path is outlined in 250.4(A)(5). The *Code* does not define what is meant by *likely to become energized*. Generally, if the equipment has exposed non-current-carrying metal parts and is supplied by electric current it should be grounded by connection to an equipment grounding conductor of the supply circuit where any of the following six conditions exist at the equipment.

The six conditions are as follows:

1. Where within 2.5 m (8 ft) vertically or 1.5 m (5 ft) horizontally of ground or grounded metal objects and subject to contact by persons

2. Where located in a wet or damp location and not isolated

3. Where in electrical contact with metal

4. Where in hazardous (classified) locations as covered by Articles 500 through 517

5. Where supplied by a metal-clad, metal-sheathed, metal raceway, or other wiring method that provides an equipment ground, except as permitted by 250.86, Exception No. 2, for short sections of metal enclosures.

6. Where equipment operates with any terminal at more than 150 volts to ground.[7]

The three exceptions from this equipment grounding requirement are as follows:

Exception No. 1. Metal frames of electrically heated devices, exempted by special permission [Written approval of the authority having jurisdiction], *in which case the frames are required to be permanently and effectively insulated from ground.*

Exception No. 2. Distribution apparatus, such as transformers and capacitor cases, mounted on wooden poles, at a height exceeding 2.5 m (8 ft) above ground or grade level.

Exception No. 3. Listed equipment that is protected by a system of double insulation or its equivalent, shall not be required to be connected to the equipment grounding conductor. Where such a system is employed, the equipment shall be distinctively marked.[8]

Grounding Specific Equipment

The *Code* requires that exposed, non-current-carrying metal parts of certain specific equipment, regardless of voltage, shall be connected to the equipment grounding conductor of the supply circuit. Those items are spelled out in 250.112 and include:

Switchboard Frames and Structures

Switchboard frames and structures supporting switching equipment, except frames of 2-wire dc switchboards where effectively insulated from ground.

Pipe Organs

Generator and motor frames in an electrically operated pipe organ, unless effectively insulated from ground and the motor driving it.

Motor Frames

Motor frames, as provided by 430.242.

1. Where supplied by metal-enclosed wiring

2. Where in a wet location and not isolated or guarded
3. If in a hazardous (classified) location as covered in Articles 500 through 517
4. If the motor operates with any terminal at over 150 volts to ground.[9]

Enclosures for Motor Controllers

Enclosures for motor controllers unless attached to ungrounded portable equipment.

Elevators and Cranes

Electrical equipment for elevators and cranes.

Garages, Theaters, and Motion Picture Studios

Electric equipment in commercial garages, theaters, and motion picture studios, except pendant lampholders supplied by circuits not over 150 volts to ground.

Electric Signs

Electric signs, outline lighting, and associated equipment as provided in 600.7 [see chapter sixteen for additional information on this subject].

Motion Picture Projection Equipment

Motion picture projection equipment.

Power-Limited Remote-Control, Signaling, and Fire Alarm Circuits

Equipment supplied by Class 1 circuits shall be grounded unless operating at less than 50 volts. Equipment supplied by Class 1 power-limited circuits and by Class 1, Class 2, and Class 3 remote-control and signaling circuits, and by fire alarm circuits, shall be grounded where system is required by Part II or Part VIII of this article.[10]

Luminaires

Luminaires as required by Part V of Article 410.

Exposed Conductive Parts. Exposed metal parts shall be grounded or insulated from ground and other conductive surfaces or be inaccessible to unqualified personnel. Lamp tie wires, mounting screws, clips, and decorative bands on glass spaced at least 38 mm (1½ in.) from lamp terminals shall not be required to be grounded [see 410.42(A)].

Made of Insulating Material. Luminaires directly wired or attached to outlets supplied by a wiring method that does not provide a ready means for grounding shall be made of insulating material and shall have no exposed conductive parts [see 410.42(B)].

Luminaires with exposed metal parts shall be provided with a means for connecting an equipment grounding conductor for such luminaires (see 410.20).[11]

Skid-Mounted Equipment

Permanently mounted electrical equipment and skids shall be connected to the equipment grounding conductor jumper sized as required in 250.122.

Motor-Operated Water Pumps

Motor-operated water pumps, including the submersible type. [See 547.5(F) and 547.9 for specific requirements for grounding in agricultural buildings.]

Metal Well Casings

Where a submersible pump is used in a metal well casing, the well casing shall be connected to the pump circuit equipment grounding conductor.[12]

Grounding of Cord- and Plug-Connected Equipment

Section 250.114 covers grounding of equipment connected by cord and plug. It is mandatory under certain conditions that non-current-carrying metal parts of cord- and plug-connected equipment which are liable to become energized be connected to the equipment grounding conductor. Listed tools, listed appliances, and listed equipment that is protected by a system of double-insulation are not required to be connected to the equipment grounding conductor. This double-insulated equipment is required to be distinctively marked.

Specifically, cord- and plug-connected equipment is required to be grounded where located in:

- hazardous (classified) locations,
- if the equipment operates at more than 150 volts to ground. Exempted from this requirement are:

–motors, where guarded, and

–metal frames of electrically-heated appliances exempted by special permission, in which case the frames shall be permanently and effectively insulated from ground.[13]

The *Code* cites in 250.114(3) specific equipment of the cord- and plug-connected type that must be connected to the equipment grounding conductor in residential occupancies. In addition, 250.114(4) lists specific cord- and plug-connected equipment in other than residential occupancies that must be grounded. For tools and portable handlamps in other than residential occupancies, an exception from the requirement for grounding is provided for cord- and plug-connected equipment that is supplied through an isolating transformer with an ungrounded secondary of not over 50 volts.

Nonelectric Equipment

The grounding of nonelectric equipment is covered in 250.116. The equipment mentioned is considered as being likely to become energized and is thus required to be grounded by connection to the equipment grounding conductor as a safety measure. Included are:

- Frames and tracks of electrically operated cranes and hoists

Photo 10-2. Equipment grounding conductors connected to enclosure by listed irreversible compression connectors

- Frames of nonelectrically driven elevator cars to which electric conductors are attached
- Hand-operated metal shifting ropes or cables of electric elevators

The fine print note following this section recommends that, "where extensive metal in or on buildings may become energized and is subject to personal contact, adequate bonding and grounding will provide additional safety." [14]

Methods of Equipment Grounding

To provide the reliable and effective ground-fault return path required, it is important to connect the equipment grounding conductor recognized in 250.118 to the equipment in such a manner that the requirements for effective grounding are met.

For equipment that is fastened in place or connected by permanent wiring methods, the equipment grounding conductor must be connected to the enclosure in a proper manner.

Grounding conductors and bonding jumpers must be connected to equipment that is required to be grounded by any of the methods provided in 250.8 (see photo 10-2).

One connecting means that is not required to be listed is the exothermic welding method. It is most important that equipment to be welded be clean and dry and that manufacturer's instructions be followed to ensure a satisfactory connection.

These welds must be examined and tested after completion to be certain that a reliable connection has been made. It is common to test the welds in the field by x-ray where required by the job specifications or by striking the weld with a hammer after it has cooled for less demanding installations.

In the case of a metallic raceway being used as the equipment grounding conductor, the raceway must be connected to the enclosure by using listed fittings designed for the purpose. The fittings for various wiring methods covered in chapter 3 of the *NEC* are required to be listed. All connections must be made up tight using proper tools. This includes locknuts, bushings, conduit, and electrical metallic tubing couplings and connectors [see 250.120(A)].

Solder Connections

Connections that depend solely upon solder cannot be used for grounding connections [see 250.8 and 250.148].

The reason for this prohibition is that when equipment grounding conductors carry fault-current, they can get very hot. This elevated temperature can exceed the melting point of the solder and weaken or destroy the connection. This can create a hazard by leaving equipment at a dangerous potential above ground, creating a shock hazard to those who could contact the equipment.

It is permissible to secure the connections mechanically and then apply solder to the joint to make the electrical connection.

Special Provisions for Grounding Certain Appliances

Special requirements are set forth in the exception to 250.140 for the grounding of frames of electric ranges, electric clothes dryers and similar appliances. These provisions apply only to existing branch-circuit installations. New installations must comply with the requirements for an insulated neutral conductor as well as an equipment grounding conductor given in 250.134 and 250.138.

Appliances or equipment to which 250.140 Exception applies include:

- Electric ranges
- Wall-mounted ovens
- Counter-mounted cooking units
- Clothes dryers
- Outlet or junction boxes that are part of the circuit for these appliances

These appliances and equipment, such as the junction boxes in the circuit, are permitted to be grounded in two ways: either by use of an equipment grounding conductor; or, except for mobile homes and recreational vehicles, by the use of the grounded circuit (neutral) conductor, provided all the conditions of this section are complied with (see figures 10-4 and 10-5). If a 3-wire with ground circuit is installed, simply use the equipment grounding conductor for grounding metal equipment and install a grounding type receptacle.

The conditions that must be met for grounding the above equipment using grounded circuit conductor in existing installations are as follows:

1. The appliances are supplied by a 120/240-volt, single-phase, 3-wire circuit or by a 3-wire circuit derived from a 208Y/120-volt, 3-phase, 4-wire, wye-connected system.
2. The grounded (neutral) conductor is 10 AWG copper or 8 AWG aluminum or larger.
3. The grounded (neutral) conductor is insulated. The grounded (neutral) conductor is permitted to be uninsulated if it is part of a Type SE cable and it originates at the service equipment, and
4. Grounding contacts of receptacles furnished as part of the equipment are bonded to the equipment.[15]

It is important to note that all of these special conditions must be met before the appliances, and outlet and junction boxes that are a part of the circuit to the appliances, are permitted to be grounded using the grounded (neutral) conduct of the circuit. Note that the supply cable must have an insulated neutral

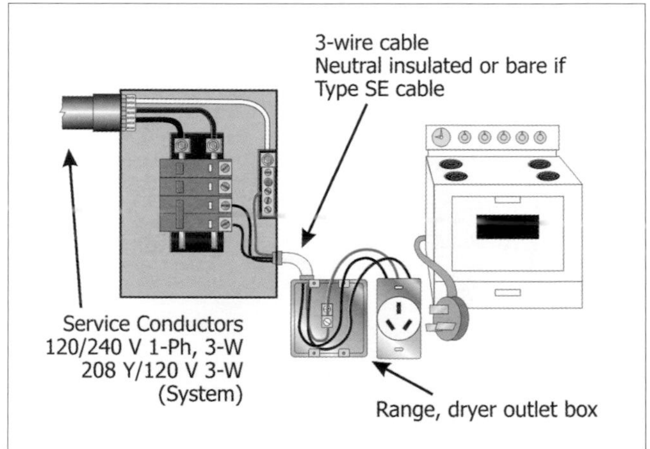

Figure 10-4. Wiring methods for appliances permitted to be grounded using the grounded (neutral) circuit conductor

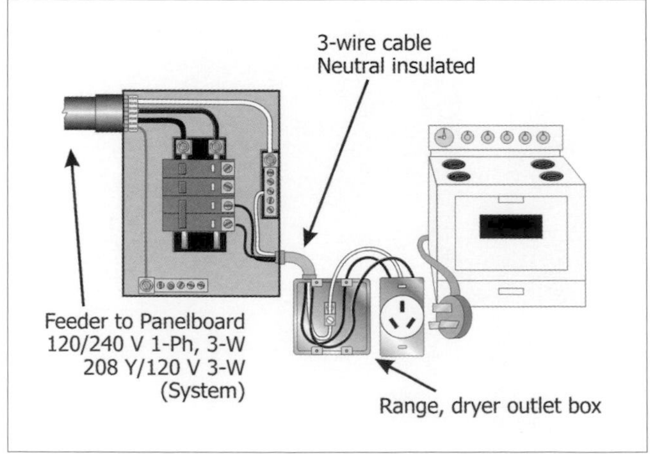

Figure 10-5. Wiring methods for appliances permitted to be grounded using the grounded (neutral) circuit conductor

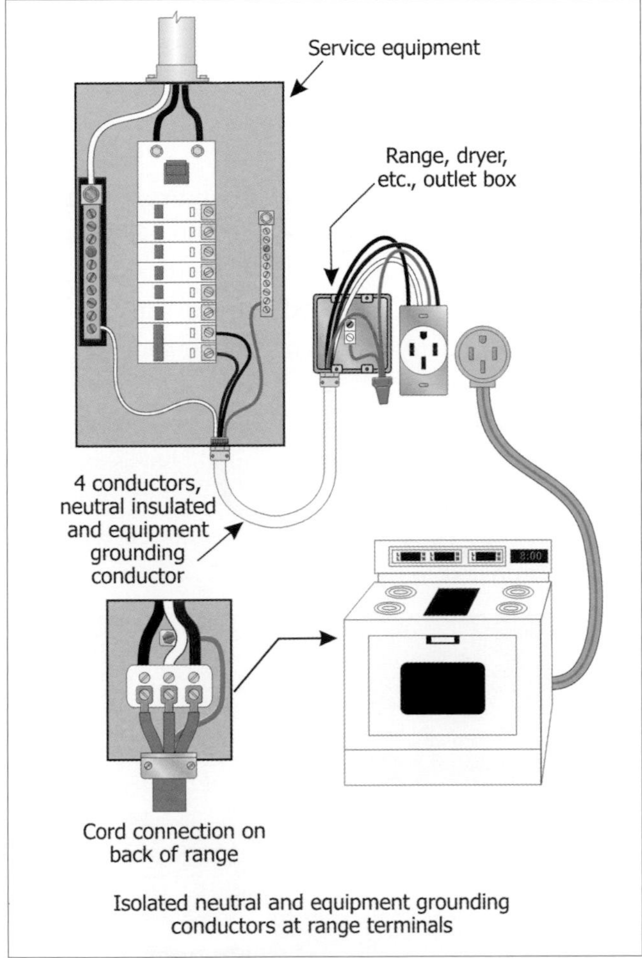

Figure 10-6. Grounding requirements for new electric range and electric dryer circuits

conductor where supplied from a panelboard on the load side of the service disconnect.

Use caution when applying the provisions of 250.140 Exception No. 3, which permits the use of a Type SE cable having a bare neutral for wiring these appliances when the circuit originates at service equipment.

Wiring for New Ranges and Dryers

New installations for ranges, dryers, and similar appliances require a *three-wire with equipment grounding conductor* circuit having an insulated neutral conductor and an equipment grounding conductor (see figure 10-6). The equipment grounding conductor is permitted to be any of those included in 250.118 including conduit and cables. Receptacles, where installed, must be of the 3-pole, 4-wire grounding type. Supply cords, where used, must be of the 3-wire with equipment grounding conductor type.

Neither the frame of the appliances nor the outlets or junction boxes that are a part of the supply to the appliances are permitted to be grounded to the circuit grounded conductor (neutral conductor). The frame of the appliances and junction boxes must be grounded by means of an equipment grounding conductor run with the branch circuit.

Care must be taken when these appliances are moved from a location employing one grounding scheme to a location having a different one. In the case where a 3-wire cable is used, the bonding jumper in the appliance junction box must be connected between the frame of the appliance and the neutral. Where a four-wire supply is used, the bonding jumper must be removed or disconnected, the neutral conductor is isolated from the appliance frame and the equipment grounding conductor connected to the frame.

Outlet, Device, Pull and Junction Boxes

Metal outlet, device, pull and junction boxes are required to be grounded and bonded in accordance with Parts I, IV, V, VI, VII and X of Article 250 [see 314.4]. It is important that the grounding of these enclosures be accomplished by the wiring method that supplies the enclosure [see 250.134].

Under no circumstances can a metal enclosure be connected to a local grounding electrode in lieu of grounding it by means of the suitable wiring method or equipment grounding conductor unless specifically permitted by *Code* rules. The metal raceway containing circuit conductors is permitted to be used as an equipment grounding conductor these enclosures if installed properly with all connections made up wrenchtight. In addition, where fittings are used for connecting cables or raceways being used for grounding, the fittings used for connecting the wiring methods must be listed. Some wiring methods have restrictions on their use as an equipment grounding conductor. These include flexible metal conduit, liquidtight flexible metal conduit and flexible metallic tubing.

Equipment grounding conductors that are supplied for grounding metallic enclosures must be sized in accordance with Table 250.122 based upon the rating of the overcurrent device protecting the circuit conductors in the raceway.

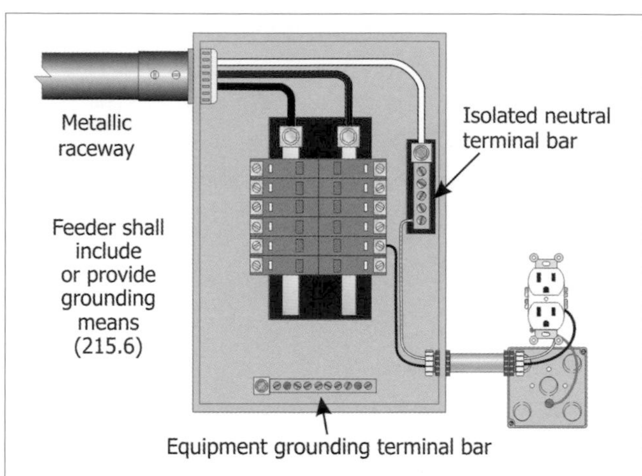

Figure 10-7. Separate equipment grounding terminal bar required in panelboard [408.40]

Section 314.40(D) requires that "a means be provided (by the manufacturer) in each metal box for the connection of the equipment grounding conductor. This means shall be permitted to be a tapped hole or equivalent." [16]

Grounding of Panelboards

Section 215.6 requires that "where a feeder supplies branch circuits in which equipment grounding conductors are required, the feeder shall include or provide an equipment grounding conductor in accordance with the provisions of 250.134 to which the equipment grounding conductors of the branch circuit shall be connected (see figures 10-7 and 10-8)." [17]

Specific requirements for grounding panelboards are contained in 408.40. All panelboard cabinets and panelboard frames if of metal are required to be in physical contact with each other and must be grounded by connection to an equipment grounding conductor. Section 408.3(C) requires that where used as service equipment, the panelboard must be provided with a main bonding jumper located inside the cabinet for the purpose of bonding the enclosure to the grounded service conductor.

"Where the panelboard is used with nonmetallic raceway or cable or where separate equipment grounding conductors are provided, a terminal bar for the equipment grounding conductors shall be secured inside the cabinet. The terminal bar shall be bonded to the cabinet and panelboard frame if of metal; otherwise it shall be connected to the equipment grounding conductor that is run with the conductors feeding the panelboard." [18] Usually, the manufacturer provides matching and tapped holes along with appropriate screws or bolts for attaching the bar to the enclosure (see photo 10-3).

Grounding and equipment grounding conductors are not permitted to be connected to the terminal bar for grounded (may be neutral) conductors unless the terminal bar is identified for the purpose. This identification will usually be on the manufacturer's label that is located within the cabinet. Section 408.41 requires that each grounded conductor (may be a neutral) be terminated in an individual terminal that is not also used for another conductor. This prohibits grounded conductors and equipment grounding conductors from being terminated in the same "hole" of a terminal bar,

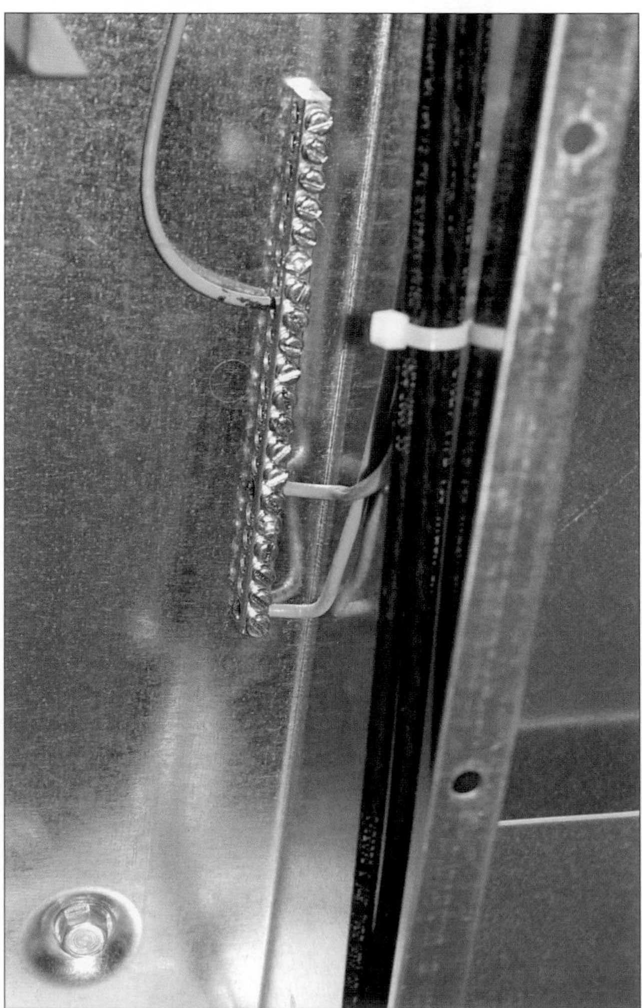

Photo 10-3. Equipment grounding terminal bar fastened to enclosure with thread-forming screws provided by the manufacturer.

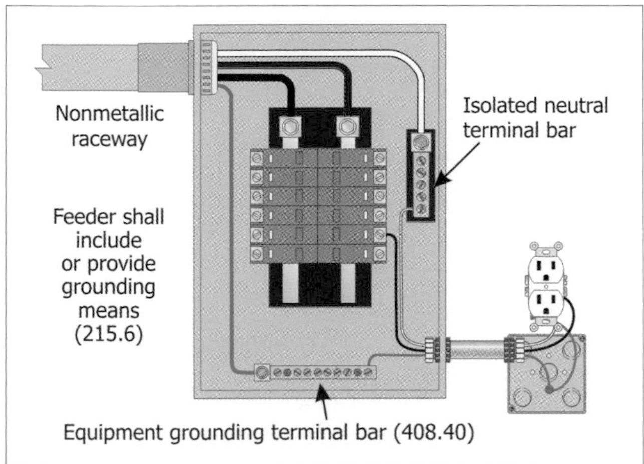

Figure 10-8. Separate the grounded (neutral) conductors and the equipment grounding conductors [250.24(A)(5) and 250.142(B)]

even if it is identified for more than one conductor (see photo 10-4). Another requirement is that the panelboard is used at a location where the grounded conductor terminal (neutral conductor) bar is connected to a grounding electrode, as permitted or required by Article 250.

These locations include: at services, at the building disconnecting means [see 225.32], and for separately derived systems [see 250.30(A)].

An exception permits an insulated equipment grounding conductor for an isolated grounding scheme to pass through a panelboard, box, wireway, or other enclosure without being connected to the equipment grounding terminal bar in compliance with 250.146(D). Generally, the grounded (may be a neutral)

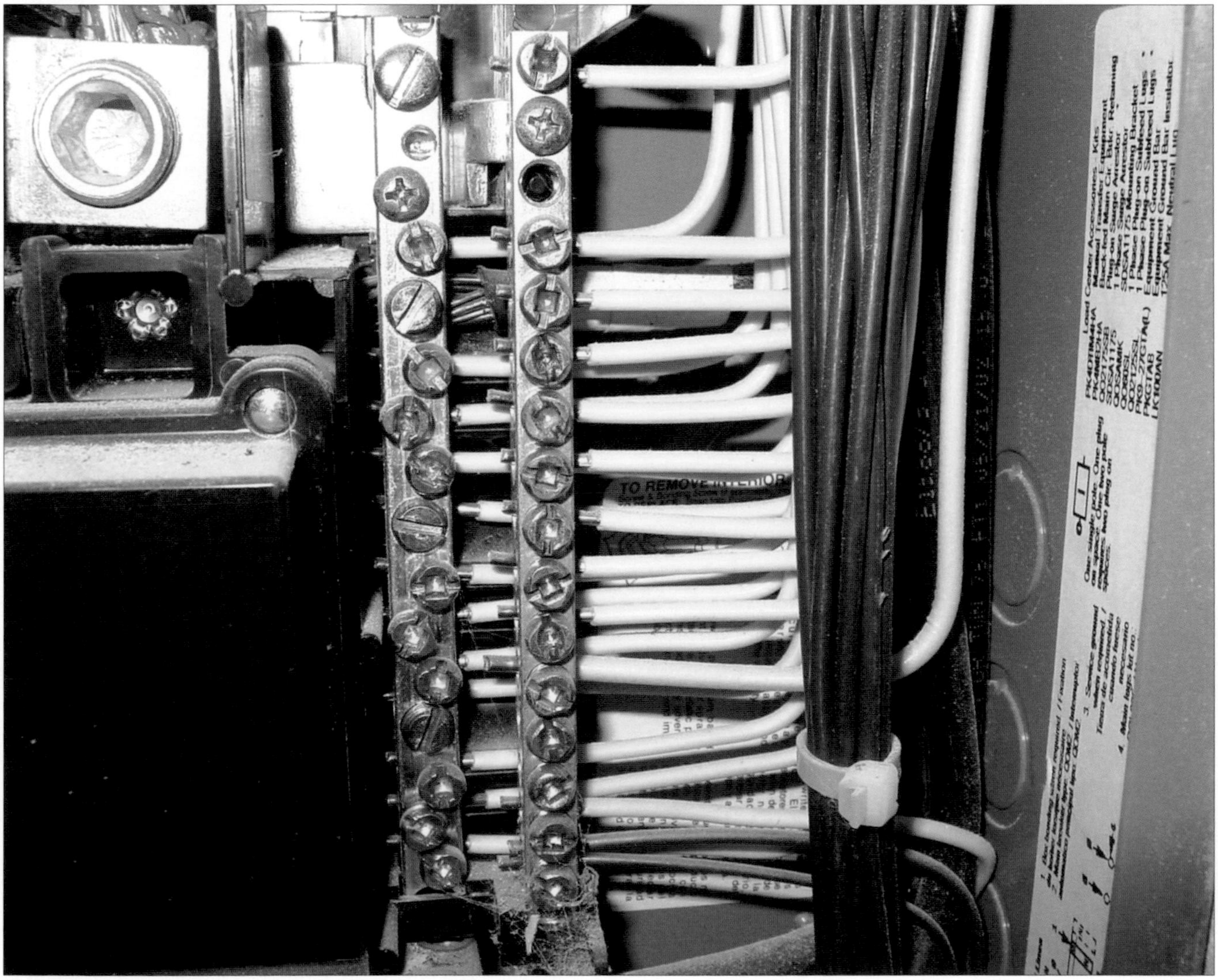

Photo 10-4. Grounded conductor terminal bar showing only one wire per terminal connected

conductor is isolated from the enclosure downstream from the service equipment [250.24(A)(5)]. Often, the term *floating neutral conductor* is used to describe the grounded conductor's relationship to the enclosure as it is insulated electrically from the enclosure. Section 250.24(A)(5) generally prohibits a grounding connection to a grounded conductor (may be a neutral) on the load side of the service disconnecting means. Exceptions to the general rule are provided for separately derived systems, at separate buildings for existing installations only, and for certain appliances such as electric ranges and clothes dryers on existing branch circuits.

See chapter twelve for additional information on separately derived systems and chapter thirteen for grounding at more than one building on the premises.

Grounding-Type Receptacles

Where grounding-type receptacles are installed, the grounding terminal of the receptacle must be connected to an equipment grounding conductor of the circuit supplying the receptacle [see 406.3(C)]. Where more than one equipment grounding conductor enters a box, they must be connected together using a suitable and listed connector. In addition, it is permitted to connect each of the equipment grounding conductors to the metal box individually using a listed clip or screw. An equipment bonding jumper must be connected to the receptacle so grounding continuity is not disturbed if the device is removed.

An equipment bonding jumper from the receptacle to the box is not required where a device with listed grounding means is installed in a metal box that is properly grounded. These devices are often referred to as *self-grounding receptacles* and are specially designed so one or more of the mounting screws are maintained in contact with the device's metal yoke (see the requirements in chapter eight for additional information on this subject).

Isolated Grounding Receptacles

Receptacles that have the equipment grounding terminal isolated from the mounting strap, and therefore from the box, are commonly installed at computer terminals and cash registers (see figure 10-9). This is permitted by

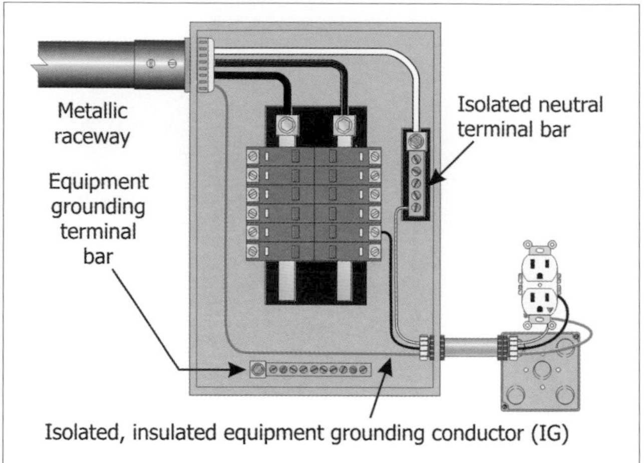

Figure 10-9. Isolated grounding receptacles [250.146(D)]

250.146(D) for the purpose of reducing electrical noise (electromagnetic interference). The grounding terminal of the receptacle must be grounded by means of an insulated equipment grounding conductor that is run with the circuit conductors.

This insulated equipment grounding conductor is permitted to pass through one or more panelboards, boxes, wireways, or other enclosures without connection to the terminal bar within the panelboard on its way back to, usually, the service disconnecting means [see 250.146(D) and 408.40, Exception and photo 10-5]. However, the insulated equipment grounding conductor must terminate within the same building at the building or structure disconnecting means or source of a separately derived system (see chapter

Photo 10-5. Panelboard with isolated grounding terminal bar (insulated) and equipment grounding terminal bar fastened to the enclosure

twelve for additional information on the subject of grounding separately derived systems).

Isolated Equipment Grounding

Section 250.96(B) permits an equipment enclosure supplied by a branch circuit to be grounded by connection to an insulated equipment grounding conductor contained within the raceway with the branch circuit that supplies the equipment. A listed nonmetallic raceway fitting at the point of connection to the equipment must be installed (see figure 10-10). This provision typically applies to listed data processing (information technology) equipment. See chapter 17 for additional information on this subject.

Underwriters Laboratories performed tests of a similar grounding scheme to determine whether isolating the metal conduit from the equipment had an adverse effect on the grounding circuit impedance. They found no appreciable increase in impedance with the metal conduit isolated from the equipment and being grounded by means of the insulated equipment grounding conductor.

Short Sections of Raceway

Where isolated sections of metal raceway or cable armor are required to be grounded, the *Code* requires in 250.132 that grounding of such sections be performed in accordance with the requirements of fixed equipment found in 250.134; in other words, they are required to be connected to the equipment grounding conductor. While the *Code* does not identify what is meant by a *short section*, perhaps this is a length less than the standard length of 10 ft. These short sections of raceways are often installed as physical protection of cables.

As mentioned previously, short sections of metal enclosures or raceways used to provide support or protection of cable assemblies from physical damage are not required to be connected to the equipment grounding conductor due to Exception No. 2 to 250.86 (see figure 10-11).

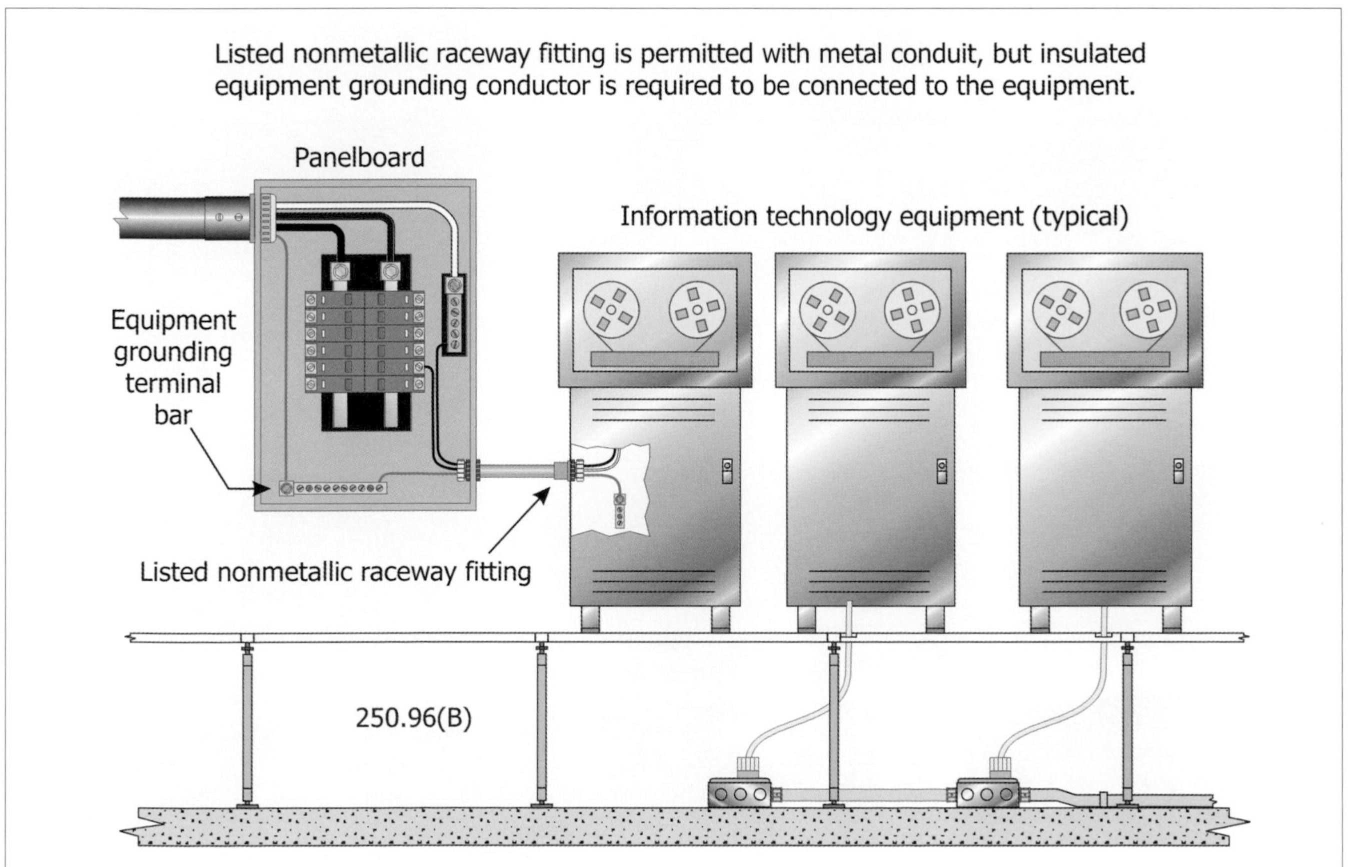

Figure 10-10. Isolated equipment grounding [250.96(B)]

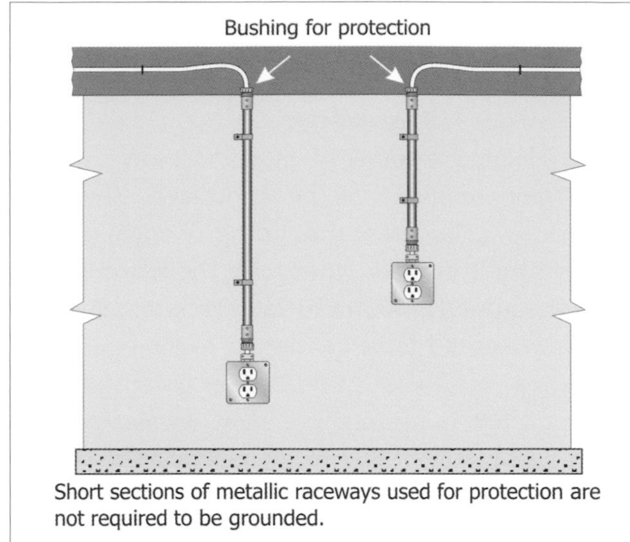

Figure 10-11. Short sections of raceway used for protection are not required to be connected to the equipment grounding conductor.

Where these short sections of raceways or cable armor are required to be grounded, 250.134 generally requires that they be grounded by connection to one of the equipment grounding conductors recognized by 250.118.

[1] NFPA 70, *National Electrical Code* 2008, Article 100, (National Fire Protection Association, Quincy, MA, 2007), p. 70-24.

[2] NFPA 70, Article 100, p. 70-24.

[3] NFPA 70, Article 100, p. 70-27.

[4] NFPA 70, Article 100, p. 70-27.

[5] NFPA 70, 250.102(E), p. 70-111.

[6] NFPA 70, 250.86, p. 70-109.

[7] NFPA 70, 250.110, 70-113.

[8] NFPA 70, 250.110, p. 70-113.

[9] NFPA 70, 430.242, p. 70-323.

[10] NFPA 70, 250.112, p. 70-113.

[11] NFPA 70, 410.20, p. 269.

[12] NFPA 70, 250.112, p. 70-113.

[13] NFPA 70, 250.114, p. 70-114.

[14] NFPA 70, 250.116, p. 70-114.

[15] NFPA 70, 250.140, p. 70-118.

[16] NFPA 70, 314.40(D), p. 70-174.

[17] NFPA 70, 215.6, p. 70-57.

[18] NFPA 70, 408.40, p. 70-264.

Review Questions

The questions included here were developed using material included in this chapter. The answers can be found by reviewing the text. It is also important that students make use of *NEC-2008*, where many answers can be found.

1. A reliable conductor to assure the required electrical conductivity between metal parts required to be electrically connected is defined as a ____.
   a. grounding electrode conductor
   b. grounded conductor
   c. bonding jumper
   d. identified conductor

2. The connection between two or more portions of the equipment grounding conductor is defined as ____.
   a. equipment bonding jumper
   b. grounded
   c. neutral
   d. main bonding jumper

3. Exposed non-current-carrying metal parts of equipment must be connected to the equipment grounding conductor where within____ feet vertically or ____ feet horizontally of ground or grounded metal objects, and subject to contact by persons.
   a. 9 - 6
   b. 8 - 5
   c. 8 - 8
   d. 9 - 9

4. Exposed noncurrent-carrying metal parts of equipment are required to be connected to the equipment grounding conductor where located in wet or damp locations and are not ____.
   a. guarded
   b. identified
   c. shielded
   d. isolated

5. Where other than short sections of metal enclosures are installed as per Section 250.86, exposed noncurrent-carrying metal parts of equipment are required to be connected to the equipment grounding conductor where supplied by any of the following wiring methods EXCEPT ____.
   a. metal-clad, metal-sheathed cables
   b. knob-and-tube wiring
   c. other wiring methods that provide an equipment ground
   d. approved metal wireways

6. Exposed noncurrent-carrying metal parts of equipment not required to be connected to the equipment grounding conductor include metal frames of electrically heated devices unless exempted by ____ in which case the frames must be permanently and effectively insulated from ground.
   a. the product instructions
   b. the local government
   c. special permission
   d. the code

7. Exposed non-current-carrying metal parts of equipment not required to be connected to the equipment grounding conductor include distribution apparatus, such as transformers and capacitor cases, that are mounted on wooden poles at a height of more than ____ feet from the ground or grade level.
   a. 6
   b. 8
   c. 5
   d. 7

8. Exposed non-current-carrying metal parts of equipment not required to be connected to the equipment grounding conductor include ____ equipment that is distinctively marked.
   a. hospital
   b. triple insulated
   c. single insulated
   d. double insulated

9. Where added to existing installations of open wire, knob-and-tube wiring, and nonmetallic-sheathed cable without an equipment grounding conductor, a metal enclosure for conductors run in lengths not to exceed 7.5 m (25 ft), free from probable contact with ground, grounded metal, metal lath, or other conductive material, and if guarded against contact by persons are ____.
   a. required to be grounded to a grounded electrode
   b. not required to be connected to the equipment grounding conductor
   c. required to be protected by a GFCI
   d. not permitted unless approved

10. Motor-operated water pumps including the submersible type are required to be connected to an equipment grounding conductor when they operate at ____.

a. any voltage
b. 120 volts
c. 240 volts
d. 208 volts

11. All of the connecting means included below are required to be listed by a qualified electrical testing lab EXCEPT ____.

a. pressure connectors
b. lugs
c. clamps
d. exothermic welding

12. Connections that depend solely upon solder for grounding connections ____.

a. are only permitted when approved
b. cannot be used
c. are permitted on the load side of the service
d. are permitted on the line side of the service

13. Metal raceways that enclose service conductors are required to be connected to the grounded system conductor if the electrical system is grounded or to the grounding electrode conductor if the electrical system is not grounded unless it is:

a. installed on a pole
b. more than 2.5 m (8 ft) above the ground
c. an elbow in a PVC run covered by not less than 450 mm (18 in.) of earth
d. buried at a depth given in Table 300.5

14. All the following statements about grounding the frame of ranges and dryers are true EXCEPT:

a. A three-wire with equipment grounding conductor is permitted for existing installations.
b. The frame of the appliances is permitted to be grounded to the neutral conductor of the circuit in new installations.
c. A two-wire with bare neutral conductor is required for new installations.
d. New installations must be grounded to comply with Sections 250.138 and 250.140.

15. All the following statements about grounding of panelboards supplied by a feeder in the same building as the service are true EXCEPT:

a. The feeder must supply or provide an equipment grounding conductor.
b. The neutral conductor is permitted for grounding the enclosure.
c. The neutral conductor must connect to a neutral bar that is isolated from the enclosure.
d. An equipment grounding conductor must be connected to an equipment grounding terminal bar.

16. An equipment grounding conductor _______ in the same hole of a terminal bar, with a grounded (neutral) conductor, if it is identified for more than one conductor

a. is prohibited
b. is permitted
c. is required to be
d. should be

17. Isolated metal electrical equipment enclosures of grounded systems are _______to be grounded solely by connecting to a grounding electrode system.

a. permitted
b. recommended
c. not permitted
d. permitted if not greater than 25 ohms of resistance

18. Metal elbows are not required to be connected to the equipment grounding conductor where installed in nonmetallic raceways and isolated from possible contact by_______

a. a minimum cover of 450 mm (18 in.)
b. encased in not less than 50 mm (2 in.) of concrete
c. neither a. or b.
d. either a. and b.

Clearing Ground Faults & Short Circuits

Section 310.2(A) generally requires that conductors be insulated. The exception allows covered or bare conductors to be used where specifically permitted. Ungrounded (phase or hot) conductors must be insulated for the applied voltage. Typical voltage ratings of conductors are 300, 600, 2000, 5000 and 15,000. Conductors are also available with much higher rated insulations, although these are primarily used for distribution of electrical energy rather than for premises wiring.

Objectives

To understand...

- Ground faults and circuit impedance
- Fundamentals of equipment grounding, circuit design, and test procedures
- Common elements in clearing ground faults and short circuits
- Sizing of equipment grounding conductors
- Purposes served by grounded conductor on grounding systems
- Conductor withstand ratings

Chapter 11

Table 310.13 provides conductor applications and insulations. It provides the trade name, type letter such as MI, maximum operating temperature and application provisions. It also gives the conductor insulation, size in American Wire Gage (AWG), thickness in mils and outer covering, if any.

Bare or covered conductors are permitted to be used in certain cases. Grounded service-lateral conductors are permitted to be uninsulated under the conditions given in 230.30 Exception. Insulated service-entrance conductors are generally required to be used as provided in 230.41, but grounded conductors are permitted to be uninsulated in accordance with any of the exceptions to 230.41. Of course, insulated, covered or bare equipment grounding conductors are permitted to be used by 250.118(1) and 250.119. Some *Code* rules require insulated equipment grounding conductors for specific applications such as for certain electrical equipment associated with swimming pools and in patient care areas of certain health care facilities.

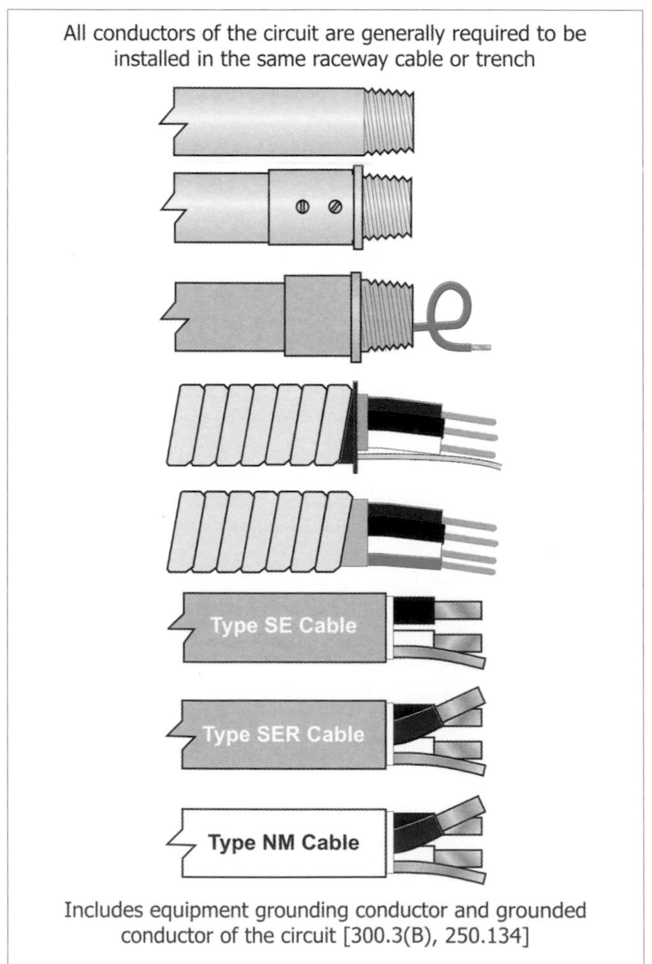
All conductors of the circuit are generally required to be installed in the same raceway cable or trench
Type SE Cable
Type SER Cable
Type NM Cable
Includes equipment grounding conductor and grounded conductor of the circuit [300.3(B), 250.134]

Figure 11-1. All circuit conductors generally are required to be together.

No hazard can exist on a distribution system unless there is an insulation failure of the ungrounded (hot) conductor, it follows that every precaution should be taken to provide the best possible insulation consistent with an economical installation. Good installation practices should be followed carefully to ensure that conductor insulation is not damaged during the installation process. Conductor insulation can be easily damaged by pulling operations, if raceways are not cleaned prior to the installation of conductors, if excessive bends are in the run, or where long or heavy pulls require the use of heavy-duty pulling equipment.

Further, if the conductor insulation breaks down, the hazard can exist only as long as the circuit remains energized. Every effort should be made to de-energize, as quickly as is practical, any circuit on which a fault to ground has developed.

When first considered, the use of a copper or aluminum equipment grounding conductor, by itself, can seem to be an ideal method of getting a low-impedance path. It will in most cases, but that is not always the best choice. Each individual circuit must be studied in relation to the size and length of the circuit, the size of its overcurrent device, and the size of the conductor enclosure before any design or engineering conclusion can be reached.

Fundamentals of Equipment Grounding Circuit Design

To get low impedance of the grounding system in an ac system, the circuit conductors and the equipment grounding conductor must generally be kept together at all times [see figure 11-1, 250.134(B) and 300.3(B)]. The equipment grounding conductor may, of course, be a copper or aluminum conductor or the metal enclosure of the conductors such as conduit or wireways, where the conduit or wireway qualifies as an equipment grounding conductor in accordance with 250.118 for such use.

In an ac electrical distribution system, whether grounded or ungrounded, inductive reactance exerts a powerful influence by directing the return current to a path closely paralleling the outgoing power conductor.

In addition to resistance, an inductive reactance value is associated with every conductor in an ac system. The inductive reactance (expressed in ohms) increases as the spacing between the conductors or the circuit is increased. This indicates that the inductive reactance of a current path closely paralleling the phase conductors offers lower impedance to the ground-fault current than any other current path regardless of a lower resistance of the other current paths. Inductive reactance will be the predominate factor in determining current division in parallel ground return paths in heavy or larger circuit constructions. This has been proven by actual tests where it is shown that there is a strong tendency for ground return currents to take a path physically close to the outgoing current power conductor.

Usually, the conduit or metallic raceway that encloses the conductors provides an excellent fault return path. The presence of magnetic material in the power conductor enclosure (conduit or raceway) introduces additional inductive effects tending to confine the return ground currents within the magnetic enclosure. Installation of an external equipment grounding conductor is generally not acceptable and in reality is quite ineffective. Connections to nearby structural building steel members, in an effort to serve as an equipment grounding means, are equally ineffective as well and are not permitted.

The intentional or accidental omission of using the metallic conductor enclosure as an equipment grounding conductor can lead to large induced voltages in nearby metallic structures, which can appear as a dangerous shock hazard or unwanted circulating currents. Only by the installation of an internal equipment grounding conductor, in parallel with the raceway, can the current carried by the raceway be reduced. Joints in conduit and raceways must be connected in a workmanlike manner and made wrenchtight, using proper tools, for the raceway to function effectively as an equipment grounding conductor and as an effective path for fault current.

Fault-Current Test Procedure

As illustrated in figure 11-2, a special installation of 2½-inch rigid steel conduit and 4/0 AWG copper conductors was made for this investigation. It was

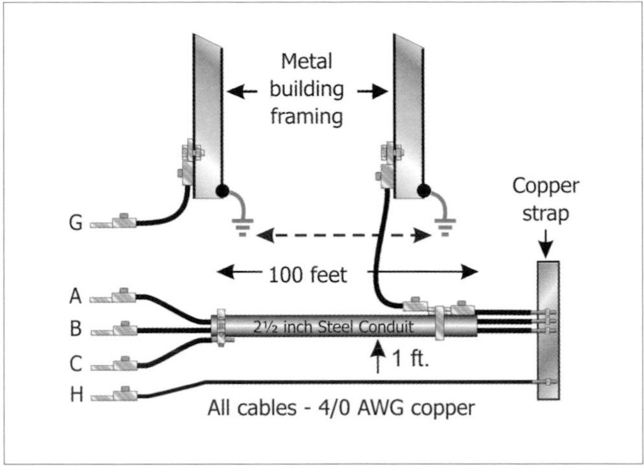

Figure 11-2. Ground current test procedure model diagram

installed in a building previously used for short-circuit testing because the building had heavy steel column construction, and all columns were tied to an extensive grounding mat composed of 250-kcmil bare copper conductors. The conduit was supported on insulators throughout the 100-foot length. The conduit was about 5 feet from a line of building columns. The external 4/0 AWG conductor was spaced about 1 foot from the conduit on the side opposite the building columns.

The setup was intended to simulate a typical electrical feeder circuit, which can be found in many commercial and industrial plants. It allowed a study of a wide variety of equipment grounding arrangements. In every case the supply current was through the "A" conductor run through the conduit. The "B" conductor served as an internal equipment grounding conductor. The "C" conductor was connected to the run of steel metal conduit, the "G" conductor was connected to the steel building columns, and the "H" conductor was the external equipment grounding conductor spaced about one foot away from and parallel to the conduit. A copper shorting bar was used at the far end of the conduit. With this arrangement it was possible to simulate various fault conditions at the right end and to verify various fault-current return paths.

Tests Performed, Low-Current

One series of tests was made at low current of 200 and 350 amperes using an ac welding transformer as a source of 60-hertz power. At these low-current magnitudes, the current could be maintained for extended periods.

Test No.	Current Flow: Out On	Current Flow: Return On	IA Total	Ic: Amperes	Ic: Per Cent of Total	Current Magnitudes: I_B	I_H	I_G	Voltage Magnitudes: E_{AC}	E_{CG}	E_{AB}	E_{AH}	E_{AG}	E_{GB}
Low-Current Tests														
A1	A	B	350	0		350	0	0	2.47		4.85			
A2	A	C	350	350	100	0	0	0	15.9	0.45	2.5			
A3	A	C	200	200	100	0	0	0	9.05	0.15	1.51			
A4	A	CH	350	340	97	0	12	0	16.0	0.05	2.55			
A5	A	CH	200	190	95	0	8	0	9.13	nil	1.55			
A6	A	CG	350	340	97	0	0	12	14.6		2.54	14.4		
A7	A	CG	200	180	90	0	0	8	9.5		1.50	9.4		
A8	A	CB	350	62	18	290	0	0	4.55	nil	4.55			
A9	A	CB	200	40	20	150	0	0	2.68	nil	2.68			
A10	A	GH	350	0	0	0	160	160	14.0	12.5	2.5	26.4	26.4	24.6
A11	A	GH	200	0	0	0	98	98	9.2	8.1	1.5	17.1	17.1	15.1
High-Current Tests														
B2	A	C	11,200	11,200	100	0	0	0	168	36*				
B3	A	C	11,070	11,070	100	0	0	0	173	38*				
B4	A	CH	11,070	11,200	101	0	1,140	0	173	18*				
B5	A	CH	11,080	11,090	100	0	1,220	0	173	17*				
B6	A	CG	10,830	10,770	99	0	0	1,080	168		71			
B7	A	CG	10,910	10,780	99	0	0	1,145	173	9*				
B8	A	CB	11,620	5,810	50	5,660	0	0		27*			155	
B9	A	CB	11,380	6,070	53	5,620	0	0	146	25*				
B10	A	CH	8,710	0	0	0	4,300	4,500		146			268	

*Distorted wave shape. Tabulated values are crest /$\sqrt{2}$

Table 11-1. Measured Electrical Current Quantites

Voltage measurements were made with high quality indicating meters. Current measurements were made with a clip-on ammeter.

Tests Performed, High Current

A second series of tests was made at high current, approximately 10,000 amperes using a 450 kVA, 3-phase, 60-hertz transformer with a 600-volt secondary as a source of power. Switching was done at the 13,800 primary voltage. An induction relay was used to control the duration of current to about ¼ second. An oscilloscope was used for all measurements of current and voltage.

Fault-Current Test Results

The results of the tests performed are shown in Table 11-1. The test number (for example, A2) is for reference purposes. The next two columns show the conductor connections used to determine the current path (for example, out on A and return on C). Next, the current values are shown, first the total input current through conductor A (350), and next the return current through the conduit (350) and its percentage of the total (100). The other columns with an "I" heading, along with a subscript indicating the circuit, show the amount of current returning over other possible paths.

The analysis conclusively confirms that only by the use of an internal grounding conductor can any sizable fraction of the return current be diverted from the raceway. In spite of the extremely low resistance of the building structural frame, it was ineffective in reducing the magnitude of the return current in the conduit (see tests A6, A7, B6 and B7).

Some interesting secondary effects were observed in the course of the tests. The first high-current test produced a shower of sparks from about half of the couplings in the conduit run. From one came a "blowtorch" stream of sparks that burned out many of the threads. Several small fires ignited nearby combustible material, which would have caused a serious fire hazard if not promptly extinguished.

The conduit run had been installed by a crew regularly engaged in such work, and they gave assurance that the joints had been tightened per normal practice and perhaps a little more. A short 4/0 copper (wire) jumper was bridged around this joint but, even so, some sparks continued to be expelled from this coupling on subsequent tests. Other couplings threw no more sparks during subsequent tests. Apparently, small tack welds had occurred on the first test.

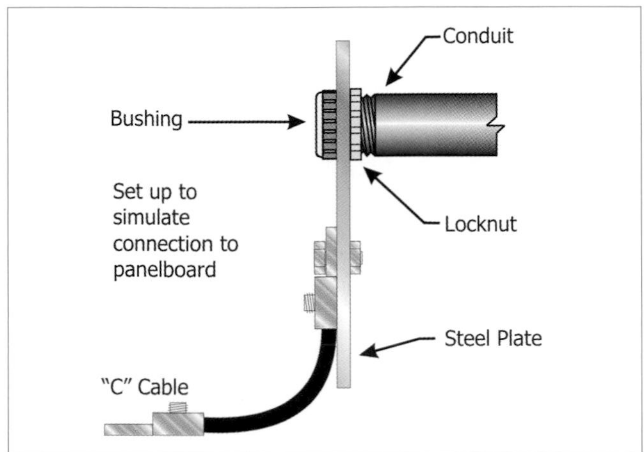

Figure 11-3. Alternate termination of "C" cable

In one high-current test, the conduit termination was altered to simulate a connection to a steel cabinet or junction box (see figure 11-3). The bushing was applied finger-tight. In one test, with about 11,000 amperes for about ¼ second, a fan-shaped shower of sparks occurred parallel to the plate. In the process, a weld resulted and the parts were separated only with considerable difficulty, with the use of wrenches and a hammer. This suggested that a repeat shot (of current) would have produced no disturbance (shower of sparks).

"During high-current test B10 (conduit circuit open), a shower of sparks was observed at an intermediate building column. Careful inspection disclosed that the origin was at a spot at which a water pipe passed through an opening cut in the web of the steel beam involved. Here is evidence of the objectionable effects of forcing the short-circuit current to seek return paths remote from the outgoing conductor. The large spacing between outgoing and returning current creates a powerful magnetic field which extends far out in space around the current-carrying conductors." [2]

Fault-Current Study Circuit Analysis

"The reactance of the circuit including the B conductor will be the lowest. Next will be the innermost tube of the conduit, followed by others in successive order until the outer tube is reached. The inductance of these tubular elements of the steel conduit assumes unusual importance because of the high magnetic permeability. Next in spacing (impedance) is the external grounding conductor (H conductor) and last, the structural members of the building frame and their interconnecting grounding conductors buried below floor level." [3]

"In test B10, both the exterior grounding conductor (H conductor) and the building frame (G terminal) were connected to provide parallel paths for the return current, but the conduit circuit was left open (C terminal not connected). The test results clearly evidence the powerful forces tending to maintain current in the conduit circuit. Note that across the open connections at the C terminal, a voltage of 146 volts (or, more significantly, over 50 percent of the impressed driving voltage) is required to force the current to return via the H conductor and the building frame in parallel. Such a voltage could be a serious shock hazard.

"Furthermore, unless the conduit was well insulated throughout its entire length (which is usually impossible or impractical in typical commercial or industrial installations), there would be a significant number of sparks at various points to constitute a serious fire hazard. It was during this test that a shower of sparks occurred between magnetic members in the building system that was caused simply by the strong magnetic field extending far out from the power conductors." [4]

Conclusions on Fault-Current Path Study

The significance of this investigation clearly points to the conclusions presented earlier. Effective use of the conduit or raceway in the equipment-grounding system is paramount. Additional work is needed to develop joints which will not "throw fire" during faults. Improving effectiveness requires greater conductivity in the conductor enclosure or the use of an internal (equipment) grounding conductor. Grounding conductors (grounding electrode conductors) con-necting the building structure to grounding electrodes (connection to earth) are needed to convey lightning currents or similar currents seeking a path to earth, but these conductors will play a negligible part in the performance of the equipment grounding system. Of course, the importance of proper equipment grounding becomes greater with the larger size feeder circuits and the availability of higher short-circuit currents. [5]

Common Elements of Fault-Current Path

Clearing faults involves one or two parts, depending on whether the service is grounded or ungrounded. In a grounded distribution system, there are two parts: (a) the grounded system conductor and (b) the equipment and conductor enclosure grounding system. In an ungrounded system, the grounding system covers only (b) the equipment and conductor enclosure grounding system.

Effective Path for Fault Current

To have an effective path for fault current for both grounded systems and ungrounded systems, the fault-current path must: (1) be electrically continuous; (2) must have ample capacity to conduct safely any fault currents likely to be imposed on it; and (3) have lowest possible impedance to limit the voltage to ground and to facilitate the operation of overcurrent protective devices in the circuit [see 250.4(A)(5) and 250.4(B)(4)]. The definition of *effective ground-fault current path* also provides this same information relative to the function of this path [see 250.2 Definition].

"Impedance sufficiently low" means, for all practical purposes, that the grounding path and the circuit conductors for an ac system must always be within the same metallic enclosure such as a conduit, wireway or cable. The metallic enclosure (such as a conduit or cable) under certain conditions may be used to provide the equipment grounding path for the circuit conductors within it. In addition, where a nonmetallic wiring method is used for an ac system, all the circuit conductors, including the equipment grounding conductor, must be in the same raceway.

Overcurrent Device - Amps	Maximum Z of Ground-Fault Circuit in Ohms Includes Fault Impedance
60	0.4
100	0.24
200	0.12
400	0.06
600	0.04
800	0.03
1200	0.02
1600	0.015
2000	0.012
2500	0.0096
3000	0.008
4000	0.006

Table 11-2.

The *Code* gives no maximum impedance of the ground-fault circuit for grounded or ungrounded systems but states in 250.4(A)(5) that it shall be sufficiently low to facilitate the operation of the circuit-protective devices. This path should have impedance no greater than will allow the circuit breaker or fuse to reach its instantaneous pickup operating range. This will cause the overcurrent device to operate quickly to remove the fault from the system. Any fault current less than the instantaneous operating value will extend or delay, by some value, the opening time of the overcurrent device.

Every manufacturer of circuit breakers and fuses publishes operating or trip-curve charts for their products (for example, see figure 11-4). These charts should be carefully reviewed to be certain the ground-fault path has an impedance low enough to allow the overcurrent device to operate quickly to reduce thermal damage to the circuit and equipment. As can be seen by reviewing these charts, circuit breakers have a different trip curve depending on whether they are single-, double- or three-pole configurations. In addition, the same family of circuit breakers will have different trip curves for different ampere-rated circuit breakers. Look for the multiple of the circuit breaker rating at which it reaches its instantaneous pickup rating which is no time delay in the operation of the circuit breaker.

A similar review of the fuse operating characteristics chart should be performed to select a ground-fault circuit that will result in instantaneous operation of the fuse.

Some engineering designs may incorporate the concept of the fault current being not less than five times the rating of the overcurrent device. For example, if the overcurrent device is 400 amperes, there needs to be not less than 2000 amperes of current in the circuit for it to open the faulted circuit in a reasonable time. For some overcurrent devices, current of five times the device rating might not reach the instantaneous trip range of the device. Generally, the longer the excessive current exists in the circuit, the greater the thermal and mechanical stress to the conductor insulation.

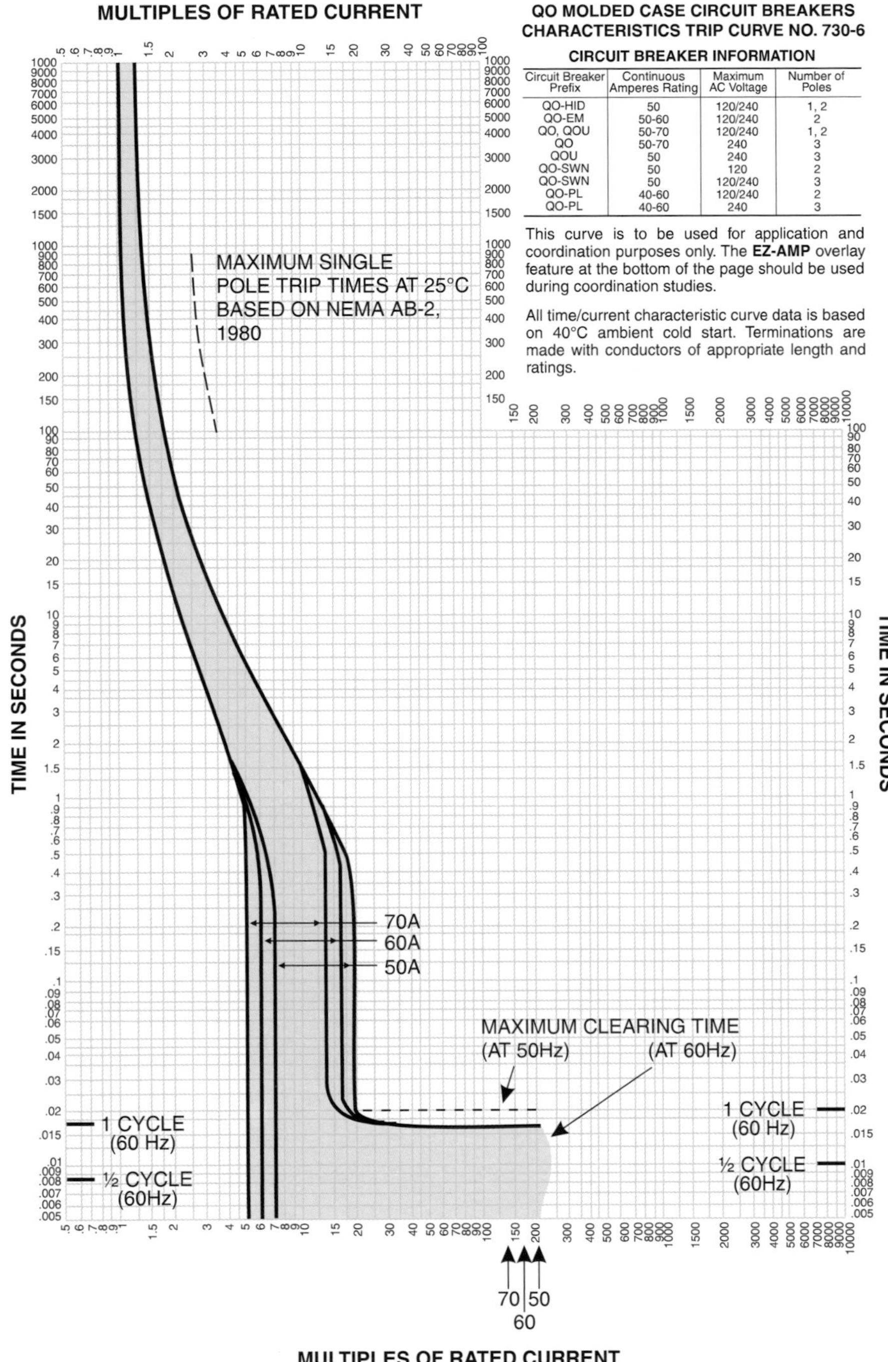

Circuit Breaker Prefix	Continuous Amperes Rating	Maximum AC Voltage	Number of Poles
QO-HID	50	120/240	1, 2
QO-EM	50-60	120/240	2
QO, QOU	50-70	120/240	1, 2
QO	50-70	240	3
QOU	50	240	3
QO-SWN	50	120	2
QO-SWN	50	120/240	3
QO-PL	40-60	120/240	2
QO-PL	40-60	240	3

Figure 11-4. Circuit Breaker Trip Curves Graph

Courtesy of Square D

Clearing Short Circuits

The method employed for clearing a short circuit is the same regardless of whether the system is grounded or ungrounded. Essentially, it involves placing an overcurrent device in series with each ungrounded system conductor (see figure 11-5).

In the event of a *short circuit*, which is a fault from conductor to conductor, the words "as quickly as is practical" means a very short period of time down to as low as a fraction of a cycle depending on the amount of short-circuit current and the characteristics of the overcurrent device. The high-interrupting capacity current-limiting fuse does that automatically within virtually all ranges of short-circuit current values. In addition current-limiting circuit breakers and circuit breaker current limiters are available from a wide variety of manufacturers. Where properly applied, these current-limiting devices significantly reduce the thermal and mechanical damage to electrical equipment.

Current-limiting fuses and their correct application are covered in Underwriters Laboratories' *Electrical Construction Materials Directory* guide card (JCQR) information. This section reads in part: "The term *current limiting* indicates that a fuse, when tested on a circuit capable of delivering a specific short-circuit current (RMS amperes symmetrical) at rated voltage, will start to melt within 90 electrical degrees and will clear the circuit within 180 electrical degrees (½ cycle).

"Because the time required for a fuse to melt is dependent on the available current of the circuit, a fuse that may be current-limiting when subjected to a specific short-circuit current (rms amperes symmetrical) may not be current- limiting on a circuit of lower maximum available current." [6]

A current-limiting circuit breaker is defined in the Underwriters Laboratories' *Electrical Construction Materials Directory* guide card (DIVQ) information as, "A current-limiting circuit breaker is one that does not employ a fusible element and that when operating within its current-limiting range, limits the let-through

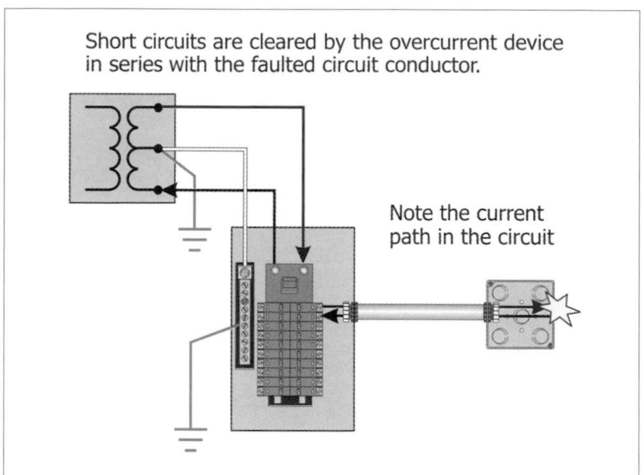

Short circuits are cleared by the overcurrent device in series with the faulted circuit conductor.
Note the current path in the circuit

Figure 11-5. Clearing short circuits

I^2t (current squared time) to a value less than the I^2t of a ½ cycle wave of the symmetrical prospective current.... Current-limiting circuit breakers are marked 'current-limiting' and are marked either to indicate the let-through characteristics or to indicate where such information may be obtained." [7]

Circuit breaker current limiters are covered in Underwriters Laboratories' *Electrical Construction Materials Directory* in guide card (DIRW). They are described as "Circuit breaker current limiters are designed to be used in conjunction with specific circuit breakers and to be directly connected to the load terminals of the circuit breakers. They contain fusible elements which function only to increase the fault current interrupting ability of the combination which is intended for use in the same manner as circuit breakers when installed at the service and as branch circuit protection. The limiters are rated 600 v or less." [8]

It is vital to carefully follow manufacturer's instructions when applying current-limiting fuses or circuit breakers.

In some cases, depending on the available fault current, this equipment might not provide current limitation. This is due to the inverse time nature of these overcurrent devices. Inverse time means that as the amount of current through the device increases, the operating time of the device reduces. Overcurrent devices see fault current at lower operating ranges as a load and react much slower than at their current-limiting range.

Some installations of fuses or circuit breakers are installed in a series combination configuration, that is, two or more fuses or circuit breakers are in series for the circuit. These devices that have been tested for their operating compatibility may be marked by the manufacturers with a series-combination rating. In this case, usually the downstream overcurrent device is rated below the fault current that is available at its line terminals, while the overcurrent device closest to the source is rated at or above the fault current that is available at its line terminals.

Where a series-rated system is installed, only the equipment that has been tested to determine its suitability can be installed in this manner. Suitability is determined by reference to manufacturer's identification of components on decals located on the equipment. For additional information, see "Interplay of Energies in Circuit Breaker and Fuse Combinations." [9]

The time-current characteristic curve of such fuses starting from a value of from five to six times the rating of the fuse indicates a clearing time of about one second and a time of well below one cycle for values of 50 times fuse rating. A further study indicates that such high-interrupting capacity current-limiting fuses have a pronounced current-limiting effect at values as high as fifty times the fuse rating.

Whether a fault current is a short circuit or a ground fault, the overcurrent devices ahead of the point of fault will see all of the fault current. Placing an overcurrent protective device at the point the conductor receives its supply, as required by 240.21, will provide the short-circuit protection that is necessary. The rating of the overcurrent device is based upon 240.4 and the ratings of standard overcurrent protective devices given in 240.6. The basic requirement is that conductors be protected in accordance with their rated ampacity in accordance with 310.15.

Clearing Ground Faults

A ground-fault circuit is different than a short circuit. In a ground fault, there can be such a high impedance of the faulted circuit that the controlling factor of current is the impedance of the ground-fault circuit. In this case, the only part the available short-circuit capacity plays is in its ability to maintain voltage. In

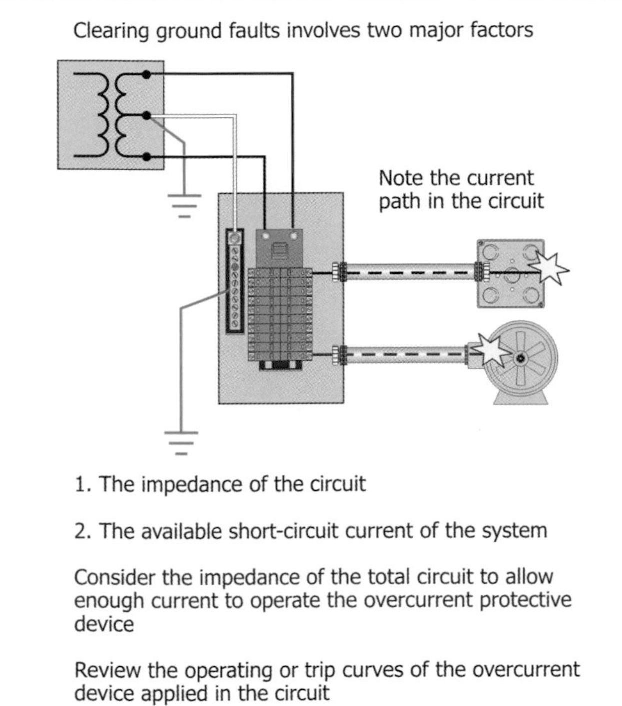

Clearing ground faults involves two major factors
Note the current path in the circuit
1. The impedance of the circuit
2. The available short-circuit current of the system
Consider the impedance of the total circuit to allow enough current to operate the overcurrent protective device
Review the operating or trip curves of the overcurrent device applied in the circuit

Figure 11-6. Clearing ground faults

some circuits, the amount of current in a ground-fault circuit is not dependent on the available capacity of the system, other than its ability to maintain full voltage during a ground fault (see figure 11-6).

When it comes to clearing a ground fault, two things control the current, these are: (1) the impedance of the circuit, and (2) the available short-circuit current of the system.

A relatively large voltage drop can be tolerated in this grounding conductor. The impedance of the complete ground-fault circuit, from the point of fault back to the transformer, should never be higher than what would permit the minimum amount of current necessary for the overcurrent device to operate within its instantaneous range [review the operating characteristics chart for the overcurrent device for the correct value] (see figure 11-12).

This will provide a factor of safety to allow for the variable impedance at the point of the fault. Some overcurrent devices operate at their instantaneous current range at about five times the rating of the overcurrent device. For example, some 50-ampere overcurrent devices have an instantaneous trip rating of about 250 amperes. In other cases, the instantaneous rating could be from six to ten times the rating of the overcurrent device.

Selective Coordination

Good electrical system designs generally involve good coordination between all levels of overcurrent protection to localize overcurrent or ground-fault conditions to the offending circuit. There are cases where the *Code* specifically requires selective coordination of the overcurrent devices. A definition of the term *selective coordination* is provided in Article 100 and reads as follows:

Coordination (selective). "Localization of an overcurrent condition to restrict outages to the circuit or equipment affected, accomplished by the choice of overcurrent protective devices and their ratings or settings."

Section 240.12 specifically requires electrical system coordination where an orderly shutdown is required to minimize hazard(s) to personnel and equipment. This system of coordination can be accomplished by either coordinated short-circuit protection or by overload indication or monitoring systems. Another requirement for selective coordination for the overcurrent devices of a system is provided in Article 700, which covers the requirements for emergency systems. Emergency system overcurrent protective devices are required to be selectively coordinated with all supply side overcurrent devices [see 700.27].

As indicated earlier in this chapter, good system design meeting the applicable requirements of the *Code* requires a careful study and selection of the overcurrent protective devices and assuring an effective ground-fault current path is provided to facilitate overcurrent device operation at the local offending circuit level. An example of a system that has not been coordinated properly would be where a ground fault on a 30-ampere branch circuit causes a 2000-ampere GFPE device in a switchboard to operate before the 30-ampere breaker can open.

Clearing Faults in Service Equipment

The service equipment enclosure is probably one of the most vulnerable locations for ground faults to occur. There is no overcurrent protection on the line side of the service conductors at their ampacity, only short-circuit protection provided by the serving utility's transformer primary overcurrent protection. Overload protection is provided by the service overcurrent device in series with the service-entrance conductors. This is one of the main reasons the *Code* requires the service disconnecting means to be located outside of the building, or, if installed inside, nearest the point of entrance of the conductors into the building. This limits the length of conductors without typical overcurrent protection inside the building.

If, on the other hand, a ground fault does not develop on the load side of the service, but at a point on the line side of the service, then that fault can only be cleared by the primary overcurrent device on the supply side of the utility transformer. This protection is the primary fuses or cutouts for the transformer which can be many times the rating of the secondary conductors. In many cases, a ground fault on the line side of the service will not clear through the primary overcurrent devices and can only clear by developing into a short circuit or by burning itself clear.

If the electrical equipment on the line side of the service is not properly bonded and a properly sized main bonding jumper installed, it is highly unlikely that there will be enough current in the path to clear the ground fault through the overcurrent devices on the line side of the utility transformer (see figure 11-7).

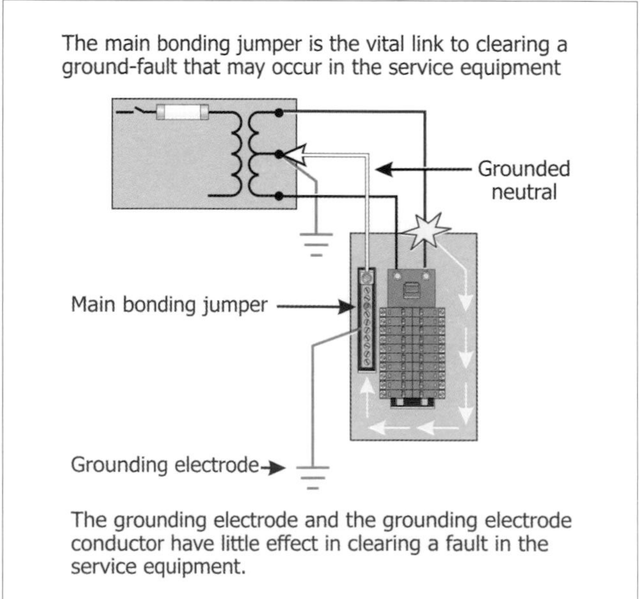

The main bonding jumper is the vital link to clearing a ground-fault that may occur in the service equipment
Grounded neutral
Main bonding jumper
Grounding electrode
The grounding electrode and the grounding electrode conductor have little effect in clearing a fault in the service equipment.

Figure 11-7. Clearing faults in service equipment

Analysis of Clearing Ground Faults

The following examples illustrate a simple electrical system and help analyze the conditions existing in the event of a ground fault. In the first case, the neutral is run from a grounded system where it is connected to the service equipment. In the second example, the grounded system (service) conductor has not been run from the system to the service disconnecting means.

Purposes Served by the Grounded Conductor (Often the Neutral Conductor) on a Grounded System

For maximum safety and to comply with 250.24(C), the neutral conductor or grounded conductor must be run from a grounded system transformer to all services and be bonded to each service disconnecting means enclosure even though all power may be utilized at line voltage only (see figure 11-7). This is required even though the service might supply only line-to-line loads [see 250.24(C) and 250.28].

The grounded conductor of any grounded system serves two main purposes. First, it permits utilization of power at line-to-neutral voltage, and therefore serves as a current-carrying conductor to carry any unbalanced current back to the source. Second, it plays a vital part in providing a low-impedance path for fault currents to facilitate the operation of the overcurrent devices in the circuit, as required by 250.4(A)(5).

The grounded conductor (neutral conductor) provides the lowest impedance return path for fault currents to the transformer neutral point as can be traced in figure 11-8. If the neutral is not needed for voltage requirements, it still must be run to the service, bonded to the service disconnecting means enclosure, and connected to the equipment grounding conductor at the service. In that event, the neutral conductor no longer serves as a neutral conductor but as a grounded conductor. If that is not done, it is difficult, if not impossible, to clear a fault on the system.

Grounded (Neutral) Conductor Installed

In figure 11-8, the neutral of the system serving as the grounded conductor is carried to the service equipment and bonded to the equipment grounding conductor to provide a low-impedance path directly to the transformer. Here it can be seen that a fault current does not have to go through the earth to complete the circuit, but will go through the grounded service conductor, which is a low impedance path. This will very likely permit sufficient current to operate the overcurrent device. The parallel path through the grounding electrode conductor and the grounding electrode and the earth still exists. The grounded service conductor, forming a relatively low impedance path, will carry most of the fault current, generally 90 percent or more in most cases.

It is obvious, then, that to get all the protection afforded by a grounded system, the grounded conductor must be run to the service and must be bonded to the disconnect enclosure and equipment grounding conductor even though the neutral conductor is not needed for serving any load. For the same reason, where the neutral is used for voltage requirements, the neutral size should be based not only on the neutral load demands [220.61] but also on the basis of the service overcurrent device and the amount of fault current necessary to operate the overcurrent devices.

Figure 11-9 shows a 120/240-volt single-phase grounded system where all the load is supplied at 240 volts. A grounding electrode conductor is properly connected to a low-resistance grounding electrode, the metal water supply system, and bonded to the equipment grounding conductor, all in accordance with the requirements of the *Code* (prior to the 1962 *Code*). However, the neutral conductor is not run to the service equipment since all power utilization is at 240 volts only. For serving the load, the neutral has no useful purpose, and it would at first appear as if it could be omitted.

From the standpoint of limiting the voltage between equipment and ground under normal conditions and from the standpoint of utilization of power at 240 volts, the circuit, as shown in figure 11-9, will function correctly, provided that the insulation remains intact and no ground faults occur on the system.

If, however, a ground fault occurs in the equipment, as illustrated in figure 11-9, due to insulation failure, the voltage between equipment and ground will rise considerably. Starting at the point of fault, first is the impedance of the fault, then the conduit itself becomes a conductor in the fault-current circuit, then the grounding electrode conductor and the grounding electrode at the service. However, the fault current must travel from the grounding electrode through the earth itself to the point where the neutral is grounded at the transformer, then through the transformer and back to the service and, finally, through the overcurrent device to the point of fault.

A study of the fault circuit as diagrammatically represented in figure 11-9 shows that it would be unlikely to have an impedance of less than 22 ohms as the sum of all the impedances shown.

That is an optimistically low value at best. The maximum current in this circuit is 5.5 amperes for

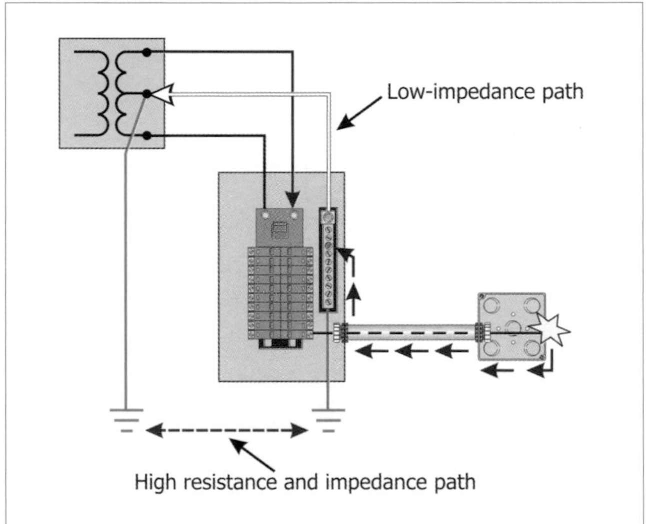

Figure 11-8. Grounded conductor run to service equipment as required by 250.24(C)

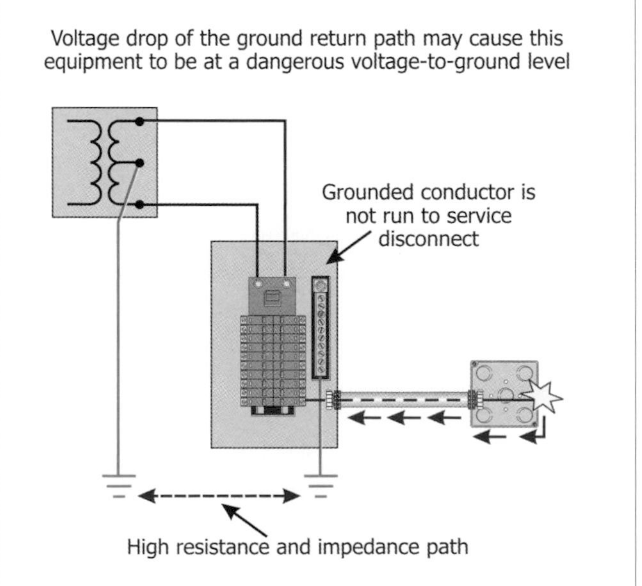

Figure 11-9. Grounded service conductor is not run to service equipment.

the 120/240-volt circuit shown (120 volts ÷ 22 ohms = 5.5 amperes). With a 100-ampere service and 20-ampere overcurrent device, it is obvious that the overcurrent device would not operate. A serious shock hazard, as well as fire hazard, will exist until the circuit is manually opened. If the circuit was opened because of being properly grounded, then the only period of time a shock hazard would exist would be for the duration of the fault. The fire hazard, as well as shock hazard, would be reduced to a relatively short period of time.

When a fault occurs on a system, it is only during the period while the fault exists that a potential hazard is present. It is of the utmost safety importance that the fault clearing time be held to the shortest practical period of time.

For the installation as shown in figure 11-9, the fault will not clear and can exist for minutes, hours or even days before it is recognized. Further, that recognition can be from observing a fire, or from noting that a victim has received a shock, which sometimes can even be fatal. It is a matter of record that many serious, and sometimes fatal, accidents have resulted from faults that were not cleared promptly because the system was not properly grounded.

If the neutral at the utility transformer were grounded to the same water pipe system as the service (may not be too likely), then the resistance of the fault path would be appreciably decreased. Because of the wide separation between the service conductor and the water pipe, the reactance and, therefore, the impedance of the fault circuit would remain high. The probabilities are that the fault current would not reach a high enough value to operate the overcurrent device. Again, fire and damage to equipment would continue until the circuit was manually opened. To improve the safety of such a system, a low-impedance path must be provided to carry enough current to clear the circuit by the overcurrent devices.

Three-Phase Services

Identical reasoning to that used for single-phase systems may be applied to any multi-wire, multi-phase grounded system. Any 3-phase power supply taken from such a grounded transformer bank must have the neutral or grounded conductor brought into the service to satisfy the requirements of 250.24(C). This is true regardless of whether or not there is neutral load at the service. The grounded conductor provides a low-impedance path for fault current to return to the source.

Open System

The above statements applying to maximum ground-fault currents would not apply to an open system. If there were no metallic enclosures, the impedance of the ground-fault circuit would be much lower. Accordingly, greater ground-fault currents can be expected in actual practice in an open or nonmetallic installation. However, since most systems of 600 volts or less are metal enclosed, there would not be such high ground-fault currents in those systems. For open systems, it is vital that the equipment grounding conductor be run with the circuit conductors to maintain a low impedance ground-fault path.

Recommended Length of Conduit for Use as Equipment Grounding Means

The *Code* currently places no restriction on the size or length of rigid metal, intermediate metal conduit or electrical metallic tubing where used as an equipment grounding conductor. Independent tests have shown that consideration must be given to both size and length of conduit.

Extensive work to determine the maximum safe length of conduit to serve as an equipment grounding conductor has been done by the School of Electrical and Computer Engineering at the Georgia Institute of Technology in Atlanta, Georgia.[10] Computer software has been developed which will allow the calculation of almost any combination of conductors and conduit for use in the ground-return path. This software can also be used for calculating the maximum length of equipment grounding conductors that are safe to use where not installed in metallic conduits, such as PVC conduit and other nonmetallic raceways.

Where a metallic conductor enclosure is used as an equipment grounding conductor, it must have continuity and the conductivity to carry enough current to facilitate the operation of the overcurrent devices. Some electrical inspection authorities require that an equipment grounding conductor be installed inside the conduit to account for poor workmanship of the raceway installation or to maintain continuity where fittings can be broken during use. This is not a minimum requirement of the *NEC*.

The engineer should examine the equipment grounding conductor (the metal enclosure) to assure it will function properly in the event of a ground fault. Where wireways, auxiliary gutters and busways have steel enclosures, there can be enough material cross section to serve as an equipment grounding conductor. It is questionable whether the electrical connections between lengths are adequate for carrying enough fault current to clear the fault. These raceways may also require a supplemental equipment grounding conductor.

The conductor enclosure, conduit, raceway, and so forth, also may be acceptable as the equipment grounding conductor in lieu of the copper or aluminum conductors given in Table 250.122. Generally, if a conduit or electrical metallic tubing meets the *Code* requirements for conductor fill, the enclosure will provide an acceptable equipment grounding conductor.

In an average busway up to 1500 amperes rating, there is enough steel in the enclosure to provide an acceptable equipment grounding conductor if proper conductivity is assured at the joints. For busways of higher rating, it is doubtful if the steel enclosure is heavy enough. For all sizes of busways, the electrical connection at the joints must be checked carefully. Many busways have aluminum enclosures. In such busways, the enclosure has sufficient conductivity. Good electrical connections at the joints also must be checked.

Long Conduit Run Designs

Consider an installation where a 3-inch conduit run is 1000 feet long and has a 400-ampere overcurrent protective device protecting the contained conductors. Assume that the minimum fault current would be 600 percent of the overcurrent device rating, or, in this case, 2400 amperes. The impedance of 3-inch conduit at 2400 amperes is approximately 0.0875 ohms/1000 feet. For a 208Y/120-volt system with zero impedance at the point of fault, the maximum current will be about 1400 amperes. That value of current would operate a 400-ampere high-interrupting capacity current-limiting fuse in about two seconds. However, such ground faults are nearly always arcing faults. This adds resistance to the circuit and reduces the fault current in the circuit. Moreover, since conduit impedance increases with decrease in fault current, the conduit impedance would be 0.129 ohms/1000 feet at a current of 1,400 amperes.

Allowing for the arc impedance in the circuit described and the increase in conduit impedance at 1400 amps (about 50 percent) and other variable factors including a further increase of conduit impedance at the still lower current, the ground-fault current is more likely to be closer to 300 amperes. With a 300-ampere fault current, the 400-ampere high-interrupting capacity current-limiting fuse obviously would not clear the fault.

For a long feeder such as this, it can be necessary to increase conductor sizes to account for voltage drop. Section 250.122 requires that the equipment grounding conductor also be increased in size in proportion to the size of the ungrounded conductors. For example, if the 3-inch conduit run was only 100 feet long, the anticipated fault current would be about 3000 amperes, which would satisfactorily operate the overcurrent device. Both fault-current values were based on the impedance of only that part of the circuit that would be within the conduit, not the impedance of the entire circuit. The fault-current values met in practice would likely be less than those given here.

What can be done to achieve maximum safety? The system must be designed to ensure that performance of the grounding and bonding in the system fulfills all the requirements as set forth in 250.4(A)(5) and 250.4(B)(4) to obtain proper or effective grounding, bonding, and an effective path for fault current. This can mean adding a supplementary equipment grounding conductor within the conduit in parallel with the conduit if the calculations indicate its need.

Where an equipment grounding conductor is added inside a raceway, it is done to increase the capacity, that is, decrease the impedance of the raceway that is serving as an equipment grounding conductor. Where a metallic system is installed, the conductor enclosure is also part of the equipment grounding conductor. The assistance of that parallel conduit path is valuable, but proper use can be made of it only if the equipment grounding conductor

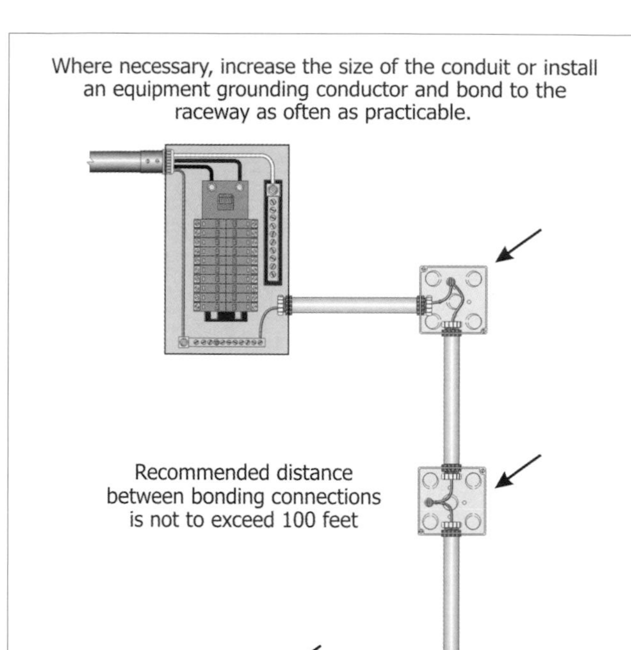

Where necessary, increase the size of the conduit or install an equipment grounding conductor and bond to the raceway as often as practicable.
Recommended distance between bonding connections is not to exceed 100 feet

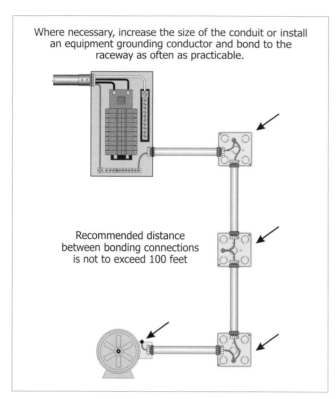

Figure 11-10. Long conduit runs can necessitate installing an equipment grounding conductor in the raceway in some cases.

and the conduit are bonded together as frequently as is practical (see figure 11-10).

If the impedance of the ground-fault circuit is higher than what will carry enough current to properly operate the overcurrent devices in a reasonable time, then a lower impedance can be obtained by adding an equipment grounding conductor within the conduit in parallel with the conduit. That conductor must be run within the conduit, that is, run with the circuit conductors. The equipment grounding conductor should never be run outside the conduit or raceway through which the conductors serving the equipment are run. Where run external to the conduit, it becomes quite ineffective in the grounding circuit, for virtually all the ground-fault current will return on the conduit. Further, the equipment grounding conductor must be run as close to the phase conductors as is practical, right to the point where it connects to the neutral conductor at the service.

By determining the impedance of the circuit involved when a ground fault occurs, it can be determined whether the metallic enclosure will make an acceptable equipment grounding conductor or whether it will be necessary to supplement the enclosure with an equipment grounding conductor. Where a steel conduit is a part of the electric circuit, as it will be where a ground fault occurs, there will be a large increase in both the resistance and reactance of the circuit, which will vary considerably with the amount of fault current.

Laboratory tests have shown that for a single-phase current through a conductor within a steel conduit, the impedance of the circuit is approximately equal to the impedance of the conduit itself. The size of the conductor within the conduit has relatively little effect on the circuit impedance. Also, despite the fact that there are many parallel paths external to the conduit, the current in all the parallel paths will be very small, and under normal conditions would be less than 10 percent of the total fault current.

Two other factors are to be taken into account in estimating the ground-fault current. They are the effects of the conduit couplings in increasing the impedance of the circuit and the voltage drop across the point of fault. If conduit couplings are installed wrenchtight, as required by *Code*, the increase in impedance of the conduit with couplings is about 50 percent more than the impedance for a straight run without couplings. This is where the value and importance of installing a properly-sized equipment grounding conductor inside the conduit and bonding it to the conduit at frequent intervals is proven for ensuring low impedance and safety.

Impedance values for conduit can be obtained from manufacturers. By using these impedance values, adjusting for the couplings and estimating a 50-volt drop across the fault, a reasonable value for the amount of fault current in the circuit can be determined.

Assume a 200-foot run of metric designator 78 (3-inch trade size) conduit with 500-kcmil conductors on a 208Y/120-volt circuit protected by 400-ampere overcurrent devices. During a ground fault, the amount of current will therefore be:

E (voltage) ÷ Z (impedance) = I (current)

With a 50-volt drop at the fault and Z (impedance) = 0.02970, the current will be about 2350 amperes. The use of a MD 78 (3-inch trade size) conduit as the equipment grounding conductor where 400-ampere overcurrent devices are used is, thus, satisfactory for this run.

A simpler method of determining if the conduit or metallic enclosure will perform satisfactorily is to first calculate the minimum desired fault-current (5 times the overcurrent device rating or better yet to reach the instantaneous portion of the time/current curve), which, in this case, is 5 x 400 or 2000 amperes. Then, on the basis of 70 volts available for a 120-volts-to-ground circuit, calculate Z (impedance), which will be found to be 0.035. A straight run of MD 78 (trade size 3) conduit has about 0.099 ohms impedance per thousand feet at 2000 amperes. To that value add 50 percent to include a factor of safety. That will give an impedance of 0.01485 ohms per 100 feet. The impedance value that will allow 2000 amperes in that circuit was found to be 0.035. Since 235 feet of conduit carrying 2000 amperes will have an impedance of 0.035, it has been determined that to have a minimum current of 2000 amperes in a ground fault, up to 235 feet of MD 78 (trade size 3) conduit can be installed for a circuit protected by a 400-ampere overcurrent device.

Table 11 in chapter twenty-one provides supporting data by the Georgia Institute of Technology relative to the maximum lengths of steel conduit or tubing that may safely be used as an equipment grounding conductor (see also tables 20-16 through 20-19).

If a metric designator 103 (trade size 4) conduit was used instead of metric designator 78 (trade size 3), and the overcurrent device rating did not change but remained at 400 amperes, the maximum length of conduit can be determined by reference to table 21-11. With some interpolation and using the same calculations, it is found that 260 feet of MD 103 (trade size 4) conduit could be installed and provide a satisfactory equipment grounding conductor for this circuit.

It can be determined for any circuit and any size conduit, for any size overcurrent device, the maximum safe length of conduit that will allow a fault current which will be sufficient to facilitate the operation of the overcurrent device. Should the circuit length exceed the maximum safe length as calculated, and then it will be necessary to add a metallic (copper or aluminum) equipment grounding conductor in parallel with the conduit or increase the size of the conduit. Minimum equipment grounding conductor sizes are provided in Table 250.122.

It would be neither practical nor desirable to substitute a copper or aluminum conductor for the conduit. Rather, add the copper or aluminum conductor, and connect it in parallel with the conduit to form an equipment grounding conductor of two conductors in parallel. The copper or aluminum conductor and the conduit should be connected together at convenient practical intervals, about every 100 feet or less. That will reduce the length of circuit through which the ground-fault current will be through the conduit alone.

The *Code* permits an equipment grounding conductor to be bare or insulated. However, if the conductor is bare, there can be some arcing between the bare conductor and the interior of the conduit at points other than the point at which the ground fault occurs. This is due to slight differences in impedance between the raceway and wire, resulting in potential differences that can result in arcing. Such arcing can damage the phase conductors without adding to the proper functioning of the ground-fault circuit. That makes a strong case for the use of insulated equipment grounding conductors where installed in a metallic enclosure.

If aluminum conduit was used instead of steel conduit for the same conditions cited above [500-kcmil copper conductors, MD 78 (trade size 3) conduit and a 400-ampere overcurrent device], the circuit run could be about 900 feet long, and the aluminum conduit would provide a satisfactory equipment grounding conductor. A MD 78 (trade size 3) aluminum conduit has a dc resistance of about 0.0088 ohms/1000 feet and 500-kcmil copper cable has a dc resistance of 0.0222 ohms/1000 feet.

Flexible metal conduit is suitable as an equipment grounding conductor for not more than 6-foot lengths in the ground-fault return path and with not more than 20-ampere overcurrent protection of the contained conductors and meeting the other conditions stipulated in 250.118(5). As such, flexible metal conduit should include an internal equipment grounding conductor. The *Code* requires that the various metal raceways shall be so constructed that adequate electrical and mechanical continuity of the complete system be secured. However, because of the various joints involved, it is important for the engineer to investigate such conductor enclosures to determine that their impedance is sufficiently low to provide an effective ground-fault current path.

Adequate Size of Conductor

In general, the minimum size of equipment grounding conductor is provided in Table 250.122 and is based on the rating of the overcurrent device ahead of the conductors. (An analysis of Table 250.122 is provided in table 7 of chapter 21.) The rule of thumb may be applied, but it should be verified by calculation that the equipment grounding conductor should not be less than 25 percent of the capacity of the phase conductors or the overcurrent device that supply the circuit. A note has been added to the table indicating that the size of equipment grounding conductor given in the table must be increased if necessary to comply with 250.4(A)(5). This adds emphasis to the heading of Table 250.122 because it indicates that equipment grounding conductors given in the table are the minimum size.

An analysis of Table 250.122 shows the relation of the equipment grounding conductor to the size of the overcurrent device (based on the continuous rating of 75°C-rated wire). It is from 50 to 125 percent of the phase conductor for overcurrent devices up to 100-ampere rating. The rating varies from 33 to 25 percent for overcurrent devices rated up to 400 amperes and is from 22 percent for 600-ampere overcurrent devices to a low of only 8 percent for an overcurrent device of 6000 amperes.

Obviously, the equipment grounding conductor must be large enough to carry that amount of current, for the amount of time necessary to clear the overcurrent device with which it is associated, and not result in extensive damage. In addition, it is vital that the equipment grounding conductor be run with the circuit conductors or enclose the circuit conductors. This was covered extensively in chapter nine.

Conductor Withstand Rating

Section 110.9 states, "Equipment intended to interrupt current at fault levels shall have an interrupting rating sufficient for the nominal circuit voltage and the current which is available at the line terminals of the equipment." [11] This includes fuses, circuit breakers, disconnect switches and similar equipment.

Section 110.10 reads in part, "The overcurrent protective devices, the total impedance, the component short-circuit current ratings, and other characteristics of the circuit to be protected shall be selected and coordinated to permit the circuit protective devices used to clear a fault to do so without extensive damage to the electrical components of the circuit. This fault shall be assumed to be either between two or more of the circuit conductors, or between any circuit conductor and the grounding conductor or enclosing raceway." [12]

Equipment grounding conductors, circuit conductors, busbars, bonding jumpers, and so forth, are not intended to break current. These conductors must be large enough to safely carry any short-circuit and ground-fault current for the time it takes the overcurrent protective device to clear the fault. This is clearly stated in 110.10, 240.1 FPN, 250.4, 250.90 and 250.96. Section 310.10 also provides details for the temperature limitations of conductors.

The integrity of equipment grounding conductors, grounding electrode conductors, main bonding jumpers and other circuit conductors must be ensured by sizing them properly. Equipment grounding conductors that are too small are of little value in clearing a fault and can, in fact, give one a false sense of security. Grounding conductors must not burn off during ground-fault conditions, leaving the equipment enclosure energized, in many cases, creating a shock hazard that can cause serious injury or even be fatal. Safety should not be compromised. In fact, the first sentence of the *Code* states that, "the purpose of this *Code* is the practical safeguarding of persons and property from hazards arising from the use of electricity." [13]

Bolted Connections

It can be calculated, using values from the Insulated Cable Engineers Association publication P32-382 (1994), that an insulated copper conductor with a bolted connection can safely carry one ampere for every 42.25 circular mils for five seconds without destroying its validity (see figures 11-11 and 11-12). That will be the short-time rating or I^2t (amperes x amperes x time) value of the conductor. Then, from the time-current characteristic curves of various approved overcurrent devices, the amount of current necessary to clear the overcurrent device in five seconds can be determined.

Using that formula, the size of the equipment grounding conductors that will be proportional to those given in Table 250.122, as analyzed in table 7 of chapter twenty, can be determined. Equipment grounding conductors are permitted to be bare, and, in most cases, are pulled into the same raceway as the insulated phase conductors. This presents a potential problem. When an equipment grounding conductor is carrying ground-fault current, an extreme rise in its temperature can cause the insulation on the adjacent phase conductors to melt, causing further damage. Again, the potential for equipment damage and electrical shock hazard to personnel is increased. It is desirable to limit the heat of the faulted circuit to reduce damage to adjacent insulated conductors.

Thus, as discussed in this chapter, for copper conductors, the clearing time and short-circuit current must be limited to:

- One ampere...
- for five seconds...
- for every 42.25 circular mils.

This can be expressed by the formula ampere squared seconds (I^2t).

For example, from Table 8 of *NEC* chapter nine, an 8 AWG conductor has a cross-sectional area of 16,510 circular mils. By dividing the circular mil area of the conductor by 42.25, the conductor's five-second withstand rating can be calculated (16,510 ÷ 42.25 = 391).

Stated another way, this conductor has an I^2t five-second withstand rating of:

391 x 391 x 5 = 764,405 ampere squared seconds.

From this five-second withstand rating value, it is easy to calculate the conductor's withstand rating for other values of time and/or for other values of current.

Example 1: How many amperes will the 8 AWG copper conductor be able to safely carry if the impedance of the circuit along with the operating characteristics of the overcurrent device protecting the circuit results in a 2-cycle (0.0333 seconds) opening time? Example 1 is solved as follows:

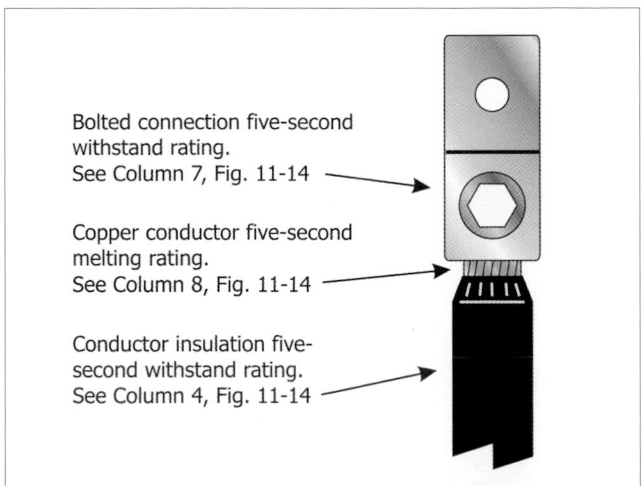

Figure 11-11. Wire, insulation and connection withstand ratings

$$I^2t = 764{,}405 \text{ ampere squared seconds}$$

$$I^2 = \frac{764{,}405}{t}$$

$$I = \sqrt{\frac{764{,}405}{0.0333}}$$

$$I = 4{,}791 \text{ amperes}$$

Example 2: How many amperes will the 8 AWG copper conductor be able to safely carry if the impedance of the circuit along with the operating characteristics of the overcurrent device protecting the circuit results in a ¼ cycle (0.0042) opening time? Example 2 is solved as follows:

$$I^2t = 764{,}405 \text{ ampere squared seconds}$$

$$I^2 = \frac{764{,}405}{t}$$

$$I = \sqrt{\frac{764{,}405}{0.0042}}$$

$$I = 13{,}491 \text{ amperes}$$

Note that in the example above, because a much faster total clearing time is achieved, the allowable fault current that the conductor will be subjected to can be increased. This is a result of substituting different time values in the I^2t formula.

Generally, where current-limiting overcurrent devices are protecting the circuit, the equipment grounding conductor sizes are determined directly from Table 250.122. Where available fault currents are high and the overcurrent protective device takes

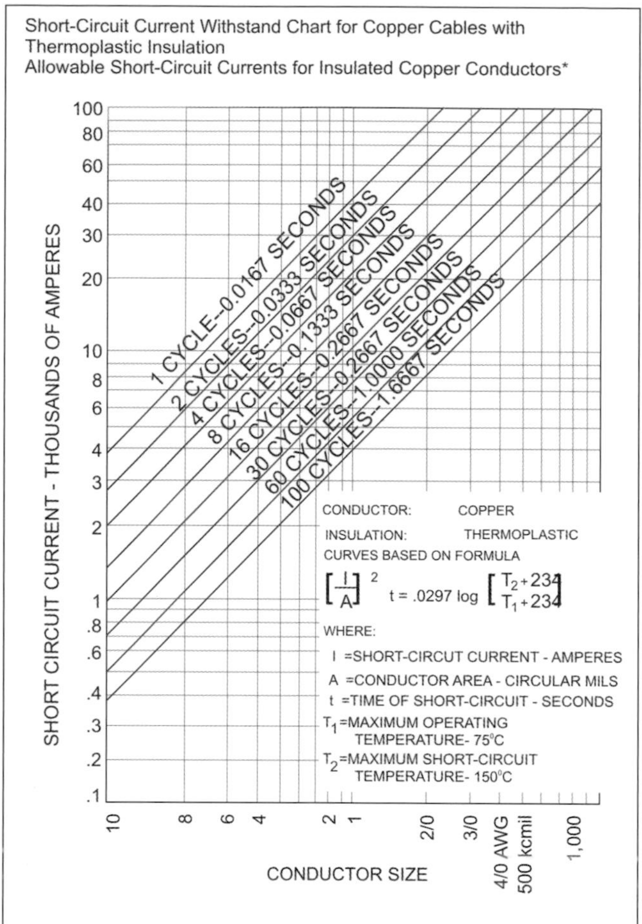

Figure 11-12. Short-circuit current withstand chart
Courtesy of the Insulated Cable Engineers Association

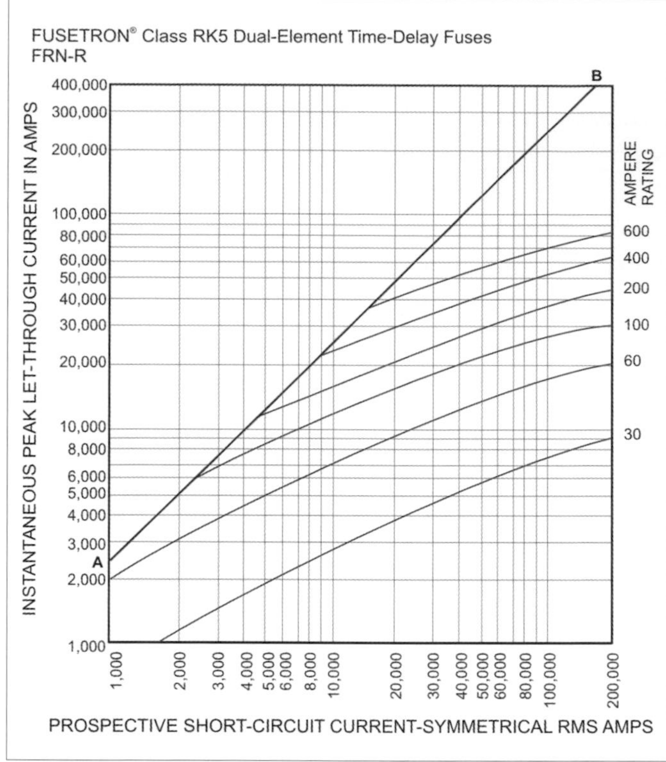

Figure 11-13. Typical time current curves for fuses. Courtesy of Copper Bussman, Cooper Industries

longer than .25 cycle to clear the fault, it is suggested that the equipment grounding conductor be sized per figure 11-12, to be on the safe side.

Figure 11-14 provides information to assist the installer in the proper selection of equipment grounding conductors. Among other information, it includes:

1. safe values for 75°C thermoplastic insulated conductors,

2. safe values for bolted connections,

3. unsafe (melting) values for the copper conductor itself.

Because the weakest link in any system is the insulation short-circuit withstand rating as found in columns 4, 5, and 6 of figure 11-14, it is recommended that column 4 be the deciding factor where selecting equipment grounding conductors. The Insulated Cable Engineers Association data, figure 11-14, column 4 calculates out to the previously discussed: *Do not exceed one ampere... for five seconds...for every 42.25 circular mils of copper conductor.*

This chart also shows the 8 AWG conductor used in the above text examples. Where you are absolutely certain that the equipment grounding conductor will not come into contact with any of the current-carrying insulated circuit conductors, the withstand rating of a bare grounding conductor, for every 29.1 circular mils of copper conductor, cannot exceed one ampere for five seconds (see column seven, figure 11-14).

This is the standard previously referred to in this text as the "Do not exceed one ampere for five seconds for every 29.1 circular mils of conductor area."

This value is only to be used where bare equipment grounding conductors are used in such a manner that they will not come in contact with insulated conductors. In this application, the limiting element of the circuit is the bolted connection of the lug.

Column 8 of figure 11-14 gives the current in amperes at which the melting temperature of copper conductors is reached. Of course, you never want to reach the current shown because the equipment grounding conductor will burn off leaving the equipment ungrounded and a possible shock hazard.

Copper 75° C. Thermoplastic Insulated Conductor, Bare Conductor, Bolted Connection Five Second Withstand Rating In Amperes and Melting of Copper Wire

1	2	3	4	5	6	7	8
Wire size	Area in circular mils	Area in square mm	ICEA Amperes	IEC Amperes	IEE Amperes	Bolted connection 250°C. Amperes	Melting of conductor 1,083°C. Amperes
14	4,110	2.080	97	107	107	141	254
12	6,530	3.310	155	170	170	224	403
10	10,380	5.261	246	271	271	357	641
8	16,510	8.367	391	430	430	567	1,020
6	26,240	13.300	621	684	684	902	1,621
4	41,740	21.150	988	1,088	1,088	1,435	2,578
3	52,620	26.670	1,245	1,372	1,372	1,808	3,251
2	66,360	33.620	1,571	1,729	1,729	2,281	4,099
1	83 690	42.410	1,981	2,181	2,181	2,876	5,170
1/0	105 600	53.490	2,499	2,751	2,751	3 629	6,523
2/0	133,100	67.430	3,150	3,468	3,468	4,574	8,222
3/0	167,800	85.010	3,972	4,372	4,372	5,767	10,366
4/0	211,600	107.200	5,009	5,513	5,513	7,272	13,071
250	250,000	126.700	5,918	6,516	6,516	8,592	15,443
300	300,000	152.000	7,101	7,818	7,818	10,310	18,532
350	350,000	177.300	8,285	9,119	9,119	12,029	21,621
400	400,000	202.700	9,467	10,425	10,425	13,747	24,709
500	500,000	253.300	11,834	13,027	13,027	17,184	30,887
600	600,000	304.000	14,201	15,636	15,636	20,621	37,064
700	700,000	354.700	16,568	18,243	18,243	24,057	43,241
750	750,000	380.000	17,752	19,544	19,544	25,776	46,330
800	800,000	405.400	18,935	20,850	20,850	27,494	49,419
900	900,000	456.000	21,302	23,453	23,453	30,931	55,596
1,000	1,000,000	506.700	23,669	26,060	26,060	34,368	61,773

Column 4 - Insulated Cable Engineers Association publication P32-382. One ampere for five seconds for every 42.25 circular mils of conductor area.
Column 5 - International Electrotechnical Commission publication 364-4-43.
Column 6 - Institute of Electrical Engineers publication 434-6.
Column 7 - Calculated from data in Electrical Engineers Handbook (75°C ambient). One ampere for five seconds for every 29.1 circular mils of conductor area.
Column 8 - Calculated from data in Electrical Engineers Handbook (75°C ambient). One ampere for five seconds for every 16.19 circular mils of conductor area.

Figure 11-14. Five-second withstand ratings for insulated conductors, bare conductors with bolted connections.

Courtesy of the ICEA

Conclusion on Equipment Grounding Conductor

For a grounded system, it is vital to safety that a low-impedance equipment grounding conductor path be provided in addition to a good grounding electrode system with as low of impedance as practical. This allows sufficient current to clear a ground fault automatically in a limited time, which would be as quickly as is practical, without undue interruption of service.

The I^2t values found in column 7 of figure 11-14 are based on the adequacy of a copper conductor and its bolted joints to carry the current values without destroying its validity. The values are obtained from an IEEE committee report in "A Guide to Safety in AC Substation Grounding." The value expressed in amperes per circular mil is one ampere for every 29.1 circular mils cross section. The time of five seconds was used to provide a safety factor and was considered a reasonable approach for distribution systems of 600 volts or less protected by high-interrupting-capacity current-limiting fuses and having equipment ground-fault protection.

As previously stated, where grounding conductors might come into contact with insulated phase conductors, use the values found in column four of figure 11-14. This column is based on one ampere for every 42.25 circular mils of conductor for five seconds. Figure 11-14 has all the calculations done and is much easier to use than performing complicated calculations.

The short-time rating of the equipment grounding conductor bears an approximately constant relation to the size of the overcurrent device. The I^2t values of the conductors given in Table 250.122 are between about 13 and 28 times their nominal continuous rating based on one ampere for every 42.25 circular mils cross section.

[1] Electric Power Distribution for Industrial Plants. The Institute of Electrical and Electronics Engineers, Inc., © 1954 AIEE, 445 Hoes Lane, P.O. Box 1331, Piscataway, NJ 08855-1331.

[2] "Some Fundamentals of Equipment-Grounding Circuit Design" by R. H. Kaufmann, Paper 54-244 presented at the AIEE Summer and Pacific General Meeting, Los Angeles, California, June 24-25, 1954, © 1954 AIEE, (now IEEE), 445 Hoes Lane, PO Box 1331, Piscataway, NJ 08855-1331.

[3] Kaufmann, ibid. 4 Kaufmann, ibid.5 Kaufmann, ibid.

[6] Fuses (JCQR), General Information for Electrical Equipment 2004, (Underwriters Laboratories, Northbrook, IL, 2004), p. 54.

[7] Circuit Breakers, Moulded-Case and Circuit Breaker Enclosures (DIVQ), General Information for Electrical Equipment 2004, (Underwriters Laboratories, Northbrook, IL, 2007), p. 88.

[8] Circuit Breaker Current Limiters (DIRW), General Information for Electrical Equipment 2007, (Underwriters Laboratories, Northbrook, IL, 2004), p. 87.

[9] 1991 IEEE Industry Applications Society Annual Meeting, Volume II. The Institute of Electrical and Electronics Engineers, Inc., © 1954 AIEE (now IEEE), 445 Hoes Lane, PO Box 1331, Piscataway, NJ 08855-1331.

[10] Modeling and Testing of Steel EMT, IMC and Rigid (GRC) Conduit School of Electrical and Computer Engineering Georgia Institute of Technology, Atlanta, Georgia 30332.

[11] NFPA 70, National Electrical Code 2008, 110.9, (National Fire Protection Association, Quincy, MA, 2007), p. 70-33.

[12] NFPA 70, 110.10, p. 70-33.

[13] NFPA 70, Front page, p. 70-1.

Review Questions

The questions included here were developed using material included in this chapter. The answers can be found by reviewing the text. It is also important that students make use of *NEC-2008*, where many answers can be found.

1. A conducting connection, whether intentional or accidental, between any of the conductors of an electrical system whether it be from line-to-line or line-to the grounded conductor, is defined as a ____.
   a. ground fault
   b. phase fault
   c. short circuit
   d. unidentified fault

2. "An unintentional, electrically conducting connection between an ungrounded conductor of an electrical circuit and the normally non-current-carrying conductors, equipment, or the earth to the electrical supply source" best defines which of the following?
   a. identified fault
   b. ground fault
   c. short circuit
   d. failure of the system

3. A short circuit may be from one phase conductor to another phase conductor, or from one phase conductor to the grounded conductor or ____.
   a. unidentified conductor
   b. enclosure
   c. neutral
   d. equipment grounding conductor

4. The grounding and bonding system must provide an electrically continuous path; must have ample carrying capacity to conduct safely any fault currents likely to be imposed on it; and must have an impedance sufficiently low to limit the voltage to ground and to facilitate the operation of the circuit protective devices in the circuit. This is best described as being ____.
   a. improved
   b. effective
   c. sufficient
   d. required

5. In general, the rule of thumb may be applied, but should be checked by calculation, that the equipment grounding conductor should not be less than ____ percent of the capacity of the phase conductors that supply the circuit.
   a. 25
   b. 20
   c. 15
   d. 18

6. There generally is no overcurrent protection on the line side of service conductors, there is generally only ____ protection on the primary of the transformer.
   a. ground-fault
   b. short-circuit
   c. overload
   d. equipment ground fault

7. The grounded service (usually a neutral) conductor must be run to each service disconnecting means and connected (bonded) to the enclosure even if the service supplies only ____.
   a. 240 volt loads
   b. 208 volt loads
   c. 480 volt loads
   d. line-to-line loads

8. An 8 AWG THW copper conductor has an allowable ampacity of 50 amperes according to Table 310.16. This conductor also has a safe five-second withstand rating of ____ amperes, but will melt if subjected to ____ amperes for five seconds.
   a. 8367 10,000
   b. 391 1020
   c. 30 500
   d. 4110 254

9. When Insulated Cable Engineers Association (ICEA) conductor short-circuit withstand rating tables are not available, it is possible to calculate the short-circuit current withstand ratings for a conductor. For example, to determine the safe withstand rating for copper conductors with 75° C insulation, use ____ ampere for every ____ seconds for every ____ circular mils of cross sectional area of the conductor.
   a. one, five, 42.25
   b. one, five, 16.9
   c. one, five, 29.1
   d. one, five, 10.2

10. Table 250.122 provides the minimum size equipment grounding conductors. Where high values of fault current are

available, the grounding conductors may have to be ____ in size to be capable of safely carrying the available fault current for the duration of time required to withstand such higher levels of current until the overcurrent device clears.

a. decreased
b. increased

11. Referring to Figure 11-12, find the minimum conductor sizes required for the fault current values indicated.

a. 10,000 amperes for one cycle. Size ____
b. 20,000 amperes for two cycles. Size ____
c. 4200 amperes for one cycle. Size ____
d. 10,000 amperes for eight cycles. Size ____

12. Emergency system overcurrent devices are required to be _______________ with all supply side overcurrent protective devices.

a. Rated at not less than 600%
b. Installed in parallel
c. Selectively coordinated
d. Located

13. The primary purpose of selective coordination of overcurrrent protective devices is to_______________.

a. Localize overcurrent conditions to restrict outages to the affected circuit or equipment
b. Ensure that the circuit always remains functional even under ground fault or short circuits
c. Enable fuses to open first
d. Enable circuiot breakers to operate first under ground fault or short circuit conditions

Grounding Separately Derived Systems

In many distribution systems of 600 volts or less for commercial or industrial occupancies, it is common to have separately derived systems at another voltage level lower or higher than the electrical system supplied by the service. Several *Code* requirements must be met when installing a separately derived system. First, if there is a separately derived system, this means there will be a new "source" established producing a voltage and providing a capacity from which current would be drawn. This capacity is usually referred to as the kVA or kW rating of the system. The system itself must have a capacity to supply the load served. Overcurrent protection sizing and the protection of conductors connected to the secondary of the separately derived system are also required. Another important element is determining which grounding and bonding rules must be applied to satisfy the requirements. This chapter focuses on those specific requirements.

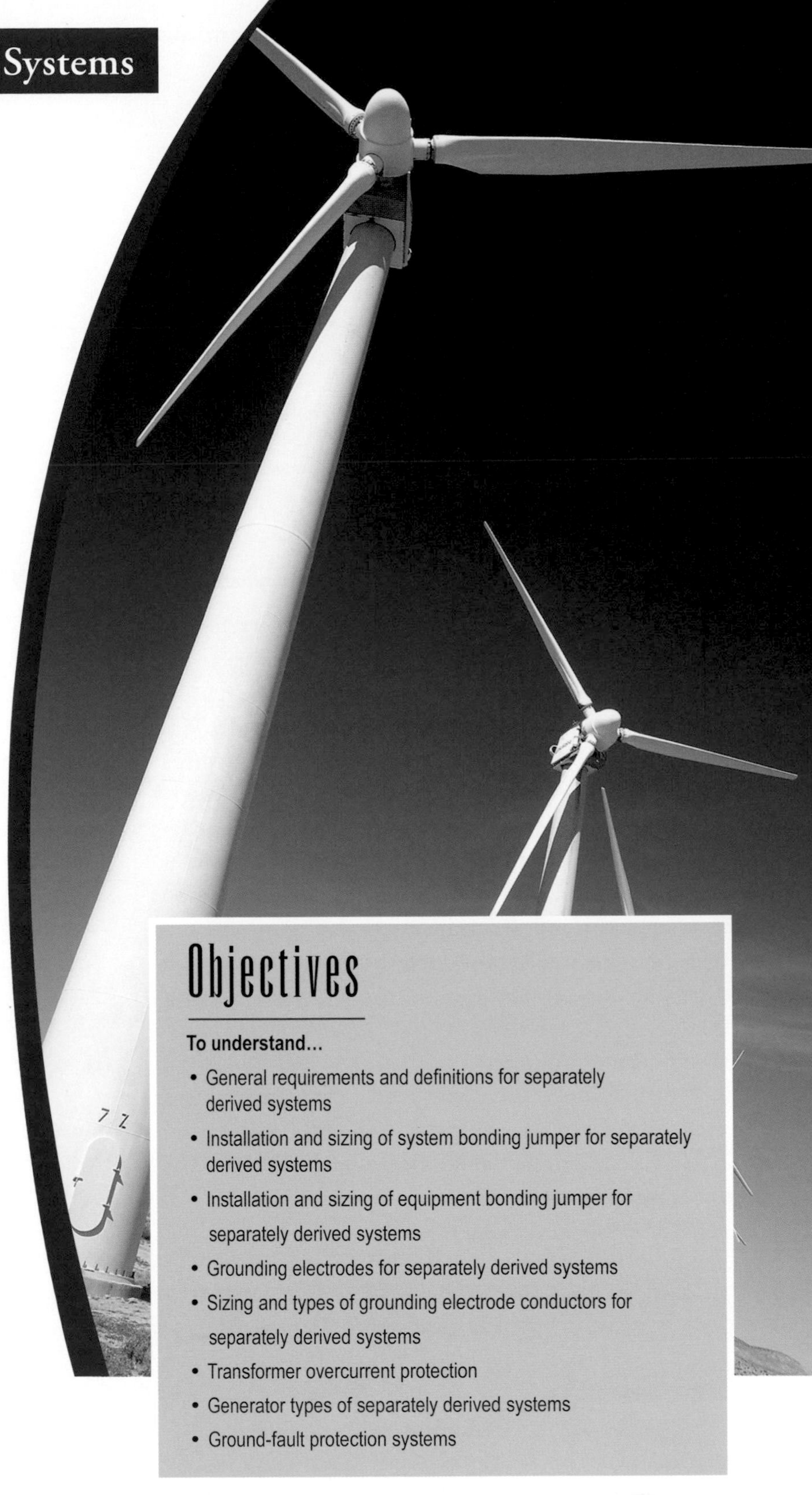

Objectives

To understand...

- General requirements and definitions for separately derived systems
- Installation and sizing of system bonding jumper for separately derived systems
- Installation and sizing of equipment bonding jumper for separately derived systems
- Grounding electrodes for separately derived systems
- Sizing and types of grounding electrode conductors for separately derived systems
- Transformer overcurrent protection
- Generator types of separately derived systems
- Ground-fault protection systems

Chapter 12

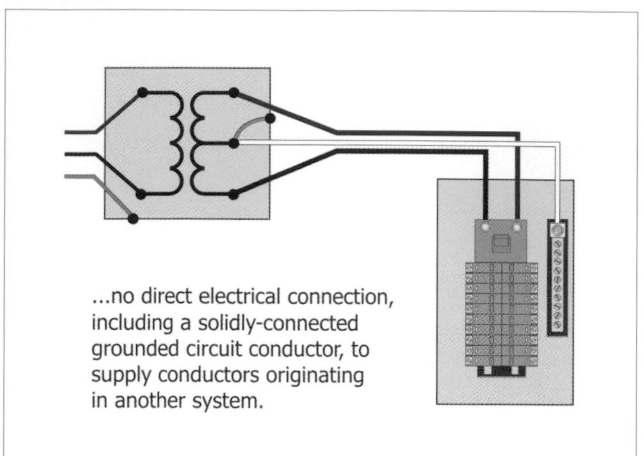

Figure 12-1. Transformer-type separately derived system

Photo 12-1. Transformer-type separately derived system

The first item that must be established is whether or not the system is separately derived. An example of a system or transformer that is not separately derived is an autotransformer. If a generator source or system is installed on a premises wiring system, it must be grounded as a separately derived system if there is a switching action in the grounded conductor through the transfer equipment. If there is no switching action in the grounded conductor through the transfer switch or equipment, then the grounded conductor remains grounded through the service grounding connection point located at the service and the generator-produced system is not grounded as required in 250.30(A) for a system that is separately derived. The *Code* requires a sign at the service equipment when the grounding electrode system at the service is also used for the grounding electrode for a generator [700.8(B), 701.9(B) and 702.8(B)]. In this case, the generator grounding and bonding is established through the equipment grounding conductor and an insulated grounded (neutral) conductor that is isolated from grounded metal parts at that location (generator) to comply with 250.24(A)(5).

Definition

Separately derived system. "A premises wiring system whose power is derived from a source of electric energy or equipment other than a service. Such systems have no direct electrical connection, including a solidly connected grounded conductor, to supply conductors originating in another system." [1]

Transformer-Type Separately Derived System

Figure 12-1 is a diagram of a transformer-type separately derived system (for simplicity, all of the grounding and bonding conductors are not shown.) The supply or primary of the transformer is at one voltage level, and the secondary is often at another voltage level, either lower or higher. As shown, there is no direct electrical connection between the transformer primary and secondary, so the installation meets the definition of a separately derived system. Where 3-phase systems are installed, the primary is usually delta connected so a neutral conductor is not needed. The secondary may be connected delta or wye as desired (see photo 12-1).

Grounding Primary Side Equipment

An equipment grounding conductor must be supplied with the primary circuit to provide a low-impedance fault-current path from the transformer case to the service or source of supply. The equipment grounding conductor can be any of the means included in 250.118 including wires and the wiring method, where appropriate. The overcurrent device on the primary of the transformer will then clear a short circuit or ground fault up to, and including, the transformer primary windings.

The equipment grounding conductor connection to the transformer enclosure, plus the system bonding jumper and equipment bonding jumper from the secondary does not constitute a direct electrical connection from the primary system to the secondary system. The equipment grounding conductor is not a system conductor as the ungrounded and grounded (neutral) conductors are.

A short circuit or ground fault on the secondary of the transformer is often seen as a load by the transformer primary winding. The amount of fault current on the secondary side will determine whether or not the primary overcurrent device will open or operate owing to the turns ratio between primary and secondary of the transformer.

Systems and Grounding Rules

Whether a separately derived system requires grounding in accordance with 250.30(A), is established in 250.20. Sometimes there is a choice as to whether the system is required, permitted to be grounded, or not permitted to be grounded. If the system can be grounded so that the voltage to ground from any of the system conductors does not exceed 150 volts, then generally it must be grounded. Sections 250.20(B)(1), (2), and (3) determine when the system produced must be grounded and a choice is not permissible. If that system is required to be grounded, or if the system is grounded by choice, in other words, if it is not required by 250.20 but is grounded anyway as permitted, then the separately derived system must be grounded according to the applicable requirements of 250.30(A). The grounding requirements are as follows in the order they appear in the *NEC*:

(1) System Bonding Jumper.
(2) Equipment Bonding Jumper Size.
(3) Grounding Electrode Conductor, Single Separately Derived System.
(4) Grounding Electrode Conductor, Multiple Separately Derived Systems.
  (a) Common Grounding Electrode Conductor Size.
  (b) Tap Conductor Size.
  (c) Connections.
(5) Installation.
(6) Bonding.
(7) Grounding Electrode.
(8) Grounded Conductor.

System Bonding Jumper

The system bonding jumper is the first component addressed in 250.30(A)(1). The system bonding jumper is the vital link for the fault-current path at the source and is the connection between the grounded conductor and the equipment grounding conductors at the source or equipment containing the first overcurrent protective device. It functions in similar fashion to the main bonding jumper at the service location.

Those in the electrical field often refer to the system bonding jumper of a separately derived system as the main bonding jumper used at the service equipment. Its function is essentially the same, but the correct terms must be used to apply *Code* requirements properly. A system bonding jumper is used at separately derived systems. This term is used in Section 250.30(A)(1) and defined in Section 250.2.

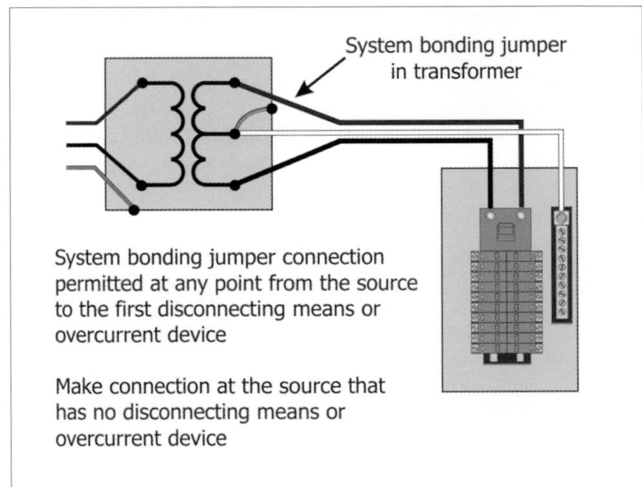

Figure 12-2. System bonding jumper located at the source enclosure

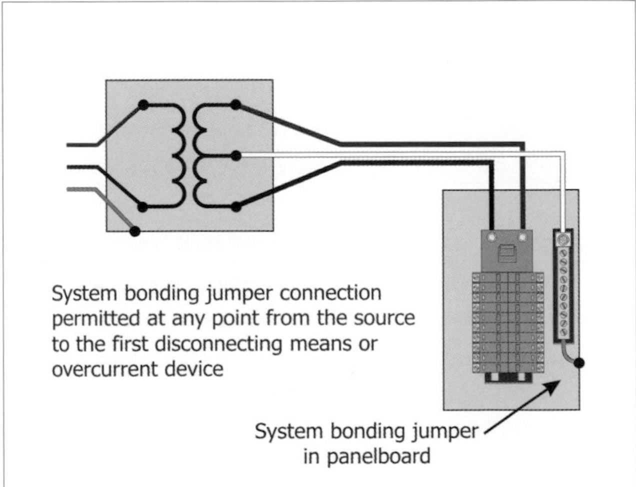

Figure 12-3. System bonding jumper located at the first disconnection means or overcurrent device enclosure

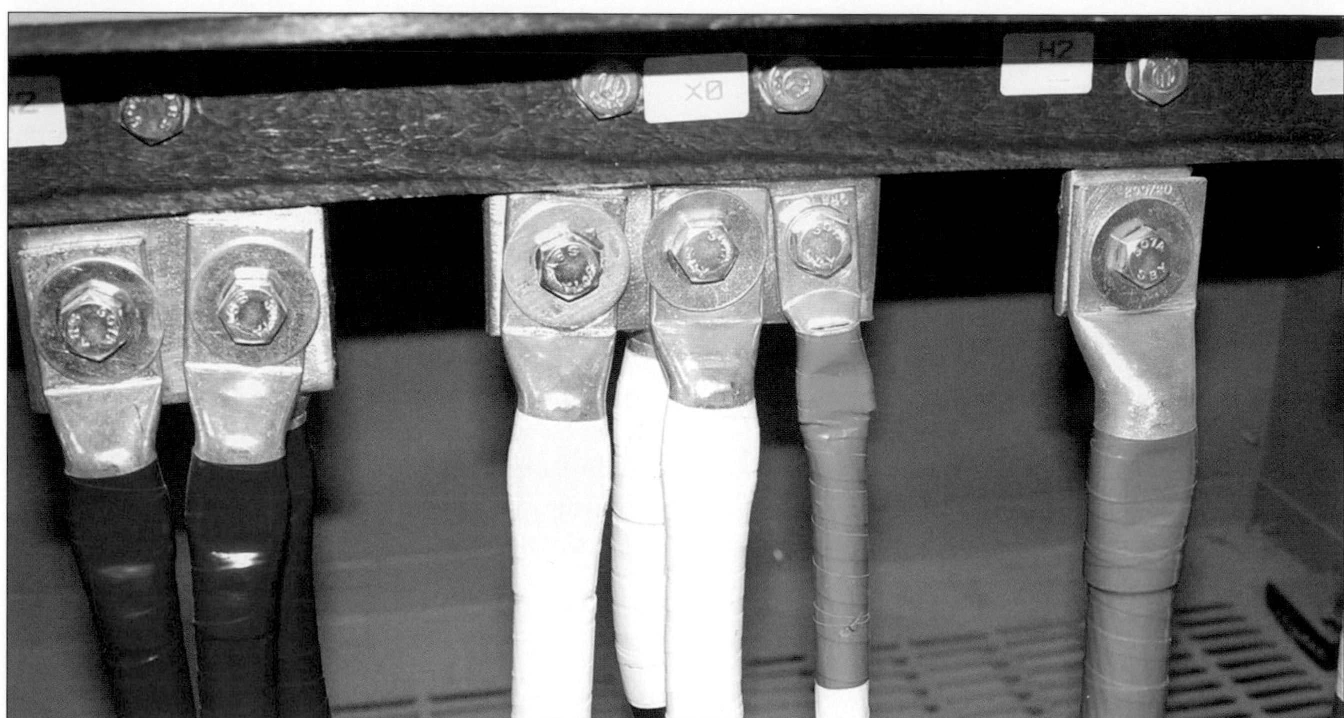
X0
H2

Photo 12-2. System bonding jumper in source enclosure

Definition

Bonding jumper, system. "The connection between the grounded circuit conductor and the equipment grounding conductor at a separately derived system."

The system bonding jumper must be installed in accordance with the requirements of 250.28(A) through (D). The sizing is to be based on the derived phase conductors supplied by the separately derived system. A brief look at the requirements in 250.28 identifies the material, construction, attachment, and sizing requirements for the system bonding jumper. It is required to be copper or other corrosion-resistant material, and is permitted to be in the form of a wire, bus, screw, or similar conductor. System bonding jumpers that are of the screw types must be identified by the color green that identifies the screw after installation. This type of system bonding jumper might be employed when the bonding jumper and grounding electrode conductor connection to the system are installed at the first system overcurrent protective device, such as a panelboard. The means of attachment of the system bonding jumper must be by the exothermic welding process, listed pressure connectors, listed clamps, or other listed means. The more common location for the system bonding jumper and grounding electrode conductor connection to the system is at the source enclosure [3] (see photo 12-2).

However, the *Code* permits this connection at any single point from the source to the first system disconnecting means or overcurrent protective device, or it must be made at the source where

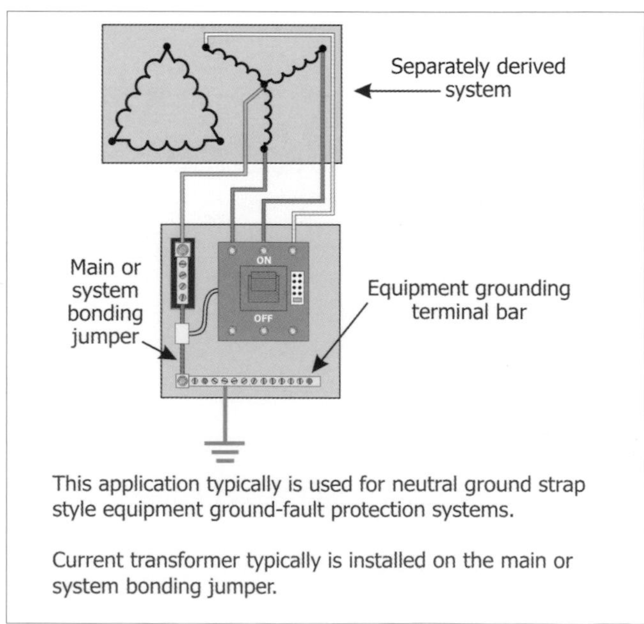
Separately derived system
ON
OFF
Main or system bonding jumper
Equipment grounding terminal bar
This application typically is used for neutral ground strap style equipment ground-fault protection systems.
Current transformer typically is installed on the main or system bonding jumper.

Figure 12-4. System bonding jumper exception

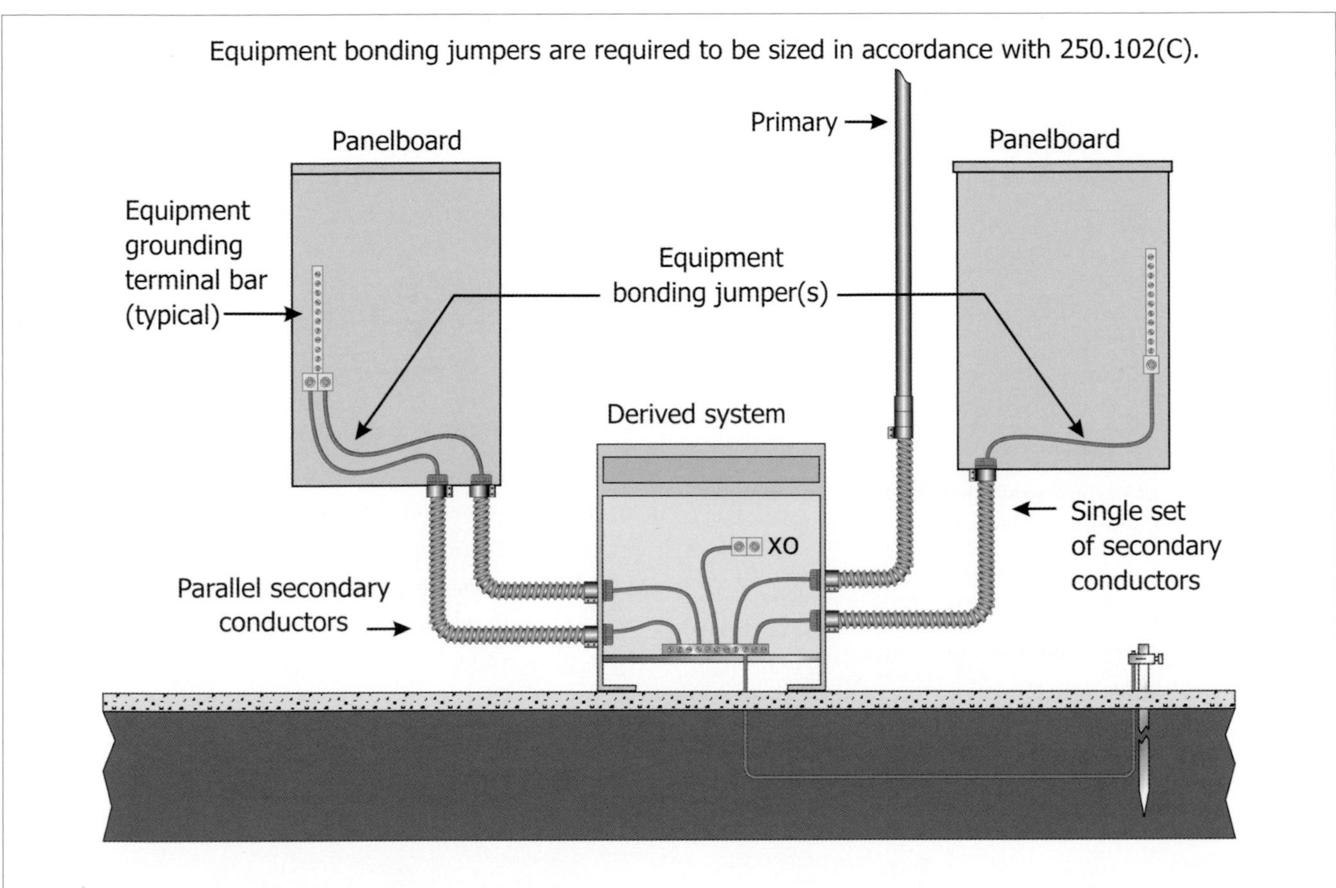

Figure 12-5. Equipment bonding jumper size [250.30(A)(2)]

there is no disconnecting means or overcurrent device located in the equipment supplied by the separately derived system (see figures 12-2 and 12-3). Section 250.28(D)(3) includes the sizing requirements for system bonding jumpers that are installed in separate enclosures, and sizing requirements for a single system bonding jumper installed at the source enclosure. Where the system supplies more than a single enclosure, the system bonding jumper in each enclosure is sized in accordance with 250.28(D)(1) based on the largest ungrounded feeder conductor serving each enclosure. Where a single system bonding jumper is installed at the source enclosure, it must be sized ini accordance with 250.28(D)(1) based on the sum of the circular mil areas of all ungrounded derived phase conductors (see photo 12-2). The system bonding jumper is required to be connected to the separately derived system at the same location as the grounding electrode conductor connection to the system.

System Bonding Jumper Location Exceptions

Section 250.30(A) Exception recognizes provisions for connecting the system bonding jumper for a separately derived system according to the requirements for a high-impedance grounded system. These rules are given in 250.36 and 250.186. These systems are grounded only through an impedance device (often a resistor) usually located at the panelboard or motor control center and not at the source enclosure. These systems are designed to limit the fault current of the first ground fault to a predetermined value and often incorporate an indication or alarm system. Additional information on high-impedance grounded systems is provided in chapter four.

Section 250.30(A)(3) Exception No. 1 permits the grounding electrode connection to be made to an equipment grounding terminal bus, rather than to the neutral bar where the main bonding jumper is a wire or busbar, as permitted by 250.24(A)(4) (see figure 12-4). This provides for residual-type

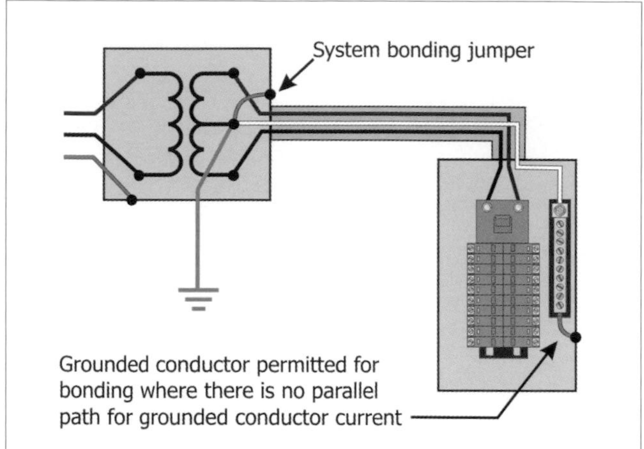
System bonding jumper
Grounded conductor permitted for bonding where there is no parallel path for grounded conductor current

Figure 12-6. Use of derived grounded conductor (neutral) for bonding is permitted by exception

equipment ground-fault protection systems where, often, a current transformer is located on the system bonding jumper to measure ground-fault currents back to the source. Another common application of this exception is in switchboards where the main or system bonding jumper is a wire or busbar.

Equipment Bonding Jumper for Separately Derived Systems

The installation and sizing requirements for equipment bonding jumpers of the wire type installed with the derived phase conductors from the source to the first disconnect or overcurrent protective are provided in 250.30(A)(2). Where the grounding and system bonding connections for separately derived systems in accordance with 250.30(A) are made at the source enclosure, most installations include the routing the derived phase conductors, the grounded (neutral) conductor of the system, and an equipment bonding jumper for bonding the metal enclosure of the derived system source to the metal enclosure at the first disconnect or overcurrent protective device are installed. This conductor is identified by the *Code* as

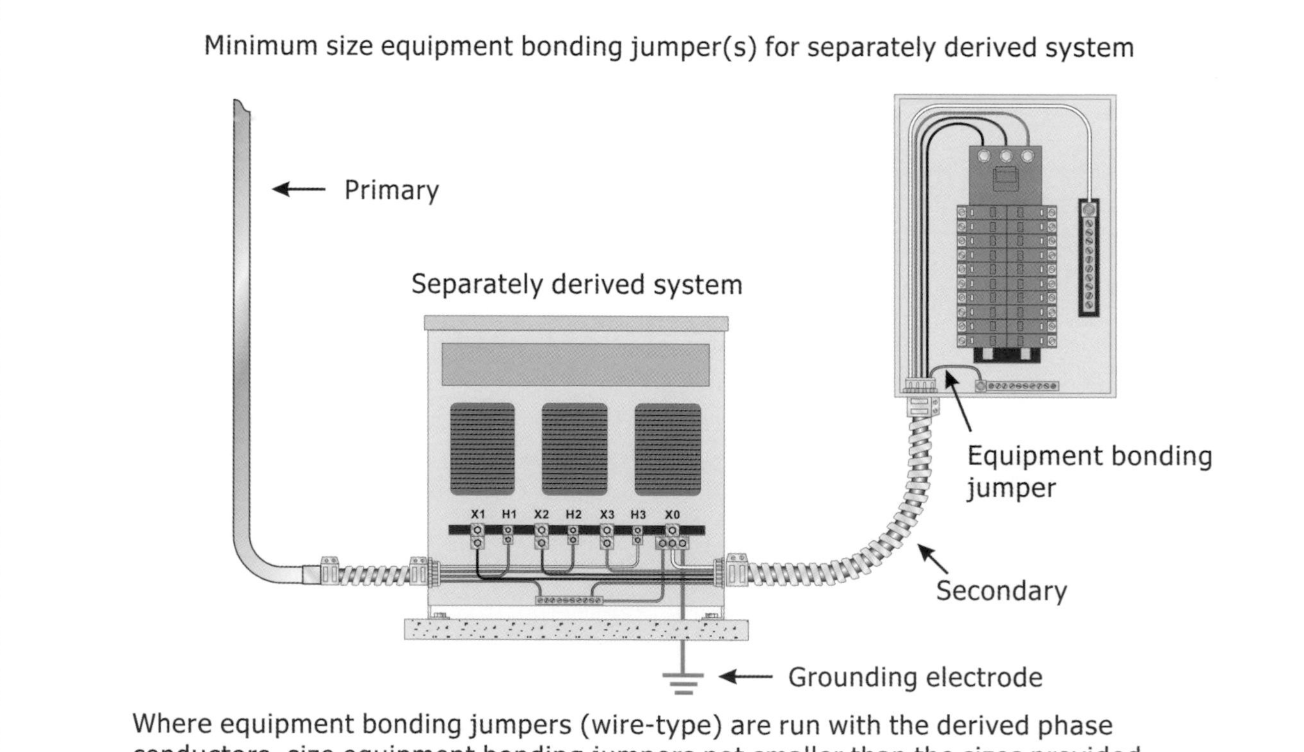
Minimum size equipment bonding jumper(s) for separately derived system
Primary
Separately derived system
X1 H1 X2 H2 X3 H3 X0
Equipment bonding jumper
Secondary
Grounding electrode
Where equipment bonding jumpers (wire-type) are run with the derived phase conductors, size equipment bonding jumpers not smaller than the sizes provided in Table 250.66
Use the 12.5% rule for where the derived phase conductors exceed 1100 kcmil copper or 250 kcmil aluminum or copper-clad aluminum

Figure 12-7. Minimum size for equipment bonding jumpers of separately derived systems [250.30(A)(2)]

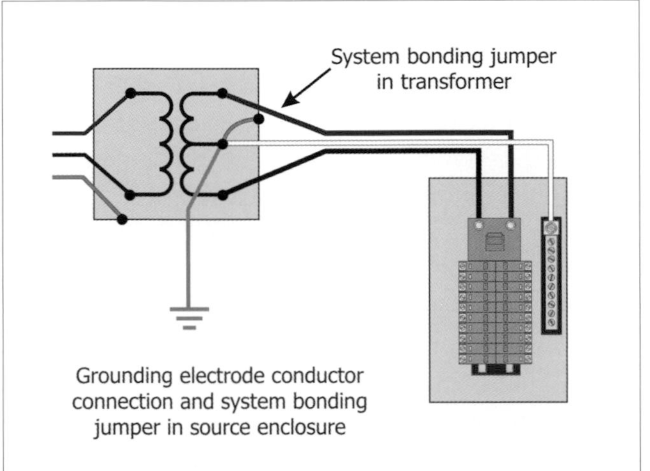

Figure 12-8. Grounding electrode conductor connections are to be made at the same location as the system bonding jumper (at source enclosure).

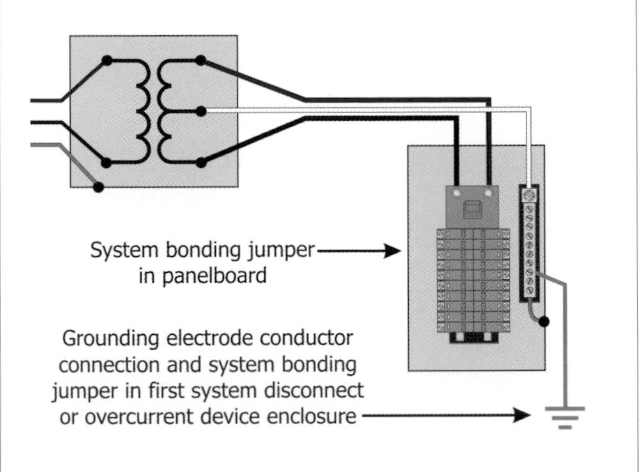

Figure 12-9. Grounding electrode conductor connections are to be made at the same location as the system bonding jumper (at first system disconnect or overcurrent device enclosure).

an equipment bonding jumper and if of the wire type must be sized in accordance with the requirements of 250.102(C) (see figure 12-5). If a raceway or other wiring method is used as permitted by 250.118, the sizing requirement does not apply but consideration should be given to ensure it provides an effective ground-fault path.

Conductors derived from the secondary of a transformer, generator or other separately derived systems are considered *unprotected* or *unfused* (line-side) conductors. They are often referred to as *tap conductors*. Tap conductors are defined in 240.2. Unless the secondary conductors are protected against overcurrent in a manner specified in 240.4(F) for transformer secondary conductors, they are not protected at their ampacity by the line-side overcurrent device, but obtain their overcurrent protection through proper application of the appropriate rules in 240.21.

The sizing requirements for the equipment bonding jumper installed from the source enclosure to the first system disconnect or overcurrent device are similar to the equipment bonding jumper sizing requirements for equipment bonding jumpers on the supply side of service equipment disconnecting means. They must be sized based on the circular mil area of the derived phase conductors and the values in Table 250.66 (see figure 12-7). Where the size of the total circular mil area of the derived phase conductors exceeds the values given in Table 250.66, the 12½ percentage rule must be applied. An example would be a 750-kVA transformer supplying a 2500-ampere switchboard. If the derived secondary phase conductors were 10 sets of four 500-kcmil copper XHHW conductors per phase, the total circular mil area of one set would equal 5,000,000 circular mils. 5,000,000 x 12.5 percent = 625,000 circular mils. Take that value back to Table 8 in *NEC* chapter nine and the next higher size is required. The minimum size for the equipment bonding jumper installed from the secondary of the derived system to the first disconnecting means or overcurrent device would be required to be not smaller than 700-kcmil copper. If the equipment bonding jumpers and phase conductors are installed in separate raceways, the equipment bonding jumper should be sized based on the size of the derived phase conductors installed in each raceway in accordance with 250.102(C).

Use of Derived Grounded (Neutral) Conductor

There is an alternative method of bonding that utilizes the grounded conductor and allows it to be bonded to the source enclosure and the first system disconnect or overcurrent device enclosure where doing so does not create a parallel path for current that would be returning to the source over the grounded conductor [see 250.30(A)(1) Exception No. 2, and figure 12-6].

Section 250.142(A)(3) permits the grounded (neutral) circuit conductor to be used for grounding equipment "on the supply side or within the enclosure of the main disconnecting means or overcurrent devices of a separately derived system where permitted by

250.30(A)(1)." As used in 250.142(A)(3), the term, *on the supply side* means "up to or within the enclosure for the disconnecting means or overcurrent devices." It is widely accepted to make bonding connections of the neutral or grounded circuit conductor within the disconnecting means enclosure for the separately derived system and upstream to the transformer or generator.

As shown in figure 12-6, the grounded (neutral) circuit conductor serves two purposes. First, it allows line-to-neutral or grounded conductor loads to be supplied, and the conductor carries the unbalanced loads from ungrounded conductors. Secondly, it serves as the equipment grounding conductor and will carry line-to-ground fault currents back to the source. This grounding scheme is similar to that used for services and for feeder(s) or branch circuit(s) to additional buildings on the premises.

Where this scheme is used, it is not necessary to install an equipment bonding jumper between the source and the disconnecting means or overcurrent protection enclosure. This requires nonmetallic raceways to be used. Where nonmetallic raceways are used, a parallel path for neutral current is not created by the wiring method.

Keep in mind that this method of bonding the grounded conductor at both ends is permitted only where doing so does not create a parallel path for grounded (neutral) conductor current. A parallel path for neutral current exists where a metal raceway is used between the source of the separately derived system and the metal enclosure for the disconnecting means or overcurrent device. A parallel path can also be established by other means such as through metal pipes, cable trays and structural metal framing members, etc. The neutral current will divide between the available paths depending upon the impedance of the paths. The lowest impedance path will obviously carry the most current.

Where a system bonding jumper is installed at both the source location and the first disconnecting means as permitted in 250.30(A)(1) Exception 2, the grounded conductor must be adequately sized to carry any fault current likely to be imposed. Section 250.30(A)(8) covers the sizing and routing requirements for grounded conductors on the secondary of derived systems connected in this manner. Basically the sizing shall be no smaller than the required grounding electrode conductor in 250.66 and where the derived phase conductors exceed 1100-kcmil copper or 1750-kcmil aluminum or copper-clad aluminum, the grounded conductor shall not be smaller than 12½ percent of the total kcmil of the largest derived phase conductor. The grounded conductor of a 3-phase, 3-wire delta-connected separately derived system shall have an ampacity not less than the ungrounded derived phase conductors.

The *NEC* in recent editions has continued to migrate away from the use of the grounded conductor for grounding non-current-carrying metal parts of equipment on the downstream side of the service grounding connection point or the grounding con-nection point for a separately derived system.

Use care when considering the use of the grounded (neutral) conductor to make grounding connections at points other than the service equipment or source of separately derived system. Current will always try to return to its source, both normal current through the grounded or neutral conductor and ground-fault current as reviewed in chapter 1.

Parallel paths for this current can create additional shock hazards for persons and could also introduce additional impedance into the path for fault current.

The Grounding Electrode Conductor(s)

Definition

Grounding electrode conductor. "A conductor used to connect the system grounded conductor or the equipment to a grounding electrode or to a point on the grounding electrode system." [4]

A separately derived system that is grounded must have a grounding electrode conductor(s) to establish the connection of the grounded system conductor and metal equipment supplied by the derived system to ground (earth).

Separately derived systems are permitted to be grounded individually with an individual grounding electrode conductors, or, under certain conditions, multiple separately derived systems are permitted to be grounded by connection to a single common grounding electrode conductor to which grounding electrode conductor taps must be connected.

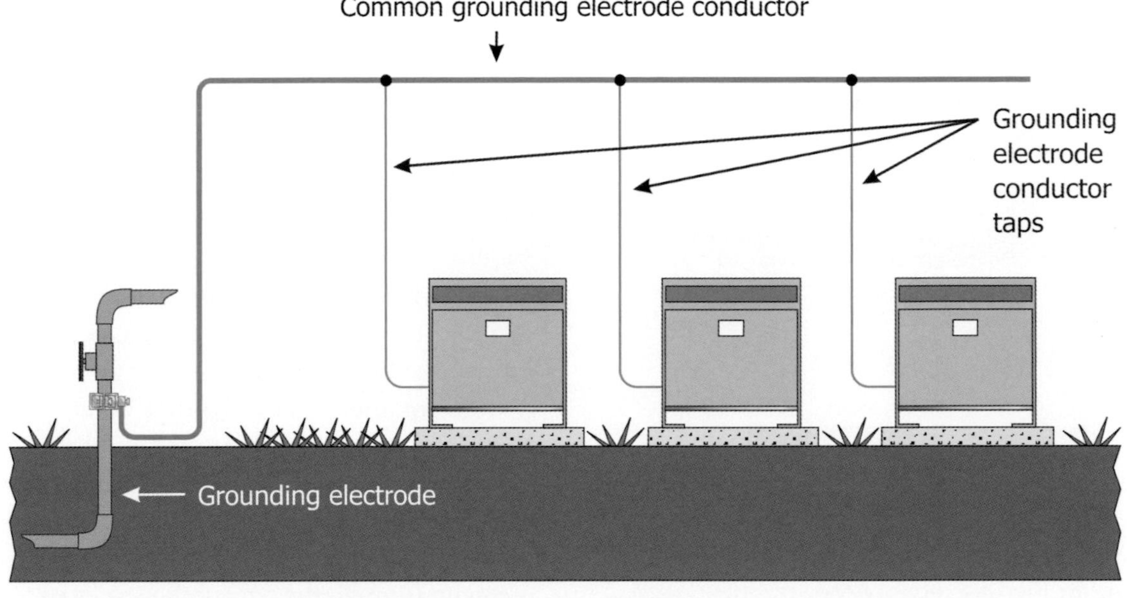

Figure 12-10. Multiple separately derived systems connected to a common grounding electrode conductor

The grounding electrode conductor for a single separately derived system in accordance with 250.30(A)(3) must be sized at a minimum in accordance with the sizes specified in Table 250.66 on the circular mil area of the derived phase conductors. This conductor must connect the grounded conductor of the derived system to the grounding electrode and must be installed in accordance with the requirements of 250.64 (see figures 12-8 and 12-9). This section includes all applicable installation requirements for grounding electrode conductors, including physical protection from damage and protection from magnetic fields. The connection of the grounding electrode conductor to the grounded (usually a neutral) conductor of the separately derived system is required to be made at the same location as the bonding jumper connection unless the installation conforms to the requirements of 250.24(A)(4). In this case, the grounding electrode conductor is permitted to be connected to an equipment grounding terminal bar installed at the source or first system disconnect or overcurrent protective device and a system bonding jumper sized in accordance with 250.28(D)(3) is installed from the grounded (neutral) conductor termination point to the equipment grounding terminal bus. This is typically done for a residual or ground strap type of equipment ground-fault protection systems [see chapter fifteen for more information on ground-fault protection systems].

Grounding Electrode Conductors for Multiple Separately Derived Systems

An alternative method for establishing and installing grounding electrode conductors for multiple separately derived systems is provided in Section 250.30(A)(4) (see figure 12-10). It is permitted by the *NEC* to install a common grounding electrode conductor to which multiple separately derived systems shall be connected by individual grounding electrode (tap) conductors that are sized based on the derived phase conductors of each individual separately derived system (see figure 12-11). An example of where this alternative method of grounding electrode conductor installation can be

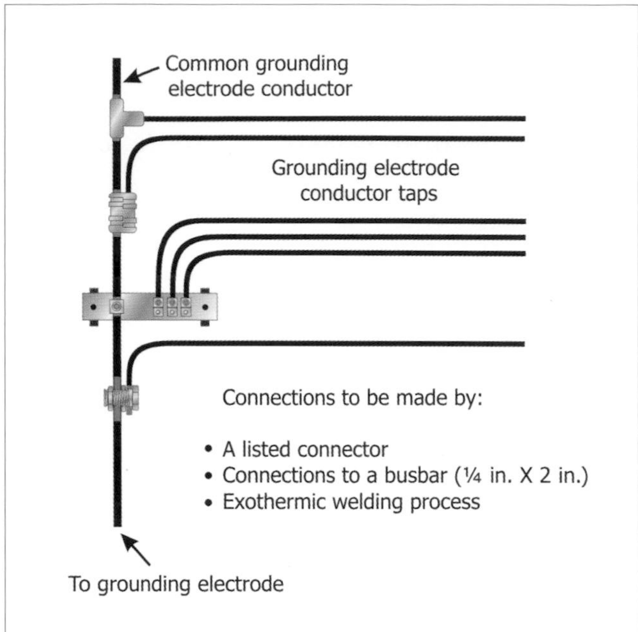
Common grounding electrode conductor
Grounding electrode conductor taps
Connections to be made by:
• A listed connector
• Connections to a busbar (¼ in. X 2 in.)
• Exothermic welding process
To grounding electrode

Figure 12-11. Grounding electrode conductor tap connections

effective is a typical high-rise building that has no structural metal grounding electrode but does have a metal water piping grounding electrode that enters the building in the basement of the structure. Section 250.30(A)(7) requires the metal water pipe grounding electrode to be used as the electrode for the separately derived systems installed in that building, and the grounding electrode conductors to be installed and connected to the water pipe at a location within 1.5 m (5 ft) of where the water pipe enters the building [250.30(A)(7)(1) and 250.52(A)(1)]. This requirement presented some challenges in the field, which led to the new alternative methods, permitted in this section.

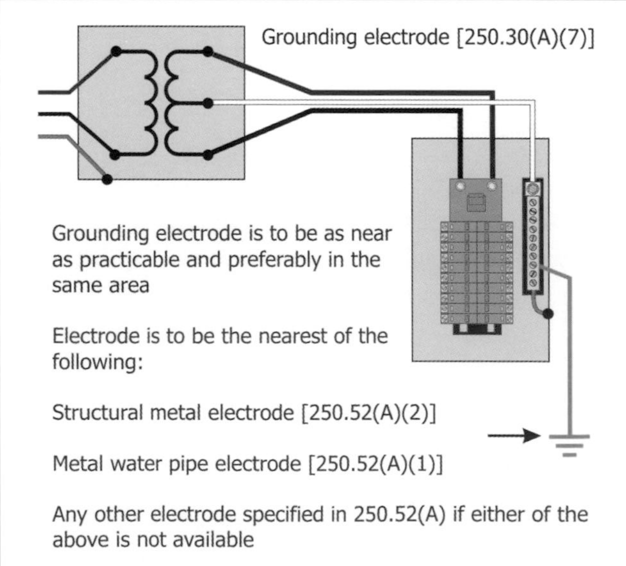
Grounding electrode [250.30(A)(7)]
Grounding electrode is to be as near as practicable and preferably in the same area
Electrode is to be the nearest of the following:
Structural metal electrode [250.52(A)(2)]
Metal water pipe electrode [250.52(A)(1)]
Any other electrode specified in 250.52(A) if either of the above is not available

Figure 12-12. Grounding electrode(s) for separately derived systems

The rules that must be followed when installing grounding electrode conductors for multiple separately derived systems in accordance with these alternative provisions are as follows:

Where more than one separately derived system is grounded to a common grounding electrode conductor, the common grounding electrode conductor shall be sized based at not less than 3/0 AWG copper or 250-kcmil aluminum [250.30(A)(4)(a)]. If longer lengths for this common grounding electrode conductor are needed, which is often the case, the common grounding electrode conductor usually is increased in size based on the conditions and the design of the system.

It is common to see high-rise buildings that have no structural metal frame grounding electrode, but the building grounding electrode system established in the basement of the structure is used and a 500-kcmil conductor (common grounding electrode conductor) is run up through the core of the structure.

Individual grounding electrode conductor taps would then be installed from each derived system to this common grounding electrode conductor. This has been an industry practice for some time to accomplish the grounding for separately derived systems in high-rise buildings.

Once the common grounding electrode conductor is installed, then the installation of the grounding electrode conductor taps is required. Sizing the individual grounding electrode conductor taps is based on the derived phase conductors supplied by each separately derived system. The grounding electrode conductor tap connects the grounded conductor of the derived system to the common grounding electrode conductor, which is connected to the grounding electrode. The grounding electrode conductor tap is required to be connected to a separately derived system at the same enclosure where the system bonding jumper is located [250.30(A)(4)]. Installation for both the common grounding electrode conductor and the grounding electrode conductor taps must be in accordance with the installation requirements for grounding electrode conductors specified in 250.64. This alternative for grounding electrode conductor taps to a common grounding electrode conductor is similar in concept to the alternative methods for grounding electrode conductor taps for services in accordance with 250.64(D).

The common grounding electrode conductor must be installed in a continuous length without a splice or joint, unless splicing is accomplished by irreversible compression connectors or by the exothermic welding process. The grounding electrode conductor taps must be connected to the common grounding electrode conductor at an accessible location by one of three methods:

1. A listed connector.
2. Listed connections to aluminum or copper busbars not less than 6 mm x 50 mm (¼ in. x 2 in.). Where aluminum busbars are used, the installation must meet the requirements of 250.64.
3. By the exothermic welding process.

The grounding electrode conductor taps must be connected to the common grounding electrode conductor in a manner that keeps it continuous without a splice or joint.

Section 250.30(A)(5) requires the grounding electrode tap conductors to comply with the installation requirements of 250.64(A), (B), (C), and (E).[6]

Grounding Electrode for the Separately Derived System

Definition

Grounding Electrode. "A conducting object through which a direct connection to earth is established."

The rules regarding grounding electrodes permitted to be used for grounding a separately derived system are located and described in Part III of Article 250.

Specific requirements apply to the allowable choices in regard to which electrode must be used for the separately derived system (see figure 12-12). The first requirement is that whichever grounding electrode is chosen, it must be located as near as practicable and preferably in the same area as the grounding electrode conductor connection to the system. The second requirement is that the grounding electrode shall be the nearest one of the following grounding electrodes.

A structural metal frame grounding electrode of the building or structure as specified in 250.52(A)(2) and a metal water pipe grounding electrode as specified in 250.52(A)(1) within 1.52 m (5 ft) from the point of entry to the building are the choices. Whichever is the closest determines which electrode must be used. If either of the grounding electrodes in 250.30(A)(7)(1) or (2) are not present for use, any of the other electrodes as specified in 250.52(A) are permitted to be used.

There is an alternative to the requirement of a connection to an metal water pipe grounding electrode within 1.52 m (5 ft) from the point of entry of the water pipe to the building provided in 250.52(A)(1) Exception. This connection shall be permitted at any point on the water pipe system in industrial and commercial buildings where conditions of maintenance and supervision ensure that only qualified persons will service the installation and the entire length of the interior metal water pipe that is being used for the grounding electrode is exposed. The term *exposed* is defined in Article 100 as "on or attached to the surface or behind panels designed to allow access."

Where separately derived systems are an integral part of listed equipment, such as in unit substations, the grounding electrode used for the service equipment, or equipment that is suitable for use as service equipment installed in accordance with the requirements of Article 225 for feeders, shall be permitted to serve as the grounding electrode for the separately derived system. The internal equipment grounding bus in such equipment is permitted to serve as the grounding electrode conductor if it is large enough, provided the grounding electrode conductor from the service or feeder to the grounding electrode is also large enough for the separately derived system (see figure 12-13).

Ungrounded Separately Derived Systems

Grounding rules are provided in 250.30(B) for ungrounded separately derived systems. Basically, a grounding electrode conductor connects the metal equipment (enclosure) of the separately derived system to a grounding electrode similar to the rules for grounded systems. However, the system itself is not grounded. Where the separately derived system itself is not grounded, the non-current-carrying metal parts of equipment and enclosures for conductors, metal raceways, etc., are required to be grounded. There must be a grounding electrode conductor

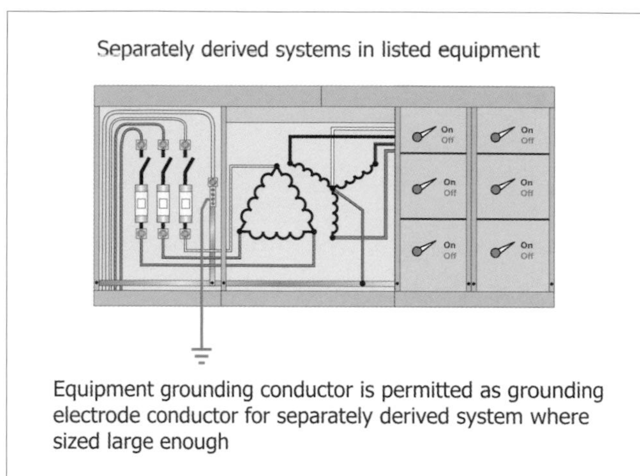
Separately derived systems in listed equipment
Equipment grounding conductor is permitted as grounding electrode conductor for separately derived system where sized large enough

Figure 12-13. Separately derived systems contained within listed equipment such as unit substations

installed from the metal enclosures or equipment supplied by the derived ungrounded system to a grounding electrode as specified in 250.30(B)(1). This grounding electrode conductor is permitted to be connected to the metal enclosures or equipment at any point on the separately derived system from the source to the first system disconnecting means. The grounding electrode for the ungrounded system must also be as near as practicable and preferably in the same area as the grounding electrode conductor connection to the enclosures or equipment. It shall be the nearest grounding electrode as specified in 250.30(A)(7).

Effective Ground-Fault Return Path

As with all electrical systems, it is vital to be certain that, for separately derived systems, an effective ground-fault return path exists from the furthermost point on the electrical system back to the source. In this case, the source is the separately derived system. Fault current downstream from the separately derived system follows the same laws of physics, as does the fault current on the line side. It attempts to return to its source, which is the secondary windings of the transformer, and primarily uses the lowest impedance path.

As can be seen in figure 12-14, fault current downstream from the separately derived system will return to the secondary windings of the transformer,

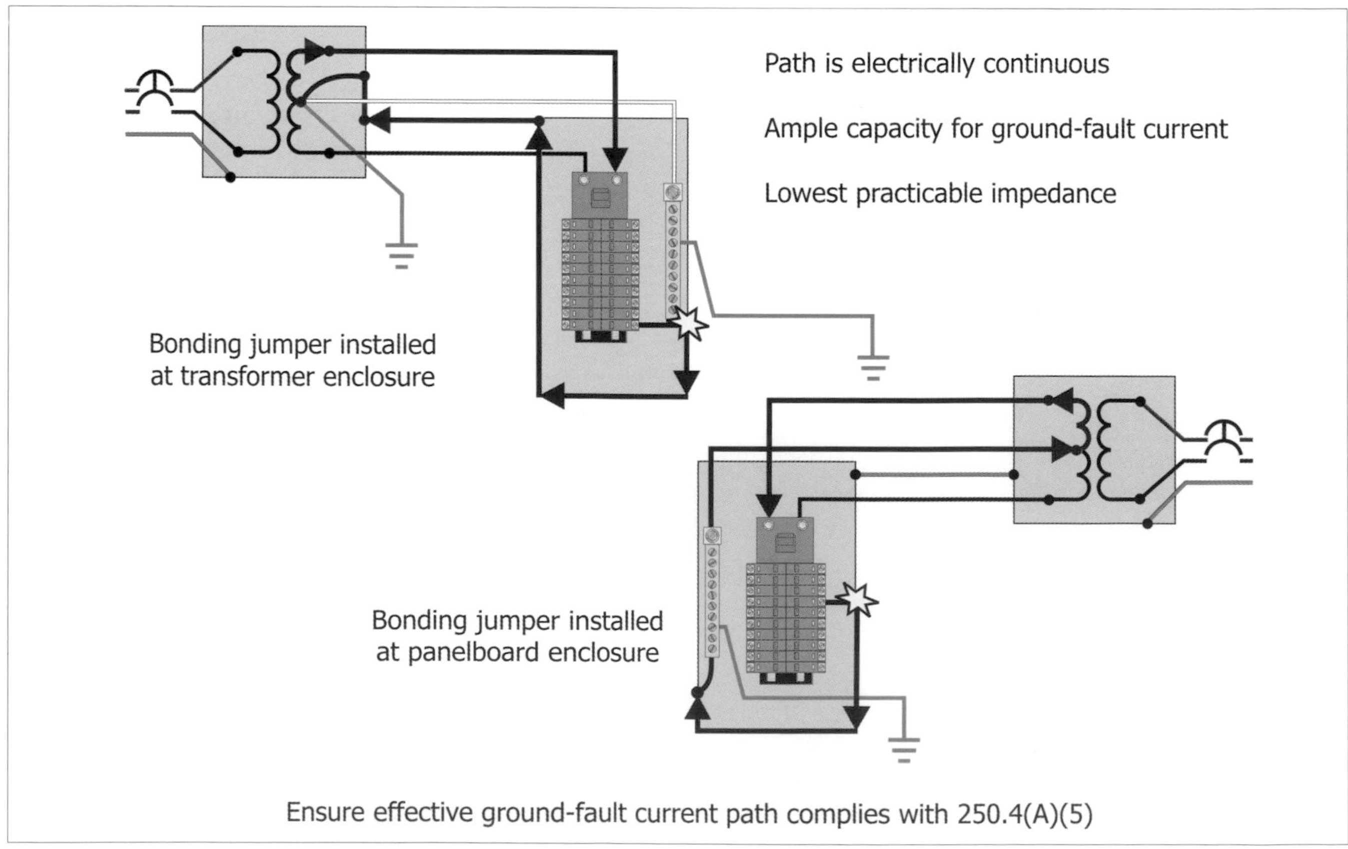
Path is electrically continuous
Ample capacity for ground-fault current
Lowest practicable impedance
Bonding jumper installed at transformer enclosure
Bonding jumper installed at panelboard enclosure
Ensure effective ground-fault current path complies with 250.4(A)(5)

Figure 12-14. Effective ground-fault current path to source

not to the primary or to the service. The transformer primary sees the fault current on the secondary as a load, and the overcurrent protection devices on the line side of the transformer will respond according to their rating.

It is necessary that an effective path for ground-fault current be electrically continuous, of adequate capacity, and low impedance from the equipment supplied back to the separately derived system. Often, it is best to make a one-line diagram of the installed system to be certain a complete and low-impedance path is provided [see 250.4(A)(5)].

If the system bonding jumper for a separately derived system is installed in the panelboard or fusible switch rather than in the transformer, and flexible metal conduit is used as the wiring method, an equipment bonding jumper must be installed between the transformer enclosure and the overcurrent device enclosure. In this case a parallel path for fault current will exist through the equipment bonding jumper and the wiring method.

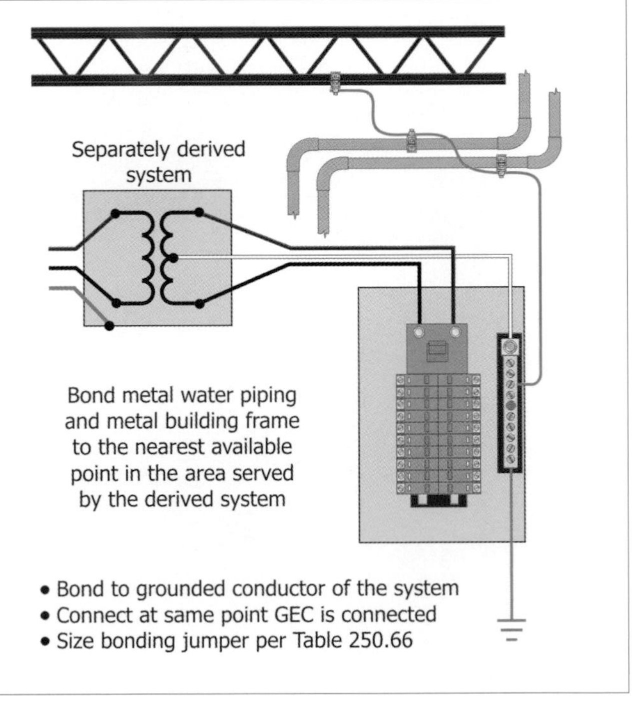
Separately derived system
Bond metal water piping and metal building frame to the nearest available point in the area served by the derived system
• Bond to grounded conductor of the system
• Connect at same point GEC is connected
• Size bonding jumper per Table 250.66

Figure 12-15. Bonding of metal water piping and structural framing members

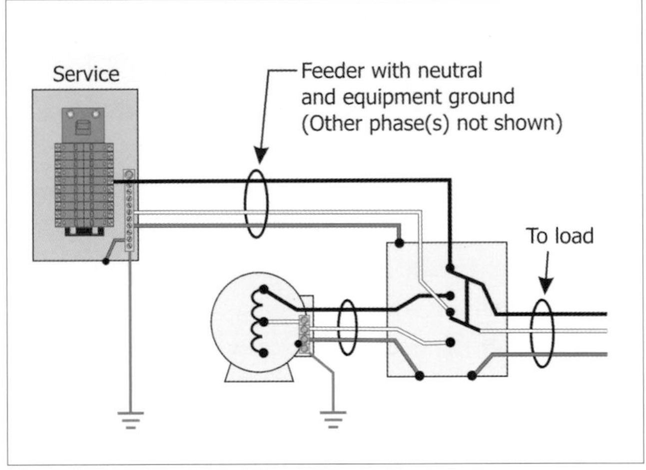
Service
Feeder with neutral and equipment ground (Other phase(s) not shown)
To load

Figure 12-16. Generator-type separately derived system using transfer equipment that switches the grounded (neutral) conductor

Equipment Grounding Conductors

Where permitted, the equipment grounding conductors of both the primary and the separately derived system can be connected together by the metal conductor enclosures, such as conduit or electrical metallic tubing or other equipment grounding conductors identified in 250.118. An equipment grounding conductor is used, if necessary, to be certain the circuit has low enough impedance to be effective. In both the primary and the separately derived system, a low-impedance path to the source grounded conductor (or neutral) of each system must be provided. This is vital to ensure that overcurrent devices will function to remove ground faults that can occur.

The important connections here require an equipment grounding conductor for the primary and equipment bonding jumper of the secondary that are separately derived systems independent of each other and terminating at the respective points of supply at their grounded conductor or neutral. Both equipment grounding conductors and secondary system equipment bonding jumpers will meet at some common point, such as the point where the system bonding jumper is connected, and they both may be connected to the common grounding electrode.

Interconnecting the equipment grounding conductor and the equipment bonding jumper of the two systems does not deeat the definition of separately derived system. This is because the equipment grounding conductor on the primary side and the

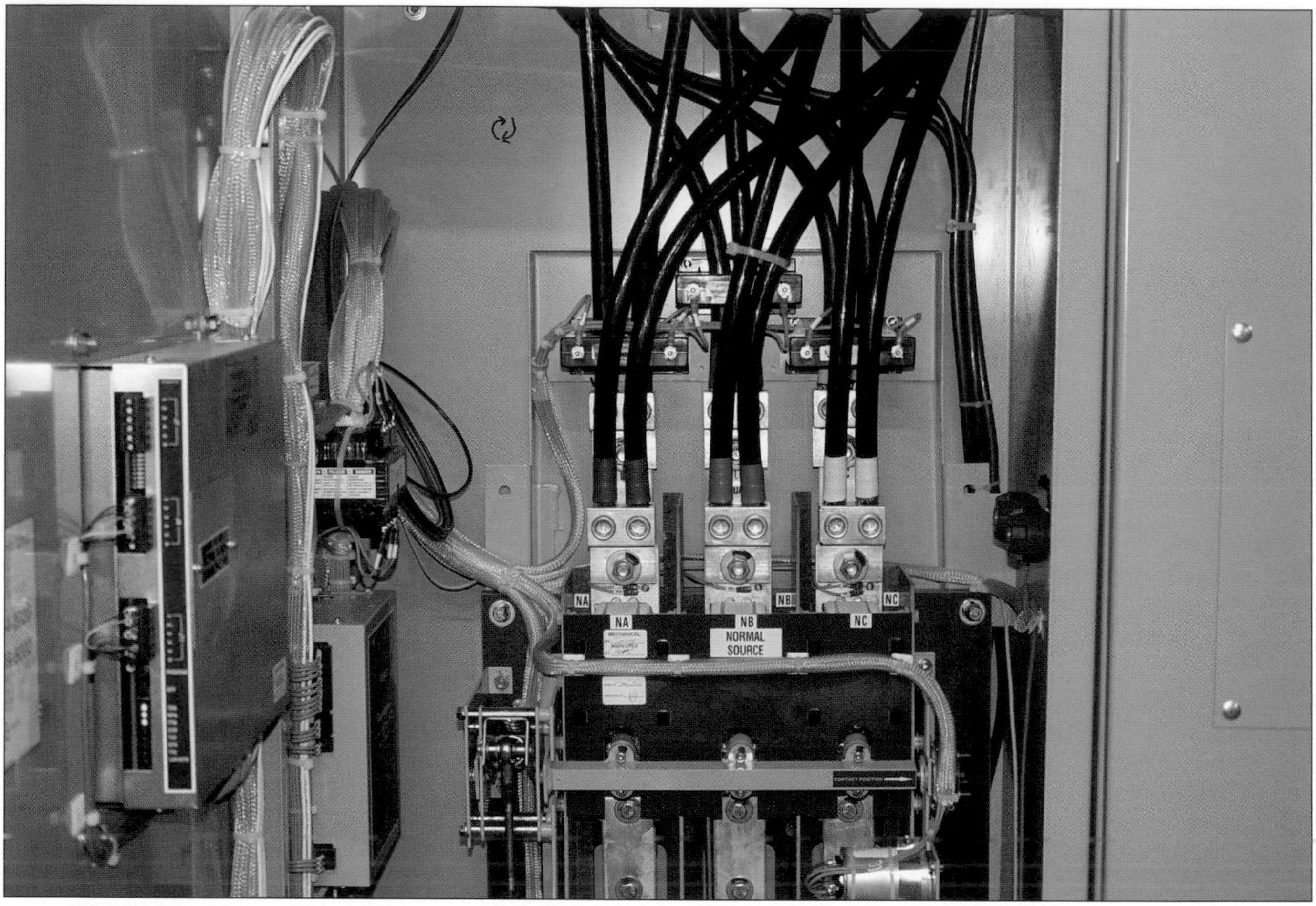

Photo 12-3. Transfer equipment that does not switch the grounded (neutral) conductor transfer equipment

equipment bonding jumpers on the secondary side are not system or power conductor(s) nor are grounded conductor(s). In addition, the equipment grounding conductor usually does not make a direct electrical connection of the neutral to ground. By following the system bonding jumper and equipment grounding conductors of the two systems, on a one-line diagram, it can be seen that the systems are, in fact, connected together by these conductors (see figure 12-14).

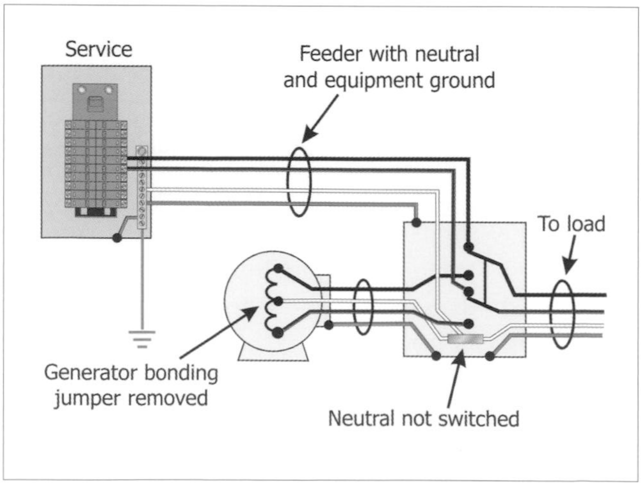

Figure 12-17. Generator system using transfer equipment that does not switch the grounded (neutral) conductor

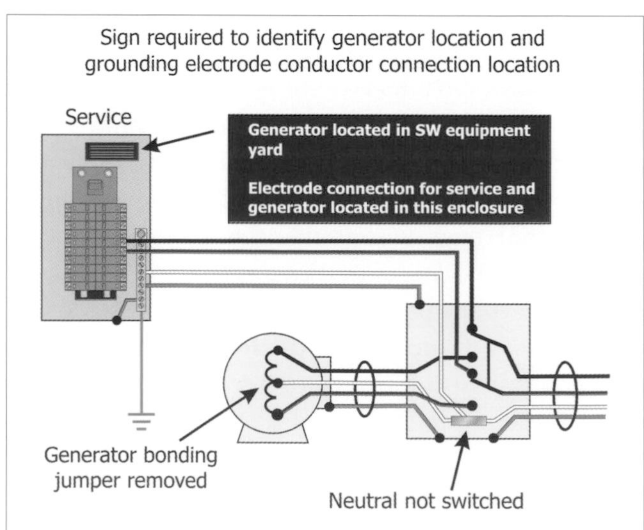

Figure 12-18. Signs are required for standby power source.

Photo 12-4. Transfer equipment that does switch the grounded (neutral) conductor

Bonding of Structural Steel and Water Piping

Where exposed structural steel that is interconnected to form the building frame or interior metal piping exists in the area served by the separately derived system, it is required to be bonded to the grounded conductor of the derived system. This effectively eliminates any possible differences of potential that can exist between the grounded conductor of the new derived source and either item, and also provides an effective path for ground-fault current directly to the source for any fault current likely to be imposed on either the steel or the water pipe. Bonding requirements for the metallic water piping system and metallic structural framing members that serve the same area as the separately derived system are provided in 250.104(D). Section 250.30(A)(6) provides a direct reference to 250.104(D) for these specific bonding rules (see figure 12-15).

Generator-Type Separately Derived System

Figure 12-16 illustrates a separately derived system from a generator. Note that there is no system connection, including that of a solidly grounded neutral conductor, between the two systems. An easy way to determine whether or not a generator must be grounded as a separately derived system in accordance with 250.30(A) is to examine the transfer equipment (see photos 12-3 and 12-4). If the grounded conductor (usually a neutral), in addition to the phase conductors, is switched at the transfer equipment, the generator is required to be grounded as a separately derived system (see figure 12-16.

If the neutral conductor is not switched by the switching action, but is solidly connected, then it is not a separately derived system (see figure 12-17). Where the generator or source is to be grounded as a separately derived system, the system bonding jumper must be installed, either at the generator or at the first disconnecting means or overcurrent device or any point between. In addition, a grounding electrode conductor must be installed

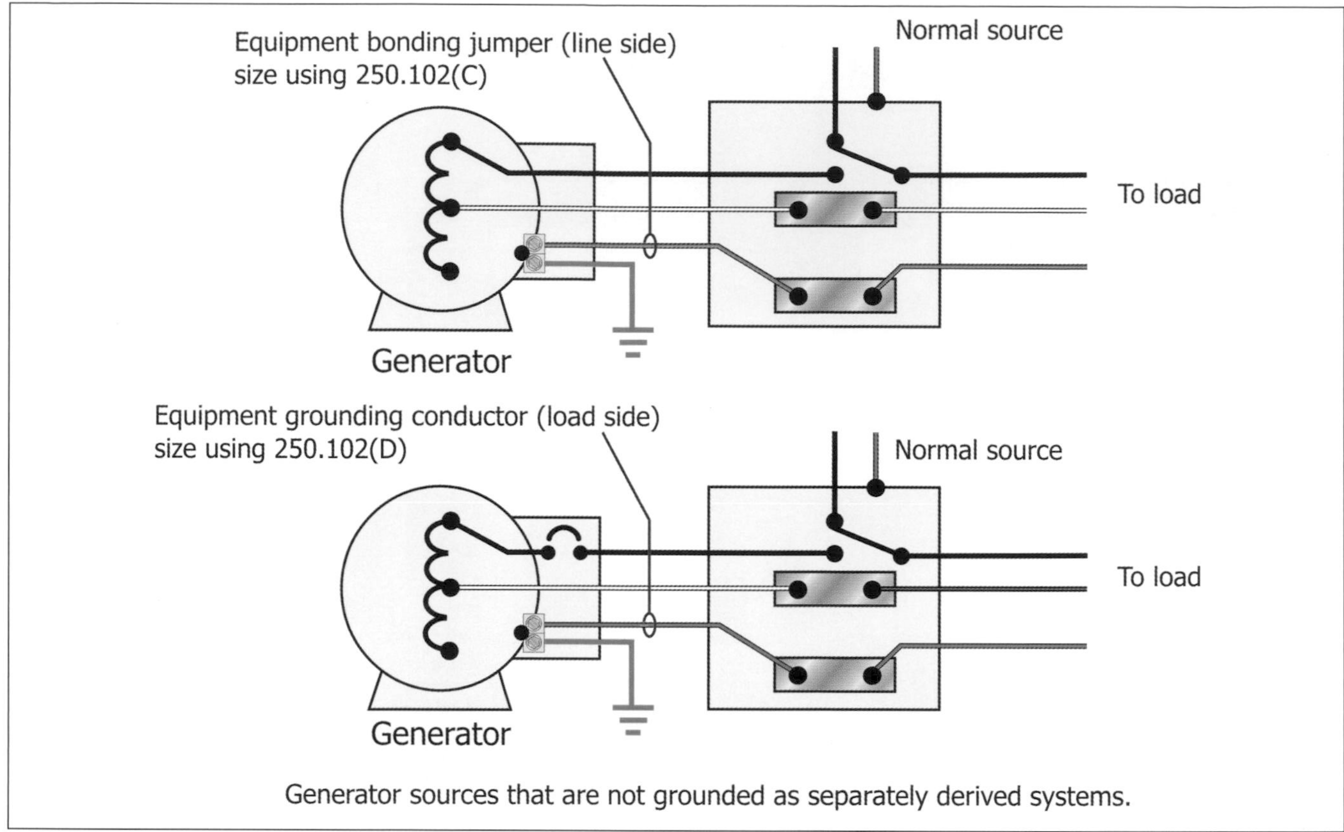

Figure 12-19. Generator is not a separately derived system

from the system grounded (neutral) conductor and a grounding electrode. The grounding electrode conductor connection to the system must be located at the same point where the system bonding jumper is installed.

Section 700.8(B), regarding grounding of emergency systems, requires that where the grounded circuit conductor connected to the emergency source is connected to a grounding electrode conductor at a location remote from the emergency source, a sign must be placed at the grounding location that shall identify all emergency and normal sources connected at that location.

Identical requirements are in 701.9(B) for legally required standby systems and in 702.8(B) for optional standby systems (see figure 12-18).

Alternate power supply systems that include grounded (neutral) conductors that are not switched through a transfer switch present a safety concern. Electricians that are unfamiliar with the system grounding scheme for that alternate source could inadvertently disconnect the grounded (neutral) conductor when working on the normal source. If the alternate or emergency system source is operating, the grounded (neutral) conductor from the transfer switch to the location where it is grounded will function as a grounding electrode conductor, and disconnecting it can present a serious shock hazard.

Section 250.35 includes requirements for providing an effective ground-fault current path between permanently installed generators and the first disconnecting means enclosure. This section addresses both separately derived systems and non-separately derived systems. Where the generator is a separately derived system as described in 250.20(D), it is required to comply with all grounding and bonding rules in 250.30(A). Where the generator is not separately derived, the grounding and bonding connections are required to be in accordance with either 250.35(B)(1) or (B)(2) depending on the location of the derived system overcurrent device. Equipment bonding jumpers on the supply side (line side) of each generator overcurrent protective device must be sized according to 250.102(C) [see figure 12-19].

Equipment grounding conductors on the load side of each generator overcurrent protective device must be sized according to 250.102(D) [see figure 12-19].

Generator Not Grounded as Separately Derived Systems

Where the generator is not a separately derived system, the neutral system bonding jumper must be removed from the generator, and the neutral must not be grounded at the generator or at any point between the generator up to the service.

In this case, the system is grounded by its solid connection to the neutral of the premises wiring system. An equipment grounding conductor or equipment bonding jumper is installed throughout the system with the circuit conductors between the service, transfer switch, and other non-current-carrying parts of the installation.

Equipment Ground-Fault Systems

There are situations where the location of the system bonding jumper is determined by the conditions of the installation and the type of equipment installed. For example, a large step-up transformer separately derived system with the voltage configuration of 240-volt 3-phase primary stepping up to 480Y/277-volt 3-phase, 4-wire secondary that supplies equipment rated at 1000 amperes or greater can require ground-fault protection for equipment [see 240.13 and 215.10].

Larger switchboards that include GFP equipment often require the system bonding jumper to be installed at the equipment and not at the source to ensure the performance of the GFP equipment and conform to the listing and installation instructions of the equipment [see 110.3(B)].

A word of caution is in order here. It is imperative that care be exercised where generators supply systems that have equipment ground-fault protection. In many cases, the designer or engineer will specify that the grounded (neutral) conductor be switched by the transfer equipment to avoid grounding connections to the grounded (neutral) conductor downstream from the ground-fault protection equipment. This is vital to prevent desensitizing the protection system. In this case, the generator is a separately derived system, and it must be grounded and bonded separately.

Section 700.26 provides that equipment ground-fault protection with automatic disconnecting means is not required for emergency systems. However, indication of a ground fault on the emergency system must be provided in accordance with the rules of 700.7(D). It is required that, where practicable, audible and visual signal devices be provided that will indicate a ground fault in a solidly-grounded wye-connected emergency system of more than 150 volts to ground and circuit protective devices rated 1000 amperes or more. A similar exception from the requirement for ground-fault protection of equipment is provided in 701.17 for legally required standby systems. See chapter 15 for a more complete discussion on the types of equipment ground-fault protection systems, how the function, and where they are required in premises wiring systems.

[1] NFPA 70, *National Electrical Code* 2008, Article 100, (National Fire Protection Association, Quincy, MA, 2007), p. 70-29.

[2] NFPA 70, 250.30, p. 70-100

[3] NFPA 70, 250.28, p. 70-100

[4] NFPA 70, Article 100, p. 70-27.

[5] NFPA 70, 250.30(A)(5) p. 70-102.

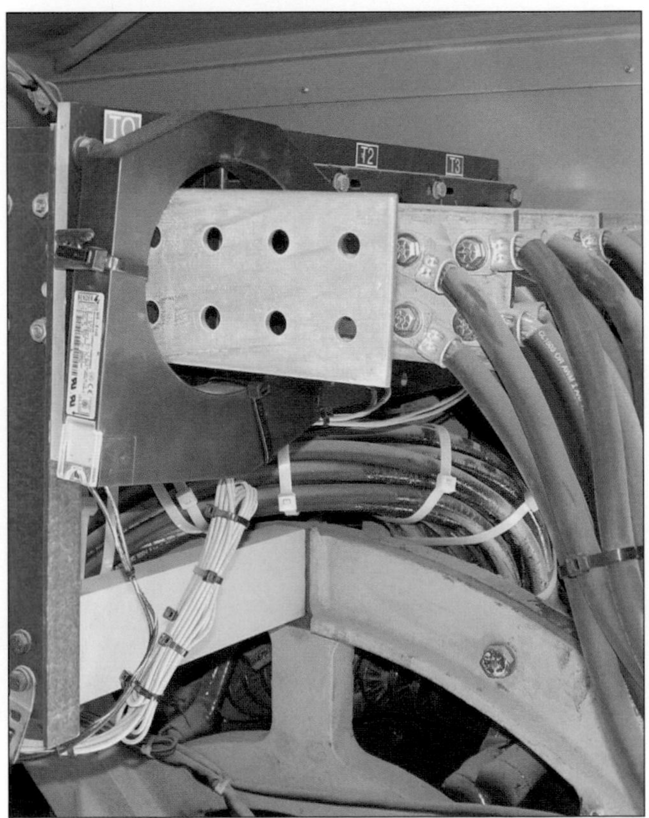

12 Review Questions

The questions included here were developed using material included in this chapter. The answers can be found by reviewing the text. It is also important that students make use of *NEC*-2008, where many answers can be found.

1. "A premises wiring system whose power is derived from a source of electric energy or equipment other than a service without any direct electrical connection, including a solidly-connected grounded circuit conductor to supply conductors originating in another system" best describes ____.
   a. medium voltage systems
   b. utility supplied systems
   c. separately derived systems
   d. high voltage systems

2. Derived phase conductors supplied by a separately derived system sized at 500 kcmil aluminum generally require a grounding electrode conductor not smaller than ____.
   a. 4 AWG copper or 2 AWG aluminum
   b. 2 AWG copper or 1/0 AWG aluminum
   c. 6 AWG copper or 4 AWG aluminum
   d. 8 AWG copper or 6 AWG aluminum

3. The connection of the system bonding jumper is permitted to be made at the source , or the ____ disconnecting means or overcurrent device or any single point between.
   a. first
   b. second
   c. any
   d. line-side

4. The grounding electrode for a separately derived system is required to be located as near as practicable to, and preferably in, the ____ as the separately derived system.
   a. same enclosure
   b. same cabinet
   c. same area
   d. same pull box

5. A grounding electrode conductor for a separately derived system is required to be connected to the nearest ____ in the structure.
   a. underground metal gas pipe
   b. structural metal grounding electrode or metal water pipe grounding electrode
   c. service equipment enclosure
   d. metal raceway

6. Where the grounding electrode(s) for a separately derived system, as in question No. 5, is or are not present and available, the grounding electrode is required to be ____.
   a. the nearest available effectively grounded metal enclosure
   b. the nearest available effectively grounded metallic gas pipe
   c. a communications system grounding electrode
   d. Any of the other electrodes identified in 250.52(A)

7. Where installed in or at the building, the same ____ is required to be used for the grounding of all ac systems.
   a. grounding electrode
   b. grounded raceway
   c. electrode enclosure
   d. bonding jumper

8. The ____ grounding conductor(s) of the primary and the equipment bonding jumper(s) of a separately derived system are permitted to be raceways or conductor enclosures, where permitted by *Code*.
   a. isolated
   b. identified
   c. system
   d. equipment

9. If the grounded (neutral) conductor and all phase conductors are switched through transfer switching equipment associated with an onsite generator, then the generator ____ required to be grounded as a separately derived system according to 250.30(A).
   a. is
   b. is not
   c. may be
   d. cannot be

10. Where a generator is grounded as a separately derived system, the ____ bonding jumper must be installed, either at the generator or at the first disconnecting means or overcurrent device.
   a. equipment
   b. system
   c. service
   d. 10 AWG copper

11. Where the generator is not a separately derived system, the ____ bonding jumper must be removed from the generator and the neutral must not be grounded. In this case, the system is grounded by its solid connection to the grounded (neutral) conductor of the premises wiring system.

a. 10 AWG
b. equipment
c. service
d. system

12. The equipment bonding jumper installed between the separately derived system enclosure and the first overcurrent device enclosure is required to be sized in accordance with ________.

a. Section 250.102(C)
b. Table 250.122
c. Section 250.122
d. Section 220.22

13. The minimum size copper grounded (neutral) conductor required for a separately derived system where the derived phase conductors are 300 kcmil copper conductors (1 per phase) and there is very little or no neutral load on the system is________

a. 6 AWG
b. 1/0 AWG
c. 2 AWG
d. 2/0 AWG

14. Where multiple separately derived systems are installed and connected to a common grounding electrode conductor, the minimum size of the copper common grounding electrode conductor must not be smaller than ________.

a. 6 AWG
b. 250 kcmil
c. 4 AWG
d. 3/0

15. Where a generator for an emergency system is provided with transfer equipment that does not switch the grounded (neutral) circuit conductor, a ____________ must be placed at the service equipment denoting the location of the generator and the grounding location and all emergency and normal sources connected at that location.

a. grounding electrode conductor
b. sign
c. bonding jumper
d. disconnect

16. Where a separately derived system is grounded by connection to a single ground rod and the derived phase conductors are sized at 750 kcmil copper, what is the minimum size grounding electrode conductor required?

a. 3/0 copper
b. 250 aluminum
c. 4 AWG copper
d. 6 AWG copper

17. The conductor used to connect the grounded circuit conductor and the equipment grounding conductor at a separately derived system is defined as the __________.

a. main bonding jumper
b. system bonding jumper
c. equipment grounding conductor
d. bonding jumper

18. Where grounding eletrode conductor taps are installed for multiple separately derived systems in accordance with 250.30(A)(4), they shall be connected to the system in which of the following locations?

a. at the service
b. within 5' of the entry of the building
c. at the source enclosure
d. in the enclosure where the system bonding jumper is installed.

19. Where a generator is a separately derived system, the line side equipment bonding jumper between the generator and the equipment grounding terminal bus of the enclosure shall be sized in accordance with which of the following *Code* sections?

a. 250.102(D)
b. 250.102(C)
c. 250.122(A)
d. none of the above

20. What is the minimum size of the water piping system bonding jumper connected to the grounded conductor of a separately derived system if the derived secondary ungrounded phase conductors are sized at 500 kcmil copper?

a. 2 AWG copper
b. 3/0 copper
c. 6 AWG copper
d. 1/0 copper

21. Where the common grounding electrode conductor tap system is used for grounding multiple separately derived systems, the grounding electrode conductor taps shall be connected in a manner that the__________ remains without a splice or joint.
    a. equipment grounding conductor
    b. grounded conductor
    c. common grounding electrode conductor
    d. equipment bonding jumper

Grounding and Bonding at More Than One Building or Structure

Section 250.32 provides requirements for grounding of electrical systems and equipment at buildings or structures that are supplied from feeders or branch circuits. Included are rules for grounding electrodes, grounded systems, ungrounded systems, and grounding at buildings where the disconnecting means are located in a separate building or structure on the same premises.

Objectives

To understand...

- Grounding at more than one building or structure
- Sizing of grounding electrode conductors
- Requirements for bonding grounding electrodes together
- Disconnecting means requirements for separate buildings
- Objectionable currents over multiple paths
- Requirements for mobile homes, recreational vehicles, and agricultural buildings

Chapter 13

Definition

Grounding electrode conductor. "A conductor used to connect the system grounded conductor or the equipment to a grounding electrode or to a point on the grounding electrode system." [1]

This definition includes the grounding electrode conductors at services, separately derived systems, and at a second or additional building or structure supplied by a common service, multiple services or other premises wiring system(s).

Grounding Electrode Is Required

The general rule in 250.32(A) is that at each building or structure served by one or more feeders or branch circuits, a grounding electrode system or grounding electrode meeting the requirements of Article 250 Part III must be connected in a manner specified in 250.32(B) or (C). Where no grounding electrodes are present at the building or structure, a grounding electrode(s) required in Part III of Article 250 (specifically 250.50) must be installed and used.

Grounding electrodes specified in 250.52 must be installed and used as required in 250.32(A) include:

"250.52(A) Electrodes Permitted for Grounding

(1) Metal Underground Water Pipe

(2) Metal Frame of the Building or Structure

(3) Concrete-Encased Electrode

(4) Ground Ring

Where none of these electrodes is available at the building or structure, one or more of the following electrodes specified in 250.52(A)(4) through (A)(7) shall be installed and used. [2]

(5) Rod and Pipe Electrodes

(6) Other Listed Electrodes

(7) Plate Electrodes

(8) Other Local Metal Underground Systems or Structures

"250.52(B) Not Permitted for Use as Grounding Electrodes. The following systems and materials shall not be used as grounding electrodes:

(1) Metal underground gas piping systems

(2) Aluminum" [3]

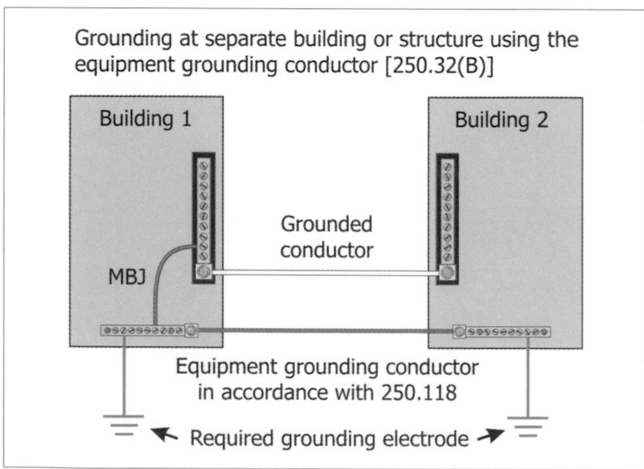

Figure 13-1. Grounding at separate buildings or structures using the equipment grounding conductor (insulate the grounded conductor from ground and grounded parts and equipment at building two).

There is a fine print note following 250.32(B) that provides clear direction for the bonding requirements for any metal gas piping systems that are installed in or attached to the building or structure. A reference to Section 250.104(B) for the rules is provided.

The exception to 250.32(A) indicates that a "grounding electrode at separate buildings or structures shall not be required where a single branch circuit supplies the building or structure and the single branch circuit including a multiwire branch circuit includes an equipment grounding conductor for grounding the conductive non-current-carrying parts of all equipment." [4] For the purposes of this section, a single branch circuit could be a multiwire branch circuit as indicated in the exception to 250.32(A). [See *NEC* 250.53 and chapter six for additional information on installing grounding electrodes].

Grounding Requirements for Grounded Systems

Rules are provided in 250.32(B) for grounding electrical systems and equipment at additional buildings or structures on the premises. Buildings or structures supplied by a feeder(s) or more than one branch circuit that include an equipment grounding conductor must comply with the rules in 250.32(B). Existing buildings or structures supplied by a feeder(s) or more than one branch circuit that does not have an equipment grounding conductor must comply with the rules in 250.32(B), Exception.

Grounding with an Equipment Grounding Conductor

In this method, an equipment grounding conductor is run with the feeder(s) or branch circuit(s), in addition to the

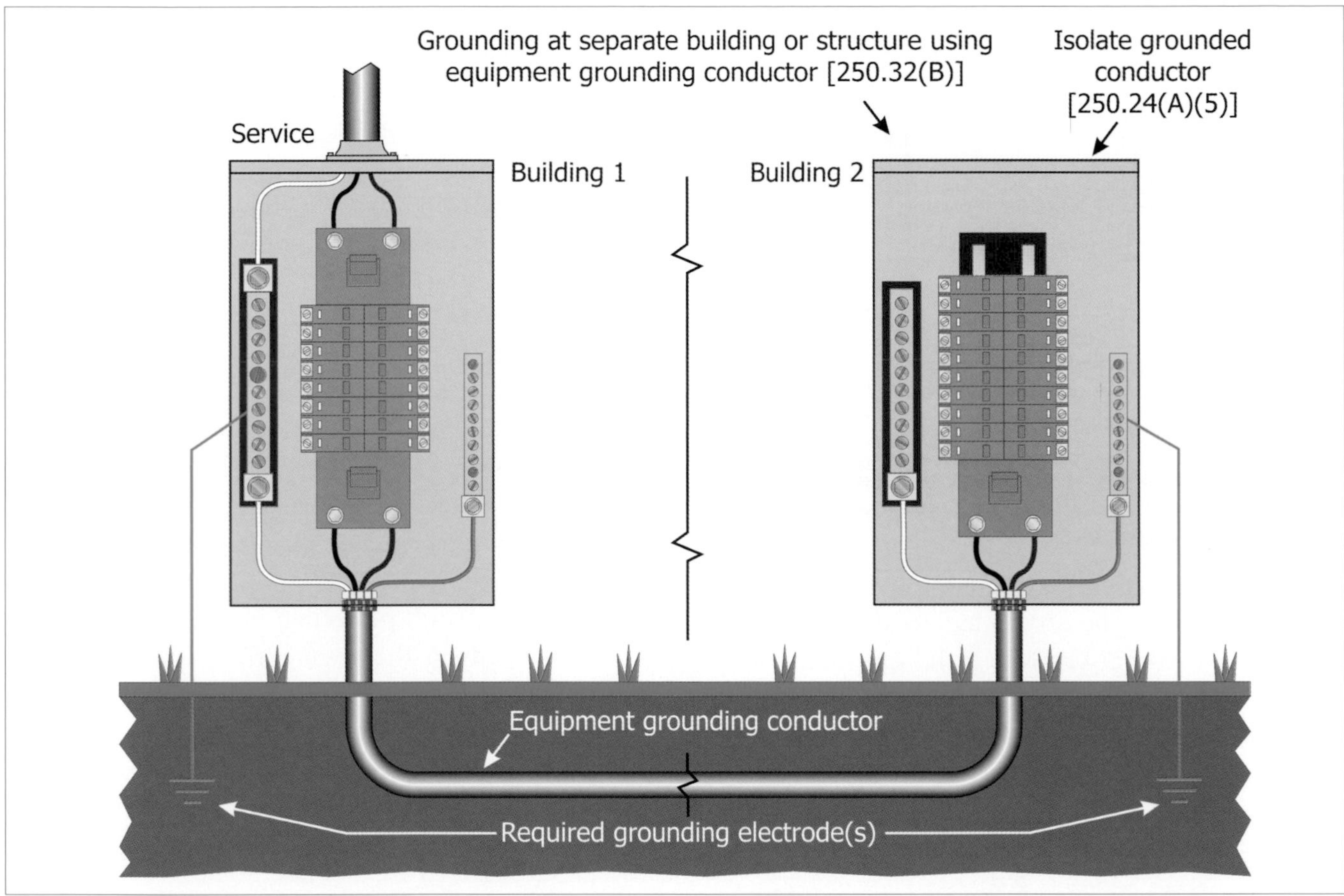

Figure 13-2. Insulate grounded (neutral) conductor from ground and grounded parts and equipment under this method.

ungrounded and grounded conductors to the building or structure. "Any installed grounded conductor shall not be connected to the equipment grounding conductor or to the grounding electrode(s)." [5] This would be in violation of 250.24(A)(5) and 250.142(B). This method is similar, or identical, to installing a feeder to a panelboard or distribution equipment that is located in the same building or service where the service is located (see figures 13-1 and 13-2).

The equipment grounding conductor (which could be any of those specified on 250.118) that is run from the building or structure where the feeder or branch circuit(s) originate is connected to a terminal bar on the supply side of, or inside, the building or structure disconnecting means. This disconnecting means (or equipment) is grounded by connecting the required grounding electrode at the building or structure to the equipment grounding terminal bar of the disconnecting means enclosure. Equipment grounding conductors of the type included in 250.118 are acceptable and include wires as well as some conduits or other raceways.

The number of conductors for various systems that must be installed is summarized in a handy reference, Table 13-1.

The grounded (often a neutral) conductor, installed as a part of the feeder or branch circuit from the building or structure where the service is located to the second or additional building or structure, must be an insulated conductor [see 310.2(A)].

By using this method, the grounded (often a neutral) conductor(s) must be connected to the terminal bar for the

Table 13-1
Equipment Grounding Conductor Installed

System	Grounded	Ungrounded	Equip. Ground
120V	1	1	1
120/240V	1	2	1
208Y/120V	1	3	1
480Y/277V	1	3	1

Table 13-1. Equipment Grounding Conductor Installed

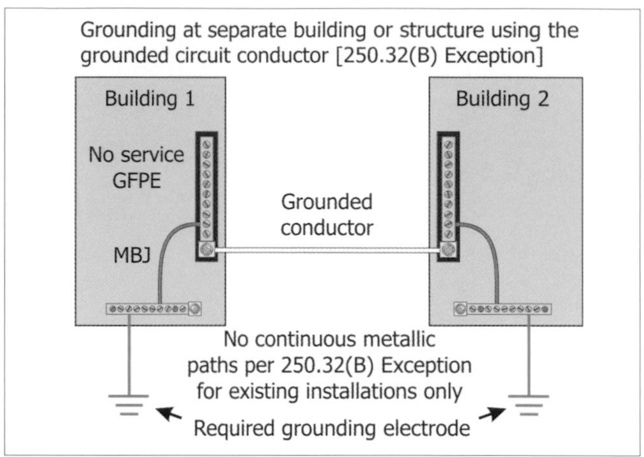

Figure 13-3. Grounding at separate buildings or structures using the grounded (neutral) conductor as allowed in 250.32(B), Exception [Existing Installation]

grounded conductors, and the equipment grounding conductor(s) must be connected to the equipment grounding terminal bar where the grounding (electrode) is connected inside the building or structure disconnecting means enclosure.

The equipment grounding conductor that is run to the additional building or structure must be sized from Table 250.122 based on the rating of the overcurrent device protecting the feeder or branch circuit.[6]

Existing Installations by Exception

The exception to 250.32(B) applies only to existing installations and permits the grounded circuit conductor (often a neutral) to be grounded again at the disconnecting means for the building or structure only if all the following conditions are complied with:

1. "an equipment grounding conductor is not run with the supply to the building or structure,
2. "there are no continuous metallic paths bonded to the grounding system in each building or structure involved, and
3. "ground-fault protection of equipment has not been installed on the common ac service…." [7] (see figure 13-5).

In this case, the electrical system at the additional

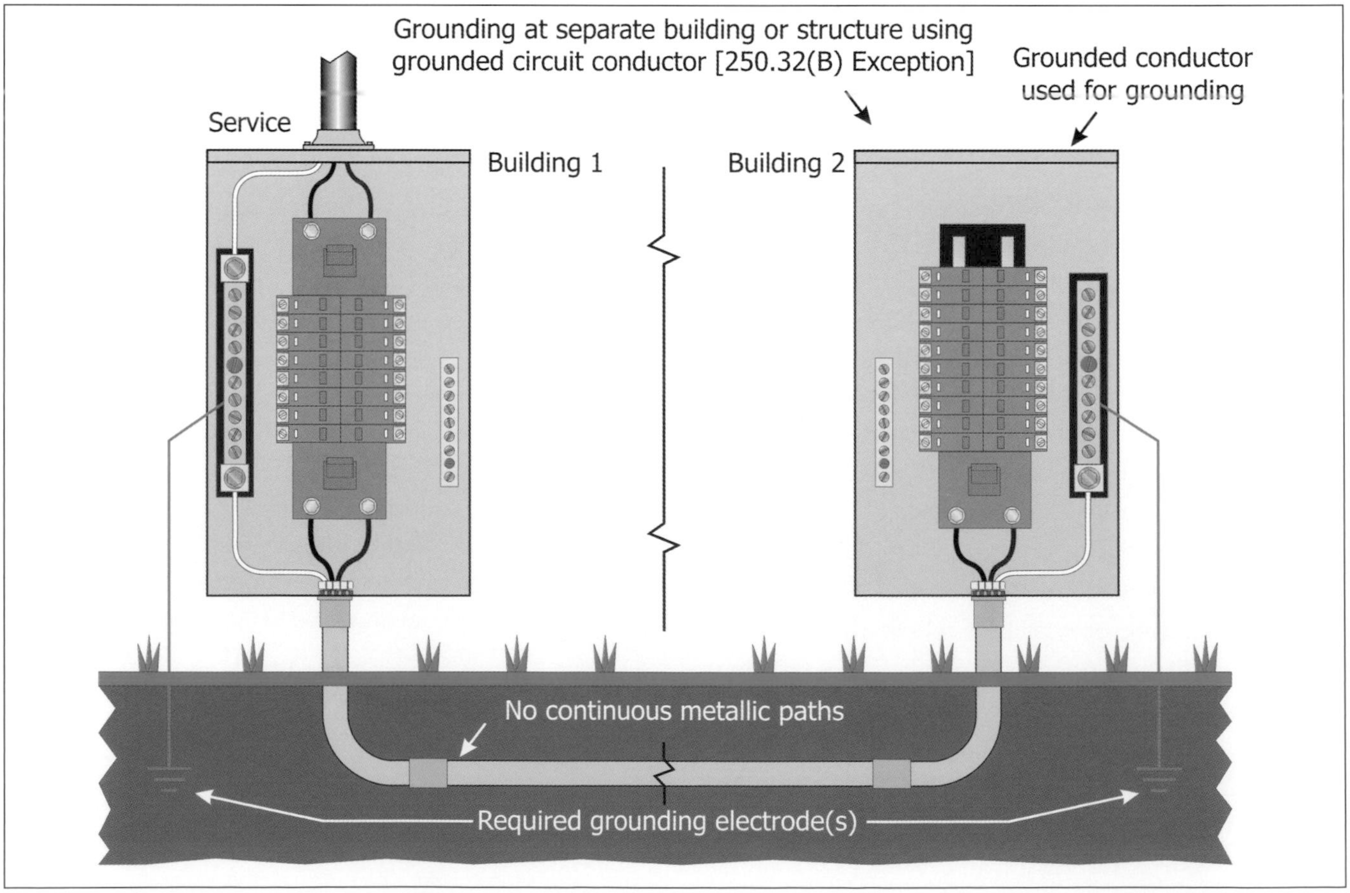

Figure 13-4. Grounding at separate buildings using the grounded (neutral) conductor by exception for existing premises wiring systems

System	Grounded	Ungrounded	Equip. Ground
120V	1	1	0
120/240V	1	2	0
208Y/120V	1	3	0
480Y/277V	1	3	0

Table 13-2
Equipment Grounding Not Conductor Installed

Table 13-2. Equipment Grounding Not Conductor Installed by Exception for Existing Installations

building or structure is treated like a service for grounding purposes, although the building or structure is actually supplied by a feeder or branch circuit. There are also some different requirements regarding bonding provisions and minimum conductor sizes. The grounded circuit conductor is bonded to the disconnecting means enclosure for the building or structure. The grounding electrode connection to the grounded (often a neutral) conductor must be made on the supply side of or within the building or structure disconnecting means.

In this method provided in the exception to 250.32(B), both the grounded circuit conductor (usually a neutral) and the non-current-carrying equipment are connected to a grounding electrode (system) at the additional building or structure. The number and type of conductors that must be taken, from the building or structure where the feeder or branch circuit originates, to the second building or structure is summarized in Table 13-2 (see also figures 13-3 and 13-4).

Both the grounded (usually a neutral) conductor(s) and the equipment grounding conductor(s) are permitted to be connected to the terminal bar where the grounding electrode conductor connects to the grounded conductor.

Note that the grounded conductor supplied with the feeder or branch circuit(s) to the second or additional building or structure must generally be an insulated conductor [see 310.2(A)]. Section 338.10(B)(2) permits the bare conductor of Type SE cable to be used only as an equipment grounding conductor for feeders or branch circuits but generally not as a grounded (neutral) conductor between buildings [see 338.10(B)(2) and the exception].

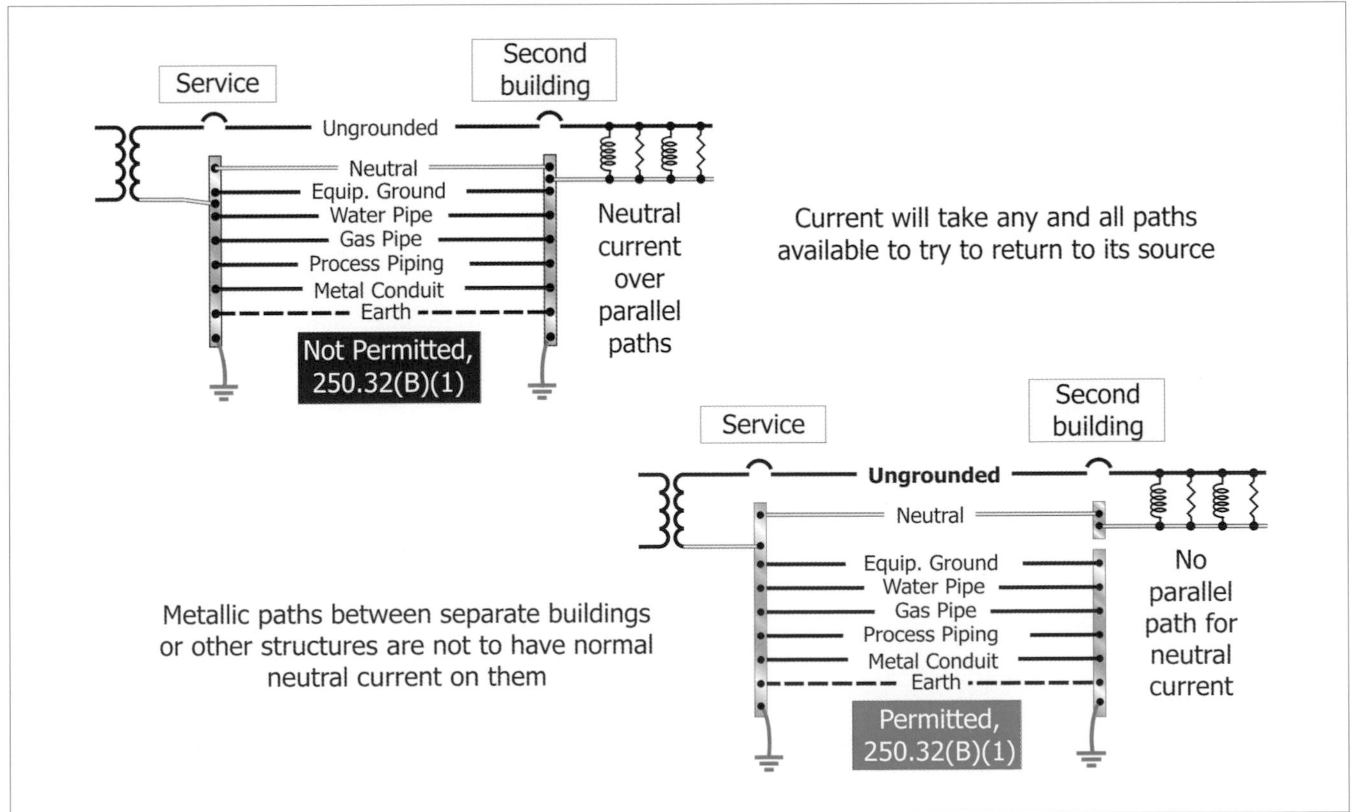

Figure 13-5. Objectionable currents—Paths

Size of Grounded Conductor

For existing installations where an equipment grounding conductor is not run to the additional building or structure in accordance with 250.32(B) exception, the grounded circuit conductor must be sized no smaller than an equipment grounding conductor from Table 250.122. In this case, the grounded circuit conductor serves three purposes: first, to permit line-to-neutral loads to be utilized; second, to carry unbalanced loads back to the source; and, third, it must function as an equipment grounding conductor in the event of a ground-fault condition. As such, it must be sized for the calculated load according to 220.61 and not smaller than the minimum size equipment grounding conductor required from Table 250.122. Keep in mind that feeders or branch circuit(s) that supply separate buildings or structures are often in extensive lengths, and voltage drop considerations can have an impact on the minimum sizes required for the equipment grounding conductor [see 250.122(B) and the note to Table 250.122].

Grounding Electrode Conductor and Connections

A grounding electrode conductor, sized in accordance with 250.66 based on the size of the largest ungrounded conductor supplying the building or structure, is used to connect the grounded circuit conductor or equipment grounding conductor and equipment to the grounding electrode system that exists or is installed at the additional building or structure (see figure 13-6). As covered above, where no grounding electrodes are present at the additional building or structure, one or more grounding electrodes must be installed in compliance with Part III of Article 250.

This connection must be made inside the building or structure disconnecting means to the equipment grounding terminal bar or to the terminal bar for the grounded conductor as appropriate. The grounding electrode conductor connections are required to meet the requirements of 250.8.

Sizing the Grounding Electrode Conductor

"The size of the required grounding electrode conductor to the grounding electrode(s) at a separate building or structure shall not be smaller than given in 250.66, based on the largest ungrounded supply

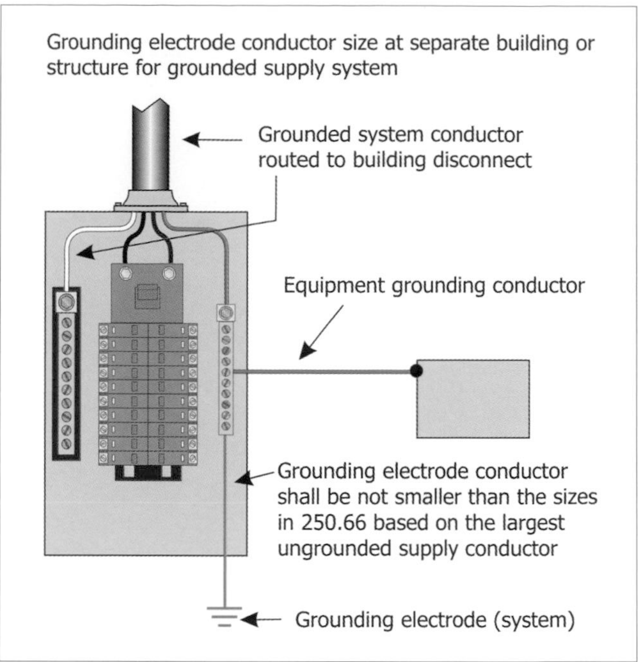

Figure 13-6. Sizing grounding electrode conductor(s) at separate buildings or structures

conductor. The installation shall comply with Part III of this article." [9]

Bonding Grounding Electrodes Together

Section 250.58 does not require that grounding electrodes at separate buildings or structures be bonded back to the service. A study of the system will show that there is no real electrical separation between two separate grounding electrode systems. The grounding electrode system for the electrical system at the individual buildings or structures is connected together either by the grounded conductor or by the equipment grounding conductor routed with the feeder or branch circuit(s) to that additional building(s) or structure(s) (see figures 13-1 and 13-3).

Ungrounded Electrical System(s)

"Where an ungrounded ac system is connected to a grounding electrode in or at a building as specified in 250.24 and 250.32(C), the same electrode shall be used to ground conductor enclosures and equipment in or on that building." [10] A grounding electrode system in compliance with Part III of Article 250 must be installed and used at the building or structure where none exists (see figure 13-7). The size of the grounding electrode conductor is based on the size of the largest

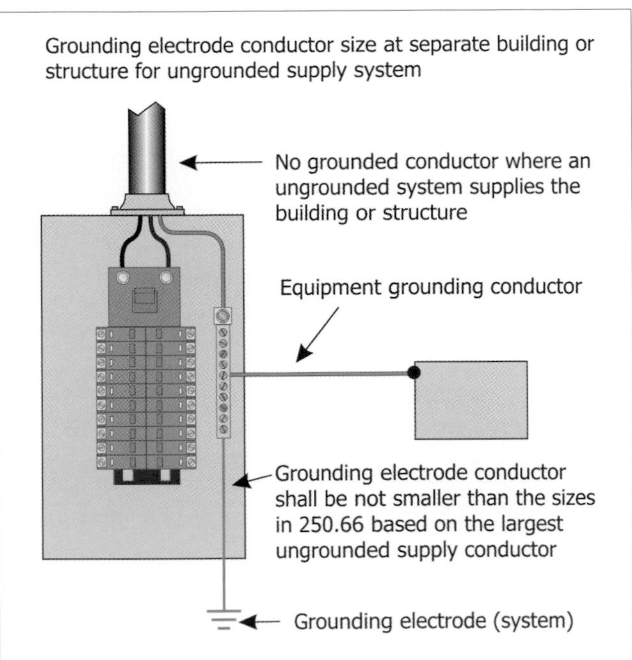

Figure 13-7. Sizing grounding electrode conductor at separate buildings that are supplied by an ungrounded system

ungrounded phase conductor in the feeder supplying the separate building or structure using Table 250.66.

Disconnecting Means Located in Separate Building or Structure on the Same Premises

Special rules have been provided for large capacity, multibuilding industrial installations under single management in 250.32(D) (see figure 13-8). These occupancies have trained and qualified personnel and have established procedures for safe switching of electrical feeders or circuits. As a result, the disconnecting means are permitted to be at other locations on the premises rather than at the building served. Often, the switching is managed by automatic or manual means from a control room or station.

The special rules for grounding the electrical system at these separate buildings or structures are as follows:

1. "The connection of the grounded circuit conductor to the grounding electrode, to normally non-current-carrying metal parts of equipment, or to the equipment grounding conductor at a separate building or structure shall not be made.

2. "An equipment grounding conductor for grounding and bonding any normally non-current-carrying metal parts of equipment, interior metal piping systems, and building or structural metal frames is run with the circuit conductors to a separate building or structure and connected to existing grounding electrode(s) required in Part III of this article, or, where there are no existing electrodes, the grounding electrode(s) required in Part III of this article shall be installed where a separate building or structure is supplied by more than one branch circuit.

3. "The connection between the equipment grounding conductor and the grounding electrode at a separate building or structure shall be made in a junction box, panelboard, or similar enclosure located immediately inside or outside the separate building or structure." [11]

Mobile Homes and Recreational Vehicles

The basic requirement in 550.32(A) is that mobile home service equipment be located remote from the structure but within sight from and not more than 9.0 m (30 ft) from the mobile home (see figure 13-9). There, the grounded (usually a neutral) conductor is connected to a grounding electrode(s). Feeders consisting of four insulated and color-coded conductors, one of which is an equipment grounding conductor, must be run to the mobile home distribution panelboard [see 550.33(A)].

"The service equipment shall be permitted to be located elsewhere on the premises, provided that a disconnecting means suitable for use as service equipment is located in sight from and not more than 9.0 m (30 ft) from the exterior wall of the mobile home it serves." [12] As such, this exception permits the two options for grounding to be used as described earlier in this chapter.

A three-conductor feeder is permitted to be run from the service located remote from the mobile home disconnecting means where the neutral is grounded, and a feeder consisting of four insulated conductors in compliance with 550.33(A) are then run from the mobile home disconnecting means to the mobile home panelboard. This permits the mobile home service equipment to be located at a common location, such as a separate laundry or utility building as is required by some serving electric utilities. The grounding of the grounded (usually neutral) circuit conductor at the disconnecting means must comply with 250.32(B) as covered above.

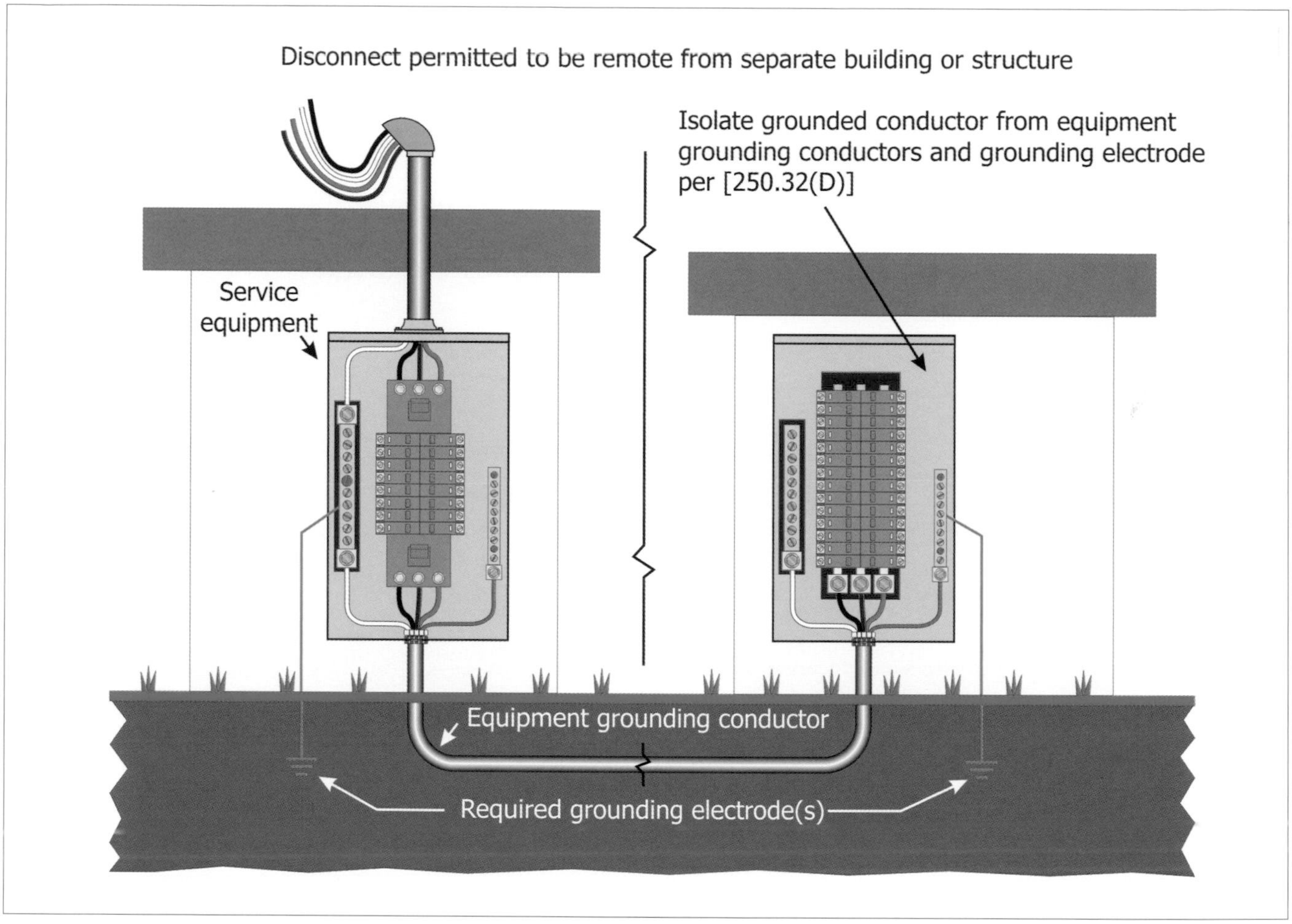

Figure 13-8. Disconnecting means located remote from separate buildings or structures according to 250.32(D)

Section 550.32(B) permits service equipment to be installed directly on the manufactured home (not mobile home — see the definitions in 550.2) under seven conditions.

1. "The manufacturer shall include in its written installation instructions information indicating that the home shall be secured in place by an anchoring system or installed on and secured to a permanent foundation.

2. "The installation of the service equipment shall comply with Part I through Part V of Article 230.

3. "Means shall be provided for the connection of a grounding electrode conductor to the service equipment and routing it outside the structure.

4. "Bonding and grounding of the service shall be in accordance with Part I through Part VII ofArticle 250.

5. "The manufacturer shall include in its written installation instructions one method of grounding the service equipment at the installation site. The instructions shall clearly state that other methods of grounding are found in Article 250.

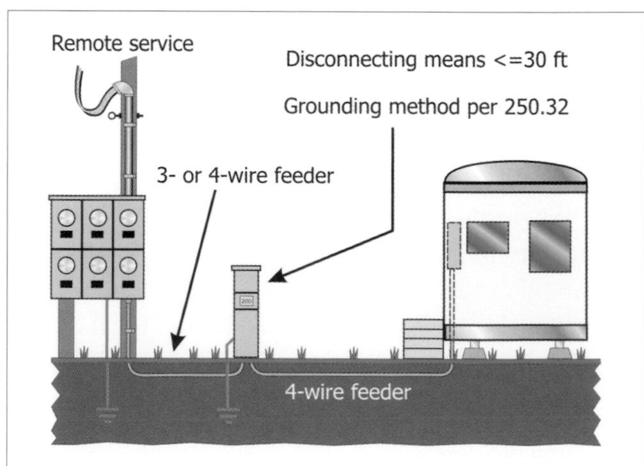

Figure 13-9. Service equipment located remote from mobile/ manufactured home

6. "The minimum size grounding electrode conductor shall be specified in the instructions.

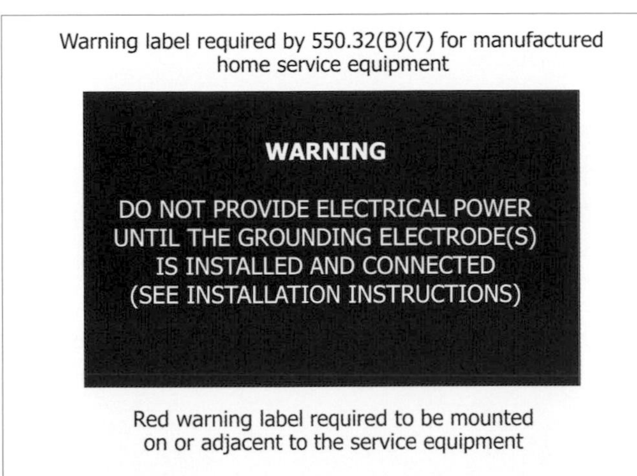

Figure 13-10. Warning label at service equipment of manufactured home

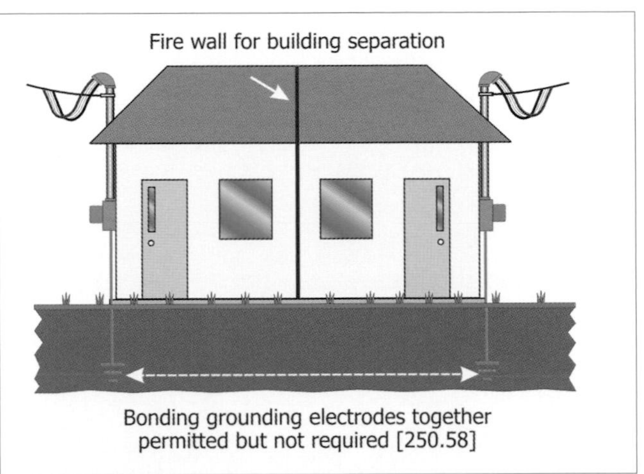

Figure 13-11. Two buildings (by definition) in one structure or on a common foundation or slab

7. "A red warning label shall be mounted on or adjacent to the service equipment. The label shall state the following:

Warning
Do not provide electrical power
until the grounding electrode(s)
is installed and connected
(see installation instructions)." [See figure 13.10].

Where this concept is chosen, the service is grounded to the grounding electrode at the manufactured home, and no service disconnecting means remote from the manufactured home is required.

For additional information, see Part 3280, Manufactured Home Construction and Safety Standards, of the Federal Department of Housing and Urban Development for requirements related to manufactured homes.

If the service equipment is not installed in or on the unit, then the installation is required to meet the other provisions of Section 550.32.

Agricultural Buildings

The grounding of agricultural buildings is covered in Article 547. Sections 547.5(F), 547.9(A)(5), 547.9(B)(3), 547.9(C), and 547.10 all include grounding and bonding requirements that amend or add to the general requirements of Article 250. Where a building or structure houses livestock, an insulated or covered copper conductor must be installed where the equipment grounding conductor is run underground from one building or structure to another. See chapter sixteen for additional information on this subject.

Two Buildings in One Structure

Figure 13-11 shows two buildings in one structure that are separated by a fire-rated wall. The two buildings, by definition (on the same slab or foundation), may be served by a single service or by multiple services in accordance with 230.2(A), (B), (C) or (D).

As previously discussed, 250.58 requires that all grounding electrodes at a building be bonded together. "Where separate services … supply a building and are required to be connected to a grounding electrode(s), the same grounding electrode(s) shall be used. Two or more grounding electrodes that are effectively bonded together shall be considered as a single grounding electrode system in this sense." [14] The connecting conductor, or specifically, a bonding jumper in accordance with 250.53(C), must be properly sized according to 250.66.

Since the fire-rated wall creates two buildings by definition, bonding of the electrodes together is desirable but not required by *Code*. The *Code* only requires that multiple electrodes for multiple services that might be allowed in accordance with 230.2 on the same building or structure be bonded together to meet the requirements in 250.58. The electrodes can be bonded together by connecting them together by a properly sized bonding jumper (see figure 13-11).

[1] NFPA 70, *National Electrical Code* 2008, Article 100, (National Fire Protection Association, Quincy, MA, 2007), p. 70-27.

[2] NFPA 70, 250.50, p. 104.

[3] NFPA 70, 250.52, p. 104.

[4] NFPA 70, 250.32(A), p. 102.

[5] NFPA 70, 250.32(B), p. 102.

[6] NFPA 70, 250.32(B), p. 102.

[7] NFPA 70, 250.32(B) Exception, p. 102.

[8] NFPA 70, 250.32(E), p. 103.

[9] NFPA 70, 250.32(E), p. 103.

[10] NFPA 70, 250.58, p. 70-106.

[11] NFPA 70, 250.32(D), p. 70-103.

[12] NFPA 70, 550.32(A), p. 70-474.

[13] NFPA 70, 550.32(B), p. 70-474.

[14] NFPA 70, 250.58, p. 106.

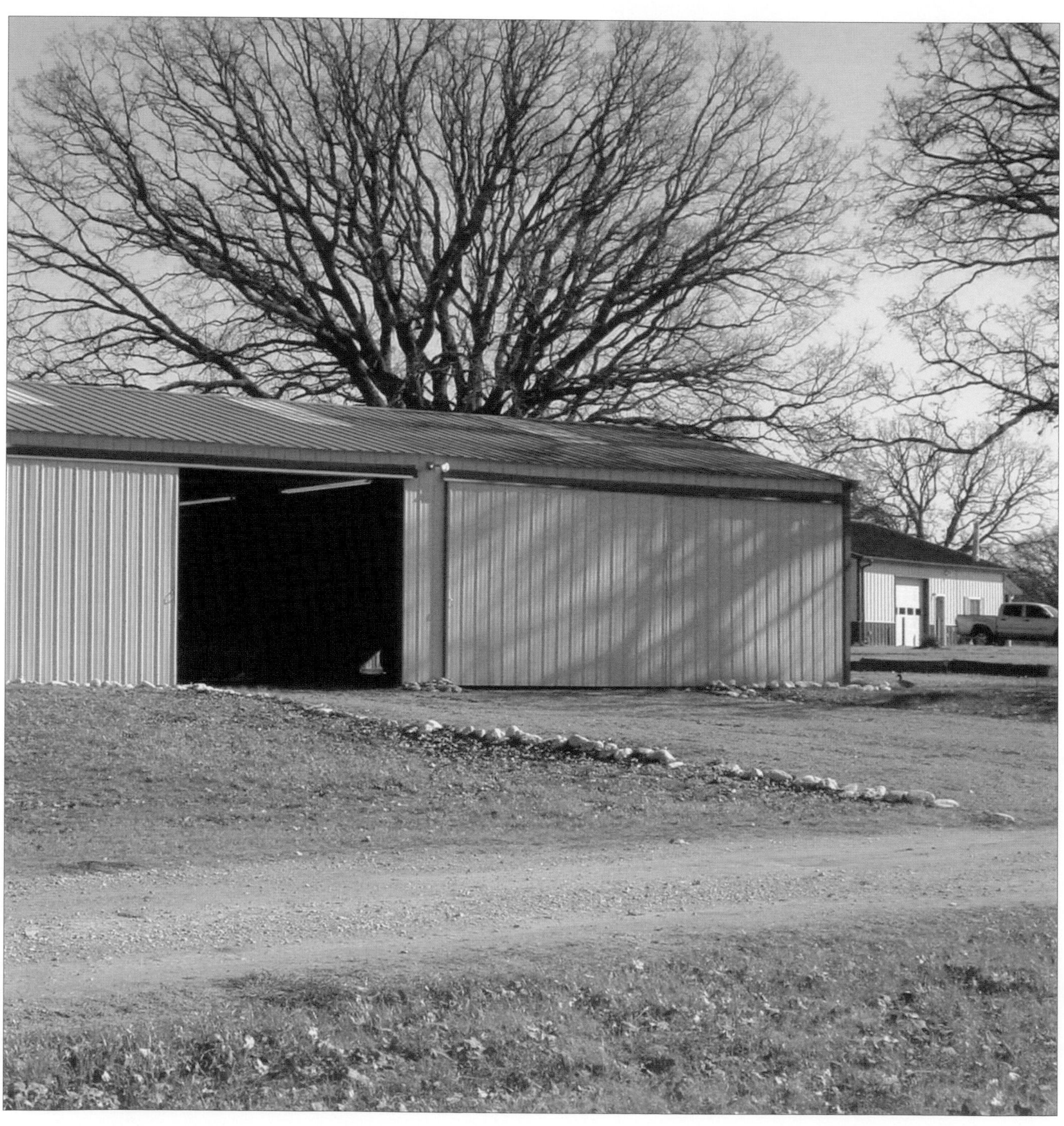

Review Questions

The questions included here were developed using material included in this chapter. The answers can be found by reviewing the text. It is also important that students make use of *NEC-2008*, where many answers can be found.

1. "A conductor used to connect the system grounded conductor or the equipment to a grounding electrode or to a point on the grounding electrode system" best defines which of the following?
   a. main bonding jumper
   b. grounding electrode conductor
   c. grounded conductor
   d. neutral conductor

2. A grounding electrode is required at a separate building or structure under which of the following conditions?
   a. one grounded system feeder is installed
   b. one ungrounded system feeder is installed
   c. more than a single branch circuit is installed
   d. all of the above

3. Which of the following methods is required for grounding at separate buildings or structures in new installations?
   a. grounding using the grounded system conductor
   b. grounding using the equipment grounding conductor
   c. grounding using the ungrounded conductor
   d. installing ground detectors

4. Where a grounding electrode(s) is not present at a building or structure supplied by a feeder(s) or branch circuit(s), ___.
   a. one or more must be installed.
   b. installing one or more is optional.
   c. an equipment ground must be installed
   d. a grounded conductor must be installed

5. The general rule for grounding at separate buildings requires the installation of an ____ where it is run with the feeder or circuit ungrounded (hot) conductors.
   a. approved conductor
   b. identified bonding jumper
   c. equipment grounding conductor
   d. acceptable conductor

6. All of the following are acceptable as a grounding electrode at a separate building or structure EXCEPT
   a. Metric designator 16 (trade size ½) conduit
   b. 16 mm (5/8 in.) iron or steel not less than 8 feet long
   c. Metric designator 21 (3/4-in.) galvanized conduit
   d. plate electrode

7. Where an equipment grounding conductor is run with the feeder(s) or branch circuit(s) from a first building to a separate building, the minimum size of the equipment grounding conductor is based upon the ampere rating of the overcurrent protective device ____ the feeder or branch circuit.
   a. on the load side of
   b. in the transformer of
   c. on the line side of
   d. in the service disconnect of

8. A grounding electrode conductor is used to connect the ____ circuit conductor, or equipment to the grounding electrode system at the separate building(s) or structure(s).
   a. feeder
   b. main bonding
   c. grounded
   d. branch

9. For existing premises wiring systems only, the grounded circuit conductor ____ to be used for grounding equipment on the line (supply) side of the disconnecting means for separate buildings.
   a. is permitted
   b. is not permitted
   c. is permitted by special permission
   d. is not required

10. Currents that introduce noise or data errors in electronic equipment, such as data processing equipment, are ____ considered objectionable currents.
   a. not to be
   b. considered to be
   c. always to be
   d. sometimes

11. Where the additional building or structure houses livestock, an insulated or covered ____ conductor must be installed where run underground.
   a. bonding
   b. copper clad
   c. aluminum
   d. copper

12. Where a single branch circuit or multiwire branch circuit is installed to supply a separate structure, a grounding electrode is ___________________.

   a. not required
   b. not permitted
   c. always required
   d. never installed

13. Where a 400 ampere feeder containing 600 kcmil copper phase conductors is installed to supply a separate building on the premises, the minimum size equipment grounding conductor required for the feeder shall; not be less than ____________.

   a. 2 AWG copper
   b. 4 AWG copper
   c. 3/0 aluminum
   d. 3 AWG copper

Chapter 14

Ground-Fault Circuit Interrupters

The Underwriters Laboratories requirement for Class A ground-fault circuit interrupters (GFCIs) is that tripping shall occur when the continuous 60-hertz differential current exceeds 6 mA, but it shall not occur at less than 4 milliamperes (5 mA ± 1 mA) (see figure 14-1). Some people contend that 5 mA is too low and should be increased to 10 mA or higher.

Objectives

To understand...

- Requirements for replacement of ungrounded receptacles
- Required locations for GFCI-protected receptacles
- Ground-fault circuit interrupter principles of operation
- Ratings of GFCI devices
- Markings for GFCI devices
- GFCI application and consideration
- GFCI requirements in special occupancies
- Ground-fault protection for equipment in comparison

Several eminent investigators, including C. F. Dalziel, F. P. Kouwenhoven, O. R. Langworthy and others, have prepared papers on the dangers of electric shock hazards. They define *let-go current* as the maximum current at which a person is able to release a conductor by commanding those muscles directly stimulated by the shock. Currents over the let-go levels are said to freeze the victim to the circuit. [1]

C. F. Dalziel's paper, titled "Electric Shock Hazard," published in the Institute of Electrical and Electronics Engineers (IEEE) *Spectrum*, Vol. 9, February 1972, summarizes the studies that estimate shock currents based on the effective impedance of the body under various conditions. According to Dalziel, the reasonably safe electric current for normal healthy adults is the let-go current from which 99.5 percent of the population can extricate themselves from the circuit by releasing the conductor. On page 44, Dalziel states, "so far, it has been impossible to obtain reliable [let-go] values for children; they just cry at the higher values." However, the IIT Research Institute report cites a value for children evidently based on engineering judgment by Dalziel and others. The following summarizes the let-go currents for 99.5 percent of the population for:

Children	4.5 mA
Women	6.0 mA
Men	9.0 mA

Percentage of the Population Estimated to Be Protected Against Inability to Let Go for Several Levels of Shock Current

Level of Shock Current	6mA(rms)	10mA(rms)	20mA (rms)	30mA(rms)
Men	100%	98.5%	7.5%	0%
Women	99.5%	60%	0%	0%
Children*	92.5%	7.5%	0%	0%

*half of let-go threshold for men

Table 14-1. Let-go thresholds at various current levels

Based on information provided in the report by IIT Research Institute, 4.5 milliamperes for children can be the appropriate GFCI trip level relative to the 6 mA and 9 mA thresholds for women and men, but the rationale for this selection is not very evident in Dalziel's or other publications.

Based on Dalziel's data, UL's estimates of the percentage of the population that would be protected against the inability to let-go at various current levels is shown in table 14-1.

For general purposes, a let-go limit for GFCIs higher than 6 mA appears inappropriate because too large a fraction of the population would be left unprotected. UL has decided to con-tinue to designate 5 mA (± 1 mA) as the limit based on tolerated reaction and physiological effects. Moreover, UL retains the 5 mA limit because the 4.5 mA level is only an estimate that is not based on data and because the 5 mA limit has withstood the test of time with no evidence of being inadequate and has appeared in a number of standards for many years.[2]

Photo 14-1. Ground-fault circuit interrupter (breaker type)

The first sensation of electricity can be felt by most people at currents considerably less than 0.5 mA, 60 Hz (frequency in Hertz). Current near 0.5 mA can produce an involuntary startled reaction such as to cause a person to drop a skillet of hot grease or cause a worker to fall from a ladder. As the current increases, involuntary muscular contractions increase, accompanied by current-generated heat.[3]

Implied Safe Voltage[4] Based on Several Published Values of Body Resistance and Selected Body Current Safety Criteria as Published by Dalziel				
Criterion	Body Resistance			
Ohms:	300	500	1500	3000
"let-go" 4.5 mA for children	1.35 V	2.25 V	6.75 V	13.5 V
"let-go" 9 mA for adult males	2.7 V	4.5 V	13.5 V	27 V
fibrillation at 23 mA 5 sec. pulse of 60-Hz current for 18 kg children	6.9 V	11.5 V	34.5 V	69 V
fibrillation at 52 mA 5 sec. pulse of 60-Hz current for 50 kg adult	15.6 V	26 V	78 V	156 V

Table 14-2. Implied safe voltages

Higher currents of longer duration than one second can cause the heart to go into ventricular fibrillation, considered the most dangerous effect of electric shock. Once fibrillation begins, it practically never stops spontaneously. Death is almost certain within minutes. The rhythmic contractions of the heart become disordered, its pumping action stops, and the pulse soon ceases altogether. Fibrillation in adults can occur at 52 mA and in children at 23 mA. According to V. G. Biegelmeier, the onset of fibrillation in a 50-kilogram (110-pound) adult occurs within the range of 50 mA to 200 mA when the duration of the shock exceeds two seconds.[4] Table 14-2 shows implied safe voltages based on these values.

The table's 3000-Ω (ohm) body resistance column, for example, indicates that 156 volts would result in a shock current of 52 milliamperes.

Extensive work on this subject has been reported in "Effects of Current on Human Beings and Livestock" IEC 479-1 by the International Electrotechnical Commission.[5] Several charts and graphs with background material are provided. Measurements were made on 50 living persons at a touch voltage of 15 volts and on 100 persons at a touch voltage of 25 volts. The total body impedance of one living person was measured with touch voltages of up to 200 volts. Measurements were also made on a large number of corpses.

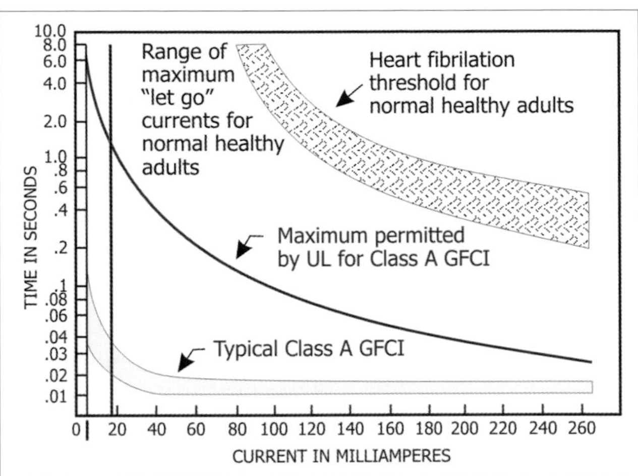

Figure 14-1. GFCI tripping characteristics graph

Three-Wire Grounded System vs. GFCI

For many years, grounding and bonding were emphasized as the primary means for the protection of the electrical system, equipment and personnel from fires and injury, as well as for operating and maintenance advantages. The grounded neutral as a protective element was recognized more and more with each succeeding edition of the *NEC*. The *Code* has placed great emphasis on the importance of an effective ground-fault current path in order to ensure that wiring faults to ground became overcurrents as required to activate the overcurrent device. The concept that wiring and equipment were designed to be protected, and the thought that the grounded system also provided adequate protection for people became ingrained and accepted.

The trend toward grounding equipment and appliances has been gradual, characterized as deliberate but cautious. The belief that grounding provided adequate protection against electric shock and fire hazards became so ingrained that consumers generally have not recognized its limitations, and they find it difficult to accept other more effective means of protection from electric shock and electric fire hazards.

Ground faults occur when an insulation failure causes electrical current in a circuit to return through: (1) the equipment grounding conductor, (2) conductive material other than the electrical system ground (metal, water, plumbing pipes, etc.), (3) a person, or (4) a combination of these ground return paths.

If a person becomes a path for electrical current to ground or a grounded object, the person will incur an electrical shock, can be seriously injured or can be electrocuted depending on the:

1. amount of current,
2. duration or time the current exists,
3. size of the person, and the
4. pathway the current follows through the body.
5. the current frequency (60 HZ, 400 HZ, etc.)

On the other hand, arcing ground faults can occur just about anywhere on an electrical system, resulting in a fire. In either case, the ground faults can be of too low a magnitude to open or operate the overcurrent device and interrupt the circuit.

A person can become a path in an electrical circuit in one of two ways: in series contact or in parallel contact. In series contact, the person is the only current path to ground and the equipment grounding conductor is not involved in the circuit. There are many ways in which contact can occur. One example of series contact was that of an infant that stuck a hairpin into a receptacle slot while sitting on a floor-heating vent. The infant was electrocuted. Section 406.11 now requires all receptacles in areas specified in 210.52 to be listed tamper-resistant types. This should help minimize possibilities of electric shock or electrocution of unsuspecting infants and children. Another case involved a man operating a metal-encased electric drill that had a 3-wire grounded cord. He used a two-to-three wire adapter but did not connect the adapter pigtail to the wall plate grounding screw. Inadvertently, the pigtail touched the plug blade, thus energizing the drill case and electrocuting the man. In this incident, the equipment grounding conductor contributed to the electrocution by providing a current path to energize the drill case. In both cases, the 3-wire grounded system was totally ineffective since the equipment grounding conductor was not involved in the current path, and the current through the body was not large enough to trip the overcurrent protective device.

In parallel contact, the victim becomes a path to ground in parallel with the equipment grounding conductor. One scenario of parallel contact occurs when the metal case of an appliance becomes energized (charged electrically) by some internal fault condition resulting in current leakage to ground via the equipment grounding conductor. The leakage to ground, however, might not be sufficient to activate the branch-circuit overcurrent protective device. A person who touches the charged case and, at the same time, contacts a grounded surface such as a water faucet or pipe will be subjected to an electric shock. In such parallel contact situations, the effectiveness of the equipment grounding conductor in preventing electrocution of the victim is dependent on several variables, including the following:

1. whether or not the ground-fault current reaches the instantaneous trip level of the overcurrent protective device (which is relatively high—over 15 amperes),
2. how fast the overcurrent device reacts,
3. the voltage level from faulted enclosure to ground, and
4. the impedance of the grounding path (composed of connections, contacts and the ground wire).

An effective ground-fault current path in the grounding and bonding system depends upon the integrity of many series connections, which must be properly made and maintained. The higher the resistance or impedance of the grounding (earth) path, the less effectual will be the protection provided by the 3-wire grounded system. Higher impedances can be due to long wire lengths, small wire sizes, loose and/or corroded ground wire clamps and connections and other causes. A detailed analysis of the relationship of the impedance ratio of the line circuit to ground circuit to shock current levels is described in the previously cited Consumer Product Safety Commission paper titled "Three-Wire Grounding Systems vs. GFCI" and the cited IITRI report.[6]

A similar study by Mr. R. H. Lee of Dupont Company corroborates IITRI's analysis of circuit impedances.[7] His paper deals with the hazard vs. safeguard of a 3-phase grounded power distribution system.

An assessment of the effectiveness of the 3-wire grounded system for providing protection against electric shock hazards was conducted by Mr. A. W. Smoot of Underwriters Laboratories.[8] He analyzed 164 fatal electric shock accidents occurring over a 3¾ year period in and around homes. His study indicated the limitations of the 3-wire system and was submitted to the *National Electrical Code* Committee to support proposed amendments to the 1971 *NEC*.

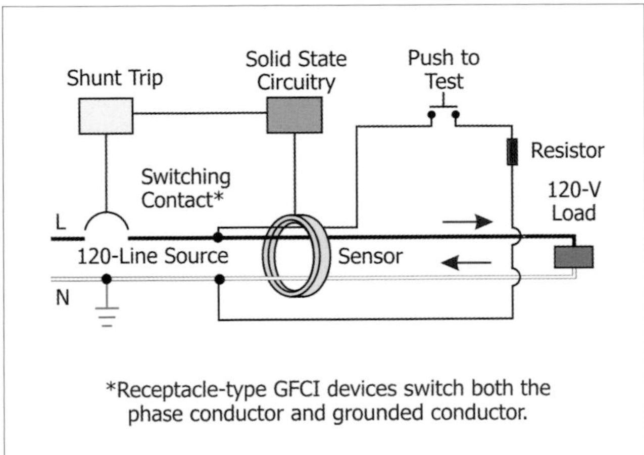

Figure 14-2. GFCI principles of operation

Principles of GFCI Operation

"The GFCI sensing system continuously monitors the current balance in the ungrounded (hot) conductor and the grounded (neutral) conductor. If the current in the grounded conductor becomes less than the current in the ungrounded conductor, a ground fault could exist. A portion of the current returns to the supply by some path other than the grounded conductor. With a current imbalance as low as 4–6 mA, the GFCI will interrupt the circuit and this will be shown by a trip or off indicator on the device (see figure 14-2).

"The GFCI does not limit the magnitude of the ground-fault current. It limits the time that a current of given magnitude exists. The trip level-time combinations are based on physiological data established for avoiding injury to normal healthy persons. These trip level-time combinations can be too high for persons with heart problems, such as those wearing a pacemaker or under treatment in health care facilities." [9]

The principle of operation of ground-fault circuit interrupters provides a significant advancement in safety for both equipment that is grounded by an equipment grounding conductor as well as for equipment that is ungrounded. Since the GFCI detects an imbalance of current in both the supply and return paths, it protects equipment supplied by both a 2-wire circuit and a 2-wire with ground circuit. This is the reason some *NEC* sections will allow a grounding-type receptacle to be supplied on a 2-wire circuit that has GFCI protection.

Several kitchen appliances, as well as portable heaters, are manufactured with 2-wire supply cords and, therefore, do not have their housings or enclosures grounded. A significant advancement in safety is realized where these appliances are supplied from receptacle outlets that have GFCI protection.

GFCI Required for Replacement Receptacles

The general requirement in 406.3(A) is that "receptacles installed on 15- and 20-ampere branch circuits shall be of the grounding type. Grounding-type receptacles shall be installed only on circuits of the voltage class and current for which they are rated, except as provided in Table 210.21(B)(2) and Table 210.21(B)(3)." [10]

"Ground-fault circuit-interrupter protected receptacles shall be provided where replacements are made at receptacle outlets that are required to be so protected elsewhere in this *Code*." [11]

This requires that installers be aware of the rules that call for GFCI-protected receptacles in areas covered by 210.8(A) for dwellings, 210.8(B) for nondwellings, as well as many other locations in the *Code* for other facilities (see table 14-3 for a non-inclusive list of GFCI requirement locations). This includes 15- and 20-ampere, 125-volt receptacles installed in dwelling unit kitchens, bathrooms, garages, outdoor receptacles, and so forth. Also, receptacles that are replaced in locations where GFCI protection is required applies to dwellings and to other than dwelling units such as commercial repair garages, elevators and elevator pits, health care

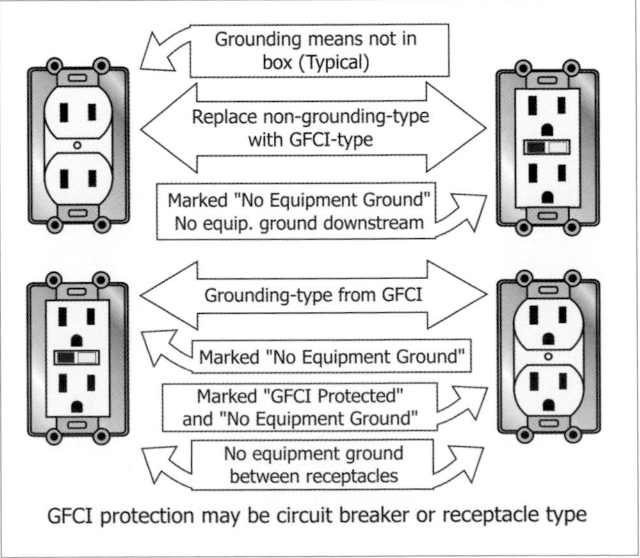

Figure 14-3. Replacement of nongrounding-type receptacles with GFCI receptacles and marking requirements

Location	*NEC* Location
Accessory building dwellings	210.8(A)(2)
Agricultural buildings	547.5(G)
Aircraft Hangars	513.12
Unfinished basements, dwelling units	210.8(A)(5)
Bathrooms, dwellings	210.8(A)(1)
Bathrooms, nondwellings	210.8(B)(1)
Dwelling Unit Boathouses	210.8(A)(8)
Dwelling Boat Hoists	210.8(C)
Carnivals, Circuses, etc.	525.23
Construction sites	590.6
Crawl spaces, dwellings	210.8(A)(4)
Electric Vehicle Charging	625.22
Elevators, etc.	620.85
Existing	406.3(D)(2)
Feeders	215.9
Fountains	680.51, 680.58
Garages, dwellings	210.8(A)(2)
Garages, commercial	511.12
Health care facilities	517.20(A), 517.21
Heating, AC, Ref. Outlet	210.63, 210.8(B)(4)
Hydromassage bathtubs	680.71
Kitchens in dwellings	210.8(A)(6)
Kitchens in nondwellings	210.8(B)(2)
Laundry, Utility, Wet Bar Sinks	210.8(A)(7)
Marinas and Boatyards	555.19(B)(1)
Mobile Homes	550.13(B) 550.32(E)
Outdoors, dwellings	210.8(A)(3)
Outdoors	210.8(B)(4)
Park Trailers	552.41(C)
Rooftops, nondwellings	210.8(B)(3)
RV Parks	551.41(C)
Signs, Outdoor Portable	600.10(C)(2)
Signs, with fountains	680.57(B)
Spas and Hot Tubs	680.43(A)(2) & (3) 680.44
Swimming Pools	680.5 680.6(6) 680.22(A)(1) & (4) 680.22(B), (C)(4) 680.23(A)(3) 680.32
Therapeutic Pools and Tubs	680.62(A)
Vending Machines	422.51

Table 14-3. GFCI requirement locations (non-inclusive)

facilities and bathrooms in commercial and industrial facilities, as provided in 210.8(B).

Ground-fault circuit interrupters of various types, the most common of which are the circuit-breaker and receptacle-types, are permitted to be used to provide the protection required unless the particular type of device is specified. For example, 620.85 requires that the ground-fault circuit-interrupter protection in pits, on elevator car tops, and in escalator and moving walk wellways shall be of the receptacle-type.[12] So GFCIs of the circuit-breaker type would not be acceptable in those applications. The goal is to provide the service person the convenience of resetting the GFCI local to the elevator car or pit where the work is being performed.

Specific marking requirements exist if non-grounding-type receptacles are replaced with grounding-type receptacles where an equipment grounding conductor does not exist at the outlet (see figure 14-3). These receptacles must be marked "No Equipment Ground" so the user will be informed that even though the receptacle has an equipment-ground slot, an equipment grounding conductor is not connected to the device. An equipment grounding conductor is not per-mitted to be connected from the GFCI-type receptacle to any outlet supplied from the GFCI receptacle.

Where grounding-type receptacles are installed on a circuit that does not include an equipment grounding conductor but are protected on the line side by a ground-fault circuit-interrupter device, the receptacle(s) must be marked "GFCI Protected" and "No Equipment Ground." This will inform the user that even though the receptacle has a grounding terminal, it is not in fact grounded by a connection to an equipment grounding conductor, but is provided with GFCI protection. The grounding-type receptacle is protected by the GFCI device, but an equipment grounding conductor is not present as can be required for some sensitive electronic equipment such as computers. An equipment grounding conductor is not permitted to be connected between the grounding-type receptacles as this would present incorrect information to the users. As explained earlier, the GFCI device will protect the user from line-to-ground faults even though an equipment grounding conductor is not connected to the devices.

Section 250.130(C) provides a method for installing an equipment grounding conductor for receptacles that are being replaced at a location where an equipment grounding conductor is desired but not present in the outlet box.

"The equipment grounding conductor of a grounding-type receptacle or a branch-circuit extension shall be permitted to be connected to any of the following:

1. "Any accessible point on the grounding electrode system as described in 250.50
2. "Any accessible point on the grounding electrode conductor
3. "The equipment grounding terminal bar within the enclosure where the branch circuit for the receptacle or branch circuit originates
4. "For grounded systems, the grounded service conductor within the service equipment enclosure
5. "For ungrounded systems, the grounding ter-minal bar within the service equipment enclosure" [13]

Receptacles in Bathrooms

Section 210.8(A)(1) requires that all 125-volt, single-phase, 15- and 20-ampere receptacles installed in dwelling unit bathrooms have ground-fault circuit-interrupter protection (see photo 14-2).

The word *bathroom* is defined in Article 100 as "an area including a basin with one or more of the following: a toilet, a tub, or a shower." It is important to recognize that the definition does not use the word *room* but rather *an area* to describe the location where these receptacles must be provided with GFCI protection. This makes it clear that GFCI protection of these receptacles is required, even though the receptacle(s) may actually be in separate rooms but in the same area. For example, a wall with a door or doorway separates a basin from the area where the toilet and tub or shower is installed. No distance measurement is given beyond which GFCI protection of these receptacles is not required. This definition applies to all bathrooms regardless of type of facility, due to the requirement in 210.8(B) for other than dwelling units (see figures 14-4 and 14-5 for a graphic illustration of typical bathroom areas).

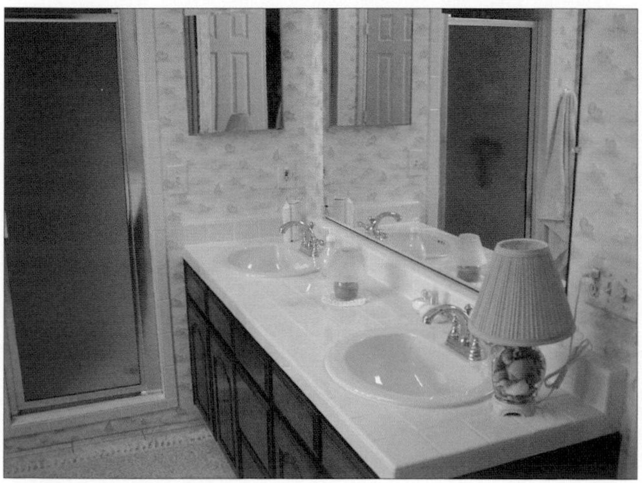

Photo 14-2. Receptacles in dwelling unit bathrooms require GFCI protection.

Receptacles in Dwelling Unit Garages and Accessory Buildings

Section 210.8(A)(2) generally requires GFCI protection for all 125-volt, single-phase, 15- and 20-ampere receptacles installed in garages and in accessory buildings that have a floor located at or below grade level, that are not intended as habitable rooms but are used for storage or work areas (see figure 14-6).

Dwelling Unit Outdoor Receptacles

Section 210.8(A)(3) generally requires all 125-volt, single-phase, 15- and 20-ampere receptacles installed outdoors to have GFCI protection for personnel (see figure 14-7). This requirement applies to these receptacles that are installed outdoors, for any purpose and at any height above grade or platform. The receptacles do not have to be accessible from grade level before the requirement applies.

The exception for 210.8(A)(3) excludes from the GFCI requirement those "receptacles that are not

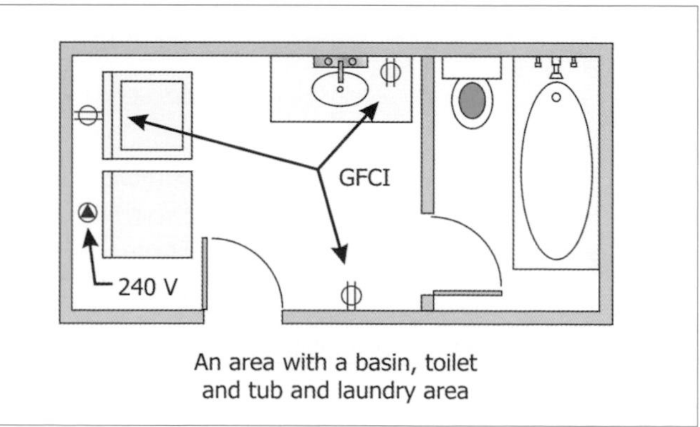

Figure 14-4. GFCI protection required in bathrooms

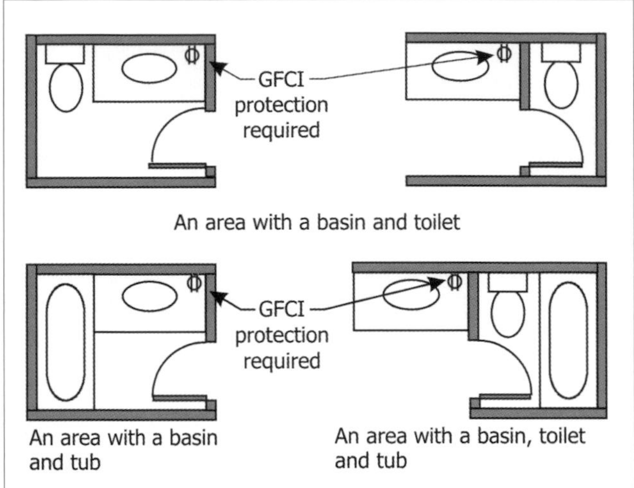
GFCI protection required
An area with a basin and toilet
GFCI protection required
An area with a basin and tub
An area with a basin, toilet and tub

Figure 14-5. GFCI protection required in bathrooms

readily accessible and are supplied by a dedicated branch circuit for electric snow-melting or deicing equipment shall be permitted to be installed in accordance with 426.28." Section 426.28 requires ground-fault protection of equipment (GFPE), and not GFCI protection for personnel for resistance heating elements used for electric snow-melting or deicing unless the cable is "mineral-insulated, metal-sheathed cable embedded in a noncombustible medium."

Dwelling Unit Crawl Space Receptacles

Section 210.8(A)(4) requires that receptacles that are of the 125-volt, single-phase, 15- and 20-ampere type and located in crawl spaces that are at or below grade level have GFCI protection for personnel (see figure 14-8).

These receptacles are often installed for servicing equipment such as heating, air-conditioning or refrigerating equipment that is located in crawl spaces.

Dwelling Unit Unfinished Basements

Section 210.8(A)(5) requires that receptacles that are of the 125-volt, single-phase, 15- and 20-ampere type and located in unfinished basements have GFCI protection for personnel (see figure 14-9).

"Unfinished basements — for purposes of this section, unfinished basements are defined as portions or areas of the basement not intended as habitable rooms and limited to storage areas, work areas, and the like." [14] It is likely that building codes will be consulted for determining whether rooms in basements are permitted to be classified as habitable rooms.

The exception to (5) allows a receptacle for a permanent burglar alarm system or fire alarm system to be installed and used without GFCI protection.

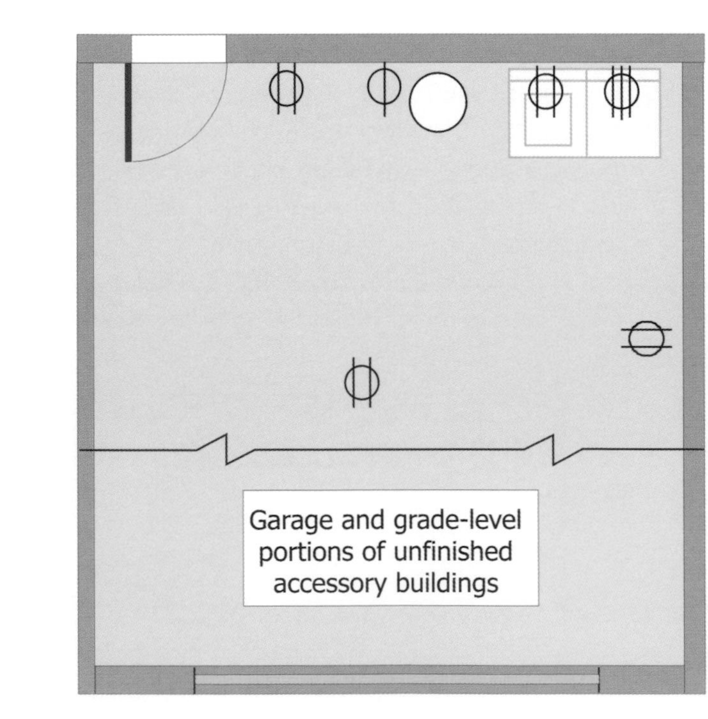
Garage and grade-level portions of unfinished accessory buildings
1. Duplex receptacles generally require GFCI protection
2. Single receptacles
3. Receptacle for laundry requires GFCI
4. GFCI required for receptacles not readily accessible
5. Additional receptacles generally require GFCI protection

Figure 14-6. GFCI requirements in garages and accessory buildings [210.8(A)(2) and (A)(5)]

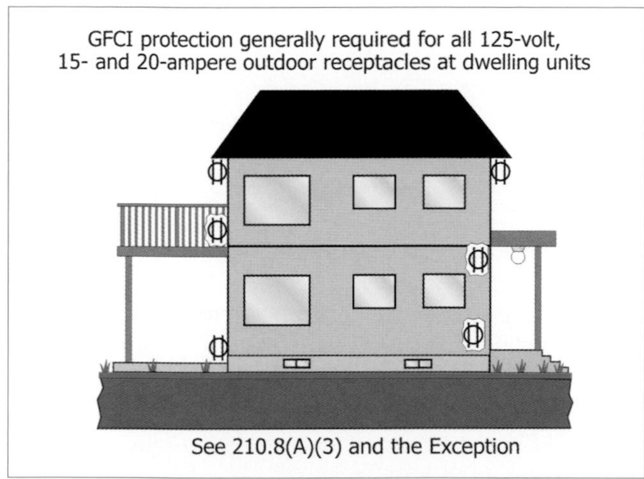

GFCI protection generally required for all 125-volt,
15- and 20-ampere outdoor receptacles at dwelling units
See 210.8(A)(3) and the Exception

Figure 14-7. GFCI requirements for outdoor receptacles

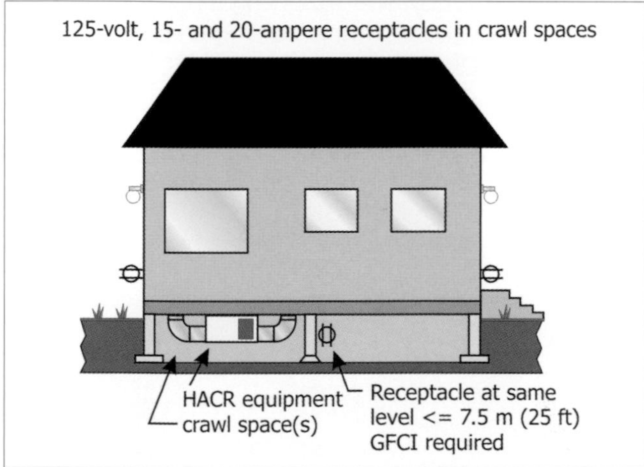

Figure 14-8. GFCI requirements for receptacles in dwelling unit crawl spaces

The receptacles installed to meet the exception to Section 210.8(A)(5) do not satisfy the requirements of 210.52(G).

Ground-fault circuit interrupter protection is required for outlets for boat hoists installed in dwelling applications. This GFCI requirement applies to all outlets not exceeding 240 volts [see 210.8(C)].

Dwelling Kitchen Counter Receptacles

All 125-volt, single-phase, 15- and 20-ampere receptacles that serve countertop surfaces in dwelling kitchens are required by 210.8(A)(6) to have ground-fault circuit-interrupter protection for personnel (see figure 14-10 and photo 14-3).

This requirement does not apply to those receptacles that do not serve countertop surfaces but are installed for other purposes such as for a waste disposal, a trash compactor or a dishwasher. Also excluded from this requirement are wall-mounted receptacles that are installed for equipment, such as electric clocks, or receptacles that are installed for appliances, such as refrigerators, where the receptacle is located behind the appliance and does not serve the counter surface.

Kitchen, Bathroom, and Outdoor Receptacles at Other Than Dwelling Units

All 125-volt, single-phase, 15- and 20-ampere receptacles installed in bathrooms are required to have ground-fault circuit-interrupter protection.

Commercial and institutional kitchens are also required to have ground-fault circuit-interrupter protection for personnel (see figure 14-11). One significant difference between the kitchen receptacle GFCI requirement in other than dwelling units is that the requirement applies to all the 125-volt 15- and 20-ampere receptacles in the kitchen, and not just the ones intended to serve the countertop surfaces as covered in 210.8(A)(6) for dwellings.

All 125-volt, single-phase, 15- and 20-ampere receptacles installed on rooftops and in outdoor locations accessible to the public are required to be GFCI-protected as well [210.8(B)(3) and (4)]. Any receptacle in the configuration stated above installed to meet the requirements of 210.63 (receptacle for servicing heating and air-conditioning or refrigeration equipment) must also be provided with GFCI protection for personnel [210.8(B)(4)].

Dwelling Unit Laundry, Utility, and Wet Bar Sinks

Section 210.8(A)(7) requires that all 125-volt, single-phase, 15- and 20-ampere receptacles located within 1.8 m (6 ft) of the outside edge of a laundry, utility, or wet bar sink are required to be protected by GFCI (see photos 14-4 and 14-5 and figure 14-12).

Section 210.8(B)(5) generally requires all 15- and 20-ampere, 125-volt single-phase receptacles installed within 1.8 m (6 ft) of the outside edge of sinks in other than dwelling units. There are two exceptions to this requirement. Exception No. 1 relaxes the GFCI requirement for receptacles in industrial laboratories where interruption of power would introduce greater hazards. Exception No. 2 correlates with the GFCI protection restrictions for health

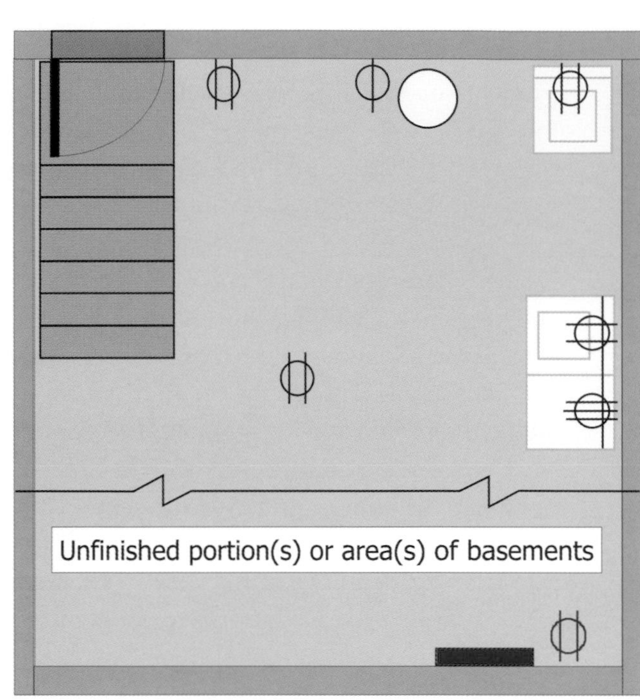

Unfinished portion(s) or area(s) of basements
1. Duplex receptacles generally require GFCI protection for receptacles installed in each unfinished portion or area
2. Single receptacle requires GFCI
3. Receptacle for laundry requires GFCI
4. GFCI required for receptacles not readily accessible
5. Single or duplex receptacles require GFCI protection for two cord-and-plug -connected appliances
6. A single receptacle supplying a burglar or fire alarm system is not required to have GFCI protection

Figure 14-9. GFCI requirements in dwelling unit basements (unfinished)

care facilities other than those covered in 210.8(B)(1). [*NEC* 210.8(B)(5) and Exceptions].

Rooftop Heating, Air-Conditioning and Refrigeration Equipment Receptacle Outlet

Receptacles installed on rooftops must have ground-fault circuit-interrupter protection for personnel in accordance with 210.8(A)(3) and (B)(3) (see figure 14-13). Note that a receptacle outlet for servicing this equipment is required for both dwellings and other than dwelling units. The receptacle must have GFCI protection if installed at these locations. This receptacle is for the purpose of servicing heating, air-conditioning and refrigeration equipment and is required to be located on the same level as the equipment, within 7.5 m (25 ft) of the equipment, and must not be connected to the load side of the equipment disconnecting means.

For other than dwelling units, the exception to 210.8(B)(3) and (4) excludes from the GFCI requirement those "receptacles that are not readily accessible and are supplied from a dedicated branch circuit for electric snow-melting or deicing equipment shall be permitted to be installed in accordance with the applicable provisions of Article 426."

Photo 14-3. Receptacles that serve countertop surfaces in dwelling kitchens require GFCI protection.

Section 426.28 generally requires that "ground-fault protection of equipment (GFPE) ... be provided for fixed outdoor electric deicing and snow-melting equipment, except for equipment that employs mineral-insulated, metal-sheathed cable embedded in a noncombustible medium."

GFCI Requirements in Special Occupancies or for Special Equipment

Section 210.8 provides general information relating to

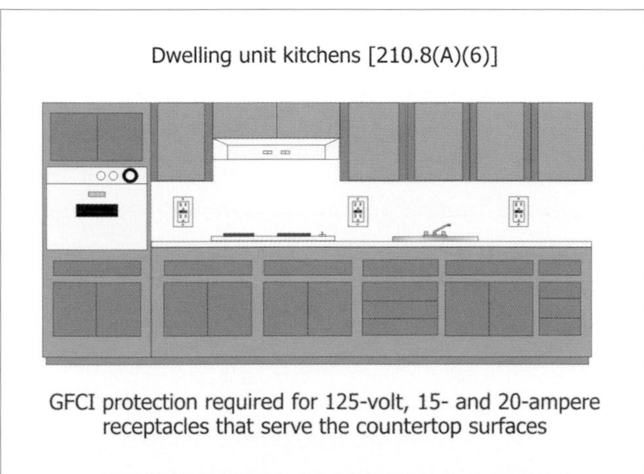

Figure 14-10. GFCI requirements for dwelling unit kitchens

Photo 14-4. Receptacles within six feet of laundry room sink require GFCI protection.

requirements for GFCI protection for dwelling units and GFCI protection for other than dwelling units. It is important to recognize the expanding requirements for GFCI protection in the *NEC* for protection of personnel. The following text provides some additional information

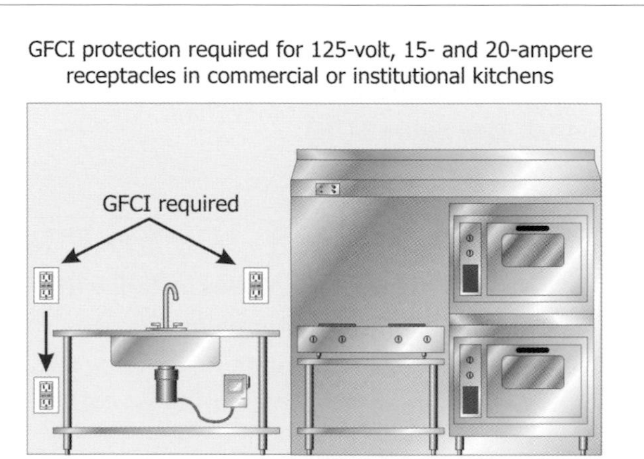

GFCI protection required for 125-volt, 15- and 20-ampere receptacles in commercial or institutional kitchens
GFCI required

Figure 14-11. GFCI requirements for commercial and institutional kitchens as defined in 210.8(B)(2)

about some of the additional GFCI requirements found in chapters 5 and 6 of the *NEC* and is not intended to be all-inclusive.

Commercial Garages

Section 511.12 requires "all 125-volt, single-phase, 15- and 20-ampere receptacles installed in areas where

Photo 14-5. Receptacles within six feet of a wet bar sink require GFCI protection.

electrical diagnostic equipment, electrical hand tools, or portable lighting equipment are to be used to provide ground-fault circuit-interrupter protection for personnel" (see figure 14-14).

Aircraft Hangars

Section 513.12 requires "all 125-volt, 50/60 Hz, single-phase, 15- and 20-ampere receptacles installed in areas where electrical diagnostic equipment, electrical hand tools, or portable lighting equipment are to be used to provide ground-fault circuit-interrupter protection for personnel" (see figure 14-15).

Health Care Facilities (Wet Procedure Locations)

The term *wet procedure locations* is defined under the Patient Care Area of 517.2 as "those spaces within patient care areas where a procedure is performed and that are normally subject to wet conditions while patients are present. These include standing fluids on the floor or drenching of the work area, either of which condition is intimate to the patient or staff." This definition of wet procedure locations modifies for the purposes of Article 517 the similar definition of *wet location* in Article 100.

The general requirement in 517.20 is that "all receptacles and fixed equipment within the area of the wet procedure location shall have ground-fault circuit-interrupter protection for personnel if interruption of power ... can be tolerated, or be served by an isolated power system if such interruption cannot be tolerated." [16]

The exception to 517.20 provides that "branch circuits supplying only listed, fixed, therapeutic and diagnostic equipment shall be permitted to be supplied from a normal grounded service, single- or 3-phase system, provided that

"(a) Wiring for grounded and isolated circuits does not occupy the same raceway, and

"(b) All conductive surfaces of the equipment are connected to an equipment grounding conductor." [17]

It is common that operating rooms and some special procedures rooms are designated as wet procedure locations by the governing body of the facility, particularly in hospitals. Many complex and

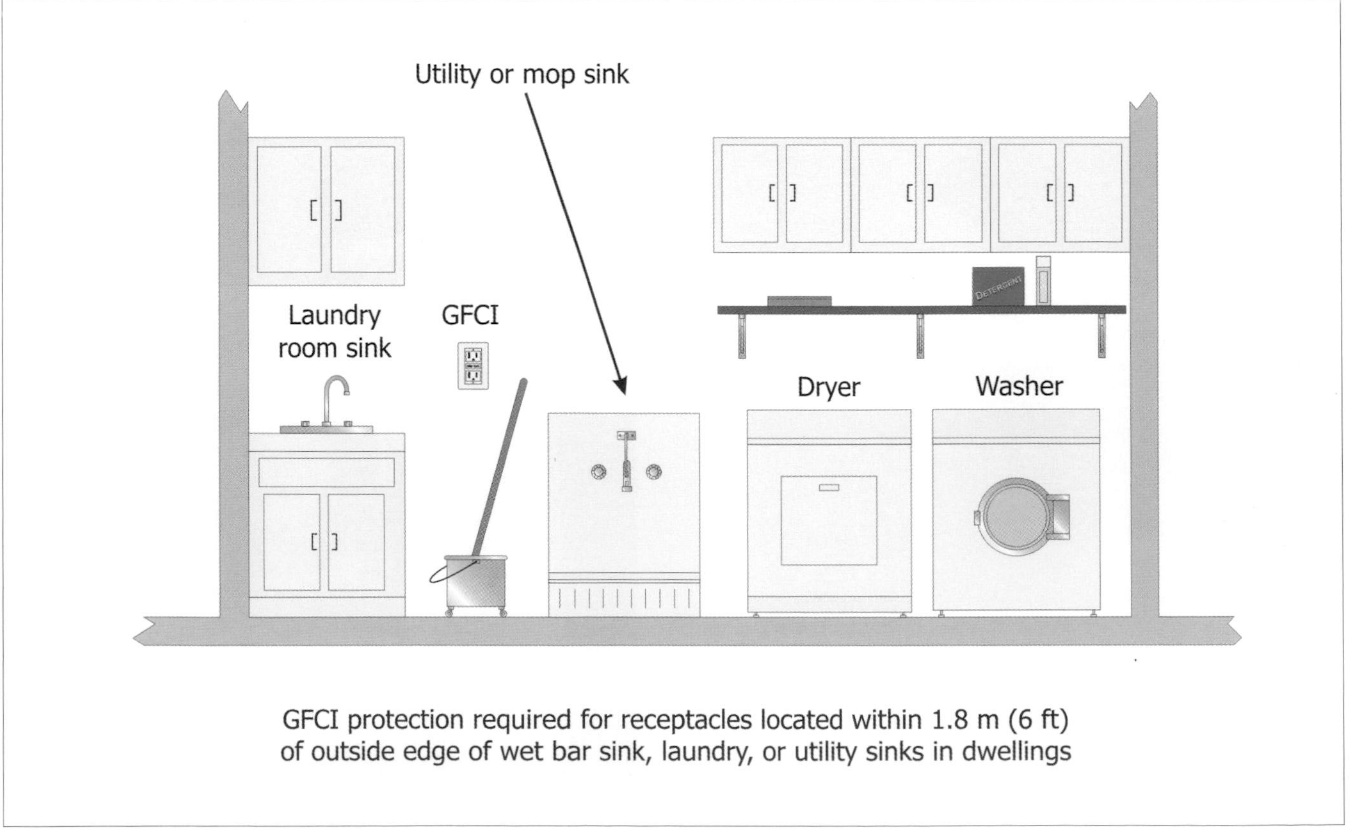
Utility or mop sink
Laundry room sink
GFCI
Dryer
Washer
GFCI protection required for receptacles located within 1.8 m (6 ft) of outside edge of wet bar sink, laundry, or utility sinks in dwellings

Figure 14-12. GFCI protection required at laundry, utility, and wetbar sinks

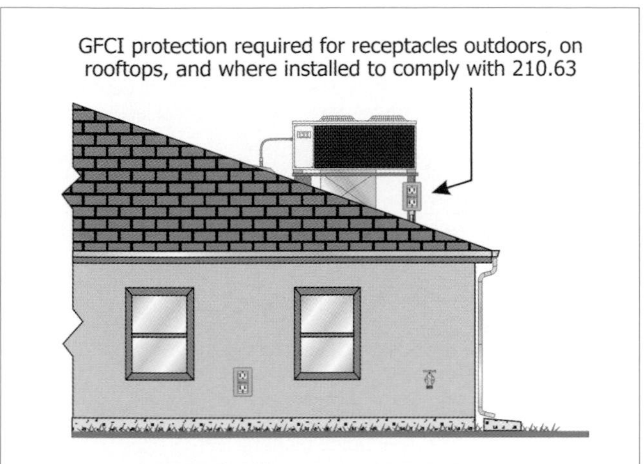

Figure 14-13. GFCI protection required for receptacles installed to comply with 210.63

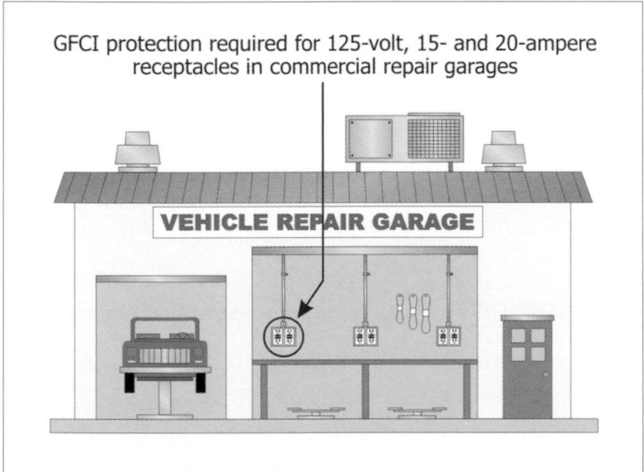

Figure 14-14. GFCI requirements in commercial repair garages in accordance with 511.12

lengthy procedures are performed in these rooms, which cannot be interrupted by the operation of a ground-fault circuit interrupter. As a result, isolated power systems are often installed.

Where an isolated power system is utilized, the equipment shall be listed for the purpose and be installed so that it meets the provisions of and is in accordance with 517.160.[18] This section provides the rules for isolated power systems.

Ground-Fault Circuit-Interrupter Protection for Personnel in Health Care Facilities

The general requirement for providing ground-fault circuit interrupters of all 125-volt, single-phase, 15- and 20-ampere receptacles in bathrooms is contained in 210.8(B)(1). GFCI protection of these receptacles is

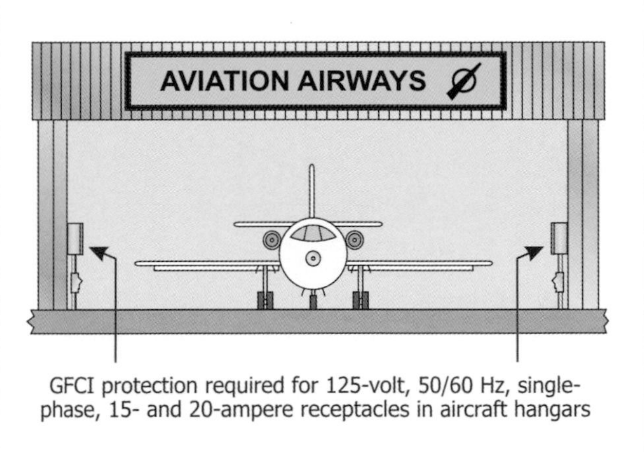

Figure 14-15. GFCI requirements in aircraft hangars in accordance with 513.12

generally required. Note that the definition of the term *bathroom* is contained in Article 100, Part I and reads, "An area including a basin with one or more of the following: a toilet, a tub, or a shower." Generally, receptacles of the type mentioned above that are installed in areas in health care facilities that meet this definition are required to have ground-fault circuit-interrupter protection for personnel.

"Ground-fault circuit-interrupter protection for personnel shall not be required for receptacles installed in those critical care areas where the toilet and basin are installed within the patient room" [19] [*NEC* 517.21]. This section essentially provides an exception to the general requirement for GFCI protection of receptacles in 210.8(B)(1). Toilets of the traditional configuration are not generally installed in critical care patient rooms, as these patients are often too ill to get out of bed. Some special configurations of patient care equipment may include a patient toilet in the critical care room. GFCI protection of the receptacles is not required where, for example, only a basin is installed in the patient room. In this configuration, the room does not meet the definition of bathroom in Article 100 (see figure 14-16 and photo 14-6).

GFCI protection of the receptacles identified above is required in other patient care rooms that meet the definition of bathroom, and typically have a basin and a toilet and/or a shower. Note that the definition of bathroom uses the term *area* rather than *room*. As a result, GFCI protection is typically required for receptacles at the basin and in the room where

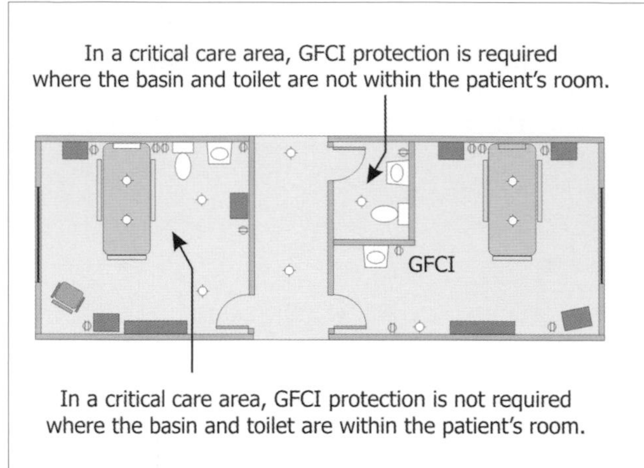

In a critical care area, GFCI protection is required where the basin and toilet are not within the patient's room.
GFCI
In a critical care area, GFCI protection is not required where the basin and toilet are within the patient's room.

Figure 14-16. GFCI requirements in health care facilities

Photo 14-6. Temporary construction power panel that provides GFCI protection

the toilet and/or shower are installed and not for the receptacles that are not located adjacent to the basin.

GFCI Requirements on Construction Sites

Section 590.6 contains requirements for ground-fault protection for personnel using temporary power at buildings, structures or on equipment or similar activities while performing any of the following tasks: "construction, remodeling, maintenance, repair, or demolition of buildings, structures, equipment, or similar activities."

GFCI protection required if used for temporary power during construction, remodeling, maintenance, repair, or demolition of buildings, structures, equipment or similar activities

GFCI protection required for all 125-volt, 15-, 20- or 30-ampere receptacles

For receptacles not covered above, the "Assured Equipment Grounding Conductor Program" is permitted to be used as provided in 590.6(B).

Figure 14-17. GFCI requirements on construction sites

Ground-fault circuit-interrupter protection is generally required for all 125-volt, single-phase, 15-, 20-, and 30-ampere receptacles used for these purposes (see figure 14-17). The rule is the same regardless of whether the power is from a temporary construction service, from an existing permanent power source or from one or more receptacles installed as a part of the permanent wiring of the building or structure).

Section 590.6 requires ground-fault circuit interupter protection for all 125-volt, single-phase, 15-, 20-, and 30-ampere receptacles where supplied from a utility source or service or other on-site-generated power source (see figure 14-18).

A cord set incorporating listed ground-fault circuit-interrupter protection that is identified for portable use is permitted to be used by personnel in lieu of permanently or temporarily installed receptacle outlets. This may be a very practical means of complying with this requirement, especially for maintenance and repair activities [see 590.6(A) and photo 14-7].

The exception to 590.6 states that "in industrial establishments only, where conditions of maintenance

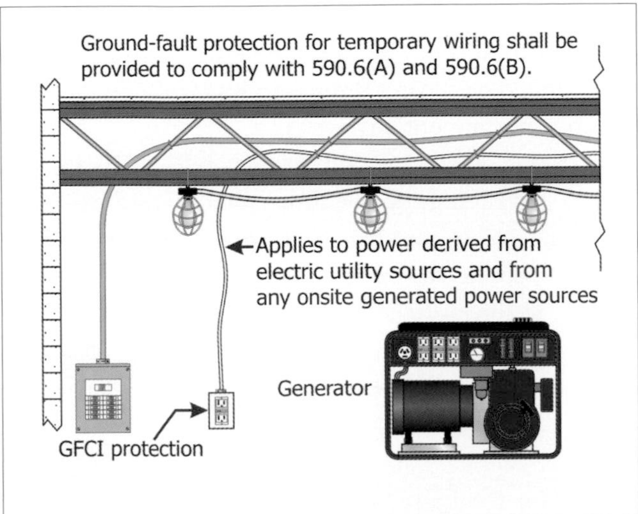
Ground-fault protection for temporary wiring shall be provided to comply with 590.6(A) and 590.6(B).
Applies to power derived from electric utility sources and from any onsite generated power sources
Generator
GFCI protection

Figure 14-18. GFCI is required for receptacles supplied by utilitiy source or on-site-generated source of power.

Photo 14-7. Cord set incorporating listed ground-fault circuit-interruption protection identified for portable use

and supervision ensure that only qualified personnel are involved, an assured equipment grounding conductor program as specified in 590.6(B)(2) shall be permitted for only those receptacle outlets used to supply equipment that would create a greater hazard if power was interrupted or having a design that is not compatible with GFCI protection."

For receptacles like 250-volt, 15- or 20-ampere or 125/250-volt, 30- or 50-ampere, the assured equipment grounding conductor program, as provided for in Section 590.6(B)(2), is permitted only if the requirements and conditions of 590.6(A)(1) Exception are complied with. This program requires a written procedure enforced by one or more individuals at the construction site. Detailed requirements for testing of the equipment grounding conductor continuity on a specific time schedule are provided. A record of this testing program must be maintained and provided to the authority having jurisdiction.

GFCI Requirements for Receptacles in Agricultural Buildings

Section 547.5(G) requires that "all 125-volt, single-phase, 15- and 20-ampere general-purpose receptacles installed in the following locations shall have ground-fault circuit-interrupter pro-tection for personnel:

"(1) In areas having an equi-potential plane

"(2) Outdoors

"(3) Damp or wet locations

"(4) Dirt confinement areas for live-stock" [20] (See figure 14-19).

Note that Article 547 does not include a definition of *damp location* or *wet location* so the definitions of these terms in Article 100 apply. While the term *general purpose receptacles* is not defined, it no doubt refers to those receptacles that would be used for portable tools and portable power cords and not to receptacles installed for a specific purpose such as for fixed equipment like brooders, incubators, etc.

GFCI Requirements for Swimming Pools and Similar Installations

Section 680.5 contains specific requirements relative to ground-fault circuit interrupters for swimming pools. They are permitted to be self-contained units, circuit-breaker types, receptacle types, or other approved types.

"Conductors on the load side of a ground-fault circuit interrupter or of a transformer, used to comply with the provisions of 680.23(A)(8), shall not occupy raceways, boxes, or enclosures containing other conductors"[even if the other conductors are for pool-related equipment such as for a water pump.] [21] Grounding conductors are permitted in these raceways and enclosures.

Conductors protected by ground-fault circuit interrupters are permitted in a panelboard that contains circuits protected by other than ground-fault circuit interrupters (see figure 14-20).

Supply conductors to a feed-through-type ground-fault circuit interrupter are permitted in the same enclosure with the conductors supplied by the GFCI device.

GFCI Requirements for Receptacles

Section 680.22(A)(2) generally requires that receptacles on the property be located at least 1.83 m (6 ft) from the inside walls of a pool (outside spa or hot tub) or fountain. A receptacle to supply power for a water-pump for a permanently installed pool or fountain, as permitted in 680.22(A)(1), must be installed between 1.83 m (6 ft) and 3.0 m (10 ft) from the inside walls of the pool or fountain. Where so located, the receptacle must be of the single and of the locking and grounding type and be protected by a GFCI device (see figure 14-20).

Section 680.22(A)(3) requires "where a permanently installed pool is installed at a dwelling unit(s), no fewer than one 125-volt 15- or 20-ampere receptacle on a general-purpose branch circuit shall be located not less than 1.83 m (6 ft) from and not more than 6.0 m (20 ft) from the inside wall of the pool. This receptacle shall be located not more than 2.0 m (6 ft 6 in.) above the floor, platform, or grade level serving the pool."[22] [Where the word pool is used, it includes outdoor spas and hot tubs.]

Section 680.22(A)(4) notes that "all...125-volt receptacles located within 6.0 m (20 ft) of the inside walls of a pool shall be protected by a ground-fault circuit interrupter." Section 680.22(A)(5) provides information on determining the distances from the pool. It reads, "In determining the dimensions in this section addressing receptacle spacings, the distance to be measured shall be the shortest path the supply cord of an appliance connected to the receptacle would follow without piercing a floor, wall, ceiling, doorway with hinged or sliding door, window opening, or other effective permanent barrier."[23]

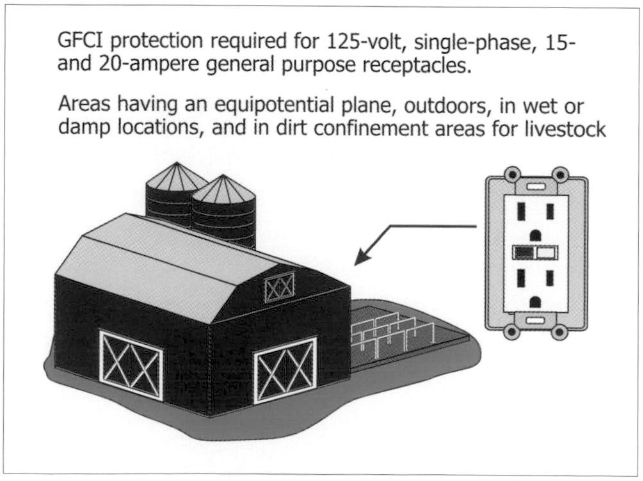

125-volt or 240-volt, single-phase, 15- or 20-ampere outlets supplying pool pump motors require GFCI protection. Note that this requirement applies to cord-and-plug-connected motors and those that are wired directly to the branch-circuit outlet [680.22(B)].

The GFCI requirements for motors associated with permanently installed pools are found in Part II of Article 680, specifically 680.21.

Spa or Hot Tub Installed Outdoors

"A spa or hot tub installed outdoors shall comply with the provisions in Part I and II of this article, except as permitted in 680.42(A) and 680.42(B), that would otherwise apply to pools installed outdoors."[24] This requirement includes those for GFCI protection of receptacles and luminaires as well as required clearances from receptacles and luminaires.

Section 680.42(A) permits listed packaged units (spas or hot tubs) utilizing a factory-installed or assembled control panel or panelboard to be connected

GFCI protection required for 125-volt, single-phase, 15- and 20-ampere general purpose receptacles.

Areas having an equipotential plane, outdoors, in wet or damp locations, and in dirt confinement areas for livestock

Figure 14-19. GFCI requirements at agricultural buildings

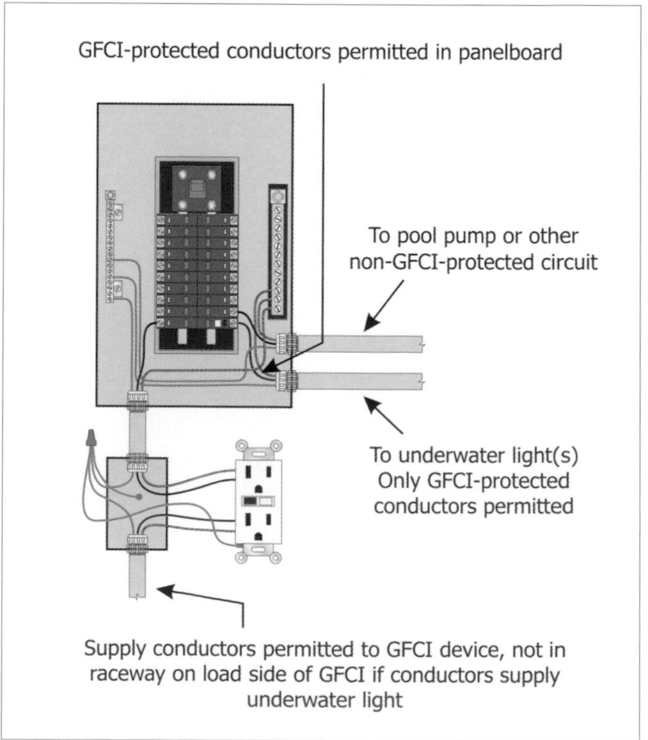
GFCI-protected conductors permitted in panelboard
To pool pump or other non-GFCI-protected circuit
To underwater light(s) Only GFCI-protected conductors permitted
Supply conductors permitted to GFCI device, not in raceway on load side of GFCI if conductors supply underwater light

Figure 14-20. GFCI wiring in raceways and enclosures with other conductors

with not more than 1.8 m (6 ft) of liquid-tight flexible conduit (metallic or nonmetallic) or be cord-and-plug connected with a cord not longer than 4.6 m (15 ft) if protected by a ground-fault circuit interrupter.

Section 680.42(B) permits bonding to be accomplished by metal-to-metal mounting on a common frame or base. This is a common construction of a packaged spa or hot tub equipment assembly, which is commonly referred to as a skid pack.

Metal bands or hoops that are used to secure wooden staves must be bonded to a common bonding grid.

Spa or Hot Tub Installed Indoors

A spa or hot tub installed indoors must conform to the requirements of Article 680, Parts I and II, except as modified by 680.43. They must be connected by wiring methods of *NEC* chapter 3. The exception to 680.43 reads, "Listed spa and hot tub packaged units rated 20 amperes or less shall be permitted to be cord-and-plug connected to facilitate the removal or disconnection of the unit for maintenance and repair."

Section 680.43(A) states that "at least one 125-volt, 15- or 20-ampere receptacle on a general-purpose branch circuit shall be located not less than 1.83 m (6 ft) from, and not exceeding 3.0 m (10 ft) from, the inside wall of a spa or hot tub.

"1) Location. Receptacles shall be located at least 1.83 m (6 ft) measured horizontally from the inside walls of a spa or hot tub.

"2) Protection, General. Receptacles rated 125 volts and 30 amperes or less and located within 3.0 m (10 ft) of the inside walls of a spa or hot tub shall be protected by a ground-fault circuit interrupter."[25]

GFCI Protection of Spa or Hot tub Installed Indoors or Outdoors

Section 680.44 contains GFCI requirements that apply to outdoor and indoor spa and hot tub equipment. GFCI protection of the outlet is generally required if it supplies:

- self-contained spa or hot tub;
- packaged spa or hot tub equipment assembly; and

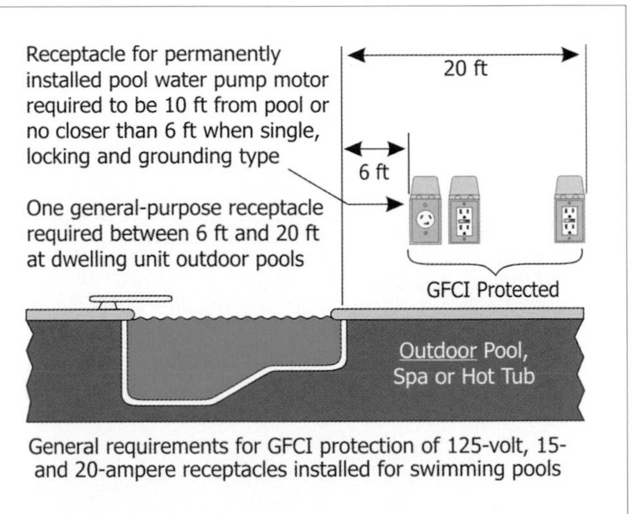

Figure 14-21. GFCI requirements at swimming pools at dwellings

- a field-assembled spa or hot tub.

GFCI protection of the outlet is not required for a listed self-contained unit or listed packaged equipment assembly that is marked to indicate that integral ground-fault circuit-interrupter protection is provided for all electrical parts within the unit or assembly (pumps, air blowers, heaters, lights, controls, sanitizer generators, wiring, etc.).

The electrical supply for a field-assembled spa or hot tub is not required to be protected by a ground-fault circuit interrupter if the supply is rated greater than 250 volts or rated 3-phase or if the heater load is more than 50 amperes. A combination pool/hot tub or spa assembly commonly bonded need not be protected by a GFCI.[26]

[1] A. Albert Biss, Ground-Fault Circuit-Interrupter (GFCI) Technical Report, (Washington, D.C.: U.S. Consumer Product Safety Commission, February 28, 1992).

[2] Walter Skuggevig, 5-Milliampere Trip Level for GFCIs, (Northbrook, IL: Underwriters Laboratories, March 1989).

[3] IITRI report, the voltage and current values are assumed to be "root-mean-square" RMS values.

[4] Electric Shock Prevention, p. 25.

[5] Technical Report IEC 479-1, Effects of current on human beings and livestock. International Electrotechnical Commission. Available from Global Engineering Documents, 15 Inverness Way East, Englewood, Colorado 80112.

[6] A. Albert Biss, Three-Wire Grounding Systems vs. GFCI, (Washington, D.C.: U.S. Consumer Product Safety Commission, December 1985).

[7] Ralph H. Lee, "Electrical Grounding: Safe or Hazardous?" *Chemical Engineering*, July 28, 1969.

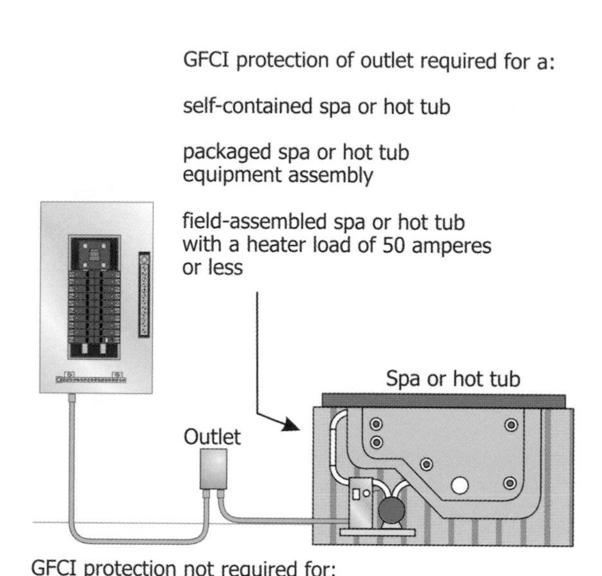
GFCI protection of outlet required for a:
self-contained spa or hot tub
packaged spa or hot tub equipment assembly
field-assembled spa or hot tub with a heater load of 50 amperes or less
Spa or hot tub
Outlet
GFCI protection not required for:
Listed self-contained unit or listed packaged equipment assembly marked to indicate integral GFCI protection is provided for all electrical parts within the unit or assembly
Field-assembled spa or hot tub rated greater than 250 volts or 3-phase
A combination pool/hot tub commonly bonded

Figure 14-22. GFCI protection requirements for spas and hot tubs

8 A. W. Smoot, Analysis of Accidents, (Northbrook, IL: Underwriters Laboratories, 1971).

9 Application Guide for Ground-Fault Circuit Interrupters, Standards Publication No. 280-1990, National Electrical Manufacturers Association, 1300 North 17th Street, Suite 1847, Rosslyn, VA 22209. Reprinted with permission.

10 NFPA 70, *National Electrical Code* 2008, 406.3(A), (National Fire Protection Association, Quincy, MA, 2007), p. 70-259.

11 NFPA 70, 406.3(D)(2), p. 70-259.

12 NFPA 70, 620.85, p. 70-527.

13 NFPA 70, 250.130(C), p. 70-113.

14 NFPA 70, 210.8(A)(5), p. 70-97.

15 NFPA 70, 210.8(A)(7), p. 70-97.

16 NFPA 70, 517.20, p. 70-430.

17 NFPA 70, 517.20(A), Exception, p. 70-430.

18 NFPA 70, 517.20(B), p. 70-430.

19 NFPA 70, 517.21, p. 70-430.

20 NFPA 70, 547.5(G), p. 70-463.

21 NFPA 70, 680.23(F)(3), p. 70-563.

22 NFPA 70, 680.22(A)(3), p. 70-562.

23 NFPA 70, 680.22(A)(6), p. 70-562.

24 NFPA 70, 680.42, p. 70-569.

25 NFPA 70, 680.43(A), p. 70-569.

26 NFPA 70, 680.44, p. 70-570.

Review Questions

The questions included here were developed using material included in this chapter. It is also important that students make use of *NEC*-2008, where many answers can be found.

1. Where new receptacles are installed in locations that are required by the *Code* to be GFCI protected, they must be ___.
   a. of the grounding type
   b. GFCI protected
   c. on dedicated circuit
   d. polarized

2. Circuit breaker or receptacle-type ground-fault circuit-interrupters ____ be used to protect the receptacle, where nongrounding types of receptacles are replaced with grounding type receptacles where an equipment ground does not exist in the receptacle enclosure.
   a. are permitted to
   b. cannot
   c. must always
   d. by special permission can

3. Where used for a single cord- and plug-connected appliance that occupies a dedicated space in a residential garage, GFCI protection is required where a ____ receptacle is installed and dedicated for that appliance.
   a. duplex receptacle (20-ampere, 125-volt, single-phase)
   b. single receptacle (20-ampere, 125-volt, single-phase)
   c. isolated receptacle (20-ampere, 125-volt, single-phase)
   d. all of the above

4. Portions or areas of a dwelling unit basement that are not intended to be used as habitable rooms and limited to storage areas, work areas and the like, defines:
   a. a basement
   b. a finished basement
   c. an unfinished basement
   d. a storage space

5. Where installed in dwelling unit kitchens, all 125-volt, single-phase, 15- and 20-ampere receptacles that are installed ____ must be ground-fault circuit-interrupter protected.
   a. within 6 feet of the sink
   b. on the dishwasher circuit
   c. to serve countertop surfaces
   d. to serve refrigeration equipment

6. The required receptacle on a rooftop, including rooftops at one- and two-family dwellings, must be located to be not further than ____ feet from the heating, air-conditioning and refrigeration equipment.
   a. 10
   b. 15
   c. 25
   d. 50

7. Where installed in ____ of commercial and institutional type occupancies, all 125-volt, single-phase, 15- and 20-ampere receptacles must be ground-fault circuit-interrupter protected.
   a. kitchens
   b. basements
   c. closets
   d. office spaces

8. All 125-volt, single-phase, 15- and 20-ampere receptacles must be ground-fault circuit-interrupter protected where installed in which of the following locations.
   a. outdoors at one family dwelling units
   b. outdoors at two family dwelling units
   c. in dwelling unit garages
   d. all the above

9. With a current imbalance as low as ____ milliamperes, a GFCI will interrupt the circuit and this will be shown by a trip or "off" indicator on the device.
   a. 3
   b. 4
   c. 5
   d. 6

10. A GFCI of the Class ___ type is designed so that it will automatically trip if the neutral conductor is grounded on the load side of the sensor.
   a. A
   b. B
   c. C
   d. D

11. The basic GFCI consists of a ground-fault detecting means that is coupled with a circuit-interrupting means. This detecting means measures the ground-fault current as the difference between outgoing and incoming load current on the ____ conductors.
   a. protected

b. neutral
c. bonded
d. service

12. A condition which could result in injury or electrocution to a person, caused by electrical current through the body, is defined as:
a. a high resistance ground
b. an electrical shock hazard
c. a short circuit
d. ground fault

13. To determine compliance with specified mechanical and electrical life requirements of a GFCI, ___ is made.
a. an endurance test
b. a ground fault test
c. an operational test
d. an overcurrent test

14. GFCI protection is required for receptacles in which of the following locations.
a. family room receptacles in a dwelling
b. commercial office receptacles
c. restaurant kitchens
d. basements of commercial office buildings

15. All 125-volt receptacles located within ____ feet of the inside walls of a pool are required to be protected by a ground-fault circuit-interrupter.
a. 30
b. 25
c. 22
d. 20

16. Ground-fault circuit-interrupters used around pools are permitted to be installed in a ____ that contains circuits protected by other than ground-fault circuit-interrupters.
a. cabinet
b. metal device box
c. panelboard
d. nonmetallic box

17. Ground-fault circuit-interrupters used around pools are required to be self-contained units, circuit-breaker types, receptacle-types or other ____ types.
a. listed
b. acceptable
c. approved
d. labeled

18. A ground-fault circuit-interrupter is required to be installed in the branch circuit supplying underwater luminaires operating at more than ____ volts, so that there is no shock hazard during relamping.
a. 6
b. 10
c. 12
d. 15

19. The installation of the ground-fault circuit-interrupter installed to protect an underwater luminaire must be such that there is no shock hazard with any likely fault-condition combination that involves a person in a conductive path from any ungrounded part of the ____ or the fixture to ground.
a. service
b. system
c. feeder
d. branch circuit

20. A pool pump motor rated at 240-volts, single-phase shall be required to be protected by which of the following?
a. arc-fault circuit-interrupter protection
b. ground-fault protection for equipment
c. overload protection
d. ground-fault circuit-interrupter protection

Chapter 15

Objectives

To understand...

- Requirements for ground-fault protection of equipment
- Types of ground-fault protection systems
- Requirements for feeder ground-fault protection
- Ground faults in an ungrounded system
- Testing of equipment ground-fault protection

Ground-Fault Protection for Equipment

Ground-fault protection for equipment is required for solidly grounded wye electrical services of more than 150 volts to ground but not exceeding 600 volts phase-to-phase for each service disconnect rated 1000 amperes or more [see 230.95 and figure 15-1]. Similar requirements for feeders exist in Sections 215.10 and 240.13 where ground-fault protection is not provided on the main overcurrent device ahead of the feeder. As can be seen, this protection is required for nominal 480Y/277 or 600Y/347-volt, three-phase, 4-wire wye connected systems where the circuit breaker or fused switch rating is 1000 amperes or more. These provisions do not apply to fire pumps or continuous industrial processes where a nonorderly shutdown would introduce additional or increased hazards.

"The maximum setting of the ground-fault protection shall be 1200 amperes, and the maximum time delay shall be one second for ground-fault currents equal to or greater than 3000 amperes." [1]

The magnitude of current in the event of a line-to-ground fault on a grounded system is determined by the overall impedance of the circuit from the source to the point of the ground-fault and back to the source. The major elements making up this impedance include the internal reactance of the grounded source, the resistance and reactance of the lines or conductors leading to the fault and the resistance and reactance of the ground return path(s) including any intentional grounding resistance or reactance. For interconnected systems, calculation of the current can be rather complicated. For simpler cases, the available ground-fault current can be calculated or a close approximation of the available fault current may be obtained.

The requirement to provide specific equipment ground-fault protection is due to a history of destructive burn downs of electrical equipment operating at this voltage level. The most destructive of these phase-to-ground faults are the arcing type faults as opposed to a bolted type fault. An electric arc, which generates a tremendous amount of heat and ionizes the surrounding air at the arc point, is readily maintained with a supply voltage of 277 volts or greater to ground. In addition, an arcing type ground-fault has the current additionally limited by the resistance of the arc. This often results in insufficient current in the circuit to cause the typical overcurrent device ahead of the fault to open. With a substantial amount of the fault energy concentrated at the point of the arc, a great deal of damage is done to the electrical equipment while the arc is burning (see photos 15-1 and 15-2).

Ground-fault protection for equipment required

Solidly grounded wye system over 150 volts to ground
Not greater than 600 volts from phase-to-phase
Each service disconnect rated 1000 amperes or more

Figure 15-1. Ground-fault protection is required.

Case History

As can be seen in photos 15-1 and 15-2, this equipment was extensively damaged by a ground-fault event that became a phase-to-phase short circuit. This service equipment was supplied by a utility source delivering considerably high levels of short-circuit current (approximately 42,000 amperes). This 2500-ampere 480Y/277-volt, 3-phase, 4-wire service included four service disconnects that were all rated at less than 1000 amperes. In accordance with 230.71(A) and 230.95, this meets the minimum requirements without having

Photos 15-1 and 15-2. This equipment endured an arcing event that resulted in extensive destruction (an arcing burn down) of the equipment. GFPE could have prevented, or at least limited, this amount of damage.

ground-fault protection for the equipment. The ground-fault event happened on the line-side busbars in the enclosure ahead of the service disconnects and quickly escalated into an arcing short-circuit fault which destroyed the equipment. The only protection on the line side of the service busbars was on the primary side of the serving utility transformer. The photos clearly show the result of substantial electrical forces and heat due to this ground fault in the equipment. This equipment had to be totally replaced and the building it served suffered considerable downtime as a result of this event. Ground-fault protection provides considerable protection for equipment from these types of events and could have helped prevent this damage. It should be noted that even if the four service disconnects had ground-fault protection, a ground fault on the line side of them will not be detected and can result in the same destruction. It is very important for the safety of electrical workers to understand the rules for electrical safety in the workplace. It is always the best plan to put electrical equipment into an electrically safe work condition, including verifying it through testing before working on it. See NFPA 70E - 2004 and OSHA 1910.331 to 1910.335 for additional information about this subject.

Photos 15-3 and 15-4. GFP equipment (bolted pressure switch with EGFP)

Definition

Ground-fault protection of equipment. "A system intended to provide protection of equipment from damaging line-to-ground fault currents by operating to cause a disconnecting means to open all ungrounded conductors of the faulted circuit. This protection is provided at current levels less than those required to protect conductors from damage through the operation of a supply circuit overcurrent device." [2]

Ground-Fault Protection System Types

There are basically two types of equipment ground-fault protection systems in use, although these systems may have different names in the industry. The most common types are known as *zero-sequence (residual)*, which may have more than one form, and *neutral ground strap-type*, which is sometimes referred to as *ground-strap*, or *ground-return type* (see figure 15-2). Both types are designed to protect equipment downstream of the ground-fault sensor from destructive arcing burn downs. Note that this equipment will not protect equipment or the system on the line side of the sensor from line-to-ground faults because the fault current will not pass through the ground-fault sensing equipment.

Neutral Ground Strap-Type System

The neutral ground-strap type of equipment ground-fault protection system consists of a current sensor, control relay, and, usually, a shunt trip circuit breaker or shunt trip fused disconnect switch. The unique design feature of this type system is that the main bonding jumper passes through the current sensor as shown in figure 15-2.

One advantage of this system is that it is the least

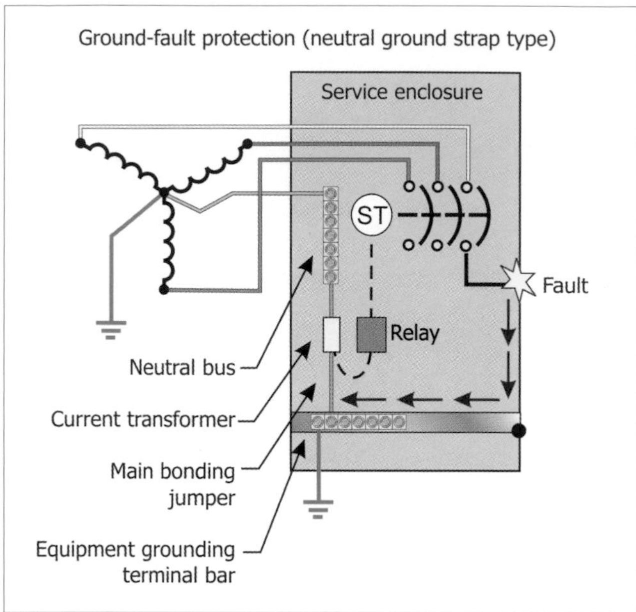

Figure 15-2. EGFP System – Neutral ground strap-type

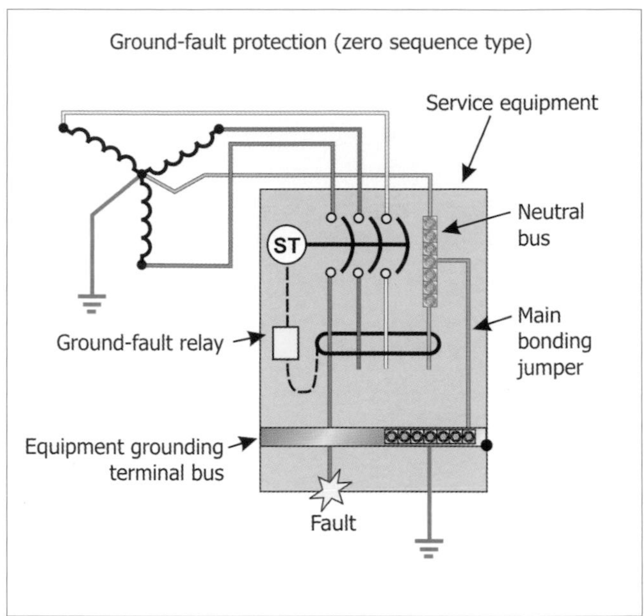

Figure 15-3. Zero-sequence EGFP system

expensive, but the big disadvantage is that it is limited to application only at the main service or supply source. It generally cannot be used downstream of the main bonding jumper at the service or system bonding jumper at a separately derived system. Additionally, it is critical that all equipment grounding conductors and earth grounding electrode connections be to the equipment ground bar so there is only one path for fault current—through the main bonding jumper or the system bonding jumper.

With the connections as shown, the maximum ground-fault current will be through the main bonding jumper and will, therefore, be recognized by the ground-fault current sensor. Although there are many parallel bypass circuits for ground-fault current, the relative impedance of the main circuit is so low compared to all the parallel bypass circuits that a minimum of 90 percent of the total ground-fault current will be through the main bonding jumper and, therefore, be seen by the current sensor.

Examination of the diagram in figure 15-2, describing the ground-fault protective device, will show that the main ground-fault current path is from the transformer to the service, through the feeder to the fault, back to the service over the equipment grounding conductor, through the main bonding jumper where it returns to the transformer by the neutral conductor. The parallel circuit from the grounding electrode at the service to the grounding electrode is both a high-resistance and a high-reactance circuit. As a result, little current on the order of less than 5 percent of available ground-fault current will return through the earth.

Some small amount of ground-fault current will be carried by the building structural metal framing if it is in the circuit. It is preferable to adequately bond the structural metal framing using the same grounding electrode as used for the service. This bonding, connected as indicated above, still captures this current for the ground-fault sensor. The bonding prevents the building structural metal frame from rising to a dangerous potential above ground and often serves as a grounding electrode as provided by 250.52(A)(2). Even though the building structural metal framing represents a parallel path for fault current, most of the current will return to the transformer through the neutral because of the lower reactance of that path as compared to the reactance of the other available parallel paths.

Zero-Sequence Ground-Fault Sensing-Type System

Probably the most popular and common type of equipment ground-fault protection system is the zero-sequence type. This system is shown in figures 15-3 and 15-4. It consists of one current sensor(s) that is placed around all the current-carrying conductors of the circuit, including the grounded (neutral) conductor, a ground-fault relay, and a shunt trip circuit breaker or

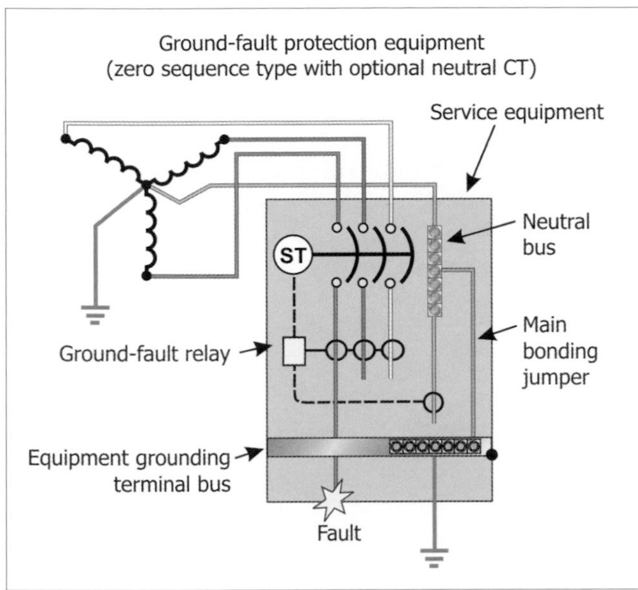

Figure 15-4. Zero-sequence EGFP system with optional neutral sensing window

shut trip fused disconnect switch. Optionally, there may be one sensor around all the phase conductors and a second sensor for the grounded (neutral) conductor. As shown, the ground-fault current sensor must be placed around the neutral downstream from the main bonding jumper connection point. The equipment grounding conductors and the main bonding jumper or system bonding jumper do not pass through the window (see photos 15-5 and 15-6).

The general difference between the types of zero-sequence equipment shown in figures 15-3 and 15-4 is the current sensors are generally built into the circuit breaker shown in figure 15-4, with an external ground-fault current sensor through which the neutral passes (see photo 15-6). In some cases, this neutral sensor may be field-installed. Often, these same current sensors are used by the circuit breaker as a part of its internal overcurrent protection operating system. The equipment ground-fault pro-tection system shown in figure 15-4 is sometimes referred to as a residual-type as the ground-fault relay or system vectorally sums up the current through all four sensors and considers any excess current as residual. Generally, the current sensor through which all conductors of the circuit pass, as

Photo 15-5. Zero-sequence GFP equipment (breaker-types) in a switchboard

Photo 15-6. Zero-sequence sensing current transformers on respective neutrals in switchboard

shown in figure 15-3, is installed by the manufacturer of the switchboard. Cables for feeders are field-installed.

Under normal operation, the vector summation of all phase and neutral (if used) currents approaches zero. This is due to the canceling effect of the currents in the conductors. Under ground-fault conditions, not all the current going from the source to the fault location and loads returns on the phase and neutral conductors. The sensor around the phase and neutral conductors detects the current imbalance and sends the resultant current signal to the ground-fault relay. The imbalance happens because of the current passing outside the current sensor "window(s)" on the ground-fault current path.

The output of the sensor is proportional to the magnitude of the ground-fault current. This output is fed to a ground-fault relay. The relays are usually field-adjustable. Pickup ranges of from 4 to 1200 amperes are common, depending on the relay selection. Typical ranges for service equipment ground-fault relays are 100 to 1200 amps. Time delay settings are available from instantaneous (1.5 cycles) to 1-second (60 cycles) delay. When the ground-fault current exceeds a pre-selected level for the set time delay, the relay will activate the circuit-interrupting device, which usually is a shunt trip circuit breaker or shunt trip fused switch, to open the circuit (see photo 15-7).

Ground-Fault Protection Not Required

The exception to Section 230.95 requires that ground-fault protection provisions of 230.95 "not apply to a service disconnect for a continuous industrial process where a nonorderly shutdown will introduce additional or increased hazards." This is a mandatory exception.

Likewise, ground-fault protection of equipment is not permitted for fire pump services [see 695.6(H)]. For an emergency system, 700.26 excludes the alternate source of power from the requirement to have "ground-fault protection of equipment with automatic disconnection means." This applies to fire pumps that are classified as an emergency system in accordance with 700.1. Indication of a ground fault of the emergency source is required to indicate a ground fault in solidly grounded wye emergency systems of more than 150 volts to ground and circuit-protective devices rated 1000 amperes or more [700.7(D)]. The sensor for the ground-fault signal devices shall be located at, or ahead of, the main system disconnecting means for the emergency source, and the maximum setting of the signal devices shall be for a ground-fault current of 1200 amperes. Instructions on the course of action to be taken in event of indicated ground fault shall be located at or near the sensor location." [3]

A similar exclusion from the ground-fault protection of equipment mandatory requirement is provided for legally required standby systems in 701.17. No such exclusion is provided for optional standby systems installed in accordance with Article 702.

Selective Coordination

Where GFPE systems are installed, they should be designed such that they are selective so the offending ground-fault event opens the closest device on the upstream side of the fault. A definition of the term *selective coordination* is provided in Article 100 of the *Code* and works hand in hand with the new requirements for selective coordination of overcurrent devices for emergency systems specified in 700.27. These overcurrent protective devices may or may not provide GFPE. Generally, if an overcurrent device of larger size is part of an emergency system, it will not be equipped with GFPE.

System Coordination

System coordination is easily accomplished when applying ground-fault protective devices (see figure

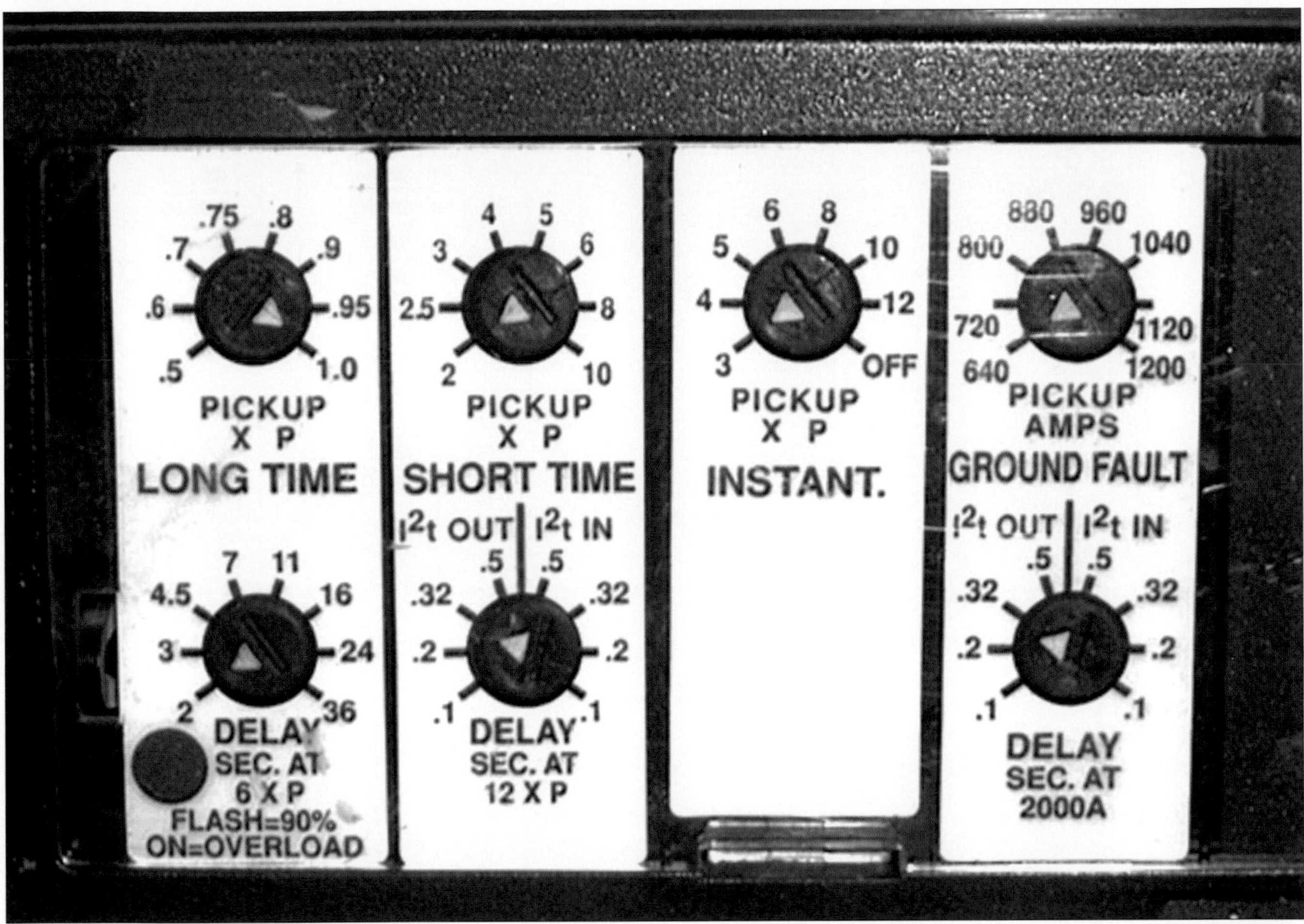

Photo 15-7. GFPE setting adjustments (breaker-type)

15-5). The two most common types of coordination schemes are the *cascading time delay scheme* and the *zone interlocking system*. The time delay system will have ground-fault protective devices cascaded where the economics of the design warrant doing so. The time-delay settings may become lower and lower as the device gets further from the service, so that the furthest ground-fault protective device in the system may even be set for instantaneous trip. The advantage is that this is easily set up and does not require any interconnecting signal conductors. A disadvantage is that the long time delay at the service can allow significant damage to happen before the relay actuates. Using the zone interlocking system, similar coordination also can be obtained by locking out the relay or relays upstream from the sensor that sees the fault first. This permits keeping the time delay for a fault on the main source ground-fault relay lower and minimizes damage. The zone interlocking systems require signal wiring to be installed, which adds complexity to the overall system.

Because it can't be predicted where a fault-to-ground or fault-to-enclosure will originate in the system, and because it is desirable to coordinate the protection and clear the fault nearest to its point of origin, it is necessary to delay the action of tripping the main overcurrent device. Therefore the ground-fault relay is set with a time delay that provides proper coordination and maximum continuity of service consistent with safety.

When a fault-to-ground or fault-to-enclosure occurs, the possibility should be considered that the fault can reach a very high value. If so, the normal main overcurrent device should be set at a time value to allow it to function and clear the fault through that means. Even though the ground-fault relay should be set at as low a value as possible, enough time delay should be set on the main overcurrent device to permit a standard

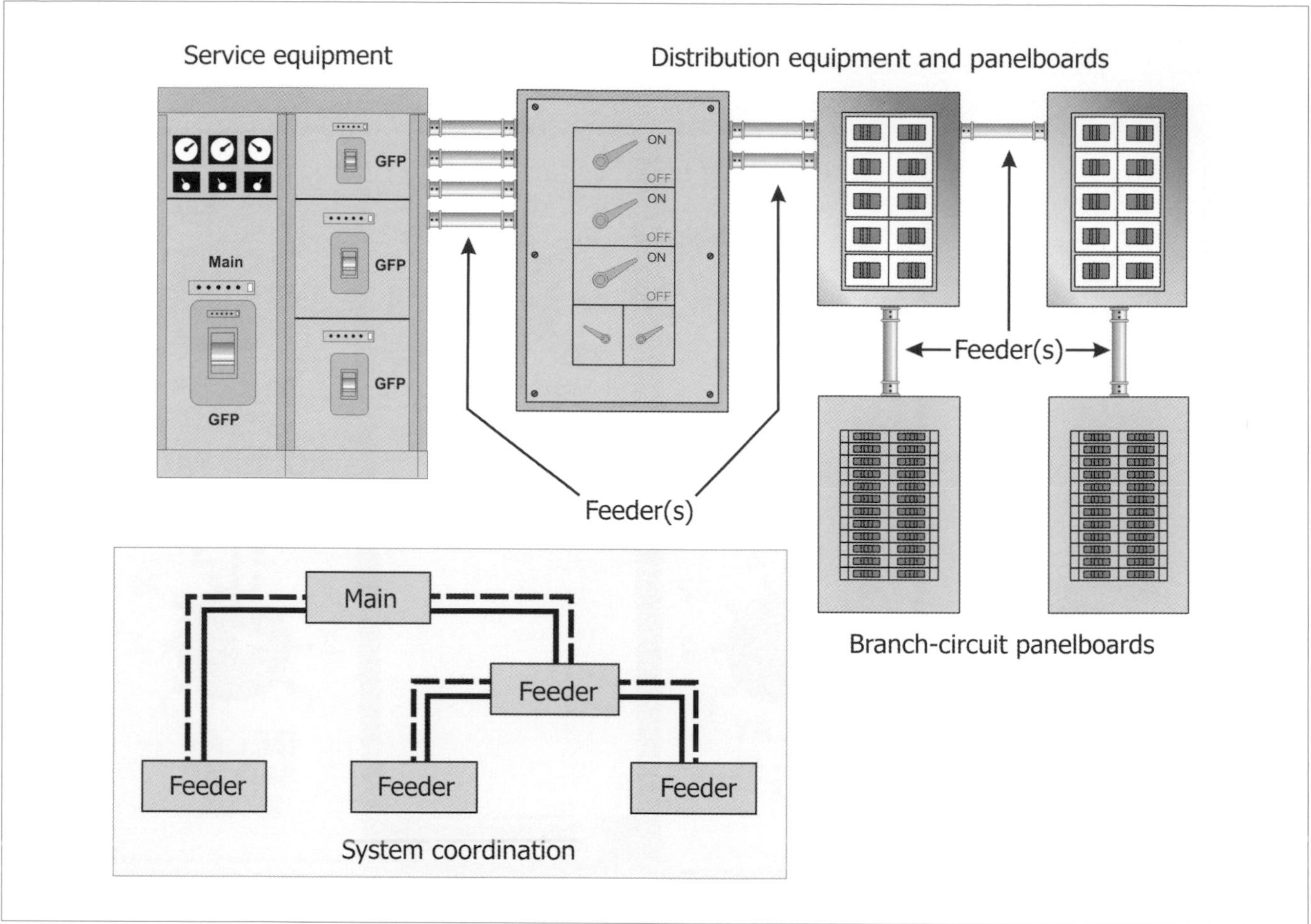

Figure 15-5. System coordination

overcurrent device nearest the point of origin of the fault to clear first. This ground-fault protective device is an adjunct to, but does not replace, the main overcurrent device. It gives added protection below the rating of the main overcurrent device. Being affected only by ground-fault currents, it does not interfere with normal operation of the main overcurrent device. In effect then, the ground-fault system will sense ground faults at currents lower than the normal overcurrent device and allow them to exist for only a limited time, which is selected based on experience, to give maximum safety without undue interruption of power.

Second Level Required in Health Care Facilities

Special rules for ground-fault protection of electrical systems apply to hospitals and other buildings (including multiple occupancy buildings) with critical care areas or utilizing electrical life support equipment as well as facilities providing the required essential utilities or services for the operation of critical care areas or electrical life support equipment [*NEC* 517.17]. Section 517.17(B) requires that "where ground-fault protection is provided for operation of the service disconnecting means or feeder disconnecting means as specified by 230.95 or 215.10, an additional step of ground-fault protection shall be provided in all next level feeder disconnecting means downstream toward the load" (see figure 15-7).

This rule clarifies that regardless of whether or not the health care facility is supplied by a service or feeder, where equipment ground-fault protection is provided at the service or feeder disconnecting means, a second level is required downstream. The provision also does not limit the circuit breaker or fused switch to ratings of 1000 amperes or more; it applies to all feeders regardless of rating (see photos 15-8 and 15-9).

The additional levels of ground-fault protection shall not be installed in the following locations or applications (see figure 15-6):

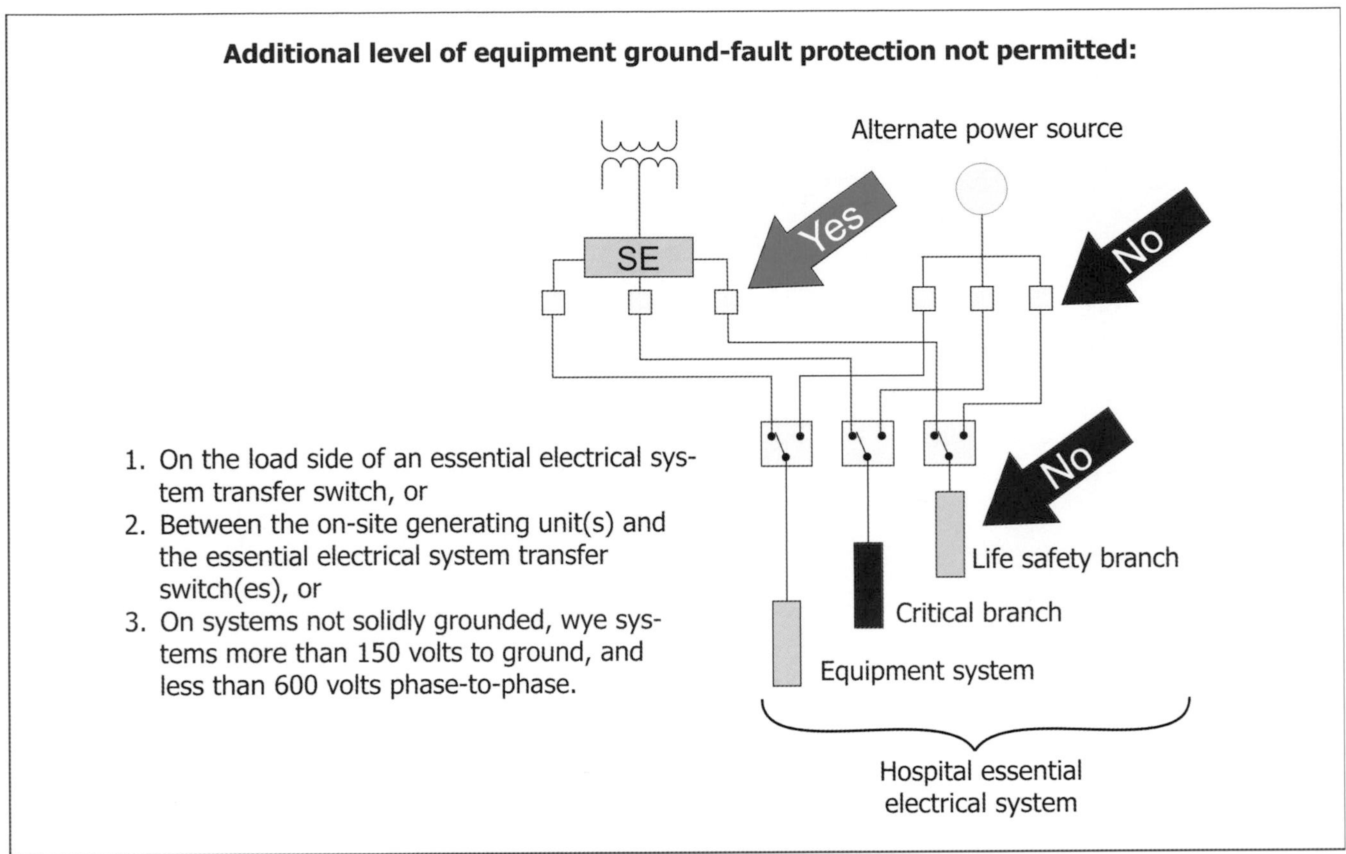

Figure 15-6. Where GFPE is not permitted

- On the load side of an essential electrical system transfer switch
- Between the on-site generating unit(s) described in 517.35(B) and the essential system transfer switch(es) (ground-fault protection systems with indication, but not tripping still may be required)
- On electrical systems that are not solidly grounded wye systems with greater than 150 volts to ground, but not exceeding 600 volts phase-to-phase [4]

This feeder protection is required to be 100 percent selective so that where a ground fault occurs downstream from the feeder overcurrent device, only the feeder overcurrent device will open and the service or feeder main will remain closed (see figure 15-8). This, of course, is to prevent the blackout of a facility caused when a main would open for a fault that can be isolated to a single feeder. To achieve this coordination, a six cycle, or greater, separation between the service and feeder tripping bands is required. Each level of the ground-fault protection system must be tested when first installed [517.17(D)]. Part of that testing must prove this selectivity and time delay separation which may require additional procedures not normally used for normal ground-fault protection system testing.

Feeder Ground-Fault Protection

Figure 15-9 illustrates a situation where equipment ground-fault protection is required for feeders. Section 215.10 requires this protection for feeders that are of the same voltage and current rating as similar services.

This is an example of a situation where the electrical system is delivered at more than 600 volts phase-to-phase and the service disconnecting means is at that voltage level. A transformer then reduces the voltage of the primary feeder to the level where equipment ground-fault protection is required. The risk of destructive arcing burn downs is the same regardless of whether the system is a service or feeder.

The same exceptions that are provided to Section 230.95 for services are also provided under 215.10 for feeders.

[However, care must be exercised to be certain that each new system created by a separately derived

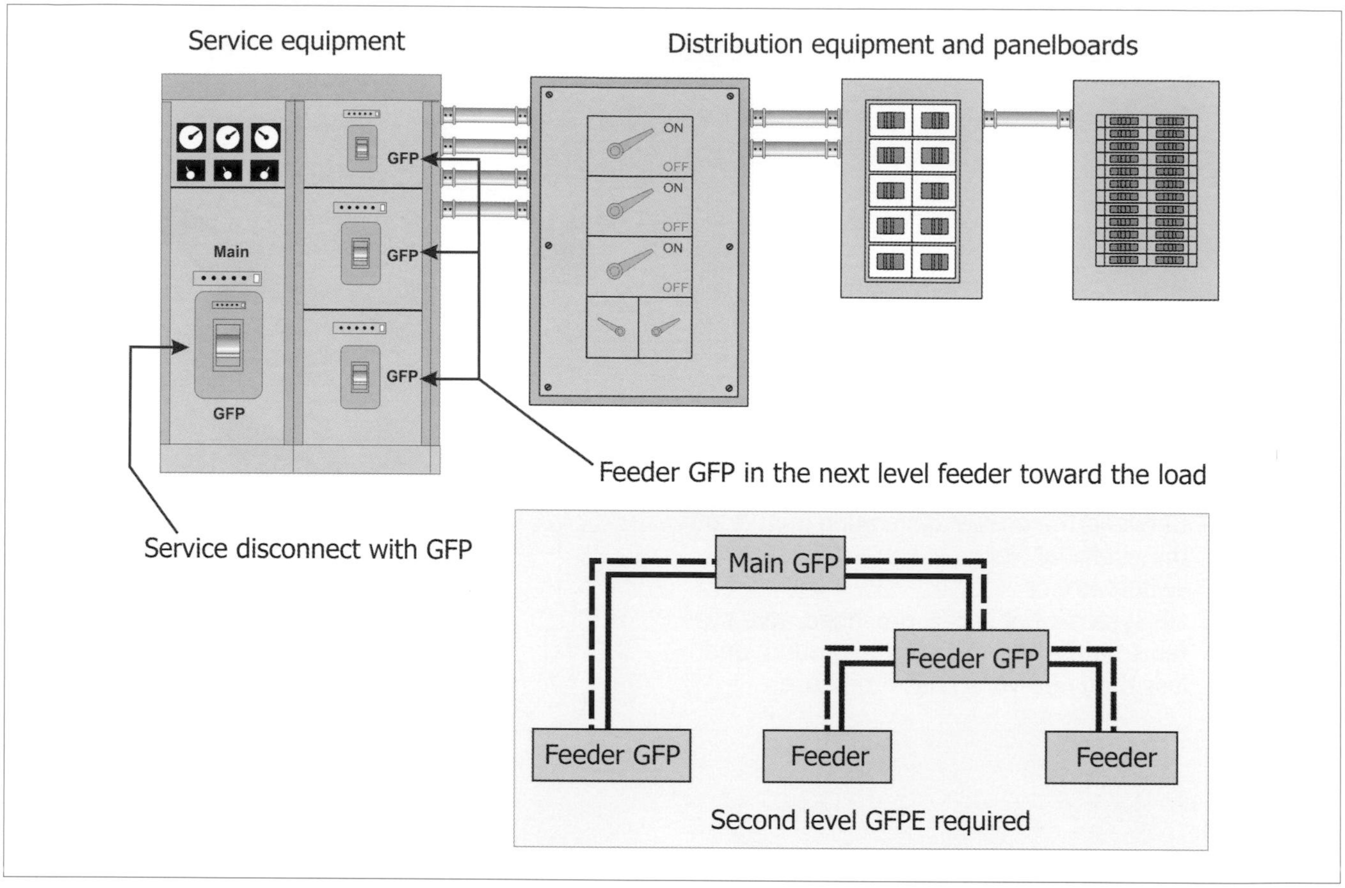

Service equipment

Distribution equipment and panelboards

ON
OFF
ON
OFF
ON
OFF

GFP
GFP
GFP
Main
GFP

Feeder GFP in the next level feeder toward the load

Service disconnect with GFP

Main GFP

Feeder GFP

Feeder GFP

Feeder

Feeder

Second level GFPE required

Figure 15-7. Second level of GFPE required [*NEC* 517.17]

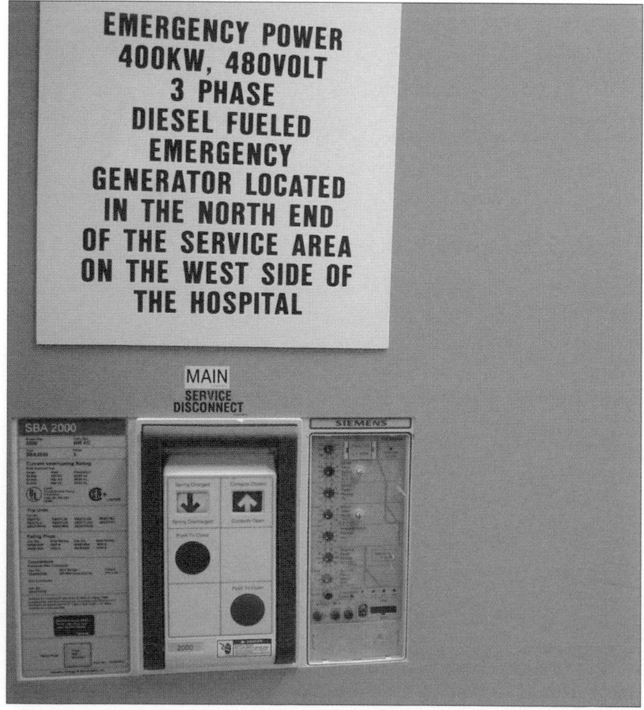

Photo 15-8. GFPE at service disconnect and in next level feeder in same the switchboard

Photo 15-9. GFPE at service disconnect and in next level feeder in the same switchboard

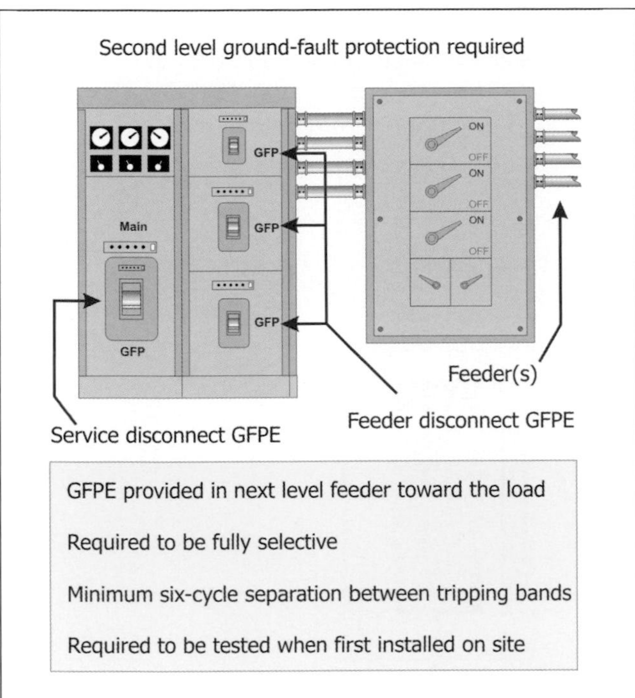

Figure 15-8. Second level of GFPE required [*NEC* 517.17]

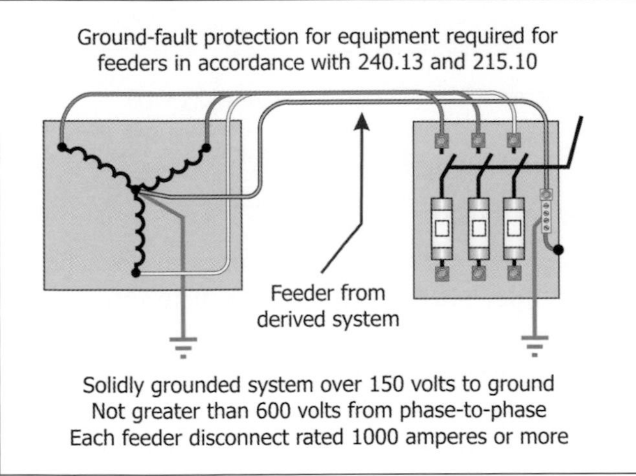

Figure 15-9. Ground-fault protection required for feeders

system has the protection required; and if it qualifies for equipment ground-fault protection by the voltage and ampacity level of the feeder and equipment, it must be provided.]

Ground-Fault Protection of Building or Structure Main Disconnecting Means

For each building or structure disconnecting means meeting the criteria, 240.13 requires ground-fault protection of equipment to be installed in accordance with 230.95 (see figure 15-10). This requirement applies for systems of identical voltage and amperage rating to that for services.

"The provisions of this section shall not apply to the disconnecting means for the following:

"(1) Continuous industrial processes where a nonorderly shutdown will introduce additional or increased hazards [see Article 685 for additional information].

"(2) Installations where ground-fault protection is provided by other requirements for services or feeders [this applies when the ground-fault protection is provided for an upstream device at the same voltage level].

"(3) Fire pumps [installed in accordance with Article 695]" [6]

It is very important that an overall engineered system approach is taken where more than one building or structure is on the premises that are fed from either a common service, feeder(s) or branch circuit(s). This is especially important where transfer switches and alternate power sources are installed in the premises wiring system. In most cases, transfer equipment that switches the grounded conductor of the system are installed to allow the system to function properly without grounding connections to the neutral or grounded conductor downstream from the service disconnecting means that incorporates ground-fault protection equipment. It is critical to the functionality of the ground-fault protection system that the grounded (neutral) conductor be completely isolated from any grounding connections downstream from the service or separately derived system neutral disconnecting means. This is a general requirement of *NEC* 110.7 and 250.24(A)(5) and is a specific requirement of the manufacturer's installation and testing instructions for the GFP equipment [*NEC* 110.3(B)].

Combination Systems

An improvement in the protection of distribution systems of 600 volts or less may be obtained by using a ground-fault protective device as a supplement to high-interrupting capacity current-limiting overcurrent protection. Fault currents in low ranges can be recognized and virtually all fault currents can be held to limited time duration with the resultant increase to the safety of such systems.

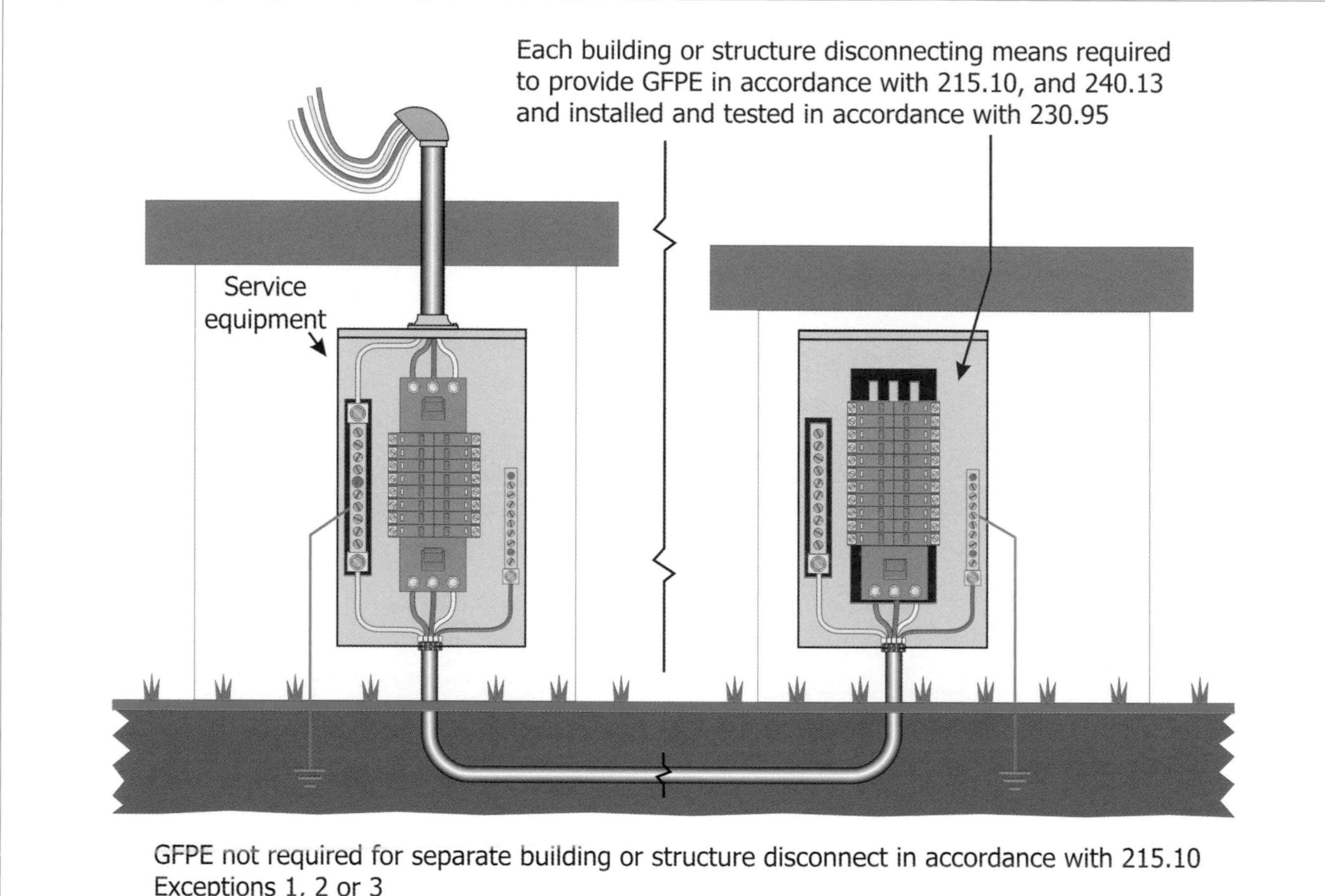
Each building or structure disconnecting means required to provide GFPE in accordance with 215.10, and 240.13 and installed and tested in accordance with 230.95
Service equipment
GFPE not required for separate building or structure disconnect in accordance with 215.10 Exceptions 1, 2 or 3

Figure 15-10. Equipment ground-fault protection for feeders

If circuit breakers are used as the disconnecting means and overcurrent protection and a ground-fault reaches a high enough value, the circuit breaker will open and clear all three phases, thus preventing a feedback into the ground fault from the two unaffected phases. When a circuit breaker is used, a ground-fault protection device also may be applied to automatically open the circuit breaker in a reasonable and limited time if a ground fault should develop and not be cleared by some overcurrent device nearer the point of the ground fault.

If a manually-operated, fused load-break switch is used as the disconnecting means and overcurrent protection, and if a ground fault reaches a high enough value to clear one fuse, an alarm should be provided to indicate a blown fuse so all three phases can be cleared manually.

If an electrically operated, fused load-break switch is used, it can be provided with an automatic blown fuse indicator which will immediately open the switch, and thus, all three phases, if any one or more fuses should open. An electrically operated switch also permits the application of a ground-fault protective device which will automatically open the switch in a reasonable and limited time if a ground fault should develop and not be cleared by an overcurrent device near the point of the fault.

For systems served with large values of available short-circuit current, and which require the best degree of protection, the electrically-operated switch or circuit breaker can provide full protection where it is equipped with high-interrupting capacity current-limiting fuses, a ground-fault protective device and an automatic blown-fuse indicator.

Testing of EGFP System

Section 230.95(C) requires that service ground-fault protection systems be performance-tested when first installed on site to ensure that they will operate

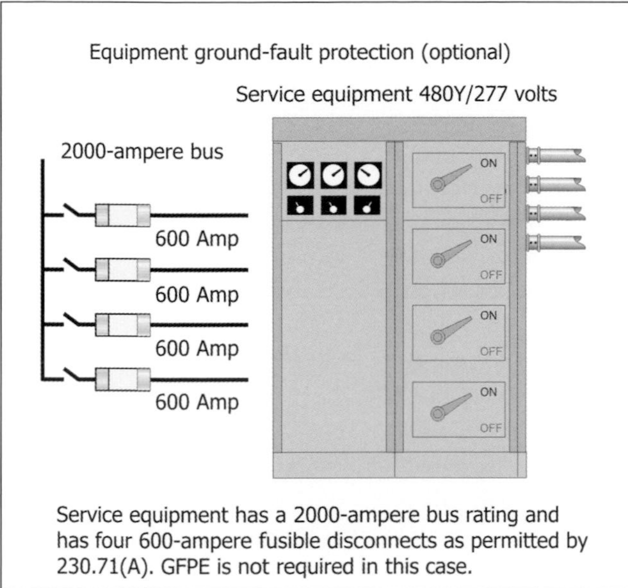

Figure 15-11 Equipment ground-fault protection is required for each disconnect rated 1000 amperes or more (it is optional as shown in the diagram).

properly. Experience has shown that the majority of these systems do not operate properly, or at all, when first installed. This is most often due to improper field wiring of the system. The two biggest problems found are undersized main bonding jumpers and the incorrect installation of grounding connections downstream of the neutral disconnect and/or the ground-fault sensor.

The test must be performed in full compliance with the manufacturer's written instructions [see 110.3(B)]. These instructions must be furnished with the equipment. A vital part of the test is to remove the neutral disconnect link in the distribution equipment and test the neutral with a continuity tester or meg-ohm meter to be certain that it is clear from any grounding connections downstream from main bonding jumper at the service. This test must be of all the neutral conductor system including the bus, feeders, and branch circuit wiring. Accidental or intentional grounding connections to the grounded (neutral) conductor past the service equipment and GFP neutral sensor can desensitize ground-fault protection systems and render them ineffective. These connections also provide a parallel path for normal neutral currents on the grounding system and thus can cause nuisance tripping of the ground-fault protection system.

"A written record of this test shall be made and shall be available to the authority having jurisdiction." [7] System safety requires that the GFP equipment be properly installed, as does the threat of lawsuit where negligence of the installer or other responsible party can be proven. Many jurisdictions now require this test to be completed before they will authorize the equipment to be energized.

The GFPE testing requirements apply to services and feeders where it is first installed.

Optional Protection

Figure 15-11 illustrates an installation where equipment ground-fault protection is not required even though the system voltage is the same as that where it is required. The difference is the rating of the disconnecting means is not 1000 amperes or greater, even though the combined rating exceeds 1000 amperes. For example, a service with a single 2000-ampere main requires equipment ground-fault protection at the stated voltage and ampere rating. If a service with four, 600-ampere overcurrent devices is installed rather than the single main, equipment ground-fault protection is not required, but the same damaging levels of ground-fault energy are still available.

Obviously, the system can benefit from equipment ground-fault protection even though it is not required.

In the case where a 900-ampere fuse is installed in a 1200-ampere switch, the requirements are different. Equipment ground-fault protection is required since the fuse holder will accept a 1000-ampere or larger fuse. Where adjustable circuit breakers are used, the rating, for purposes of 230.95, is the actual overcurrent device installed in the circuit breaker or the maximum rating that it can be adjusted to.

[1] NFPA 70, *National Electrical Code* 2008, 230.95(A), (National Fire Protection Association, Quincy, MA, 2007), p. 70-81.

[2] NFPA 70, Article 100, p. 70-27.

[3] NFPA 70, 700.7(D), p. 70-597.

[4] NFPA 70, 517.17(A), p. 70-428.

[5] NFPA 70, 215.10, p. 70-57.

[6] NFPA 70, 240.13, p. 70-86.

[7] NFPA 70, 230.95(C), p. 70-81.

15 Review Questions

The questions included here were developed using material included in this chapter. The answers can be found by reviewing the text. It is also important that students make use of *NEC*-2008, where many answers can be found.

1. A system intended to provide protection of equipment from damaging line-to-ground fault currents, by operating to cause a disconnecting means to open all ungrounded conductors of the faulted circuit, is defined as ____.
   a. ground-fault circuit interrupter
   b. ground-fault protection of equipment
   c. leakage current detector
   d. saturable core reactor

2. The *NEC* requires ground-fault protection of all solidly-grounded wye electrical services of more than 150 volts to ground, but not exceeding 600 volts phase-to-phase, for each service disconnect rated ____ amperes or more.
   a. 800
   b. 1000
   c. 600
   d. 400

3. Ground-fault protection of equipment is not applicable to ____ or continuous industrial processes where a nonorderly shutdown would introduce additional or increased hazards.
   a. electronically-actuated fuses
   b. phase converters
   c. fire pumps
   d. industrial buildings

4. Where a service is protected by a system of GFPE, the maximum setting of the GFPE devices is ____ amperes and the maximum time delay is one second for ground-fault currents equal to or greater than ____.
   a. 1300 - 2000
   b. 1200 - 3000
   c. 1400 - 1500
   d. 1600 - 1000

5. For feeders rated 1,000 amperes or more in a solidly-grounded wye system with greater than 150 volts to ground, but not exceeding 600 volts phase-to-phase, ground-fault protection ____ required.
   a. is generally
   b. is not
   c. sensors are
   d. indication is

6. The most common types of equipment ground-fault protection equipment is:
   a. three-phase, three wire
   b. residual and bypass relaying
   c. zero-sequence
   d. isolation transformer and relay

7. One of the most critical elements of a ground-fault protection system is:
   a. the grounded conductor (neutral) must be isolated from ground downstream
   b. a choke coil is installed downstream
   c. capacitors are installed downstream
   d. all the above

8. The equipment ground-fault protection system must be performance tested:
   a. by pushing the "push to test" button
   b. on an annual basis
   c. within 30 days of installation
   d. when first installed on site

9. A second level of equipment ground-fault protection is required
   a. for fire pump installations
   b. for hospitals and other buildings with critical care
   c. for legally required standby systems
   d. for optional standby systems

10. Equipment ground-fault protection is generally required for solidly grounded wye electrical systems of more than 150 volts to ground and less than 600 volts phase-to-phase:
   a. for feeders of less than 800 amperes
   b. at a building disconnecting means rated at 200 amperes or more
   c. for feeders rated more than 1000 amperes
   d. for a continuous industrial process

11. Where a 480-volt, 3-phase, 3-wire 2500 ampere feeder from an ungrounded system supplies a unit of equipment in a commercial building, which of the following is required?
   a. Ground-fault protection of equipment
   b. A grounding electrode
   c. A grounded conductor
   d. Ground detectors

Grounding and Bonding for Special Locations and Conditions

This chapter takes a look at grounding and bonding requirements for special occupancies and special conditions found in the *NEC*. General information is also provided about grounding and bonding for static electricity protection that is beyond the scope of the *NEC*, but is directly related to similar safety concerns. A review of some general requirements from NFPA 77-2007 Recommended Practice on Static Electricity is provided; however, it is not inclusive.

Objectives

To understand...

- Static electricity protection practices
- Bonding in hazardous (classified) locations
- Agricultural buildings
- Health care facilities
- Swimming pools, hot tubs and spas
- Electric signs and outline lighting

Chapter 16

Hazardous (Classified) Locations Grounding and Bonding

The *Code* has some special requirements for grounding and bonding in hazardous (classified) locations (see figure 16-1). These requirements can be found in 501.30 for Class I locations, 502.30 for Class II locations, and 503.30 for Class III locations.

For Class I, Zones 0, 1 and 2 hazardous (classified) locations, see 505.25 which requires compliance with the grounding and bonding rules in 505.25(A) and (B).

Grounding and bonding is essential for electrical safety in nonhazardous locations as well as in hazardous locations. In hazardous locations it is vital to have effective grounding and bonding to prevent an explosion. Under fault-current conditions when there are substantial currents through metal conduit, every connection point in the raceway system is a potential source of sparks and ignition. If there is an arcing fault to a metal enclosure in a hazardous (classified) location, the external surface temperature of the metal enclosure at the point of the arcing fault can rise to temperatures that could cause ignition of the flammable vapors or accumulations of combustible dust. Under these fault conditions it is essential that the overcurrent device operate as quickly as possible to prevent a *hot spot* on the enclosure or even arcs that can burn through the enclosure from igniting the atmosphere, dust, or fibers and flyings on the outside of the enclosure. It is extremely important that all threaded joints be made up wrenchtight to prevent sparking at those threaded joints. If joints are other than the threaded type, such as locknuts and bushings or double locknuts and bushings at boxes, enclosures, cabinets, and panelboards, it is essential that bonding be ensured around those joints in the fault-current path to prevent sparking and assure a low-impedance path for the fault current.

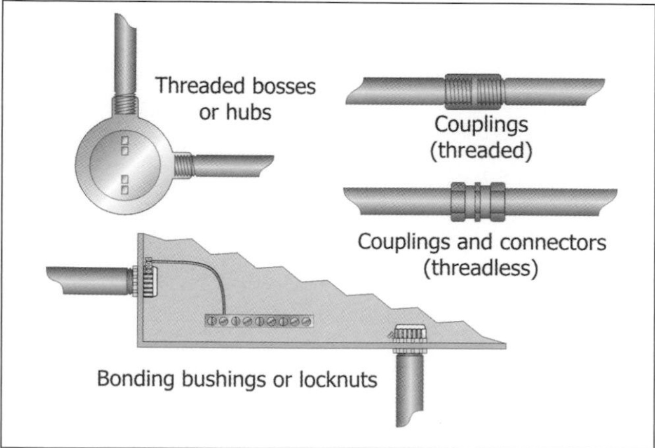

Figure 16-1. Bonding suitable for hazardous (classified) locations

Bonding requirements are found in Part V of Article 250. Section 250.90 requires that "bonding shall be provided where necessary to ensure electrical continuity and the capacity to conduct safely any fault current likely to be imposed." Section 250.100 also includes additional and more restrictive bonding requirements for hazardous locations and indicates that "regardless of the voltage of the electrical system, the electrical continuity of non-current-carrying metal parts of equipment, raceways, and other enclosures in any hazardous (classified) location as defined in 500.5 shall be ensured by any of the bonding methods specified for services in 250.92(B) specified in 250.92(B)(2) through (B)(4)." [2]

One or more of these bonding methods are required even where an equipment grounding conductor (of the wire-type) is installed in the raceway system.

By these special requirements, an effort is made to provide assured grounding and bonding to reduce the likelihood that a line-to-ground fault will cause arcing and sparking at connection points of metallic raceways and boxes or other enclosures. If such arcing or sparking were to occur in a hazardous (classified) location while a flammable gas or vapor is present in its explosive range, it is likely that the flammable atmosphere would be ignited.

Generally, these requirements provide that locknuts on each side of the enclosure, or a locknut on the outside and a bushing on the inside, cannot be used for bonding. Bonding locknuts or grounding bushings with bonding jumpers must be used to ensure the integrity of the *bonding connection* and its capability of carrying the fault current that can be imposed, hopefully, without arcing or sparking at the connections.

The bonding means required in 501.30(A), 502.30(A), 503.30(A) are generally installed from the hazardous (classified) location to the service equipment (see figure 16-2), or point of grounding of a separately derived system (see figure 16-3) that is the source of the circuit.

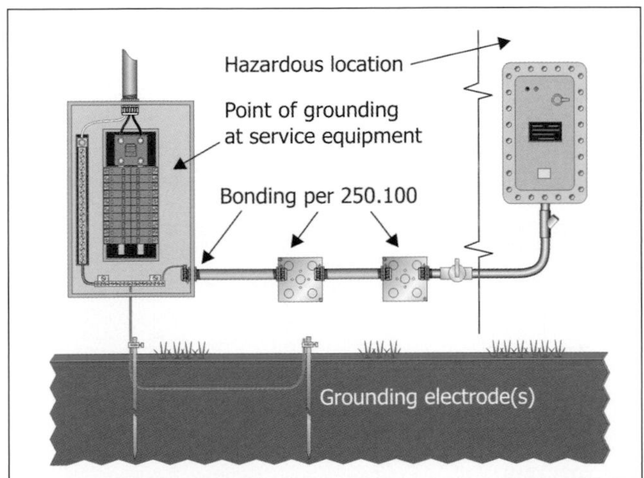

Figure 16-2. Bonding extended back to point of grounding at service equipment

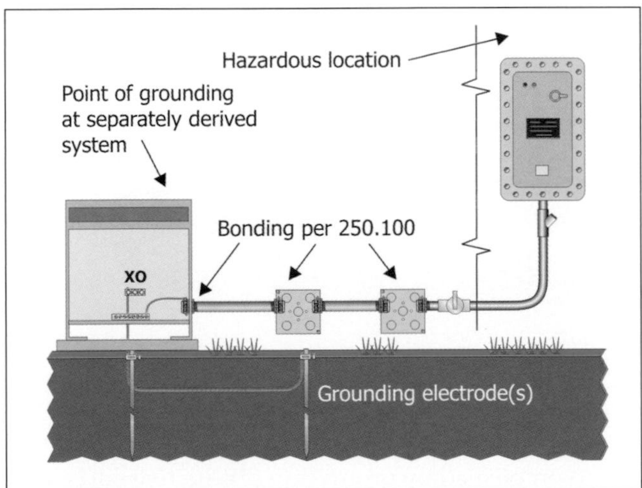

Figure 16-3. Bonding extended back to point of grounding at separately derived system

"Such means of bonding shall apply to all intervening raceways, fittings, boxes, enclosures, and so forth between Class I locations and the point of grounding for service equipment or point of grounding of a separately derived system." [3]

The goal is to ensure a substantial path for ground-fault currents to facilitate fast operation of overcurrent protective devices supplying the circuit in the hazardous locations and provide a flame path for cooling escaping hot gasses.

Sections 501.30(B), 502.30(B), and 503.30(B) do not permit flexible metal conduit or liquidtight flexible metal conduit as the sole path for ground-fault current. Where equipment bonding jumpers are installed around these flexible conduits, they must meet the requirements in 250.102.

Hazardous atmospheres can be ignited by hot temperatures on electrical enclosures. If a ground fault should occur inside an explosionproof enclosure, a hot spot at the point of fault on the enclosure could develop if the overcurrent device does not clear quickly. These more restrictive bonding rules are in *NEC* chapter 5 for these reasons. The bonding in hazardous locations is applicable to all metal raceways and enclosures in the hazardous location, and all intervening metal raceways and enclosures of the circuit extending back to the point of grounding for the applicable service or derived system. Since current and fault-current returns to the source, this method of bonding must be accomplished for the circuit to the source where the system bonding jumper is installed or service grounding point, where the main bonding jumper is installed.

The exceptions to 501.30(A), 502,30(A), and 503.30(A) clarify that "the specific bonding means shall only be required to the nearest point where the grounded circuit conductor and the grounding electrode are connected together on the line side of the building or structure disconnecting means as specified in 250.32(B), provided the branch-circuit overcurrent protection is located on the load side of the disconnecting means" (see figure 16-4).

Section 250.32(B) provides grounding and bonding requirements where more than one building or structure are on the same premises and are supplied by feeder(s) or branch circuit(s). Where the grounded circuit conductor is not grounded at the building or structure, the rule in 501.30(A), 502.30(A) and 503.30(A) requires that the bonding extend from the hazardous location back to the service even if it is in another building. This requires that the feeder raceway system be bonded if it is metallic [see chapter thirteen for additional information on grounding electrical systems at additional buildings or structures on the premises].

Flexible Conduits in Hazardous Locations

Where flexible metal conduit or liquidtight flexible metal conduit is used as permitted in 501.10(B) and is to be relied on to complete a sole equipment grounding path, it shall be installed with internal or external bonding jumpers in parallel with each conduit and

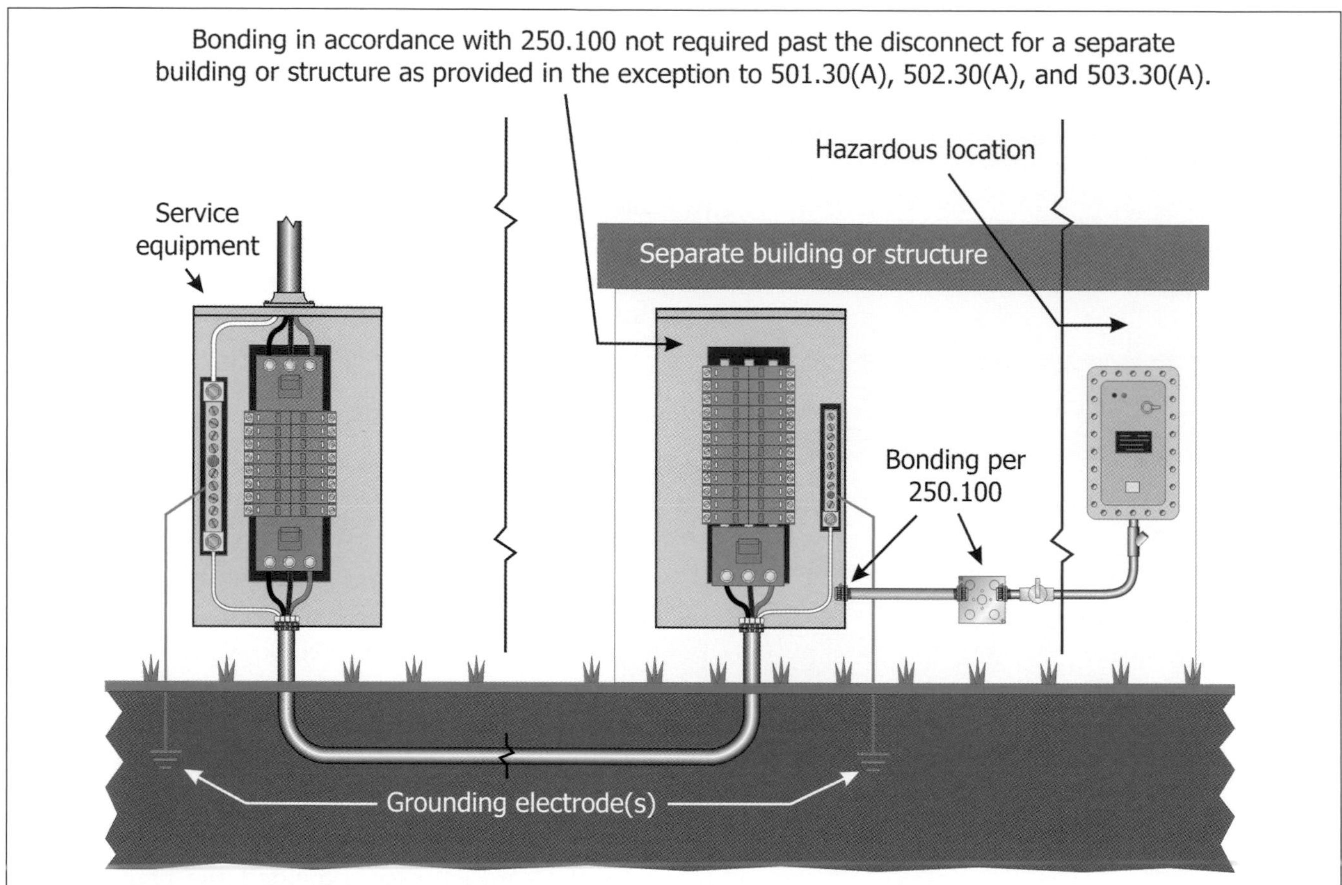
Bonding in accordance with 250.100 not required past the disconnect for a separate building or structure as provided in the exception to 501.30(A), 502.30(A), and 503.30(A).
Hazardous location
Service equipment
Separate building or structure
Bonding per 250.100
Grounding electrode(s)

Figure 16-4. Bonding in hazardous (classified) locations is not required beyond the building disconnect at a separate building or structure supplied by a feeder(s) or branch circuit(s).

complying with 250.102. If installed outside the conduit, the bonding jumper is limited to 1.8 m (6 ft) in length per 250.102(E) [see 501.30(B) and Exception].

"Exception: In Class I, Division 2 locations, the bonding jumper shall be permitted to be deleted where all the following conditions are met:

"(1) Listed liquidtight flexible metal conduit 1.8 m (6 ft) or less in length, with fittings listed for grounding, is used.

"(2) Overcurrent protection in the circuit is limited to 10 amperes or less.

"(3) The load is not a power utilization load." [4]

[See 502.30(B) for similar rules for Class II areas and 503.30(B) for Class III locations].

Static Protection through Bonding and Grounding

Effective grounding and bonding are important components in the overall electrical safety scheme. As previously discussed, the benefits of properly grounded and bonded systems and conductive parts provide protection for persons and property. Protection against electrical shock and equalizing the potential to earth are accomplished by grounding conductive parts. Fast and sure operation of overcurrent protective devices if a fault occurs is ensured by creating an effective ground-fault current path back to the source, either the applicable service or source of separately derived system. The grounding and bonding requirements in the *Code* for electrical installations in hazardous locations provide protection from such events.

In hazardous locations, electrical wiring, including the grounding and bonding circuits are extremely important for safety. Because sources of ignition are a primary concern in explosive atmospheres, it is often necessary to also provide a more enhanced protection system of handling static electricity in hazardous locations.

Humidity

Protection against static electricity is a requirement of a number of industries and establishments. The grounding of equipment is not necessarily a solution to static problems. Each static problem requires its own study and solution. Humidity plays an important part in the degree of concern. The higher the humidity is, the less the chances of a static discharge occurring. In some industries, increasing the humidity in the area of a static discharge has been found very effective. One example is in the printing industry.

While humidification does increase the surface conductivity of the material, the charge will only dissipate if there is a conductive path. The surface resistivity of many materials can be controlled by the humidity of the surroundings. At humidity of 65 percent and higher, the surface of most materials will adsorb enough moisture to ensure a surface conductivity that is sufficient to prevent accumulation of static electricity. When the humidity falls below about 30 percent, these same materials could become good insulators, in which case accumulation of charge will increase. It should be emphasized that humidification is a not a solution for all static electricity problems encountered. Some insulating materials do not adsorb moisture from the air and high humidity will not noticeably decrease their surface resistivity. Examples of such insulating materials are uncontaminated surfaces of some polymeric materials, such as plastic piping, containers, and the surface of most petroleum liquids [NFPA 77 7.4.2.3].

Concerns of Static Electricity as Ignition Source

This chapter takes the reader beyond the requirements of the electrical *Code* and looks at means of protection from static electricity and the related sources of ignition. It should be clearly understood that the primary goal in providing static protection is to eliminate the ignition source of the fire triangle. Careful consideration and planning is necessary to evaluate all known possibilities of static ignition sources relative to providing this type of protection in hazardous locations. The degree of additional protection needed is specific to each condition encountered. There are no mandatory electrical *Code* requirements to provide such protection; however, the hazards do exist and must be considered for safety. Generally the type of installation, type of explosive or flammable atmosphere (dust or gases), and the natural environment are all contributing factors to the degree or extent of static electricity as an ignition source. For a static electricity discharge to be a source of ignition, the following four conditions must exist simultaneously:

1. An effective means of separating the charge must be present.
2. A means of accumulating the separated charges and maintaining a difference of electrical potential must be available.
3. A discharge of the static electricity of adequate energy must occur.
4. The discharge must occur in an ignitible mixture [NFPA 77 – 5.3.1].

Sparks from ungrounded charged conductors, including the human body, are responsible for most fires and explosions ignited by static electricity. Sparks are typically intense capacitive discharges that occur in the gap between two charged conducting bodies, usually metal. The ability of a discharge spark to produce ignition or explosion is directly related to its energy. This will be some fraction of the total energy stored in the conductive object, which could be the human body.

Beyond the *NEC*

The *NEC* provides a reference through a fine print notes (FPN) to the recommended practice on protection from static electricity. The referenced document is titled *Recommended Practice on Static Electricity,* NFPA 77-2007. Lightning protection systems are also important considerations and provide reasonable planned protection for those natural events in the weather that produce lightning.

There is an actual industry standard for these types of protection systems and it is entitled *Standard for the Installation of Lightning Protection Systems* NFPA 780-2008. The American Petroleum Institute (API) also has produced a document that addresses protection techniques used to tackle these concerns and is titled *Protection Against Ignitions Arising Out of Static Lightning and Stray Currents* API RP 2003-1998 [*NEC* 500.4(B) FPN No. 3]. As previously discussed, the fine print notes in the *NEC* are explanatory in nature and are not mandatory requirements of the *Code* based on the structure and style of the *NEC* [90.5(C)]. However, these references provide clear direction to resources that provide specific criteria and guidelines for this enhanced protection. It is important to emphasize

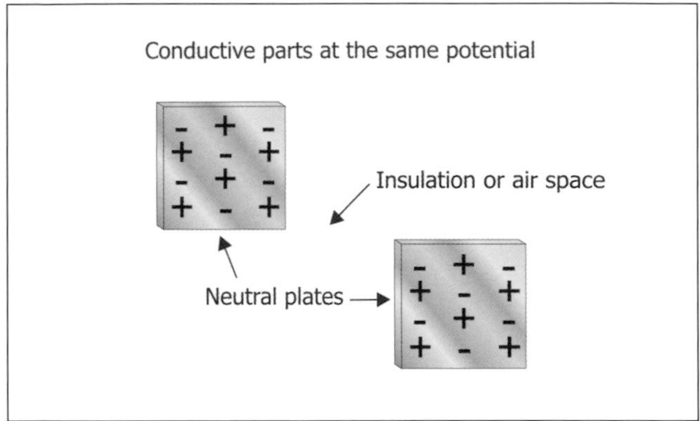

Figure 16-5. Two metal plates (conductors), each with like charges

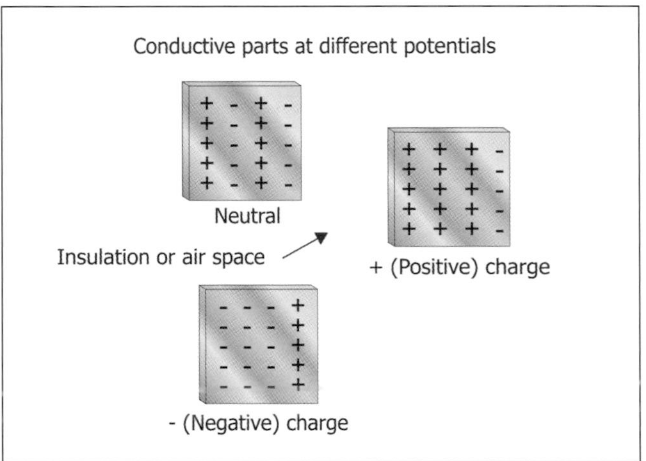

Figure 16-6. Two metal plates (conductors), with unlike charges

that these methods of protection for static electricity and static ignition sources must overlay the requirements of the *Code* and are in addition to those requirements and are never intended to substitute for those requirements.

Definitions

Static electric discharge. "A release of static electricity in the form of a spark, corona discharge, brush discharge, or propagating brush discharge that might be capable of causing ignition under appropriate circumstances" [NFPA 77 3.1.16].

Static electricity. "An electric charge that is significant only for the effects of its electrical field component and that manifests no significant magnetic field component" [NFPA 77 3.1.17].

Static Electricity Fundamentals

All matter (materials), whether liquid or solid, is made up of various arrangements of atoms. Atoms are made up of positively charged nuclear components that give them mass and are then surrounded by negatively charged electrons. Atoms are considered to be electrically neutral in their normal state. Basically this means that there are equal amounts of positive and negative charges present. The atoms can become, what is referred to as, *charged* when there is an excess or deficiency of electrons relative to the neutral state (see figures 16-5 and 16-6).

In electrically conductive materials, such as metals of the ferrous and nonferrous types, electrons move freely. In materials that are made up of insulating material such as plastic, glass, motor oil, etc., electrons are bonded more tightly to the nucleus of the atom and are not free to move. Some examples of electrically conductive materials are wire, metallic enclosures, busbars, etc, while insulating materials include such items as glass, and petroleum based products, paper, rubber, and so forth. For insulating materials in the form of fluids, an electron can separate from one atom and move freely or attach to another atom to form a negative ion. The atom losing the electron then becomes a positive ion. Ions are charged atoms and molecules.

Elimination or separation of the charge cannot be absolutely prevented, because the origin of the charge lies at the interface of materials. When materials are placed in contact, some electrons move from one material to the other until a balance (equilibrium condition) in energy is reached. This charge separation is most noticeable in liquids that are in contact with solid surfaces and in solids in contact with other solids. The flow of clean gas over a solid surface produces negligible charging [NFPA 77- 8.3.1]. This is the primary reason for the gasoline dispensing hazard warnings at motor fuel dispensers. It is important to observe and adhere to all warnings and directions relative to the transfer of gasoline to a motor vehicle or portable container. Always place portable gasoline containers on the ground when filling them; otherwise, the charging currents allow static charges to build without a path to dissipate. The possibilities of ignition or explosion of gasoline vapors during these types of operations is increased if all appropriate safety procedures are not followed. Elimination of differences of potential (voltage) between objects reduces these hazards

Static Discharge and Separation

A *capacitor* is described basically as two conductors that are separated by an insulating material. In the static electric phenomena, the charge is generally separated by a resistive barrier, such as an air gap or form of insulation between the conductors, or by the insulating property of the materials being handled or processed. In many applications, particularly those where the materials being processed are nonconductive (charged insulators), measuring their potential differences is challenging to say the least.

One is probably most familiar with the common static charge built up by walking or scuffing the feet on carpet fibers. People are conductors of electricity and therefore are capable of "holding" a static charge. The release of such static charges is also a familiar experience for most individuals. Children often are amused and entertained when this phenomenon is first realized. The electrical static charging results from rubbing materials together and is known as *triboelectric charging*. It is the result of exposing surface electrons to a broad variety of energies in an adjacent material, so that charge separation (discharge) is likely to take place. The breakup of liquids by splashing and misting or even flow, in some instances, results in a similar charge release. It is only necessary to transfer about one electron for each 500,000 atoms to produce a condition that can lead to a static electric discharge. Surface contaminants at very low concentrations can play a significant role in charge separation at the interface of materials.

Electrically conductive materials can become charged when they are in the vicinity of another highly charged surface. The electrons in the conductive material are either drawn toward or forced away from the region of closest approach to the charged surface, depending on the nature of the charge on that surface. Like charges will repel and unlike charges will attract. If the electrically conductive material that is charged is connected to ground or bonded to another object, additional electrons can pass to or from ground or the object. If contact is then broken and the conductive material and charged surface are separated, the charge on the isolated conductive object changes. The net charge that is transferred is called *induced charge*.

The basic goal when dealing with concerns of static electricity and stray voltages is to try to eliminate or at least minimize any differences of potential between electrically conductive objects and other objects and the ground. The potential difference, that is, the voltage, between any two points is the *work-per-unit charge* that would have to be done to move the charges from one point to the other. Work must be accomplished to separate charges, and there is a tendency for the charges to return to a neutral (uncharged) condition. The separation of electric charge might not, in itself, be a potential fire or explosion hazard. There must be a discharge or sudden recombination of the separated charges to create arcing to pose an ignition hazard. One of the best methods of providing protection from static electric discharge is constructing an electrically conductive or semiconductive path that will allow the controlled recombination of the charges and dissipation of charges (usually to earth). The two terms used most often when providing protection from static electricity and lightning are *grounding* or one of its derivatives, and *bonding* or one of its derivatives. Derivatives of these terms are as in the following examples:

Grounding – *Ground* or *Grounded*
Bonding – *Bond* or *Bonded*

Definitions from NFPA 70 and NFPA 77

Grounded (*Grounding*) "Connected (connecting) to ground or to a conductive body that extends the ground connection" [NFPA 70 Article 100].

Bonded (*Bonding*). "Connected to establish electrical continuity and conductivity" [NFPA 70 Article 100].

Grounding. "The process of bonding one or more conductive objects to the ground, so that all objects are at zero (0) electrical potential; also referred to as 'earthing'" [NFPA 77 - 3.3.10]. Keep in mind that the term *earthing* is not currently a defined term.

Bonding. "For the purpose of controlling static electric hazards, the process of connecting two or more conductive objects together by means of a conductor so that they are the same electrical potential, but not necessary at the same potential as sthe earth" [NFPA 77 - 3.3.2].

So for all practical purposes, when the term *grounding* is used, it should be thought of as including a connection or path to the earth to put electrically

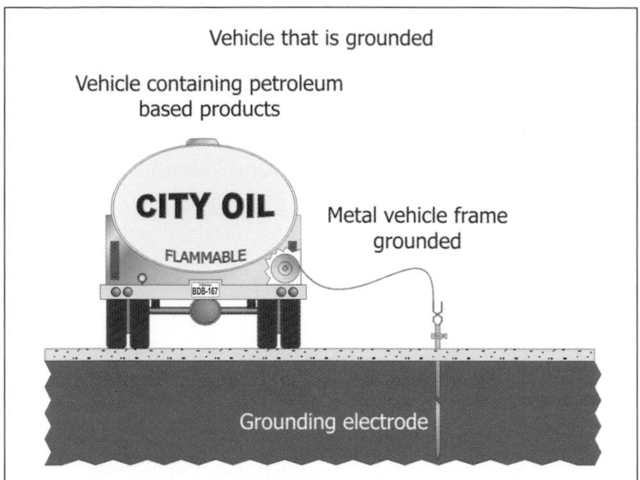

Figure 16-7. A vehicle connected to the earth (grounded)

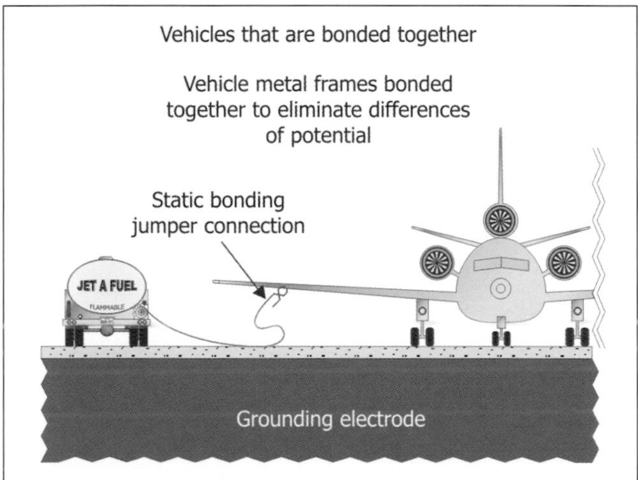

Figure 16-8. Two vehicles connected together (bonded)

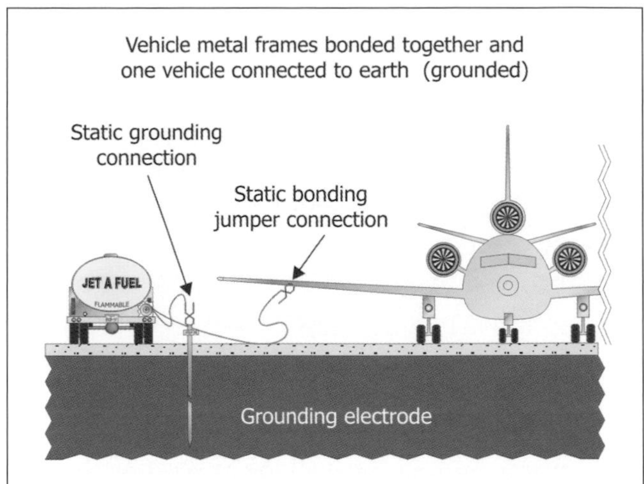

Figure 16-9. Two vehicles connected together (bonded), and one vehicle also connected to the earth

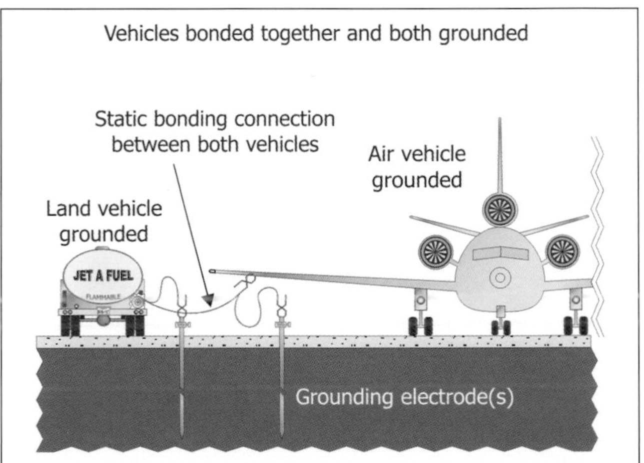

Figure 16-10. Two vehicles connected together (bonded), and also each vehicle is connected to the earth separately (grounded)

conductive materials at the same potential as the earth. When the term *bonding* is used, it should be thought of as connecting electrically conductive materials together to eliminate differences of potential between them and form one conductive mass. Note that bonding generally includes a path to the earth, but the earth is not referred to in the definitions. Figures 16-7, 16-8, 16-9 and 16-10 graphically demonstrate the differences between the two concepts and also show the two working together to provide desired protection. It can be concluded, then, that bonding conductive parts together minimizes the potential differences between them, even when the resulting system is not grounded. Grounding (earthing), on the other hand, equalizes the potential differences between the objects and the earth. The relationship between bonding and grounding is shown in figures 16-7 through 16-10.

Controlling Static Electricity Ignition Hazards

Ignition hazards from static electricity can be controlled by the following methods:

- Removing the ignitible mixture from the area where static electricity could cause an ignition-capable discharge.
- Reducing charge generation, charge accumulation, or both by means of process or product modifications.
- Neutralizing the charges.

Grounding isolated conductors and air ionization are primary methods of neutralizing charges.

Resistance in the Path to Ground

To prevent the accumulation of static electricity in

conductive equipment, the total resistance of the ground path to earth should be sufficient to dissipate charges that are otherwise likely to be present. A resistance of 1 megohm (10^6 ohms) or less is generally considered adequate. Where the bonding/grounding system is all metal, resistance in continuous ground paths will typically be less than 10 ohms. Such systems include multiple component systems. Greater resistance usually indicates the metal path is not continuous, usually because of loose connections or the effects of corrosion. A grounding system that is acceptable for power circuits or for lightning protection is more than adequate for a static electricity grounding system.

Where wire conductors are used, the minimum size of the bonding or grounding wire is dictated by mechanical strength, not by its current-carrying capacity. Stranded or braided wires should be used for bonding wires that will be connected and disconnected frequently [NFPA 77 7.4.1.4]. Grounding conductors can be insulated (e.g., a jacketed or plastic-coated cable) or uninsulated (i.e., bare conductors). Uninsulated conductors are recommended, because it is easier to detect defects in them.

Where static problems are present, workers should only be grounded through a resistance that limits the current to ground to less than 3 mA for the range of voltages experienced in the area. This method is called *soft grounding* and is used to prevent injury from an electric shock from line voltages or stray currents.

Liquids Flowing through Pipes

Charge separation occurs when liquids flow through pipes, hoses, and filters, when splashing occurs during transfer operations, or when liquids are stirred or agitated. The greater the area of the interface is between the liquid and surfaces and the higher the flow velocity, the greater the rate of charging. The charges become mixed with the liquid and are carried to receiving vessels where they can accumulate. The charge is often characterized by its bulk charge density and its flow as a streaming current to the vessel.

In the petroleum industry, for tank loading and distribution operations involving petroleum middle distillates, liquids in the semiconductive category are handled as conductive liquids. Such procedures are possible because regulations prohibit use of non-conductive plastic hoses and tanks and multiphase mixtures and end-of-line polishing filters are not involved.

Metallic Piping Systems

All parts of continuous metal piping systems should have a resistance to ground that does not exceed 10 ohms. Higher resistance could indicate poor electrical contact or connection, although this will depend on the overall system. For flanged couplings, neither paint on the flange faces nor thin plastic coatings used on nuts and bolts will normally prevent bonding across the coupling after proper tightening torque has been applied. Jumper cables and star washers are not usually needed at flanges. Star washers could even interfere with proper tightening. Electrical continuity of the bonding and grounding path should be confirmed after system is completely assembled and periodically thereafter.

Additional bonding wires (jumpers) might be needed around flexible, swivel, or sliding joints. Tests and experience have shown that resistance in these joints is normally below the 10-ohm value, which is low enough to prevent accumulation of any static charges.

Grounding Storage Tanks for Nonconductive Liquids

Storage tanks for nonconductive liquids should be grounded properly. Storage tanks on foundations constructed on the earth are considered inherently grounded, regardless of the type of foundation (e.g., concrete, sand, or asphalt). For tanks on elevated foundations or supports, the resistance to ground could be as high as 100 ohms and still be considered adequately grounded for purposes of dissipation of static electric charges, but the resistance should be verified in these cases for assurances that an adequate path to ground is achieved. The addition of grounding rods and similar grounding systems will not reduce the hazard associated with static electric charges apparent in the liquid [NFPA 77 8.5.2.2].

Basic Static Concerns with Combustible Dust

A combustible dust is defined as any finely divided solid material 420 μm or smaller in diameter (i.e., material that will pass through a U.S. No. 40 standard sieve) that can present a fire or deflagration hazard.

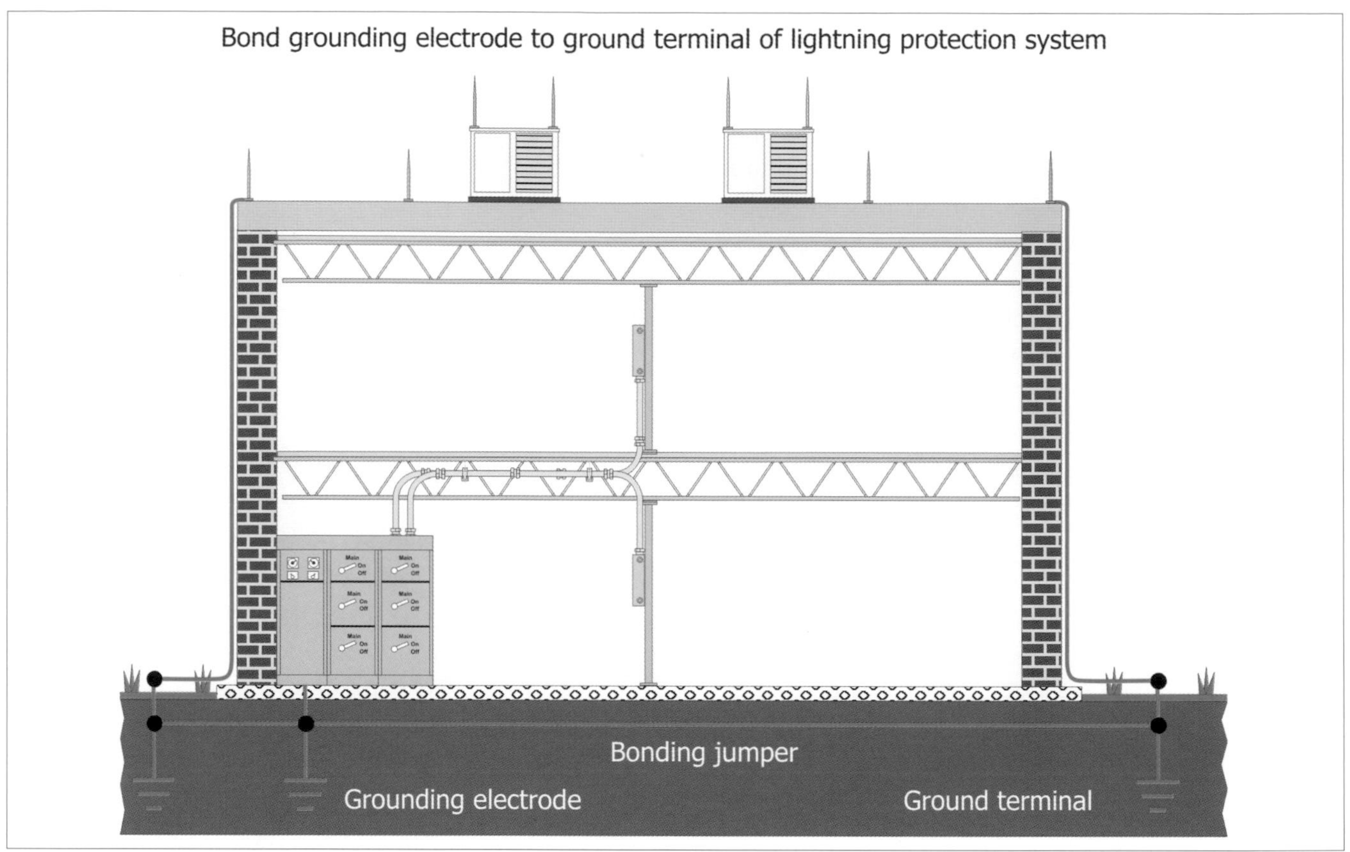

Figure 16-11. Ground terminals of lightning protection systems and grounding electrodes of power systems to be bonded together [*NEC* 250.106]

For a static electric discharge to ignite a combustible dust, the following four conditions need to be met:

1. An effective means of separating charge must be present.
2. A means of accumulating the separated charges and maintaining a difference of electrical potential must be available.
3. A discharge of the static electricity of adequate energy must be possible.
4. The discharge must occur in an ignitible mixture of the dust.

A sufficient amount of dust suspended in air needs to be present in order for an ignition to achieve sustained combustion. This minimum amount is called the *minimum explosible concentration* (MEC). It is the smallest concentration, expressed in mass per unit volume, for a given particle size that will support a deflagration when uniformly suspended in air.

For historical reasons, the ability of a solid to transmit electric charges is characterized by its volume resistivity. For liquids, this ability is characterized by its conductivity.

Powders are divided into the following three groups:

- *Low-resistivity powders* having volume resistivities in bulk of up to 108 ohm-m. Examples include metals, coal dust, and carbon black.
- *Medium-resistivity powders* having volume resistiv-ities between 108 and 1010 ohm-m. Examples include many organic powders and agricultural products.
- *High-resistivity powders* having volume resistiv-ities above 1010 ohm-m. Examples include organic powders, synthetic polymers, and quartz [NFPA 77 9.4.3].

Low-resistivity powders can become charged during flow. The charge rapidly dissipates when the powder is conveyed into a grounded container. However, if conveyed into a nonconductive container, the accumulated charge can result in an *incendive spark*.

Lightning Protection Systems

Lightning protection is an important factor at outdoor substations and at locations where thunderstorms are prevalent. Lightning discharges usually consist of very large currents of extremely short duration. Protection is accomplished by deliberately providing a path of low resistance to earth, compared to other paths. There is no guarantee that lightning will necessarily follow the lower resistance path that has been provided, but at least the low-resistance path will reduce the likelihood of damage.

"The lightning protection system ground terminals shall be bonded to the building or structure grounding electrode system" [1] (see figure 16-11). This section (250.106) no longer requires that metallic parts of electrical wiring system be bonded to the lightning protection system conductors where there is less than six feet of separation. However, specific requirements for bonding the systems together are found in NFPA 780, *Standard for the Lightning Protection Systems.*

Chapter 20 of this text provides basic information about installing lightning protection systems for pre-existing buildings and structures.

Grounding and Bonding Requirements for Agricultural Buildings

Sections 547.5(F) and 547.9 contain some specific requirements for grounding and bonding in agricultural buildings (see figure 16-12). The major concern is twofold. The first is for the integrity of the grounding path due to corrosive conditions that exist in these buildings. The second concern is for neutral-to-earth and stray voltage, which, if excessive, cause behavior responses in livestock. These behavior responses in dairy cattle can lead to loss of production and health problems.

Section 547.9(D) requires that "where livestock is housed, any portion of a direct-buried equipment grounding conductor run underground to the building or structure shall be insulated or covered copper." [6]

Equipment Grounding

Where grounded equipment is installed in an agricultural building, a copper equipment grounding conductor must be installed to ground the equipment. This applies regardless of the type of wiring method employed. "If installed underground, the equipment grounding conductor shall be insulated or covered." [7]

Motor-operated water pumps, including the submersible type are required to be grounded.

In addition to the water pump grounding requirements, where a submersible pump is used in a metal well casing, the well casing is required to be bonded to the pump circuit equipment grounding conductor [see figure 16-13 and *NEC* 250.112(L) and (M)].

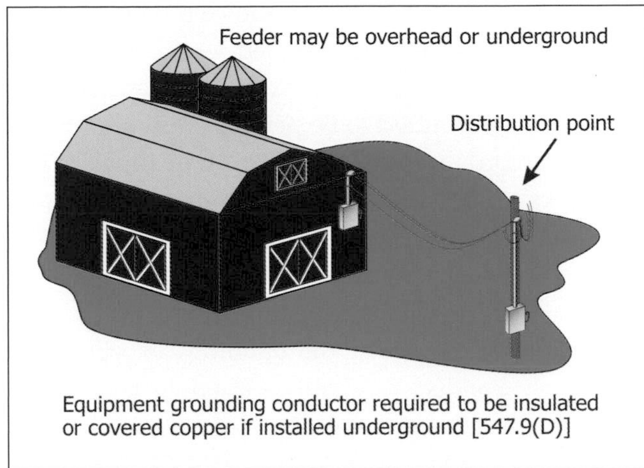

Figure 16-12. Grounding at agricultural buildings

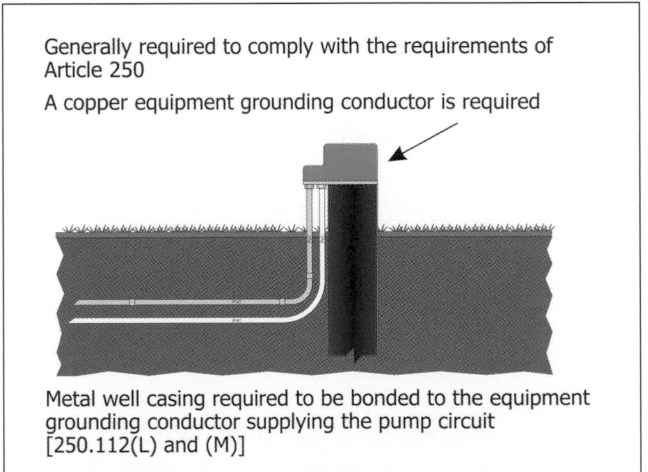

Generally required to comply with the requirements of Article 250
A copper equipment grounding conductor is required
Metal well casing required to be bonded to the equipment grounding conductor supplying the pump circuit [250.112(L) and (M)]

Figure 16-13. Equipment grounding at agricultural buildings

One of the most important elements of farm wiring systems, especially where dairy cattle are involved, is to isolate the system neutral from the equipment grounding conductor and non-current carrying metal parts of the electrical system at barns, milking parlors, and so forth. Separation of the neutral conductor will help minimize and prevent voltage drop on the feeder neutral from becoming stray voltage in the building.

Isolation of the grounded conductor (usually the neutral) at an additional building or structure on the same premises as the electrical service is required by 250.32(B). In the case of agricultural buildings, the conditions of 547.9(B) must be satisfied as well. Here two general methods are provided.

Grounding of electrical systems at agricultural buildings is permitted to be performed in one of two ways in accordance with 547.9(B)(3)(1) or (B)(3)(2). The difference in the two methods permitted depends on whether the neutral is grounded at the site distribution point or at the agricultural building.

Where the disconnecting means and overcurrent protection are located at the buildings or structures, the supply conductors shall be sized in accordance with Article 220 and installed in accordance with the requirements of Part II of Article 225. [9] [547.9(B)(1) and (2)]. Article 220 provides the rules for calculating the load on conductors, feeders and services. Article 225 includes requirements for the disconnecting means at additional buildings and structures on the premises.

Where the disconnecting means and overcurrent protection is located at the agricultural building or structure, two methods for grounding are permitted.

For each building or structure in an agricultural premises wiring system, grounding and bonding at separate buildings or structures must comply with Section 250.32 and the following two specific conditions.

1. The equipment grounding conductor is the same size as the largest supply conductor if of the same material, or is adjusted in size in accordance with the sizing columns of Table 250.122 if of different materials.

2. The equipment grounding conductor is connected to the grounded circuit conductor and the site-isolating device at the distribution point.[11]

As provided in 547.9(C), "Where the disconnecting means and overcurrent protection for each set of feeder conductors are located at the distribution point, feeders to building(s) or structure(s) shall meet the requirements of 250.32 and Article 225, Parts I and II."

Stray (Tingle) Voltage

Voltage drop on a neutral conductor supplying a building or structure housing livestock can result in elevated levels of stray voltage if it is grounded at the building or structure. It is important to balance 120-volt loads to minimize neutral loads, operate motors at 240 volts wherever practical, and size neutral conductors in service drops and feeders as large as practical.

Livestock behavioral responses, including production and health problems, can be caused by *stray voltage* (also referred to as *tingle-voltages*). These voltages can appear between various portions of grounded or ungrounded metallic systems, such as electrical equipment or piping systems, and the earth or floor such that livestock can come between two different potentials. It is not uncommon to find voltage differences between adjacent concrete slabs or between a concrete slab and the adjacent earth.

Common causes of neutral-to-earth voltages (stray voltage if it is at livestock contact points) are currents on primary distribution systems, farm secondary neutral conductors and faulty wiring on the farm. Some electric utilities have installed primary-to-secondary neutral conductor isolators on their transformers in an attempt to solve the stray voltage problem. Neutral conductors that are too small for the load and length of run and loose or corroded splices and terminations will frequently elevate the neutral-to-earth voltage to a level that can adversely affect livestock.

Another common cause of stray voltage is ground faults where the fault current gets into the earth, concrete slabs or on metal equipment that can be contacted by livestock. Ground faults that can cause stray voltage can occur in water pumps, underground wires, sump pumps, manure pumps, electrically heated livestock watering fountains, electrically operated feeders and similar equipment. Proper grounding along with bonding this equipment together usually prevents the stray voltage from occurring even where there is a high-impedance fault. In addition, it is important to have a low-impedance ground-fault return path so overcurrent devices can clear ground faults.

Equipotential Planes and Bonding of Equipotential Planes

Section 547.10(A)(1) requires that "equipotential planes shall be installed in confinement areas with concrete floors where metallic equipment is located that may become energized and is accessible to livestock."

Also, outdoor confinement areas, such as feedlots, shall have equipotential planes installed in concrete slabs where metallic equipment is located that may become energized and is accessible to livestock.

The equipotential plane shall encompass the area where the livestock stands while accessing metallic equipment that may become energized.[12] [547.10(A)(2)].

This is the best method of controlling the effects of stray voltages on livestock. This bonding does not correct or remove the faults that are causing the problem but keeps everything in the livestock area at the same potential (see figure 16-14) This, in essence, prevents the livestock from being aware of and being affected by the stray voltage.

The term *equipotential plane* is defined in 547.2 as, "an area where wire mesh or other conductive elements are embedded in or placed under concrete, bonded to all metal structures and fixed nonelectrical equipment that may become energized, and connected to the electrical grounding system to prevent a difference in voltage from developing within the plane."

Section 547.10(B) provides details on requirements for bonding of the equipotential plane. Wire mesh or other conductive elements, such as reinforcing steel rods, must be installed in the concrete floors in these areas and be bonded together, as well as to the building grounding electrode system. This is accomplished by bonding the reinforcing rods and wire mesh in the concrete together with a copper conductor that is insulated, covered or bare, not smaller than 8 AWG, and then connecting these bonding conductors to metal piping systems, stanchions and the building grounding electrode systems (see figures 16-14 and 16-15). This creates the required equipotential plane, meaning everything in the area is at the same potential. The equipotential plane prevents voltage differences between conductive bodies with which livestock can make contact. Connections must be made with pressure connectors or clamps of brass, copper, copper alloy, or an equally substantial approved means. Connections can also be made by exothermic welding. "Slatted floor sections that are supported by structures that are a part of an equipotential plane shall not require bonding." [13] These floor sections are typically pre-cast concrete sections, which by their size and mass effectively become a part of the equipotential plane when supported by the rest of the structure. In addition, the installation of bonding conductors to these floating sections has proven difficult as these sections are removed for periodic wash downs and other cleaning.

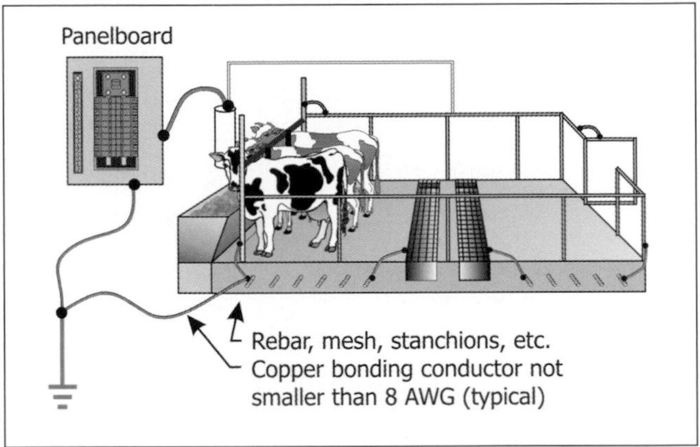

Figure 16-14. Equipotential bonding at agricultural buildings

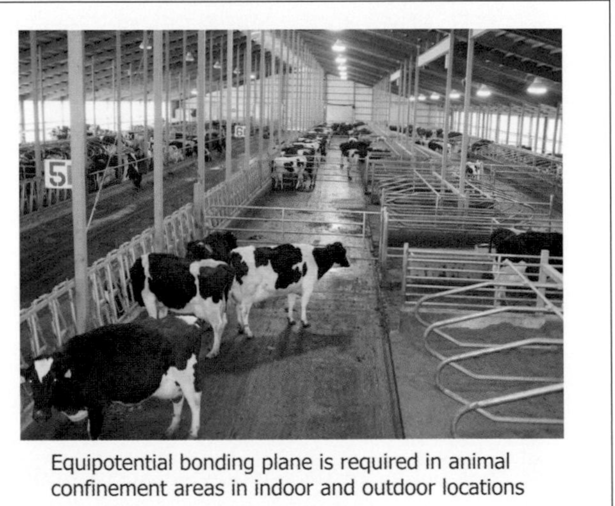

Equipotential bonding plane is required in animal confinement areas in indoor and outdoor locations

Figure 16-15. Equipotential bonding plane

Obviously, it is best to install this bonding system during the original construction of the building or portion of a building that houses livestock or serves as a milking parlor. Equipotential planes have been installed in existing buildings by sawing grooves in the concrete floors and installing bonding conductors that bond all conductive elements together. However, this is always much more expensive than making the installation before the concrete is poured.

FPN No. 1 following 547.10(B) provides information on methods for installing equipotential planes and voltage gradient ramps and reads, "Methods to establish equipotential planes are described in American Society of Agricultural and Biological Engineers (ASABE) EP473.2–2001, *Equipotential Planes in Animal Containment Areas*." [14]

Health Care Facilities

Electrical systems in health care facilities are required to comply with at least two safety standards, the *National Electrical Code* NFPA 70 and NFPA 99-2005, Standard for Health Care Facilities. These standards provide the minimum requirements for installation as well as maintenance and testing of electrical systems in health care facilities.

Depending on the type of health care facility involved and the scope of the project, several other electrical codes and standards may be involved. These include, but are not limited to NFPA 20-2007, Standard for the Installation of Stationary Pumps for Fire Protection, NFPA 72-2007, National Fire Alarm Code, NFPA 110-2005, Standard for Emergency and Standby Power Systems, and NFPA 780-2008, Standard for the Installation of Lightning Protection Systems.

Article 517 provides special requirements for grounding of equipment in certain health care facilities, particularly in patient care areas. Reasons for the extra or specialized requirements are given in the FPN following 517.11.

In a health care facility, it is difficult to prevent the occurrence of a conductive or capacitive path from the patient's body to some grounded object, because that path can be established accidentally or through instrumentation directly connected to the patient. Other electrically conductive surfaces that can make an additional contact with the patient, or instruments that can be connected to the patient, then become possible sources of currents that can traverse the patient's body. The hazard is increased as more apparatus is associated with the patient, and, therefore, more extensive precautions are needed. Control of electric shock hazards require the limitation of electric current in an electric circuit that involves the patient's body. This is accomplished by raising the resistance of the conductive circuit that includes the patient, or by insulating exposed surfaces that might become energized, in addition to reducing the potential difference that can appear between exposed conductive surfaces in the patient vicinity, or by combinations of these methods. A special problem is presented by the patient with an externalized direct conductive path to the heart muscle. The patient can be electrocuted at current levels so low that additional protection in the design of appliances, insulation of the catheter, and control of medical practice is required.[16]

Grounding in Patient Care Areas

Section 517.13(A) requires that "all branch circuits serving patient care areas …be provided with an effective ground-fault current path by installation in a metal raceway system, or cable armor…or sheath assembly. The metal raceway system, or metallic cable armor, or sheath assembly [must] qualify as an equipment grounding conductor in accordance with 250.118" (see figure 16-16).

Type AC, Type MC, and Type MI cables shall have an outer metal armor or sheath that is identified as an acceptable grounding return path.

Section 517.13(B) requires that in an area used for patient care, "the grounding terminals of all receptacles and all non-current-carrying conductive surfaces of fixed electric equipment likely to become energized that are subject to personal contact, operating at over 100 volts, shall be connected to an insulated copper equipment grounding conductor. The equipment grounding conductor shall be sized in accordance with Table 250.122 and installed in metal raceways or [metal-clad cables] with the branch-circuit conductors supplying these receptacles or fixed equipment."

Exception No. 1 provides that "metal faceplates shall be permitted to be connected to the equipment grounding conductor by means of a metal mounting screw(s) securing the faceplate to a grounded outlet box or grounded wiring device." [17]

Exception No. 2 provides that "luminaires (light fixtures) more than 2.3 m (7½ ft) above the floor and switches located outside of the patient care vicinity shall be permitted to be connected to an equipment grounding return path complying with 517.13(A)" [18] (see figure 16-17).

The term *patient care vicinity* is defined in 517.2 as "… in an area in which patients are normally cared for, the *patient care vicinity* is the space with surfaces likely to be contacted by the patient or an attendant who can touch the patient. Typically in a patient room, this encloses a space within the room not less than 1.8 m (6 ft) beyond the perimeter of the bed in its nominal location, and extending vertically not less than 2.3 m (7½ ft) above the floor." As can be seen, luminaires that

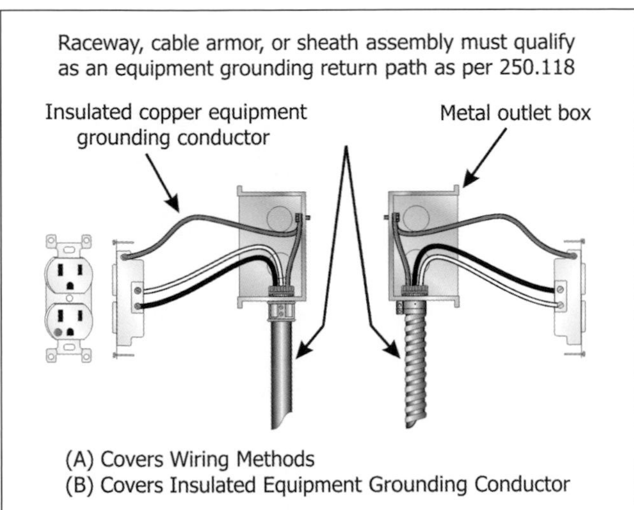
Raceway, cable armor, or sheath assembly must qualify as an equipment grounding return path as per 250.118
Insulated copper equipment grounding conductor
Metal outlet box
(A) Covers Wiring Methods
(B) Covers Insulated Equipment Grounding Conductor

Figure 16-16. Grounding of receptacles and fixed equipment in patient care areas

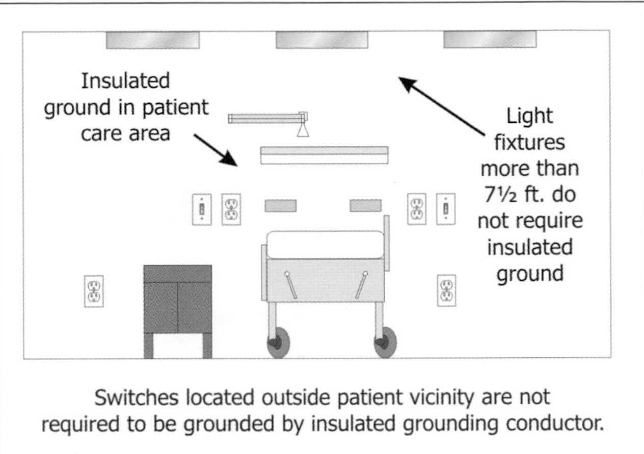
Insulated ground in patient care area
Light fixtures more than 7½ ft. do not require insulated ground
Switches located outside patient vicinity are not required to be grounded by insulated grounding conductor.

Figure 16-17. Grounding of receptacles and fixed electric equipment exception

are higher than 7½ ft from the floor do not have to be grounded by the insulated equipment grounding conductor otherwise required in the patient care area. These luminaires do have to be grounded by the wiring method used to supply the luminaires. Switches that are located either inside or outside the patient room but outside the patient vicinity are also not required to be grounded by the insulated equipment grounding conductor.

Both a primary and secondary, or redundant, means of grounding are required in patient care areas of health care facilities [see 517.13]. The metal raceway or outer metal jacket of a listed cable that is suitable for grounding provides the primary means of equipment grounding. An insulated copper equipment grounding conductor (the secondary, redundant equipment grounding means) is required to also be installed or be a part of listed cable for grounding receptacles and non-current-carrying metal portions of fixed electric equipment in the patient care areas that are subject to personal contact. The basic concept or idea is that the *Code* requires two grounding paths to be installed for electrical equipment in patient care areas so it is assured that at least one positive return path for fault current is always present if either should fail. This is often referred to as *redundant grounding* for the patient care area. The concept of this redundancy should be looked at from the standpoint that one is required to install two suitable grounding return paths, to always be assured to have at least one.

The *Code* is specific in requiring two separate types of grounding return paths in accordance with 250.118, one being the raceway or cable enclosing the circuit conductors, and the other, an insulated copper equipment grounding conductor. Installing two insulated copper grounding conductors in a raceway that does not qualify as an equipment grounding conductor does not meet the letter or intent of the *NEC* for branch-circuit grounding requirements in patient care locations. The wiring methods that are acceptable for providing the primary equipment grounding fault return path are the metal raceways or cable assembly included in 250.118 (see figure 16-18).

Of course, where these wiring methods are installed, it is critical that good work practices be followed to ensure that the wiring method provides a reliable, continuous and low-impedance path for fault current. Workmanship is important.

Section 517.13(A) places additional restrictions on the use of cable wiring methods that are used for wiring in patient care areas. The outer jacket of the cable must qualify as an equipment grounding conductor path in accordance with 250.118.

Types AC, MC and MI cables are required in 517.13(A) to have an outer armor that itself qualifies "as an equipment grounding conductor in accordance with 250.118 (see figure 16-19)." These requirements do not permit the combination of the cable outer jacket and the internal equipment grounding conductor to serve as providing the path for fault current required by this section.

Generally, the outer jacket of standard Type MC cable of the spiral-interlocking armor type is

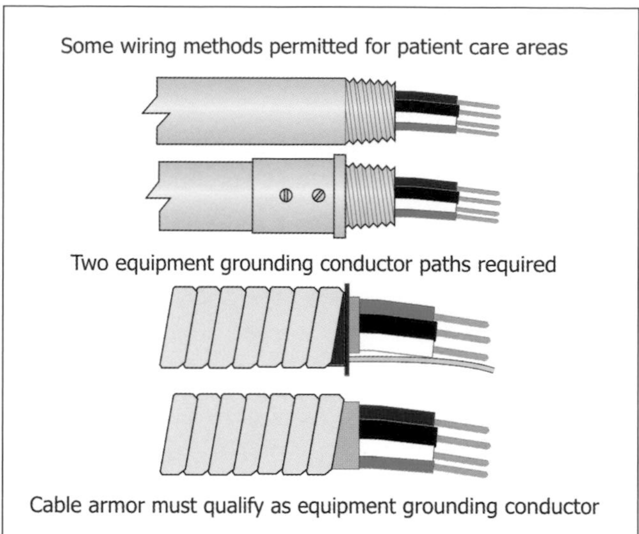

Figure 16-18. Wiring methods permitted for patient care areas

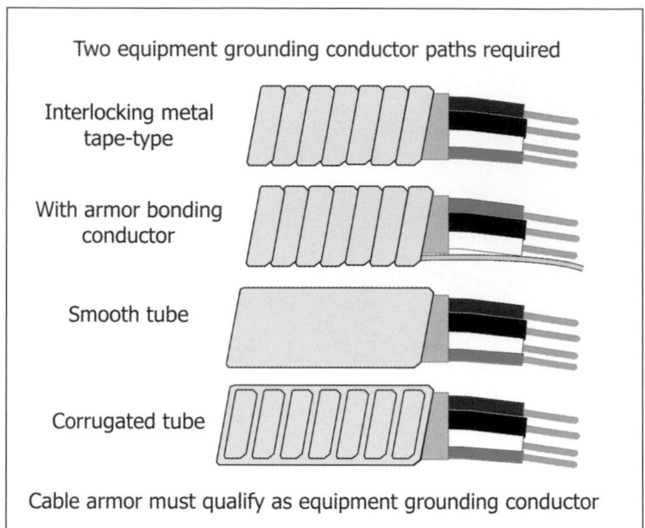

Figure 16-19. MC cable must provide two equipment grounding conductor paths where used for branch circuits in patient care areas

not suitable as an equipment grounding conductor and cannot be used in a patient care area of a health care facility that must comply with 517.13. Type MC cable with a smooth or corrugated continuous tube is suitable for these patient care areas if it contains a green-insulated equipment grounding conductor with a yellow stripe or surface marking or both to indicate that it is an additional equipment (or isolated) grounding conductor. This acceptable type of MC cable would not contain a supplemental bare or unstriped green-insulated grounding conductor. A supplemental bare or unstriped green-insulated equipment grounding conductor in Type MC cable of the smooth or corrugated continuous type is an indication that the outer armor by itself does not qualify as an equipment ground. There is also a new type of MC cable of the interlocking metal tape-type construction that incorporates an armor and bare bonding conductor combination that does qualify as an equipment grounding conductor. This type of MC cable is readily identified as providing the two needed ground-fault current paths required by 517.13 (see photo 16-1).

By its construction and listing by a qualified electrical testing laboratory, Type AC cable with a spiral interlocking metal jacket and a bonding strip or wire in intimate contact with the outer jacket is suitable as an equipment grounding conductor (see figure 16-20). Type AC cable with this outer cable armor assembly plus the internal insulated equipment grounding conductor constitutes two ground return paths and qualifies the cable for use in wiring in patient care areas.

Copper, Type MI cable, by nature of its construction, has an outer jacket that is suitable as an equipment grounding conductor. This cable is manufactured with bare conductors that are physically separated from each other by the mineral compound inside the metal sheath. The individual conductors are insulated by slip-on insulation at the time the cable is terminated.

While recognized in 250.118 as equipment grounding conductors, severe restrictions are placed on the use of flexible metal conduit and liquidtight flexible metal conduit due to their limited capabilities of carrying current and functioning as a ground-fault return path.

Additional restrictions are placed on the wiring of the emergency system in hospitals. While these

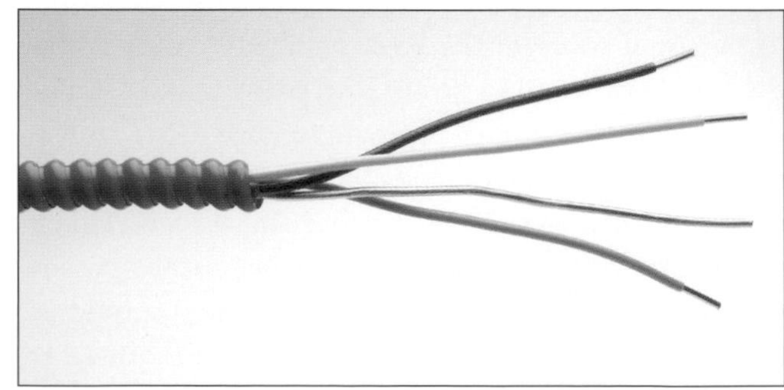

Photo 16-1. MC "AP" Cable providing two grounding paths
Courtesy of Southwire Company

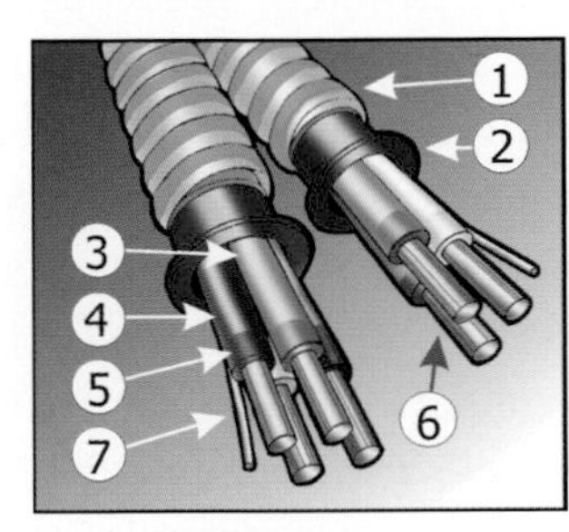

Figure 16-20. AC cable with insulated equipment grounding conductor meets the requirements of 517.13

requirements relate to mechanical protection of the circuits and not directly to grounding, the installer should be aware of these rules. [See 517.30(B)(2) for designation of these branches, 517.2 for definitions, and 517.30(C)(3) for wiring requirements].

Testing of Grounding System in Patient Care Areas

NFPA 99, *The Standard for Health Care Facilities,* requires that the integrity of the grounding path provided by the wiring method in patient care areas of health care facilities be tested before acceptance of the initial installation and after any alterations or replacement of the electrical system is made [see NFPA 99: 4.3.3.1 and figure 16-21]. Both voltage and impedance measurements must be made of exposed conductive surfaces including the grounding contacts of receptacles in the patient-care vicinity.

Excluded from the testing requirements are small, wall-mounted conductive surfaces not likely to become energized, such as surface-mounted towel and soap dispensers, mirrors and the like.

Also exempt are large, metal conductive surfaces not likely to become energized, such as window frames, door frames, and drains. A note following this section indicates that the grounding system, including both metallic raceway and equipment grounding conductor, is to be tested as an integral system. Removing equipment grounds from receptacles or equipment is not required or recommended.

The voltage and impedance measurements must be measured against a reference point. The reference point is permitted to be one of the following: (1) a reference grounding point; (2) a grounding point in the room under test that is electrically remote from the equipment under test, such as a metal water pipe; and (3) the grounding contact of a receptacle that is powered from a different branch circuit from the receptacle being tested.

The criteria for new construction that must be met are as follows: (1) the voltage limit is 20 mV; (2) the impedance limit is 0.1 ohm; and (3) for quiet ground systems the impedance limit is 0.2 ohm.[19] [NFPA 99: 4.3.3.1.6].

Receptacle Testing in Patient Care Areas

In wet locations, fixed receptacles, equipment connected by cord and plug, and fixed electrical equipment must be tested: (1) when first installed, (2) where there is evidence of damage, (3) after any repairs, or (4) at intervals not exceeding 6 months.

NFPA 99, Section 4.3.3.1 gives requirements for the testing interval for receptacles in patient care areas. It is required that:

- Testing be performed after initial installation, replacement, or servicing of the device.
- Additional testing be performed at intervals defined by documented performance data.

Receptacles that are not listed as hospital-grade must be tested at intervals not exceeding 12 months.

Record keeping requirements are found in NFPA 99: 4.3.4.2. A record must be maintained of the tests required by the chapter and associated repairs or modification. At a minimum, this record must contain the date, the rooms tested, and an indication of which items have met or have failed to meet the performance requirements of this chapter.

Where an isolated power system is installed, NFPA 99: 4.3.3.3.2 requires that a permanent record be kept of the results of each of the tests.

Tests that are required to be performed for each receptacle include:

- Visual inspection to confirm physical integrity
- Continuity of the grounding circuit
- Correct polarity
- Retention force (except locking type) to be not less than 115 g (4 oz) [20]

Bonding of Panelboards in Patient Care Areas

Section 517.14 requires that the "equipment grounding terminal busses of the normal and essential branch-circuit panelboards serving the same individual patient vicinity...be connected together with an insulated

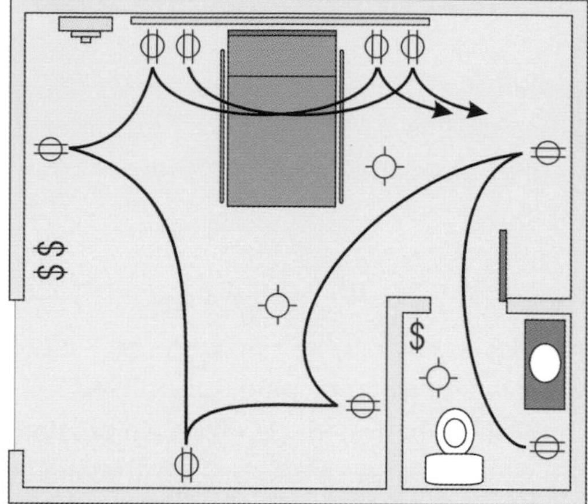

Voltage and impedance measurements from reference point:
1. reference grounding point
2. remote grounding point in room
3. grounding contact on receptacle from different branch circuit.
Criteria to be met for new construction:
1. voltage max. 20 mV
2. impedance max. 0.1 ohm
3. quiet ground max 0.2 ohm.

Figure 16-21. Testing of grounding system in patient care areas

continuous copper conductor not smaller than 10 AWG" (see figure 16-22).

"Where two or more panelboards serving the same individual patient care vicinity are served from separate transfer switches on the emergency system, the equipment grounding terminal buses of those panelboards shall be connected together with an insulated continuous copper conductor not smaller than 10 AWG." This conductor shall be continuous from panel to panel but "shall be permitted to be broken in order to terminate on the equipment terminal bus in each panelboard." [21]

Typically, the circuits supplying receptacles serving the patient bed locations are supplied from different separately derived systems. This bonding is intended to ensure that little if any potential difference exists between exposed non-current-carrying metal portions of equipment in the patient-care area. It is not required that this bonding conductor be installed in a raceway between panelboards where it is protected by normal building construction such as studs and gypsum board.

This bonding is provided to equalize the potential of electrical equipment in the patient vicinity that might be supplied from two different panelboards. This is especially important should a line-to-ground fault occur in one piece of equipment while another piece of equipment in the same patient vicinity is in a stable condition. There might be a slight voltage rise on the faulted circuit until the overcurrent device on the line side of the fault clears. The 10 AWG bonding conductor helps keep the electrical equipment in the patient care vicinity at the same potential thus reducing the shock hazard.

Patient Equipment Grounding Point

The patient equipment grounding point is an option for improving electrical safety in patient care areas. Though an optional feature [see 517.19(C)], a patient equipment grounding point continues to be provided by many hospital equipment manufacturers and is specified by consulting engineers. Where installed, it consists of a grounding terminal bus and can include one or more grounding jacks. "An equipment bonding jumper not smaller than 10 AWG copper shall be used to connect the grounding terminals of all grounding-type receptacles to the patient equipment grounding point." [22] Again, this is done to reduce any potential difference between conductive surfaces in the patient care vicinity (see photo 16-2).

Panelboard Grounding, Critical Care Areas

Section 517.19(D) requires that where metal raceways or Type MC or MI cables are used for feeders, grounding

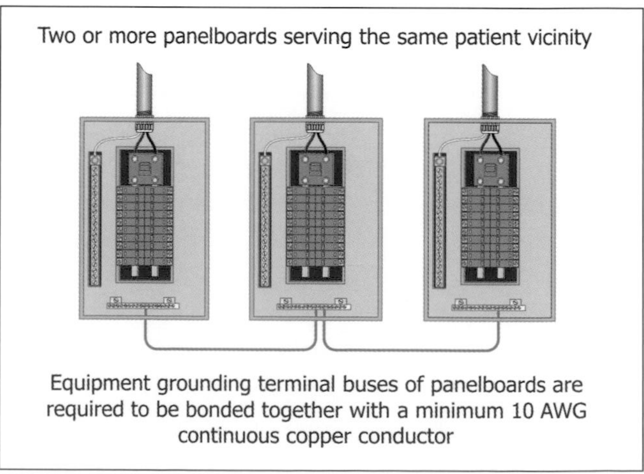

Figure 16-22. Equipment grounding terminals of panelboards serving the same patient care vicinity required to be bonded together. Also where supplied by separate transfer switches, which is usually the case.

and bonding of panelboards and switchboards serving critical care areas must be assured (see figure 16-23). Acceptable methods include as follows:

"(1) A grounding bushing and a continuous copper bonding jumper, sized in accordance with 250.122, with the bonding jumper connected to the junction enclosure or the ground bus of the panel

"(2) Connection of feeder raceways or Type MC or MI cable to threaded hubs or bosses on terminating enclosures

Note that Type MC or MI cable used in panelboard feeders that supply branch circuits in critical care areas must qualify as an equipment grounding conductor in accordance with 250.118 [517.19(D)].

"(3) Other approved devices such as bonding-type locknuts or bushings." [23]

Isolated Power Systems Permitted as Protection Technique

Isolated power systems are ungrounded systems, and "shall be permitted to be used for critical care areas, and, if used, the isolated power system equipment shall be listed [for the purpose and the] system...designed and installed [so that it meets the provisions of and is] in accordance with 517.160." [24] The exception to 517.19 states that "the audible and

Photo 16-2. Patient equipment grounding point equipment

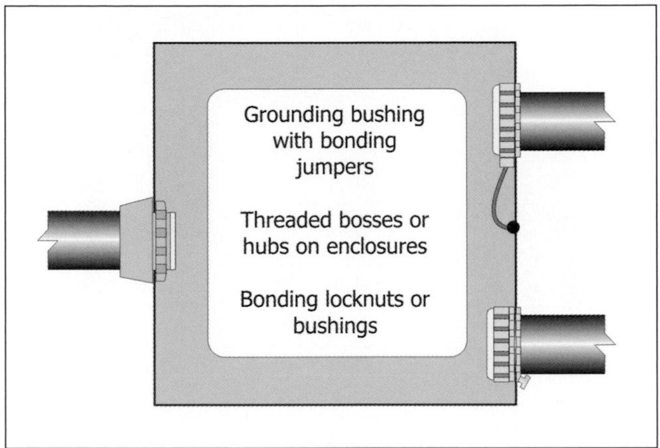

Figure 16-23. Panelboard bonding, critical care feeders installed in metallic raceways

visual indicators of the line isolation monitor shall be permitted to be located at the nursing station for the area being served."

An isolated power system usually consists of a transformer that has an ungrounded secondary. A protective shield is placed between the primary and secondary windings. A line isolation monitor with visual and audible alarms is provided to warn of excessive line-to-ground leakage as well as to indicate that a line-to-ground fault has occurred. The primary purpose of an isolated power system is to allow equipment to function with the first line-to-ground fault without opening an overcurrent device. Faulty equipment can be repaired or replaced at the earliest opportunity but will remain in service until a second line-to-ground fault occurs which then becomes a line-to-line fault. This would cause the overcurrent device to operate and take the faulted equipment off the line. Reference grounding terminal bars are included in isolated power systems enclosures for termination of equipment grounding conductors of the branch circuits supplied by the isolated power system. These systems must be tested prior to being put into service and periodically thereafter (see photo 16-3).

There are specific requirements regarding the types of insulation on the branch-circuit conductors, length of the branch circuits, and color code for these conductors. Wire pulling compound is not permitted to be used for installation of these circuit conductors in raceways as it can have an adverse affect on the dielectric characteristics of the conductors and the system operation. It is important to follow the manufacturer's installation instructions

Photo 16-3. Isolated power systems equipment

to ensure compliance with the equipment listing requirements of 110.3(B) in addition to the requirements of 517.160. Other specific requirements for isolated power systems are in 517.160.

Special Purpose Receptacle Grounding

Section 517.19(F) states that "where an isolated ungrounded power source is used and limits the first-fault current to a low magnitude, the grounding conductor associated with the secondary circuit shall be permitted to be run outside of the enclosure of the power conductors in the same circuit."

The FPN reads, "Although it is permitted to run the grounding conductor outside of the conduit, it is safer to run it with the power conductors to provide better protection in case of a second ground fault." [25]

NEC 517.21 provides that "ground-fault circuit-interrupter protection for personnel shall not be required for receptacles installed in those critical care areas where the toilet and basin are installed within the patient room."

Swimming Pools, Fountains, and Similar Installations

It is important to understand the organization of Article 680 for proper application of its requirements. Article 680 is divided in several parts, namely Part I through Part VII. Part I includes general requirements and applies to all equipment and installations covered in Article 680. Part II covers permanently installed pools (see photo 16-4). Other parts cover the equipment or installation identified in the title of the part and do not necessarily apply to all equipment included in Article 680. For example, the rules in Part II for grounding panelboards do not apply to the wiring of a hydromassage bathtub covered in Part VII (see photo 16-5).

However, be aware of the requirements for wiring spas and hot tubs. Section 680.40 reads, "electrical installations at spas or hot tubs shall comply with the provisions of Parts I and IV of this article." Generally, a spa or hot tub installed outdoors must be in accordance with 680.42 and is required to meet all the rules for swimming pools, while a spa or hot tub installed indoors is required to meet the rules in 680.43, which are different from those for swimming pools.

"680.1 Scope. The provisions of this article apply to the construction and installation of electrical wiring for and equipment in or adjacent to all swimming, wading, therapeutic, and decorative pools; fountains; hot tubs; spas; and hydromassage bath tubs, whether permanently installed or storable, and to any metallic auxiliary equipment, such as pumps, filters, and similar equipment. The term body of water used throughout Part I applies to all bodies of water covered in this scope unless otherwise amended." [26]

Grounding Requirements

The following swimming pool and outdoor spa and hot tub equipment is required to be grounded:

"1) Through-wall lighting assemblies and underwater luminaires, other than those low-voltage lighting products listed for the application without a grounding conductor

"2) All electric equipment located within 1.5 m (5 ft) of the inside wall of the specified body of water

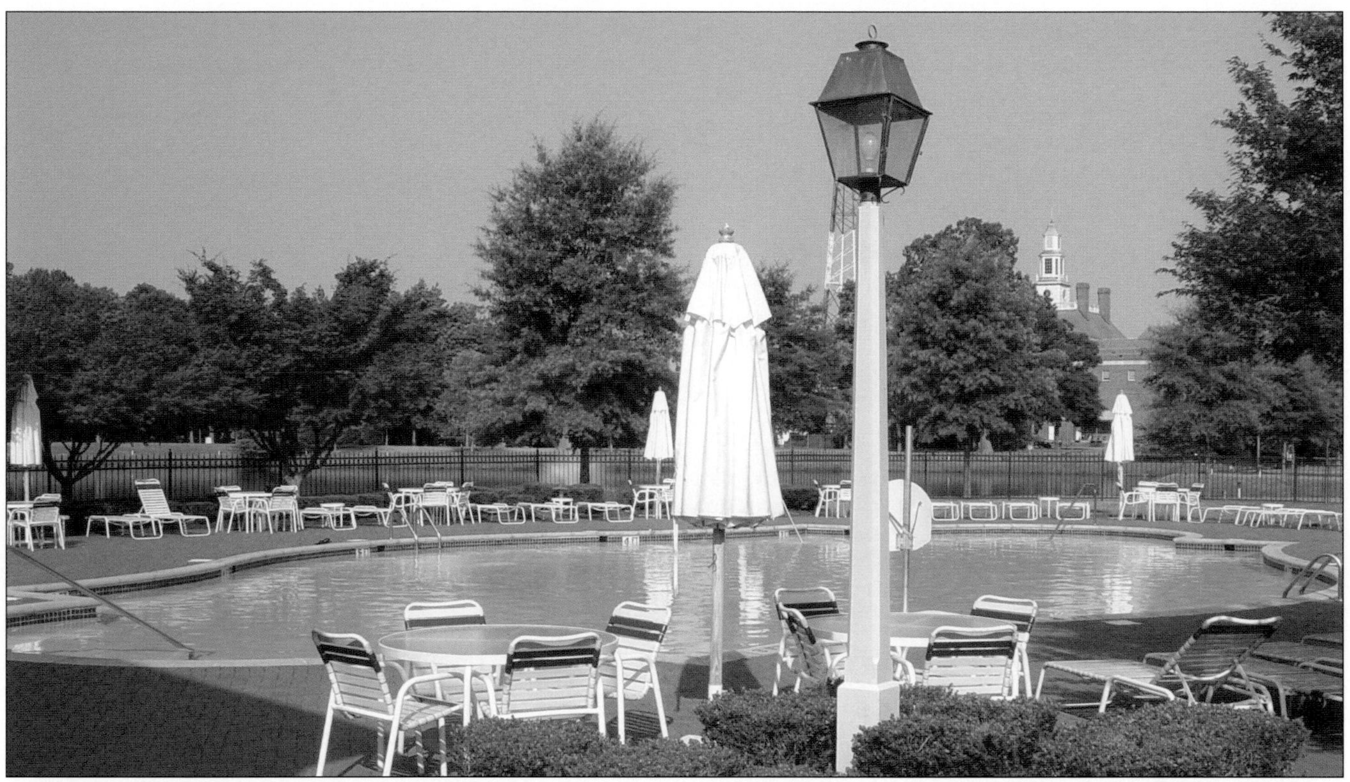

Photo 16-4. Permanently installed swimming pool

"3) All electric equipment associated with the recirculating system of the specified body of water

"4) Junction boxes

"5) Transformer enclosures

"6) Ground-fault circuit interrupters

"7) Panelboards that are not part of the service equipment and that supply any electric equipment associated with the specified body of water." [31]

Cord-Connected Equipment

Where fixed or stationary equipment is connected with a flexible cord and plug to facilitate removal or disconnection for maintenance, repair, or storage as provided in 680.7, the equipment grounding conductor must "be connected to a fixed metal part of the assembly. The removable part must be mounted on or bonded to the fixed metal part" [see 680.7(C)].

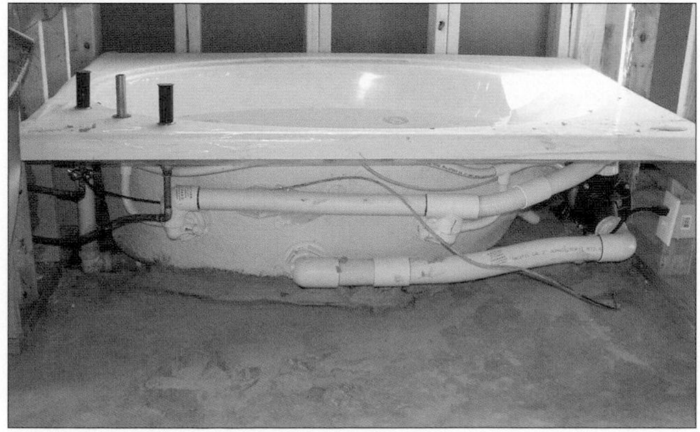

Photo 16-5. Hydromassage Bathtub

All other electrical equipment must be grounded in accordance with Article 250 and be connected by an approved wiring method covered in *NEC* chapter 3.

Methods of Grounding

Section 680.23(F)(2) reiterates that "through-wall lighting assemblies, wet-niche, dry-niche, and no-niche luminaires shall be connected to an insulated copper equipment grounding conductor installed with the circuit conductors. The equipment grounding conductor shall be installed without joint or splice except as permitted in (F)(2)(a) and (F)(2)(b). The equipment grounding conductor shall be sized in accordance with Table 250.122 but shall not be smaller than 12 AWG." The equipment grounding conductor installed "between the wiring chamber of the secondary winding of a transformer and a junction box [must] be sized in accordance with the overcurrent device in this circuit" (see Table 250.122). The equipment grounding conductor must be an insulated copper conductor and generally be installed with the circuit conductors in rigid metal conduit, intermediate metal conduit, liquidtight flexible nonmetallic conduit, or rigid nonmetallic conduit.

Electrical metallic tubing is permitted to be used for these conductors where installed on or within buildings. Electrical nonmetallic tubing is permitted to be used to protect circuit conductors where installed within buildings as ENT is generally limited to installation inside buildings as it is not suitable for direct-sunlight exposure [see 680.23(F)(1)].

Where connecting to transformers for pool luminaires, liquidtight flexible metal conduit or liquidtight flexible nonmetallic conduit Type B (LFNC-B) is permitted to be used. Any one length is limited to not more than 1.8 m (6 ft) feet and a total of not more than 3.0 m (10 ft) is permitted. Liquidtight flexible nonmetallic conduit, Type B (LFNC-B), is permitted in lengths longer than 1.8 m (6 ft).

"The junction box, transformer enclosure, or other enclosure in the supply circuit to a wet-niche or no-niche luminaire and the field-wiring chamber of a dry-niche luminaire shall be connected to the equipment grounding terminal of the panelboard. This terminal shall be directly connected to the panelboard enclosure." [32]

"a) If more than one underwater luminaire is supplied by the same branch circuit, the equipment grounding conductor, installed between the junction boxes, transformer enclosures, or other enclosures in the supply circuit to wet-niche luminaires, or between the field-wiring compartments of dry-niche luminaires, shall be permitted to be terminated on grounding terminals.

"b) If the underwater luminaire is supplied from a transformer, ground-fault circuit interrupter, clock-operated switch, or a manual snap switch that is located between the panelboard and a junction box connected to the conduit that extends directly to the underwater luminaire, the equipment grounding conductor shall be permitted to terminate on grounding terminals on the transformer, ground-fault circuit interrupter, clock-operated switch enclosure, or an outlet box used to enclose a snap switch." [33]

Section 680.23(B)(3) requires "wet-niche luminaires that are supplied by a flexible cord or cable shall have all their exposed non-current-carrying metal parts grounded by an insulated copper equipment grounding conductor that is an integral part of the cord or cable. This grounding conductor shall be connected to a grounding terminal in the supply junction box, transformer enclosure, or other enclosure. The grounding conductor shall not be smaller than the supply conductors and not smaller than 16 AWG." [34]

Pool-Associated Motors

All pool-associated motors must be connected to an equipment grounding conductor sized in accordance with Table 250.122 [see 680.21(A)]. This equipment grounding conductor must be a copper conductor and cannot be smaller than 12 AWG. "The branch circuits for pool-associated motors shall be installed in rigid metal conduit, intermediate metal conduit, rigid polyvinyl chloride conduit, reinforced thermosetting resin conduit, or Type MC cable listed for the location." [35] This Type MC cable will have an overall outer jacket of PVC material and an insulated equipment grounding conductor not smaller than 12 AWG. Electrical metallic tubing is permitted to be used to protect conductors where it is installed on or within buildings.

Where flexible connections are necessary at or adjacent to the motor, liquidtight flexible metal or liquidtight flexible nonmetallic conduit with approved fittings are permitted.

"In the interior of dwelling units, or in the interior of accessory buildings associated with a dwelling unit, any of the wiring methods recognized in Chapter 3 of this *Code* that comply with the provisions of this section shall be permitted. Where run in a raceway, the equipment grounding conductor shall be insulated. Where run in a cable assembly, the equipment grounding conductor shall be permitted to be uninsulated, but it shall be enclosed within the outer sheath of the cable assembly." [36]

Permanently installed pools are permitted to be provided with listed cord-and plug connected pool pumps that are protected by a system of double insulation. They are to be provided with a means for grounding only the internal and nonaccessible non-current-carrying metal parts of the pump [680.21(A)(5)].

Panelboard Grounding

A panelboard and, where installed, a disconnecting means that are not part of the service equipment or source of a separately derived system are required to have an equipment grounding conductor installed between its equipment grounding terminal and the equipment grounding terminal located at the service equipment or source of a separately derived system. This conductor must be sized in accordance with Table 250.122 and be an insulated conductor of copper, aluminum or copper-clad aluminum and can never be smaller than 12 AWG. This conductor must generally be installed with the feeder conductors in a rigid metal conduit, intermediate metal conduit, liquidtight flexible metal conduit, or rigid nonmetallic conduit.

Electrical metallic tubing is permitted to be used to protect conductors where installed on or within buildings in accordance with Article 358. Electrical nonmetallic tubing is permitted to be used to protect conductors where installed within buildings in accordance with Article 362. The equipment grounding conductor must be connected to an equipment grounding terminal of the panelboard and, where installed, to the enclosure for a disconnecting means.

Panelboards from Separately Derived Systems

There are several ways of grounding swimming pool panelboards that are supplied from a separately derived system. It is important to follow the rules for grounding the separately derived system as given in 250.30(A) and as covered extensively in chapter twelve.

Section 680.25(B) requires that the equipment grounding conductor from a separately derived system be sized in accordance with Table 250.66 based on the derived phase conductors but not smaller than an 8 AWG.

Existing Panelboards

The exception to 680.25(A) states that "in existing feeder between an existing remote panelboard and service equipment shall be permitted to run in flexible metal conduit or an approved cable assembly that includes an equipment grounding conductor within its outer sheath. The equipment grounding conductor shall comply with 250.24(A)(5)." Section 250.24(A)(5) generally does not permit a connection of grounding conductor on the load side of the service disconnecting means.

The *NEC* does not define what is meant by the word *existing*. Often, inspection authorities consider an installation to be existing after it has had final approval or after a certificate of occupancy has been issued.

Swimming Pool Panelboards in Other Buildings

Grounding at the disconnecting means for remote buildings is covered in 250.32 and in chapter thirteen where the methods for complying with the Code rules are given. Normally, one can choose between grounding the neutral conductor again at the building and not taking the equipment grounding conductor to the building, and taking the equipment grounding conductor to the building, grounding it there and not grounding the neutral but floating (isolating it from ground) it there. Specific limitations are given for re-grounding the grounded conductor (may be a neutral) at the remote building.

"2) Grounded Conductor. Where (1) an equipment grounding conductor is not run with the supply to the building or structure, (2) there are no continuous metallic paths bonded to the grounding system in both buildings or structures involved, and (3) ground-fault protection of equipment has not been installed on the common ac service, the grounded circuit conductor run with the supply to the building or structure shall be connected to the building or structure disconnecting means and to the grounding electrode(s) and shall be used for grounding or bonding of equipment, structures, or frames required to be grounded or

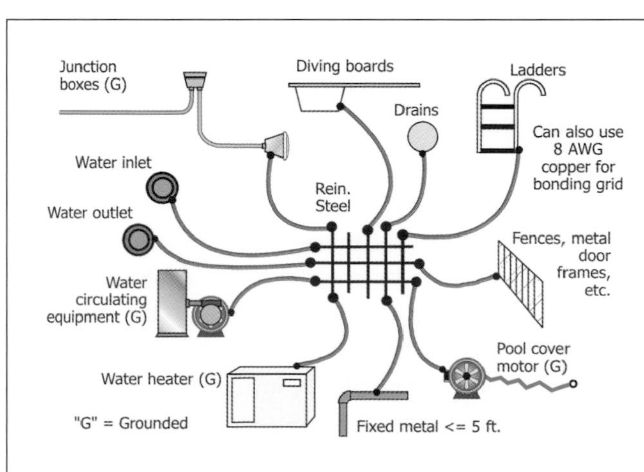

Figure 16-24. Equipotential bonding grid required

bonded."[37] Where an equipment grounding conductor is installed in accordance with 250.32(B)(1), and is included with a feeder (other than existing feeders which are covered in 680.25) to a separate building, it must be an insulated conductor in accordance with 680.25(B)(2).

Bonding Requirements (Equipotential Bonding)

Bonding requirements for pool areas are found in 680.26. Section 680.26(A) includes performance language relative to the purpose of bonding for these types of installations. The *Code* clarifies that the bonding required for the pool must "be installed to reduce voltage gradients in the pool area." [described in the article]. Swimming pools, spas and hot tubs present special shock hazards to people due to mixing water enriched with chemicals, people and electricity together. Electric shock hazards will be reduced to an acceptable level if the measures prescribed in Article 680 are carefully followed. One of these measures is the bonding together of conductive portions of the pool and metal parts of electrical equipment associated with the pool. The goal is to provide a means of equalizing the potential of all equipment and parts so there will be no current between parts. This is accomplished by connecting all the parts together by an adequatelysized and properly connected copper conductor. This concept is often referred to as equipotential bonding (see 680.26).

These requirements emphasize the purpose of bonding in a swimming pool, spa or hot tub area as that of equalizing the potential (voltage) between various parts of the pool. By keeping the potential difference as low as practicable, the shock hazard is reduced significantly.

Since several pieces of electrical equipment commonly used with pools, hot tubs or spas that are bonded together are required to also be grounded with an insulated equipment grounding conductor, an interconnection between the grounded (neutral) conductor and bonding grid exists. This interconnection is often remote from the pool and is not intended to play a part in equipotential bonding.

The *Code* requires the conductive parts identified in 680.26(B)(1) through (B)(7) to be bonded together (see figure 16-24). The bonding can be accomplished using an 8 AWG copper conductor, insulated, covered, or bare, or rigid metal conduit that is made of brass or other corrosion-resistant material. All connections of the AWG copper bonding conductor used in constructing the equipotential bonding grid are required to be accomplished using one of the methods specified in 250.8 [680.26(B)].

Section 680.26(B)(1) describes the differences between a conductive pool shell construction and pool shells that are nonconductive. This is important because bonding requirements are predicated under this differentiation.

Conductive and Nonconductive Pool Shells

Poured concrete, pneumatically applied concrete, or painted concrete block are considered conductive because of their water permeability and porosity. Of course, metal pool frames such as those for many aboveground pools are also conductive and are required to be bonded. Pools that are made of fiberglass (usually prefabricated) or made of vinyl liners are considered nonconductive and more specific bonding requirements apply.

Where structural reinforcing steel for a conductive shell pool is not encapsulated in a nonconductive compound (encapsulated), the steel is required to be bonded together using the usual steel tie wires or the equivalent. This section recognizes both steel reinforcing rods or bars and wire mesh construction methods for equipotential bonding. Where the structural reinforcing steel of a conductive pool shell is encapsulated (coated rebar), a copper conductor bonding grid meeting the requirements in 680.26(B)(1)(b) must be installed in accordance with the following criteria:

- It is required to consist of a minimum 8 AWG solid copper conductor bonded together at points of crossing.
- It shall conform to depth contour of the pool and the perimeter contour of the pool deck.
- It is required to be in a 300 mm (12 in.) x 300 mm (12 in.) pattern (mesh) with a tolerance up to 100 mm (4 in.).
- It is required to be secured within or under the pool no more than 150 mm (6 in.) from the contour of the pool shell [680.26(B)(1)(b)]

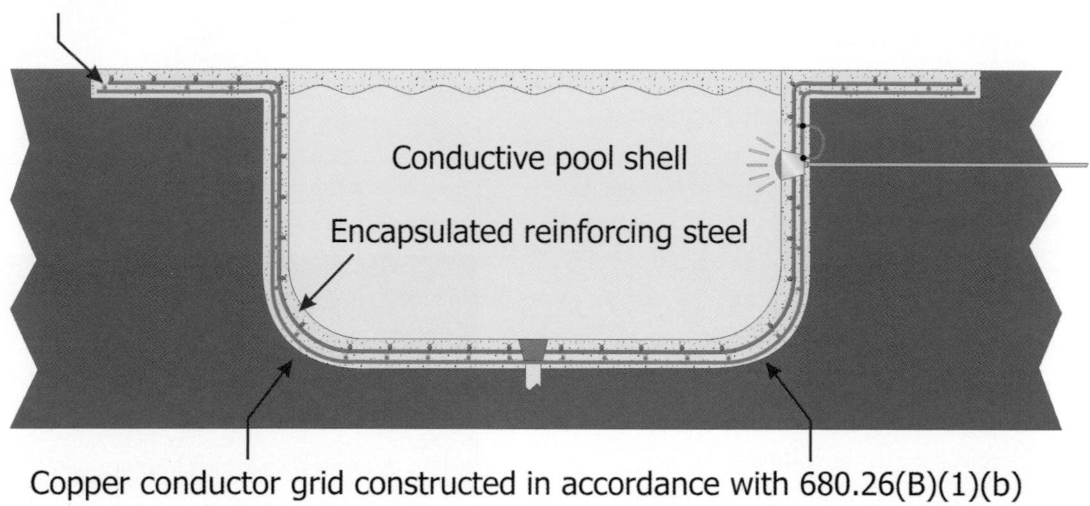

The copper conductor grid shall be constructed of minimum 8 AWG solid copper conductors bonded to each other at crossing points.
The grid shall conform to the contours of the pool shell and the pool deck.
The grid shall be arranged in a minimum 12" x 12" network (pattern) of conductors uniformly spaced with a tolerance of 100 mm (4 in.).
Conductive pool shell
Encapsulated reinforcing steel
Copper conductor grid constructed in accordance with 680.26(B)(1)(b)

Figure 16-25. Where encapsulated reinforcing steel is installed, an equipotential bonding grid structure must be installed.

Bonding Grid at Perimeter Surfaces

The surface area at the pool perimeter extending 1 m (3 ft) horizontally beyond the inside walls of the pool including, but not limited to, unpaved, conductive and poured concrete surfaces shall be included in the equipotential bonding grid requirements in accordance with either 680.26(B)(2)(a) or (2)(b). The installed bonding grid at the perimeter surface is required to be attached to the pool reinforcing steel or copper conductor at a minimum of four locations at uniformly spaced corners of the structure. Structural reinforcing steel, including wire mesh in the deck surface is required to be bonded in accordance with 680.26(B)(1)(a) and where no structural reinforcing steel is installed in the perimeter surface an alternate means is required to be installed and must meet all of the following criteria:

- It must be made up of at least a single 8 AWG solid copper conductor.
- The conductor is required to follow the contour of the pool perimeter surface.
- Any splices of this single conductor or splices of conductors connecting to this single conductor must be made with a listed means [680.26(B)(2)(b)].

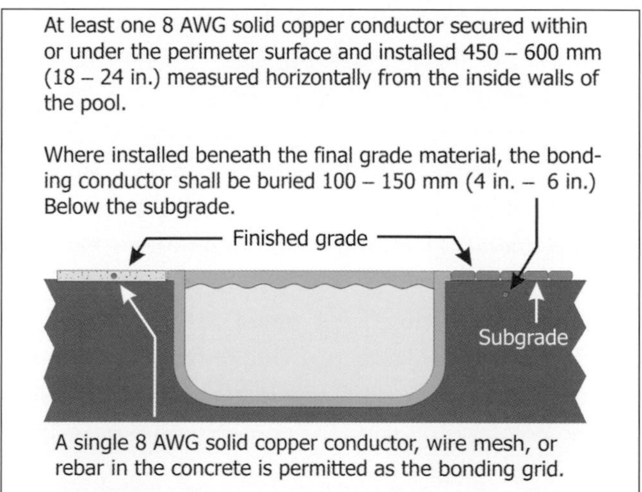

At least one 8 AWG solid copper conductor secured within or under the perimeter surface and installed 450 – 600 mm (18 – 24 in.) measured horizontally from the inside walls of the pool.
Where installed beneath the final grade material, the bonding conductor shall be buried 100 – 150 mm (4 in. – 6 in.) Below the subgrade.
Finished grade
Subgrade
A single 8 AWG solid copper conductor, wire mesh, or rebar in the concrete is permitted as the bonding grid.

Figure 16-26. Equipotential grid installation for perimeter surfaces (nonconductive pool shell shown).

The single conductor must be installed 450 – 600 mm (18 in. – 24 in.) from the inside walls of the pool and secured within or under the perimeter surface 100 mm – 150 mm (4 in. – 5 in.) below the subgrade. Note that the term *subgrade* is not specifically explained in

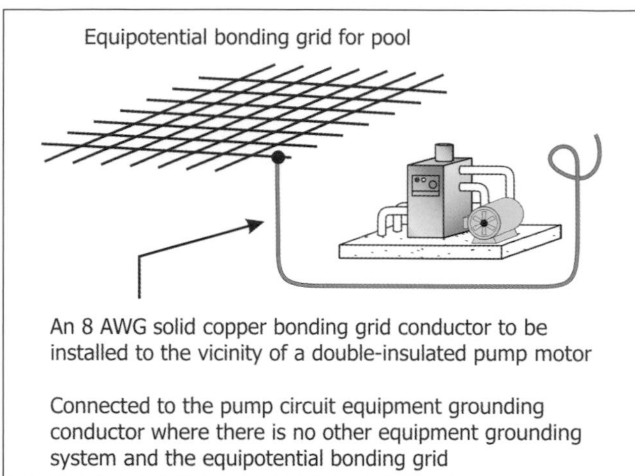

Equipotential bonding grid for pool
An 8 AWG solid copper bonding grid conductor to be installed to the vicinity of a double-insulated pump motor
Connected to the pump circuit equipment grounding conductor where there is no other equipment grounding system and the equipotential bonding grid

Figure 16-27. Bonding grid conductor routed to the double-insulated pool pump motor vicinity

Photo 16-6. A forming shell of wet-niche luminaire provides the bonding means for the pool water.

this section, but can be interpreted to mean "the earth surface that is located below a decorative surface such as pavers or decorative rock and so forth."

Section 680.26(B)(3) through (B)(7) provides additional items that are required to be bonded to the equipotential bonding gird as follows:

(3) Metallic Structural Components. All metallic parts of the pool structure, including the reinforcing metal not addressed in 680.26(B)(1)(a), shall be bonded. The usual steel tie wires shall be considered suitable for bonding the reinforcing steel (including wire mesh) together.

(4) Underwater Lighting. All forming shells and mounting brackets of no-niche luminaires shall be bonded, unless a listed low-voltage lighting system with nonmetallic forming shells not requiring bonding is used.

(5) Metal Fittings. All metal fittings within or attached to the pool structure shall be bonded. Isolated parts that are not over 100 mm (4 in.) in any dimension and do not penetrate into the pool structure more than 25 mm (1 in.) shall not require bonding.

(6) Electrical Equipment. Metal parts of electric equipment associated with the pool water circulating system, including pump motors and metal parts of equipment associated with pool covers, including electric motors, shall be bonded. Accessible metal parts of listed equipment incorporating an approved system of double insulation and providing a means for grounding internal nonaccessible, non-current-carrying metal parts shall not be bonded. Where a

Photo 16-7. A stainless steel handrail provides the bonding means for the pool water.

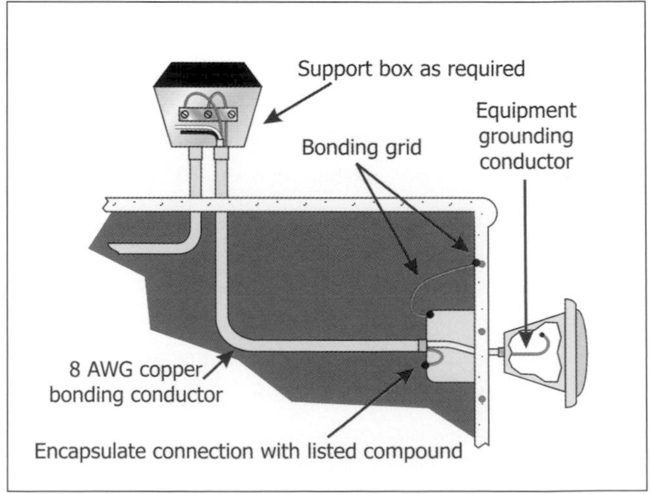

Figure 16-28. Bonding of forming shells of wet-niche luminaires (fixtures) required

double-insulated water-pump motor is installed under the provisions of this rule, a solid 8 AWG copper conductor that is of sufficient length to make a bonding connection to a replacement motor shall be extended from the bonding grid to an accessible point in the motor vicinity (see figure 16-27). Where there is no connection between the swimming pool bonding grid and the equipment grounding system for the premises, this bonding conductor shall be connected to the equipment grounding conductor of the motor circuit. Pool heating equipment rated more than 50 amperes are required to be bonded to the gird, but also must meet any specific grounding and bonding instructions that are provided by the manufacturer. In this case, only specifically designated metal parts of such equipment are required to be bonded into the equipotential bonding grid [680.26(B)(6)].

(7) Metal Wiring Methods and Equipment. Metal-sheathed cables and raceways, metal piping, and all fixed metal parts that are within the following distances of the pool except those separated from the pool by a permanent barrier shall be bonded:

1. Within 1.5 m (5 ft) horizontally of the inside walls of the pool
2. Within 3.7 m (12 ft) measured vertically above the maximum water level of the pool, or any observation stands, towers, or platforms, or any diving structures [680.26(B)(7) Exceptions 1, 2, and 3].

Section 680.26(C) requires pool water to be bonded to the equipotential bonding grid. This bonding can be accomplished by conductive parts that are in contact with the pool water such as a wet-niche luminaire forming shell or stainless steel handrails or ladders in contact with the water (see photos 16-6 and 16-7). The minimum surface area of the conductive surface in contact with the pool water cannot be smaller than 5806 mm² (9 in.²) [680.26(C)]. Where a nonconductive pool shell is installed, such as a prefabricated fiberglass pool, and no conductive parts are in contact with the pool water to bond the water to the grid, provisions must be made to bond the pool water to the gird.

Bonding of Wet-Niche Luminaires

Where mounted in a pool or fountain structure, a wet-niche luminaire must be installed in a forming shell that is designed to support a wet-niche luminaire

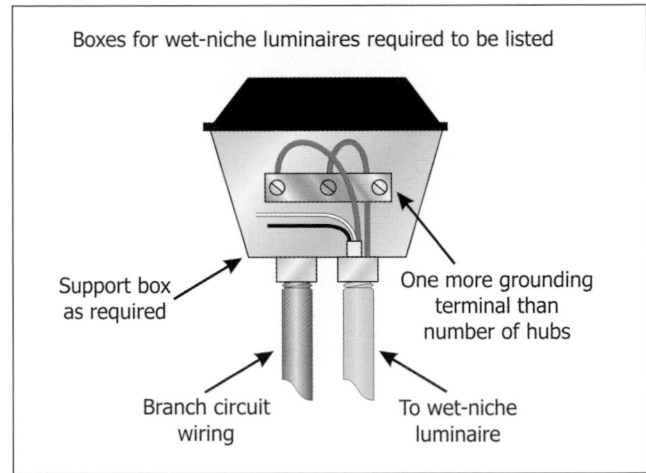

Figure 16-29. Junction boxes for wet-niche luminaires are required to be listed.

assembly. The luminaire unit will be completely surrounded by water [see definition of wet-niche luminaire in Article 680 and 680.23(B)(1)].

A forming shell must be installed for the mounting of all wet-niche underwater luminaires. The forming shell must also be equipped with provisions for threaded conduit entries.

Wiring methods permitted to be used to connect the forming shell to a suitable junction box or other permitted enclosure include rigid metal conduit, intermediate metal conduit, liquidtight flexible nonmetallic conduit or rigid PVC conduit [see 680.23(B)(2)].

Rigid metal or intermediate metal conduit used to connect the wet-niche fixture housing must be made of brass or other approved corrosion resistant metal, such as stainless steel.

Where rigid PVC or liquidtight flexible non-metallic conduit is used, an 8 AWG insulated copper conductor must be installed along with the pool flexible cord assembly so that it can be terminated on a suitable lug in the forming shell, junction box, transformer enclosure or ground-fault circuit-interrupter enclosure (see figures 16-28). This conductor is referred to as a bonding jumper in this section and the sizing must be 8 AWG copper minimum. This conductor performs as a bonding jumper to connect non-current-carrying electrical equipment together. Where a metallic method is used from the junction box to the forming shell, the bonding of the two enclosures is accomplished through the metal conduit. An equipment grounding conductor is included in the cord assembly of the wet-niche luminaire for returning any line-to-ground fault current back to the source (see photo 16-28).

At the point of termination within the forming shell, the 8 AWG bonding jumper must be covered with, or encapsulated in, a listed potting compound [680.23(B)(2)(b)]. This compound provides protection from deteriorating effects often caused by the pool water. Where a listed potting compound is not used to encapsulate the 8 AWG bonding bonding jumper inside the forming shell, brass or stainless steel conduit must be used between the pool junction box and the forming shell to eliminate the need for the bonding jumper.

Where in contact with the pool water, metal parts of the luminaire and forming shell must also be of brass or other approved corrosion-resistant metal. Forming shells used with nonmetallic conduit systems other than those that are part of a listed low-voltage lighting system not requiring grounding, include provisions for the termination of an 8 AWG copper conductor [see 680.23(B)(1)]. The end of the flexible cord jacket and flexible cord conductor terminations located within a wet-niche fixture must be covered with, or encapsulated in, a suitable potting compound. This will help to prevent the siphoning of water into the luminaire through the cord jacket or its contained conductors. This requirement is met by the manufacturer of the wet-niche luminaire. In addition, an equipment grounding conductor connection within a wet-niche fixture must be similarly treated to protect it from the deteriorating effect of pool water in the event of water entry into the luminaire.

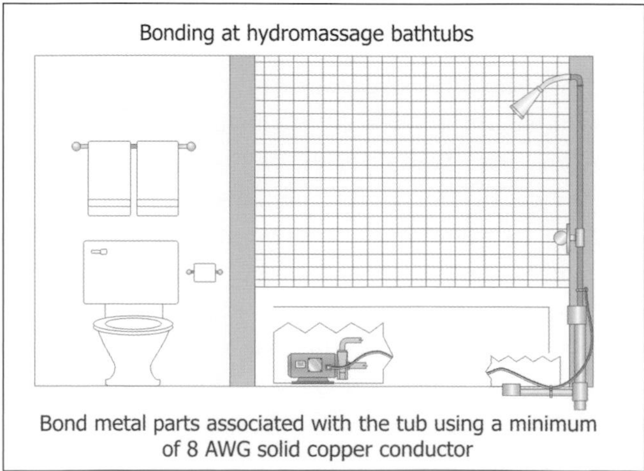

Bonding at hydromassage bathtubs
Bond metal parts associated with the tub using a minimum of 8 AWG solid copper conductor

Figure 16-30. Bonding at hydromassage bathtubs

The luminaire must be bonded to and secured to the forming shell by a positive locking device that will ensure a low resistance contact. A special tool is required to remove the luminaire from the forming shell. Bonding is not required for luminaires listed for the application, having no non-current-carrying metal parts.

Underwater wet-niche luminaires include markings that will indicate the proper housing or housings with which they are to be used, and the fixture housings are marked to indicate the fixture or fixtures with which the housings are to be used. These luminaires are provided with a factory-installed permanently-attached flexible cord that extends at least 3.7 m (12 ft) outside the luminaire enclosure. This will permit removal of the luminaire from the forming shell so that it can be lifted onto the pool or spa deck for servicing without lowering the water level or disconnecting the fixture from the branch-circuit conductors.

Luminaires with longer cords are available for installation where the junction box or splice enclosure is so located that a 12-foot-long cord will not permit the fixture to be removed from the forming shell and placed on the deck for servicing. To avoid possible cord damage, any cord length in excess of that necessary for servicing should be trimmed from the supply end rather than stored in the forming shell.

Listed Junction Boxes and Enclosures

Special requirements are contained in 680.24(A) for "a junction box connected to a conduit that extends directly to a forming shell or mounting bracket for a no-niche luminaire...

"1) Construction. The junction box shall be listed as a swimming pool junction box and shall comply with the following conditions:

"1) Be equipped with threaded entries or hubs or a nonmetallic hub

"2) Be comprised of copper, brass, suitable plastic, or other approved corrosion-resistant material

"3) Be provided with electrical continuity between every connected metal conduit and the grounding terminals by means of copper, brass, or other approved corrosion-resistant metal that is integral with the box." [29]

680.24(D) requires that junction boxes have a number of grounding terminals that shall be at least one more than the number of conduit entries. For example, where

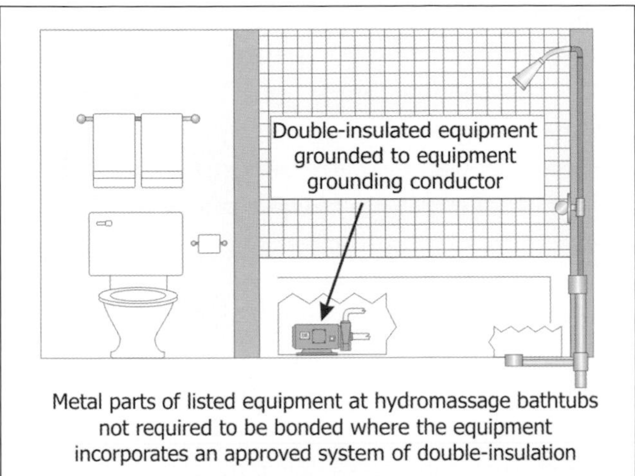

Double-insulated equipment grounded to equipment grounding conductor
Metal parts of listed equipment at hydromassage bathtubs not required to be bonded where the equipment incorporates an approved system of double-insulation

Figure 16-31. Bonding requirements for hydromassage bathtubs exclude double insulated pump motors.

there are two conduit entries, the grounding terminal bar is required to have not less than three terminals (see figure 16-29).

Installation of Luminaires over 15 Volts

The junction box or other suitable enclosure must be located not less than 100 mm (4 in.), measured from the inside of the bottom of the box or other enclosure, above the ground level or pool deck, or not less than 200 mm (8 in.) above the maximum pool water level, whichever provides the greatest elevation [see 680.24(A)(2)(a)]. It must also be located so that it is not less than 1.2 m (4 ft) from the inside wall of the pool unless it is separated from the pool by a solid fence, wall or other permanent barrier [see 680.24(A)(2)(b)]. If used on a lighting system operating at 15 volts or less, a flush deck box is permitted providing both of the following conditions are complied with:

(1) An approved [acceptable to the authority having jurisdiction, not listed or labeled] potting compound is used to fill the box to prevent the entrance of moisture.

(2) The flush deck box is located not less than 1.2 m (4 ft) from the inside wall of the pool.[30]

Bonding of Spa or Hot Tub Installed Indoors

The following parts of a spa or hot tub are required by 680.43(D) to be bonded together:

"1) All metal fittings within or attached to the spa or hot tub structure

"2) Metal parts of electrical equipment associated

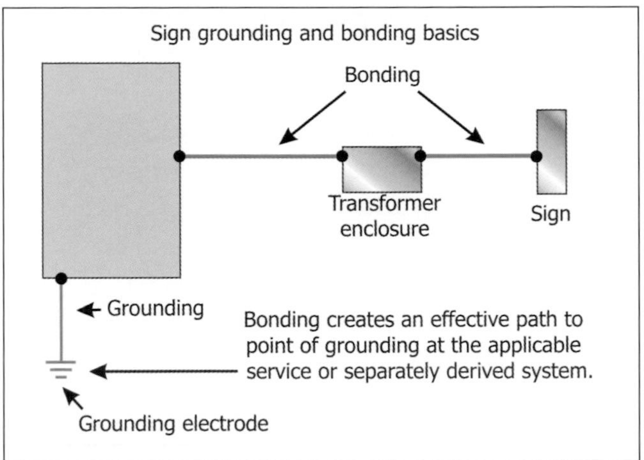

Figure 16-32. Basic concepts of grounding and bonding

with the spa or hot tub water circulating system, including pump motors

"3) Metal raceway and metal piping that are within 1.5 m (5 ft) of the inside walls of the spa or hot tub and that are not separated from the spa or hot tub by a permanent barrier

"4) All metal surfaces that are within 1.5 m (5 ft) of the inside walls of a spa or hot tub and that are not separated from the spa or hot tub area by a permanent barrier.

Exception No. 1: Small conductive surfaces not likely to become energized, such as air and water jets and drain fittings, where not connected to metallic piping, towel bars, mirror frames, and similar nonelectric equipment, shall not be required to be bonded.

"Exception No. 2: Metal parts associated with the water circulating system, including pump motors that are part of a listed self-contained spa or hot tub.

"5) Electrical devices and controls that are not associated with the spas or hot tubs and that are located not less than 1.5 m (5 ft) from such units; otherwise they shall be bonded to the spa or hot tub system."[38]

Methods of Bonding of Spa or Hot Tub Installed Indoors

"All metal parts associated with the spa or hot tub shall be bonded by any of the following methods:

"1) The interconnection of threaded metal piping and fittings

"2) Metal-to-metal mounting on a common frame or base

"3) The provisions of a solid copper bonding

jumper, insulated, covered, or bare, not smaller than 8 AWG solid." [39]

Bonding Requirements for Hydromassage Bathtubs

"Hydromassage bathtubs as defined in Article 680.2 are required to comply with only Part VII"of Article 680.[40] This part has requirements for both GFCI protection and bonding of hydromassage bathtubs,

Hydromassage bathtubs and their associated electrical components must be protected by a GFCI. This GFCI is to be of the Class A type that trips at a range of 4 to 6 mA leakage. In addition, all 125-volt, single-phase receptacles within 1.83 m (6 ft) measured horizontally of the inside walls of a hydromassage tub must be protected by a GFCI(s) (see figures 16-30 and 16-31).

Section 680.74 contains requirements on bonding of electrical equipment and metal piping systems associated with hydromassage bathtubs. It should be noted that these requirements apply to dwelling units as well as to other than dwelling units. This section requires that "all metal piping systems and all grounded metal parts in contact with the circulating water shall be bonded together using a solid copper bonding jumper, insulated covered, or bare, not smaller than 8 AWG." This bonding jumper must be connected to the terminal on the circulating system pump motor unless the motor is double-insulated (see figure 16-31). Note that this bonding jumper is not required to be routed to any panelboard, service equipment, or grounding electrode because it is for equipotential bonding purposes in the hydromassage bathtub vicinity (680.74).

Grounding and Bonding for Electric Signs and Neon Installations

It is fairly common knowledge to those in the electrical field that grounding and bonding of electrical equipment is essential for electrical safety as well as a *Code* requirement. Chapter six of the *NEC* provides minimum requirements for special equipment. The first article in this chapter is Article 600, Electric Signs and Outline Lighting. This article contains specific rules for this type of equipment that, in some cases, modify or amend the basic requirements set forth in chapters one through four. Otherwise, the basic rules

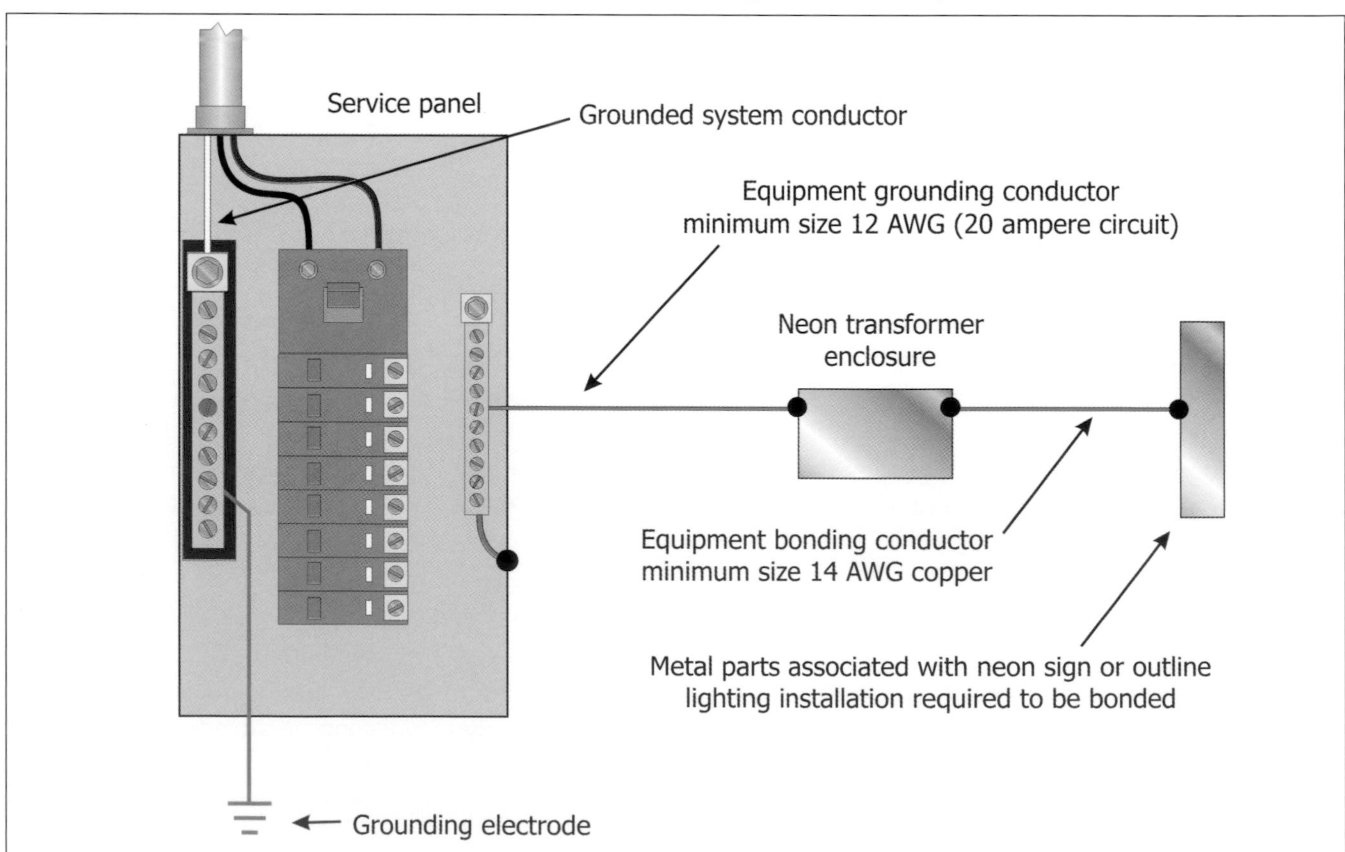

Figure 16-33. Equipment grounding conductor connected to the electric sign

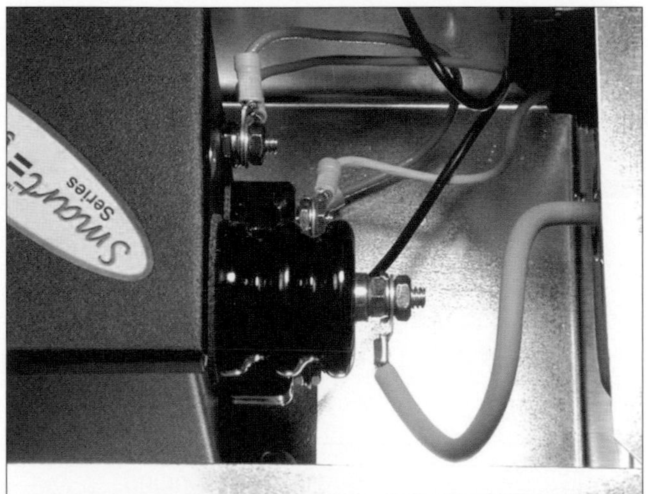

Photo 16-8. Equipment grounding conductor connected to neon transformer and enclosure

Photo 16-9. Metallic parts shown effectively bonded together

in one through four have application to this equipment as do the specific requirements of Article 600.

Electric Sign Grounding and Bonding Basics

To fully understand and apply the rules relating to grounding and bonding of this special equipment, one must grasp an understanding of the two concepts *grounding* and *bonding*. Grounding and bonding do not have to be as mysterious or complicated as many make them out to be. Let's review each of these basic concepts

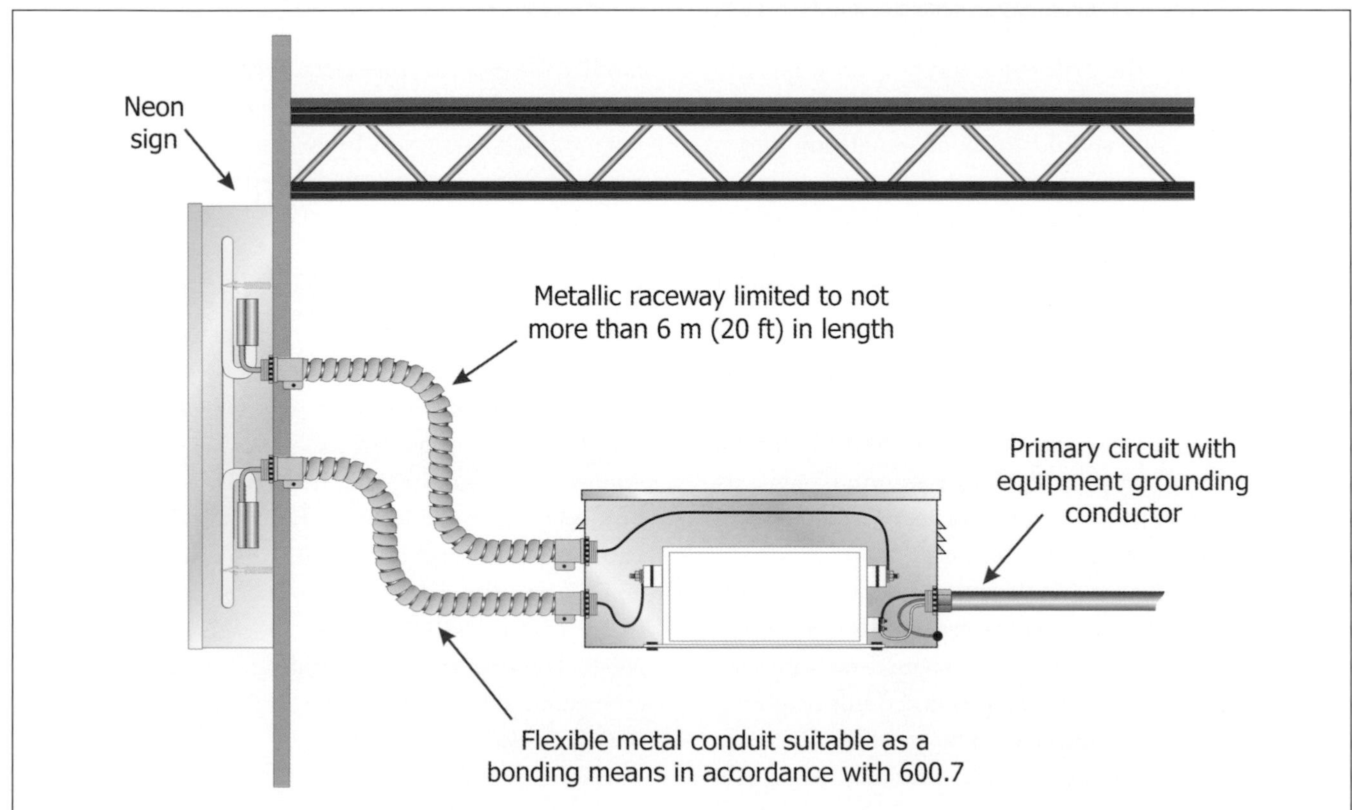
Neon sign
Metallic raceway limited to not more than 6 m (20 ft) in length
Primary circuit with equipment grounding conductor
Flexible metal conduit suitable as a bonding means in accordance with 600.7

Figure 16-34. Length of secondary circuit limited to 6 m (20 ft) in metal raceway

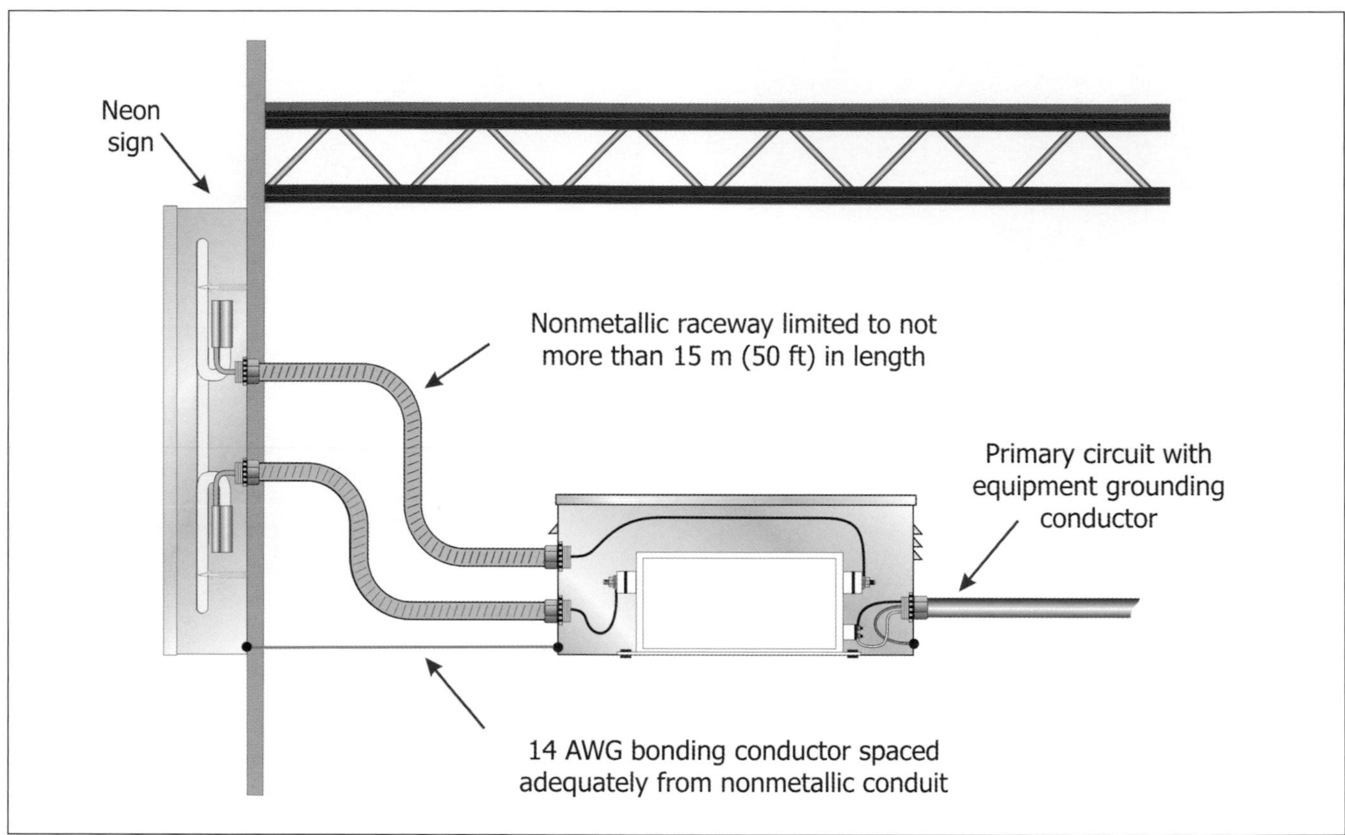

Figure 16-35. Length of secondary circuit limited to 15 m (50 ft) in nonmetallic raceway

before looking at the *Code* rules and how they apply to this special equipment. When the word *grounding* or *ground* is used, one should think of something that is connected to the earth or ground. The term *grounded (grounding)* is defined by the *Code* as "connected (connecting to ground or to a conductive body that extends the ground connection," [*NEC* Article 100]. The words *bonded (bonding)* are defined as "connected to establish electrical continuity and conductivity." So basically *bonded* means metal parts connected together and *grounded* means connected to ground (the earth) (see figure 16-32). Both concepts are essential elements and fundamentally important for electrical safety in electric signs and neon installations.

Section 600.7 contains the general requirement for grounding and bonding of electric signs and metal equipment of outline lighting systems and indicates generally that metal parts of signs and outline lighting systems must be grounded. The general rules for grounding and bonding are also provided in Article 250. When applying the *Code,* it is generally understood that the rules in chapters 5, 6, and 7 supplement or modify the general rules [*NEC* 90.3]. So when applying the *Code* to these installations, if the general rules are not modified in any way by Article 600, then grounding and bonding requirements in Article 250 would apply as well.

Equipment (electric sign and equipment) Grounding

The equipment grounding conductor and proper bonding are essential elements for safety in electrical signs and neon installations. The branch circuit supplying the equipment provides the required equipment grounding conductor for accomplishing the grounding (the connection to ground). The acceptable equipment grounding conductors are specified in 250.118 and the method of connection to the sign or metal enclosure is provided in *NEC* 250.120(A) and 250.8. The equipment grounding conductor must be connected to the equipment using methods consistent with 250.8 (see photo 16-8).

The minimum size of the equipment grounding conductor with the branch circuit must not be less than the sizes in Table 250.122 based on the branch-circuit

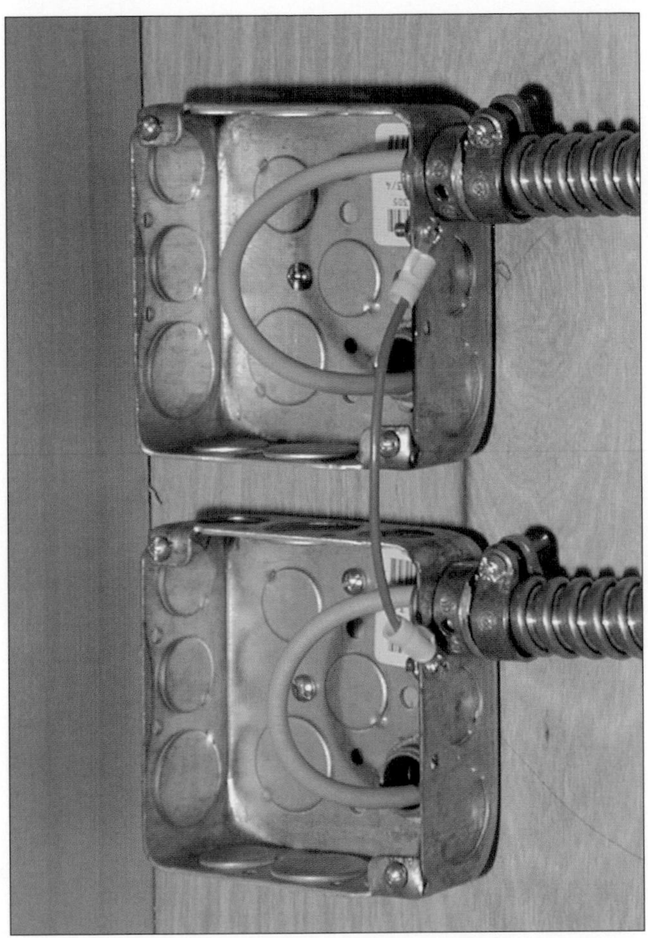

Photo 16-10. Metal parts bonded together using equipment suitable for the use

breaker or fuse rating. The equipment grounding conductor connects the equipment to ground and works to maintain it at or near earth potential (see figure 16-33). This safety component of the circuit also acts as the silent servant waiting to perform its ever-important function of facilitating the operation of the branch-circuit overcurrent protection in the event of a ground-fault circuit [the term *ground fault* is defined in Section 250.2]. Electrical installations for sign circuits and neon installations are not exempt from these basic grounding requirements and they must be grounded. It is important to note that 250.134(A) specifies in detail the required connections for the equipment grounding conductor. It must be routed with the branch circuit as indicated in 250.134(B) back to the source of a separately derived system or service grounding point. Structural metal frames of buildings are not permitted as the required equipment grounding conductor as clearly indicated in 250.136 and 600.7(B)(3).

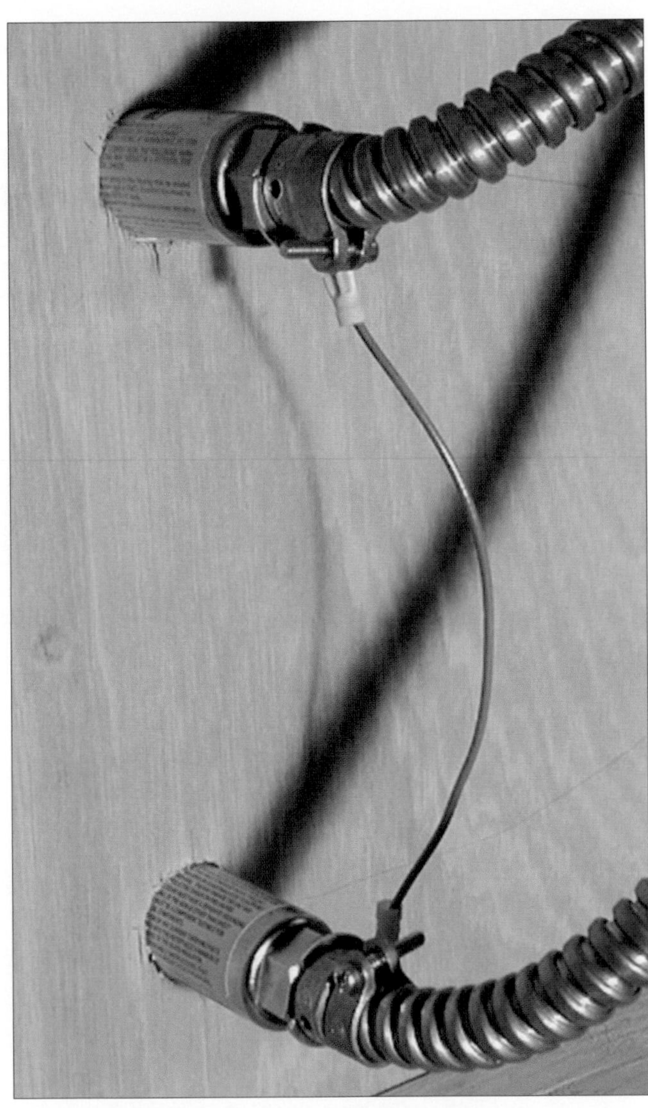

Photo 16-11. Minimum size for bonding conductor not less than 14 AWG copper

Bonding Requirements

Bonding basically means connecting metallic parts together. Section 600.7(B) includes the bonding requirements for metal parts and equipment of signs and outline lighting systems. When bonding metal equipment and parts associated with electric signs and neon lighting systems the primary objective is to establish the bonding in an acceptable manner and connect them to the equipment grounding conductor of the branch circuit supplying the sign or neon system. This accomplishes the bonding and grounding for the equipment or system. Proper grounding and bonding of metal enclosures and associated metal parts ensure that these parts remain at earth potential, and minimize differences of potential between metallic parts (see

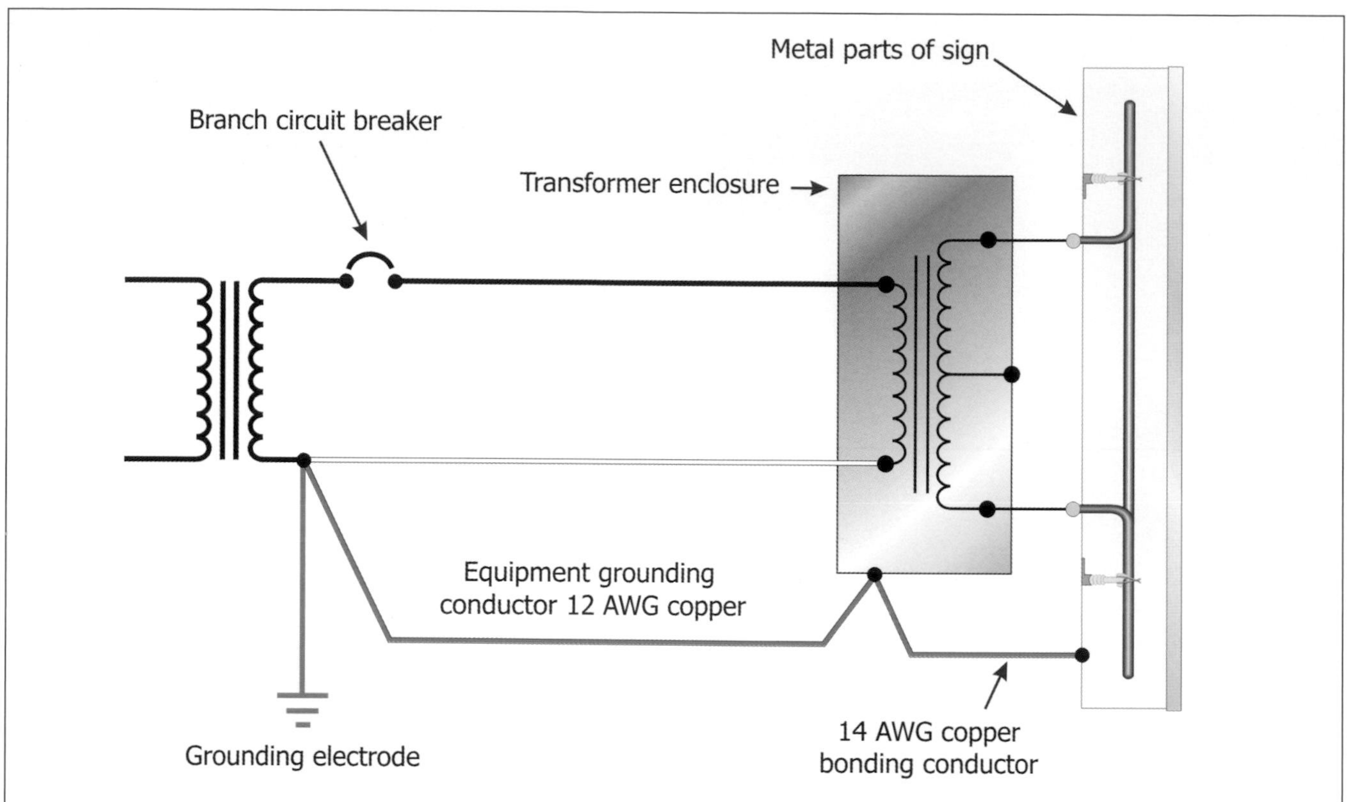

Figure 16-36. Metal parts properly bonded together

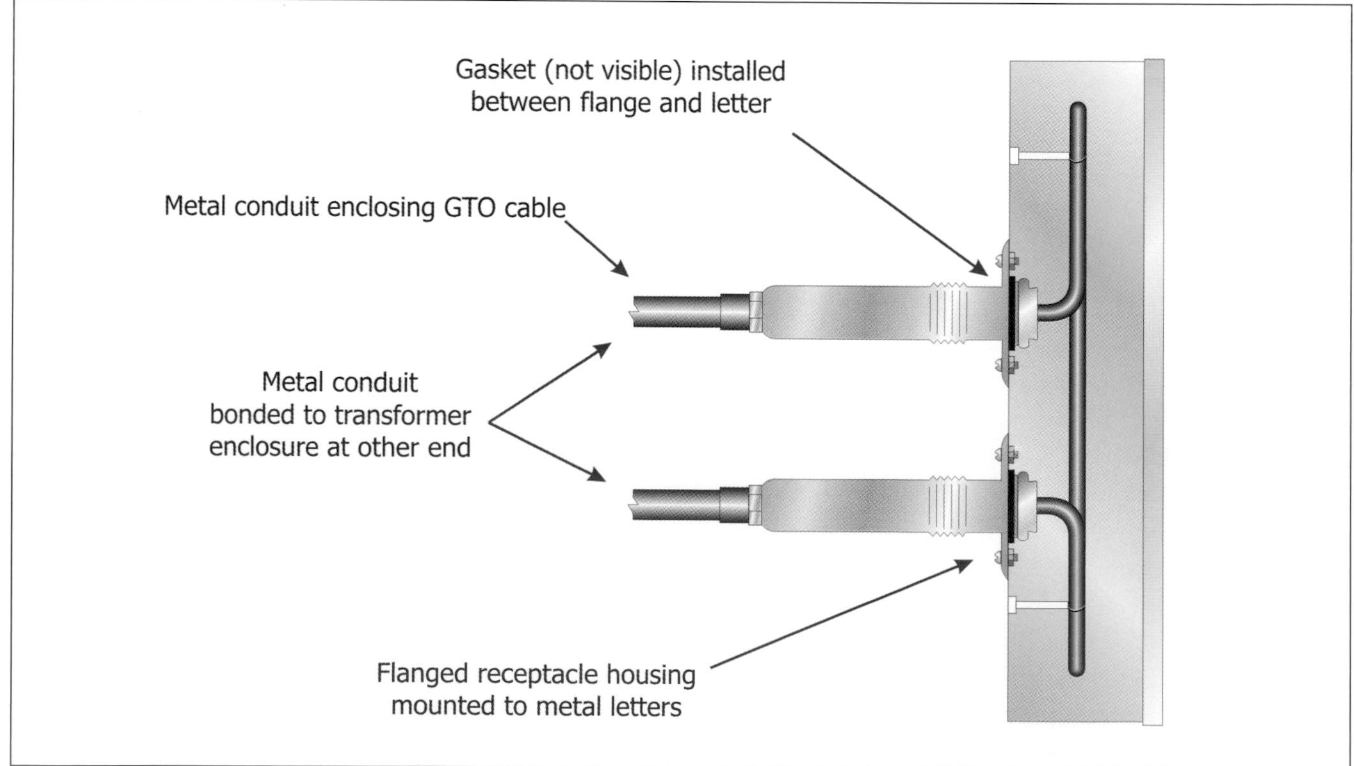

Figure 16-37. Electrode receptacle establishing bonding connection to metal channel letter by mechanical connection (Flanged electrode receptacles are available for this purpose).

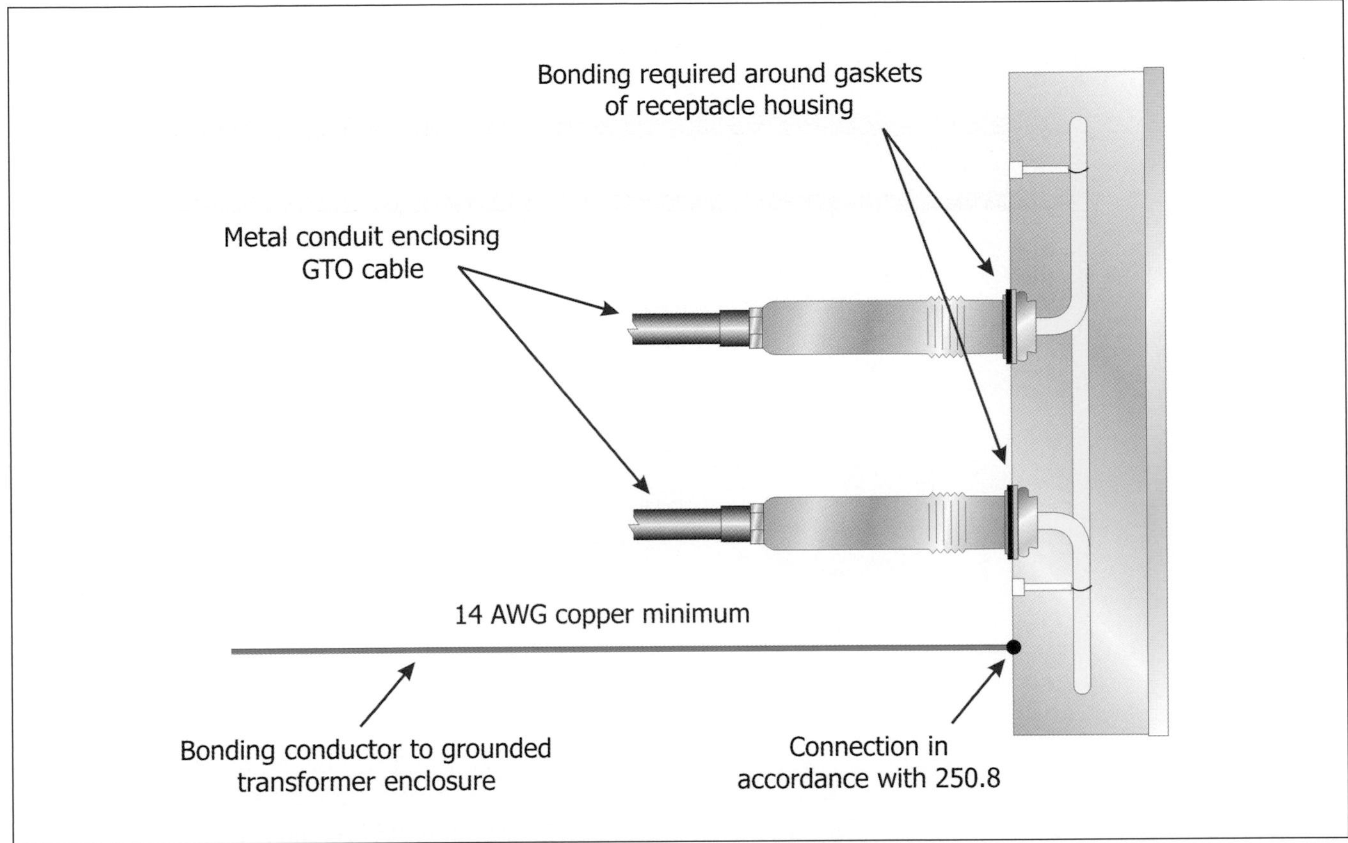

Figure 16-38. Electrode receptacle with gaskets installed requires separate 14 AWG copper bonding conductor to establish bonding connection to metal channel letter.

photo 16-9).

High voltage secondary circuits (GTO in a wiring method that extends from the transformer to the discharge tubing) for neon installations produce various levels of capacitance which is inherent to these secondary circuits and is unavoidable. Capacitance coupling in the neon or cold cathode secondary circuit can actually raise the potential (voltage) on ungrounded metal equipment and metal parts if not bonded together and connected to ground.

The *Code* allows for listed flexible metal conduit or listed liquidtight flexible metal conduit to be used as a bonding means in total accumulative lengths not exceeding 30 m (100 ft) on the secondary side of the sign transformer or power supply. These flexible metal conduits are suitable as a bonding means in lengths up to 30 m (100 ft) because the current on the secondary side of a neon transformer is in the milliampere range and the bonding provided through the conduit or bonding conductor here is not to clear an overcurrent device on the primary

Photo 16-12. Electrode receptacles (flanged and standard types shown) *Courtesy of Westrim* **[Photo is from** *Neon Lighting***]**

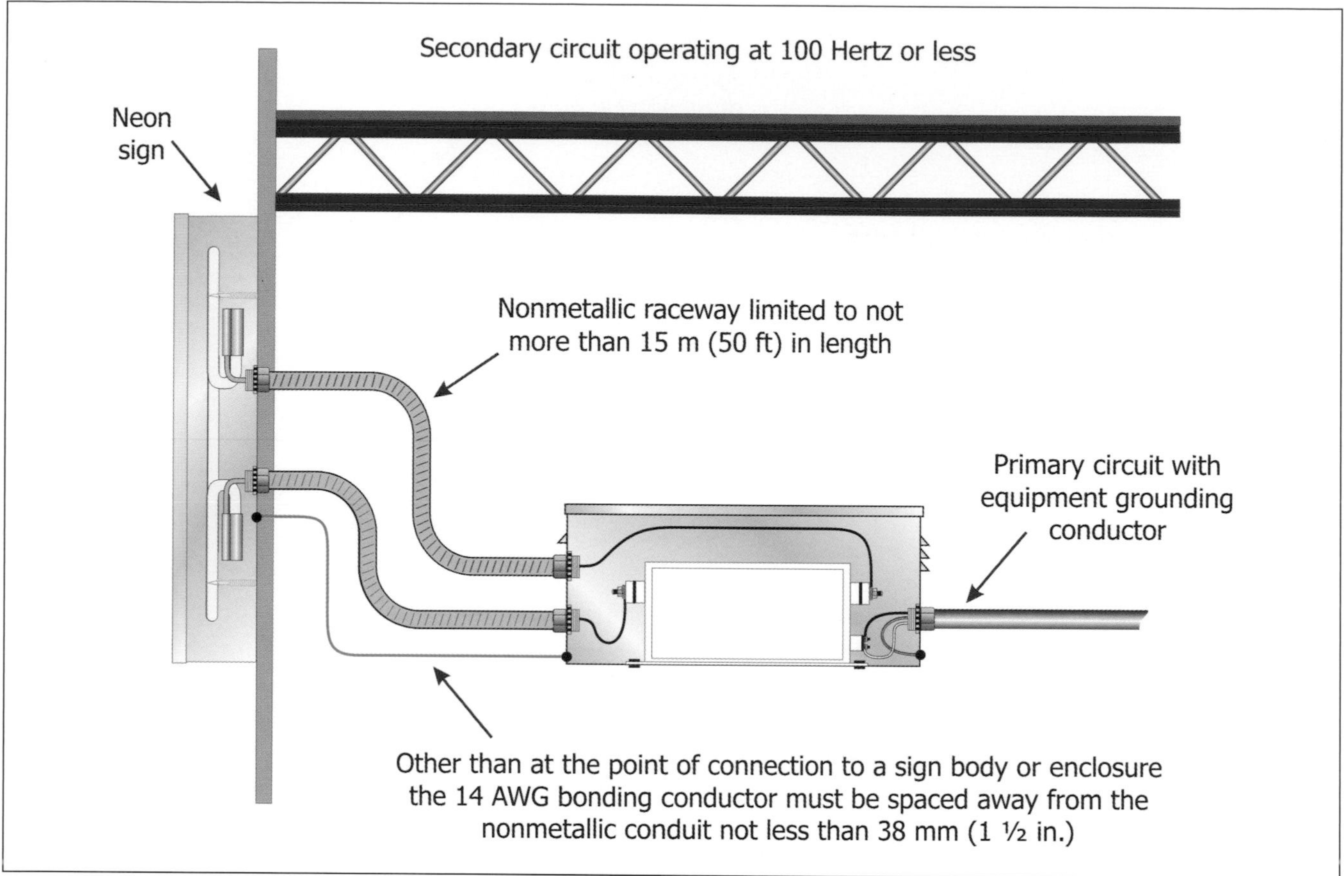

Figure 16-39. Bonding conductor spacing for circuits operating at 100 Hz or less

(supply side) of the neon transformer or power supply [600.7(B)(4)]. One should keep in mind that there is a length limitation on secondary GTO conductors of 6 m (20 ft) when installed in metallic wiring methods and 15 m (50 ft) when installed in nonmetallic wiring methods. This limitation is required to minimize the capacitance effect in the secondary that impacts both the secondary conductors (GTO cable) and the transformer [see 600.32(J) (1) and (2) and figures 16-34 and 16-35].

Bonding of electrical equipment and enclosures simply means that the enclosures will be connected together in an appropriate manner to ensure electrical continuity and conductivity.

Remember that normal current will always try to return to its source, which is the transformer or power supply secondary. The same is true for any fault current. Any metal parts or components requiring bonding must be bonded back to the source (transformer and/or combination of transformer and enclosure). When a metal conduit is connected to a metal electrical junction box with a proper conduit connector or proper fittings, the two parts are at the same potential because they are bonded together (see figure 16-36).

When metallic wiring methods are used for secondary circuits and the type of electrode connection to the neon tubing is through an electrode receptacle, particular attention to bonding must be applied. Electrode receptacles (often referred to in the industry as *PK housings*) provide gaskets that must be installed, particularly in wet locations. The installation instructions of listed parts must be followed [*NEC* 110.3(B)]. Basically this means that bonding of metallic parts must be assured around any gaskets that cause isolation between the channel letter of a sign and the metallic conduit connect to the electrode receptacle. This happens all too often in the field. This is one reason it is important to be thorough in the inspection of secondary circuits of neon signs and outline lighting systems (see figures 16-37 and 16-38).

Small associated metal parts not exceeding 50 mm (2 in.) in any dimension and not likely to become

Photo 16-13. Violation shows metal parts of a structure being used as a neon secondary (high voltage) return circuit to transformer. There are numerous other violations in the photo as well.

energized (such as the metal mounting means for tubing supports), and spaced at least 19 mm (¾ in.) from the neon tubing are not required to be bonded [*NEC* 600.7(B)(5)]. Where listed liquidtight nonmetallic conduit is used for installing the secondary high voltage GTO conductors from the transformer or power supply to the neon tubing and where there are associated metal parts that require bonding, a bonding conductor is required to be installed. This bonding conductor is required to be installed separate and remotely spaced from the nonmetallic conduit [see 600.4(B)(6)]. The wiring method referred to here is liquidtight flexible metal conduit or rigid PVC conduit.

Stress in the Secondary Circuit

Where listed nonmetallic conduit is used to enclose the secondary wiring of a transformer or power supply and a bonding conductor is required, the bonding conductor shall be installed separate and remote from the nonmetallic conduit. Bonding conductors are

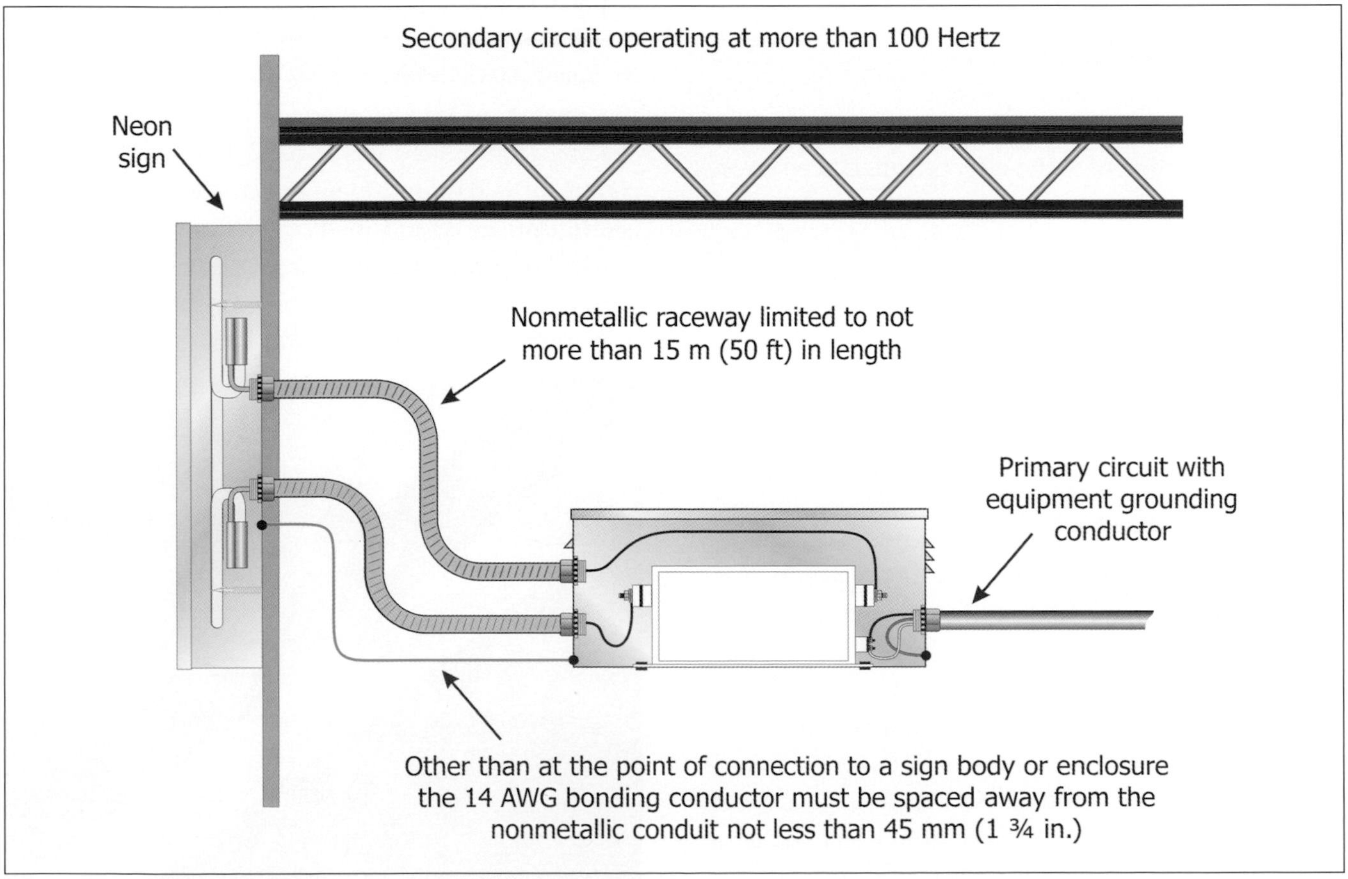

Figure 16-40. Bonding conductor spacing for circuits operating at over 100 Hz.

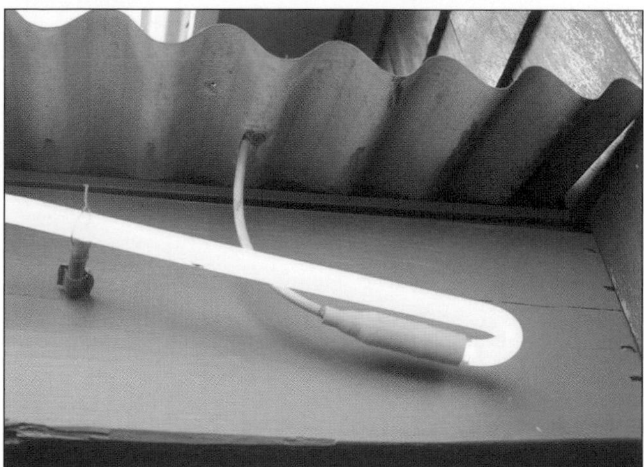

Photo 16-14. Violation shows a connection from the last tubing electrode to the metal parts of a structure used as the high voltage secondary return to the neon transformer. There are numerous other violations in the photo as well.

required to be copper and not smaller than 14 AWG. This spacing requirement is established to reduce stresses that might be imposed on the GTO conductor because the magnetic flux lines would no longer be symmetrical around the conductor.

Installing a bonding conductor or grounded and bonded electrically conductive parts remote from nonmetallic raceways and enclosures that contain high voltage secondary GTO cables is important to minimize the effects unbalanced stress on the high voltage secondary conductor installed in a nonmagnetic wiring method. Two spacing requirements in the *Code* are intended to deal with this situation [*NEC* 600.7(B)(6) and 600.32(A)(4)]. When a conductor is carrying ac current, it generates a magnetic field, which surrounds the conductor as the current flows during normal operation. This electromagnetic field produces electromagnetic lines of force.

When the conductor is installed in a metal wiring method, these electromagnetic lines of force will work to compress into the metal wiring method encircling the secondary conductor, because the metal conduit introduces less resistance to the magnetic lines of force than air. Even though the magnetic lines of force (flux) are more concentrated, they are maintained in a more symmetrical fashion around the conductor. This keeps the stress on the conductor more uniform or equal for the most part all around the GTO conductor.

When a current-carrying conductor is installed in a nonmetallic wiring method, for example liquidtight flexible nonmetallic conduit, and a bonding conductor is required to be installed for bonding metal parts associated with the neon sign or neon installation, the magnetic lines of force (flux) will try to compress into the bonding conductor or grounded metal parts on that side of the conductor and remain expanded on the other sides. As the compressed magnetic field flows in this circuit, the current is flowing with this unbalanced magnetic flux from zero to maximum voltage and back to zero with every cycle of ac current flow. The stresses on the secondary GTO conductor are much greater on one side than the other and will continue this unbalanced stress condition which can cause degradation of the high voltage secondary conductor insulation in time. By spacing this bonding conductor at least at the minimum required intervals, the amount of unbalanced stresses imposed on the high voltage secondary conductors will be significantly reduced.

Sections 600.7(B)(6) and 600.32(A)(4) both require a spacing of 38 mm (1½ in.) to be maintained between grounded metal parts or bonding conductors when the secondary circuit operates at 100 Hz or less. When the secondary circuit operates at over 100 Hz, the spacing requirement increases to 45 mm (1¾ in). This conductor is required to be not smaller than 14 AWG (see figures 16- 39 and 16-40).

Metal parts of a building or structure are not permitted to be used as a secondary return conductor or equipment grounding conductor [600.7(B)(3) and

600.32(A)(5)]. Return secondary leads cannot be connected to the metal parts of a building or structure and used as a return for a mid-point return wired secondary circuit. This is a bad situation that can lead to fire and shock hazards (see photos 16-13 and 16-14). .Proper grounding and bonding is a basic requirement in the *NEC* and is found in chapter two, which is appropriately titled, *Wiring and Protection*. The *NEC* provides the minimum requirements for essentially safe electrical installations. That means we must do at least that much. These *Code* requirements are set forth to protect persons and property from the hazards that arise from the ever-expanding use of electricity. Chapter six covers special equipment and electric signs and neon outline lighting systems. The requirements in Article 600 often amend or supplement the requirements in chapter two. Following these basic minimum requirements for grounding and bonding of signs and neon lighting installations contributes to the safe use of electricity. We hope that this brief tour through the basic grounding and bonding requirements for this type of equipment is of some help and provides some answers to questions that all too often don't get asked. More in depth information on this topic can be found in IAEI's *Neon Lighting* book.

[1] NFPA 70, National Electrical Code 2008, 250.106 (National Fire Protection Association, Quincy, MA, 2007), p. 70-113.

[2] NFPA 70, 250.100, p. 70-111.

[3] NFPA 70, 501.30(A), p.70-367.

[4] NFPA 70, 501.30(B), p 70-367.

[5] NFPA 70, 501.30(B), p. 70-367.

[6] NFPA 70, 547.9(D), p. 70-465.

[7] NFPA 70, 547.5(F), p. 70-463.

[8] NFPA 70, 250.112, p. 70-113.

[9] NFPA 70, 547.9(B)(3), p. 70-465

[10] NFPA 70, 547.9(B)(3), p. 70-465.

[11] NFPA 70, 547.9(B)(3), p. 70-465.

[12] NFPA 70, 547.10(A), p. 70-465.

[13] NFPA 70, 547.10(B), p. 70-465.

[14] NFPA 70, 547.10(B), p. 70-465.

[15] NFPA 70, 547.5(G), p. 70-464

[16] NFPA 70, 517.11, FPN, p. 70-428.

[17] NFPA 70, 517.13(B), p. 70-428.

[18] NFPA 70, 517.13(B), p. 70-428.

[19] NFPA 99, Standard for Health Care Facilities.

[20] NFPA 99, *Standard for Health Care Facilities*.

[21] NFPA 70, 517.14, p. 70-428.

[22] NFPA 70, 517.19(C), p. 70-429.

[23] NFPA 70, 517.19(D), p. 70-430.

[24] NFPA 70, 517.19(E), p. 70-430.

[25] NFPA 70, 517.19(F), FPN, p. 70-430.

[26] NFPA 70, 680.1, p. 70-558.

[27] NFPA 70, 680.26(B), p. 70-566.

[28] NFPA 70, 680.26(C), p. 70-567.

[29] NFPA 70, 680.24(A), p. 70-565.

[30] NFPA 70, 680.24(A)(2)(b), p. 565.

[31] NFPA 70, 680.6, p. 70-560.

[32] NFPA 70, 680.24(F), p. 70-565.

[33] NFPA 70, 680.23(F)(2), p. 70-564.

[34] NFPA 70, 680.23(B)(3), p. 70-563.

[35] NFPA 70, 680.21(A), p. 70-561.

[36] NFPA 70, 680.21(A)(4), p. 70-561.

[37] NFPA 70, 250.32(B)(2), p. 70-103..

[38] NFPA 70, 680.43(D), p. 70-569.

[39] NFPA 70, 680.43(E), p. 70-570.

[40] NFPA 70, 680.70, p. 70-572.

[41] NFPA 70, 680.74, p. 70-573.

[42] NFPA 70, Article 100, p. 70-27.

[43] NFPA 70, 600.7(B), p. 70-505.

[44] NFPA 70, 600.7(B), p. 70-505.

[45] NFPA 70, 600.7(B), p. 70-505.

[46] NFPA 70, 600.7(B), p. 70-505.

[47] NFPA 70, 600.7(B), p. 70-505.

[48] NFPA 70, 600.32(A)(5), p. 70-508.

16 Review Questions

The questions included here were developed using material included in this chapter. The answers can be found by reviewing the text. It is also important that students make use of *NEC*-2008, where many answers can be found.

1. Building or structure protection from lightning discharges is provided by ___.
   a. building low-rise structures
   b. building structures of wood construction
   c. providing a low-resistance path to earth
   d. providing a high-resistance path to earth

2. Ground terminals for lightning protection systems are required to be ____.
   a. connected (bonded) to the electrical power grounding electrode system
   b. bonded to the building water pipe system
   c. isolated from the building electrical system by at least six feet
   d. isolated from the building grounding electrode system

3. Generally, in hazardous (classified) locations, locknuts installed on each side of an enclosure, or a locknut on the outside and a bushing on the inside ___ be used for bonding.
   a. shall
   b. shall be permitted
   c. shall not
   d. may

4. Bonding locknuts or bonding bushings with bonding jumpers must be used in hazardous (classified) locations to ensure the integrity of the bonding connection and its capability of carrying any ____ that may be imposed.
   a. unidentified currents
   b. neutral currents
   c. objectionable current
   d. fault current

5. Where flexible metal conduit is permitted and used in a Class I, Division 2 location, it must have an internal or external bonding jumper installed to supplement the conduit. Where the jumper is installed on the outside, it is limited to ____ feet in length.
   a. 6
   b. 7
   c. 8
   d. 9

6. Where equipment is installed in an agricultural building that requires grounding, a ____ equipment grounding conductor must be installed to ground the equipment.
   a. aluminum
   b. copper
   c. copper-clad
   d. steel

7. Minimizing or elimination of stray voltages in an agricultural building is accomplished by bonding together the reinforcing rods and wire mesh in the concrete and then connecting these to the metal piping systems, stanchions and grounding electrode systems with a copper conductor not smaller than ____ AWG.
   a. 8
   b. 6
   c. 4
   d. 2

8. Which of the following statements is true about wiring in an agricultural building?
   a. An 8-ft. ground rod is required at each corner of the building.
   b. GFCI protection is required for all 125-volt, 1-phase, 15- and 20-ampere receptacles installed in damp or wet locations.
   c. Voltage-gradient ramps at entrances and exits are required.
   d. Line-to-neutral loads are not permitted.

9. A redundant means of grounding is required for branch circuits in ____ of health care facilities.
   a. limited-care facilities
   b. office areas
   c. all areas
   d. patient care areas

10. Where used in patient care areas, Type MI cable or Type ____ cable is required to have outer metal armor or sheath that (is) ____ as an acceptable grounding return path in accordance with 250.118.
   a. MV - approved by special permission
   b. SE - permitted
   c. SNM - approved
   d. MC – qualifies

11. Where metal raceways or Type MC or MI cables in accordance with 250.118 are used for feeders supplying critical care areas in health care facilities, grounding and bonding of panelboards and switchboards must be assured. Acceptable methods include all of the following EXCEPT ___
   a. grounding/bonding bushings with bonding jumpers
   b. threaded bosses

c. bonding locknuts or bushings
d. standard locknuts

12. Equipment in the vicinity of swimming pools must be bonded together with a 8 AWG or larger copper conductor. This also includes metal raceways and cables, piping and fixed metal parts within ____ feet horizontally or ____ feet vertically of the pool, observation stands, towers, platforms or diving structures.
a. 5 - 18
b. 6 - 20
c. 5 - 12
d. 9 - 14

13. A equipotential bonding grid at a pool is permitted to be the structural reinforcing steel of a concrete pool, the wall of a bolted or welded pool, or a(n) ____ AWG or larger solid copper conductor.
a. 8
b. 6
c. 4
d. 2

14. A dry-niche luminaire is required to be provided with a provision for the drainage of water and a means for accommodating ____ equipment grounding conductor(s) for each conduit entry.
a. 1
b. 2
c. 3
d. 0 or none

15. For other than storable pools, a flexible cord is not permitted to exceed ____ feet in length and must have a copper equipment grounding conductor not smaller than 12 AWG with a grounding-type attachment plug.
a. 3
b. 4
c. 5
d. 6

16. Where metal conduit is used to supply a ____ niche luminaire, it is required to be of brass or other approved corrosion-resistant metal.
a. dry-
b. wet-
c. no-
d. metal-

17. A wet-niche luminaire is required to be bonded to and secured to the forming shell by a positive locking device that assures a low-resistance contact and requires a ____ to remove the luminaire from the forming shell.
a. tool
b. permit
c. fitting
d. helper

18. Transformers used for the supply of underwater luminaires, together with the transformer enclosure, are required to be ____.
a. suitable
b. listed as a swimming pool and spa transformer
c. acceptable
d. approved

19. The transformer used to supply underwater luminaires is required to be a ____ winding type having a grounded metal barrier between the primary and secondary windings.
a. three
b. isolated
c. one
d. four

20. Electric signs and ___ shall be grounded by connection to an equipment grounding conductor.
a. all line-side equipment
b. electrode receptacle housings
c. metal equipment of outline lighting
d. electrical-discharge tubing

21. Where flexible nonmetallic conduit is used on the high voltage secondary circuit of a sign transformer, any separate bonding conductor installed to bond metal parts must be:
a. installed inside the conduit
b. not more than 6-feet long
c. installed outside the conduit and spaced not less than 38 mm (1 ½ in.) from the conduit
d. listed GTO cable

22. Listed flexible metal conduit that is used to enclose the secondary wiring of a sign transformer is permitted as the bonding means for total accumulative lengths not exceeding:
a. 6 feet
b. 25 feet
c. 50 feet
d. 100 feet

23. Which of the following methods is not permitted for connecting equipment grounding conductors or bonding conductors for electric signs and outline lighting systems?
    a. a listed pressure connector
    b. a sheet metal screw
    c. a pressure connector listed as grounding and bonding equipment
    d. machine screw-type fasstners that engage not less than 2 threads or are secured with a nut

24. Pool water of an in-ground swimming pool is required to be bonded to the equipotential bonding grid.
    a. True
    b. False

Chapter 17

Grounding and Bonding for Electronic Equipment

Article 645 covers electrical requirements for information technology equipment. Section 645.1, Scope, reads, "This article covers equipment, power-supply wiring, equipment interconnecting wiring, and grounding of information technology equipment and systems, including terminal units, in an information technology equipment room" (see photo 17-1). Articles 725 and 800 apply to interconnection of information technology equipment located outside the computer room, such as for personal computers, workstations, and printers. A fine print note to 645.1 reads, "For further information, see NFPA 75-2003, *Standard for the Protection of Information Technology Equipment.*"

Objectives

To understand...

- Requirements for grounding information technology equipment
- Isolated equipment grounding conductors
- Proper grounding methods for a data processing system
- Reducing ground loop effects
- Equalizing potential in a computer room
- High frequency effects in grounding conductors
- Surge arresters and surge protective device (SPD) usage
- Harmonic currents

As indicated, some grounding and bonding requirements for information technology equipment are contained in Article 645. Due to the rules on organization found in 90.3, the requirements for grounding in Article 250 apply to information technology equipment except as modified in Article 645.

Photo 17-1. Power distribution unit (PDU) for information technology use

Special conditions are imposed in 645.4 before the provisions of Article 645, including those for grounding, are permitted to be used for the information technology room under consideration. These requirements are as identified in (1) through (5) below:

"(1) Disconnecting means complying with 645.10 are provided.

"(2) A separate heating/ventilating/air-conditioning (HVAC) system is provided that is dedicated for information technology equipment use and is separated from other areas of occupancy. Any HVAC system that serves other occupancies shall be permitted to also serve the information technology equipment room if fire/smoke dampers are provided at the point of penetration of the room boundary. Such dampers shall operate on activation of smoke detectors and also by operation of the disconnecting means required by 645.10.

"FPN: For further information, see NFPA 75-2003, *Standard for the Protection of Information Technology Equipment*, Chapter 10, 10-1, 10-1.1, 10-1.2 and 10-1.3.

"(3) Listed information technology equipment is installed.

"(4) The room is occupied only by those personnel needed for the maintenance and functional operation of the installed information technology equipment.

"(5) The room is separated from other occupancies by fire-resistant-rated walls, floors, and ceilings with protected openings.

"FPN: For further information on room construction requirements, see NFPA 75-2003, *Standard for the Protection of Information Technology Equipment*, Chapter 5.[1]

Rooms for information technology equipment that do not meet all these conditions must be wired and the contained equipment grounded in accordance with the general requirements of *NEC* Chapters 1-4. It should also be pointed out that the provisions of Article 645 apply only within the information technology equipment room and not to other spaces, such as in general office areas, where all the features of 645.4 (1) through (5) are not provided.

Grounding requirements in 645.15 read, "All exposed non-current-carrying metal parts of an information technology system shall be bonded to the equipment grounding conductor in accordance with Article 250 or shall be double insulated." This is a general reference to Article 250 where grounding requirements for all types of equipment are found. Few additional requirements or modifications of the general grounding requirements in Article 250 are found in Article 645.

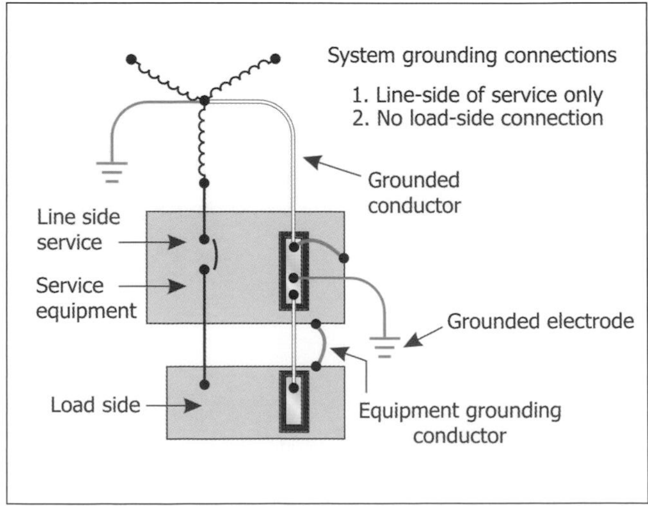

Figure 17-1. Typical power system grounding connections

In addition, 645.15 provides that, "Power systems derived within listed information technology equipment that supply information technology systems through receptacles or cable assemblies supplied as part of this equipment shall not be considered separately derived for the purpose of applying 250.20(D)." Section 250.20(D) provides the requirements for grounding of electrical systems, such as for services provided by the electric utility, as well as for systems that are separately derived. Without the modification that is provided in 645.15, electrical systems produced in power distribution units or uninterruptible power systems (UPS) that are commonly used for information technology equipment must be grounded as separately derived systems in accordance with 250.30.

Finally, 645.15 requires, "Where signal reference structures are installed, they shall be bonded to the equipment grounding conductor provided for the information technology equipment." This requirement will be explored later in this chapter.

Two fine print notes provide additional information. FPN No. 1 reads, "The bonding and grounding requirements in the product standards governing this listed equipment ensure that it complies with Article 250." FPN No. 2 states, "Where isolated grounding-type receptacles are used, see 250.146(D) and 406.2(D)" [2] (see figures 17-3, 17-4 and 17-5).

An additional reference to grounding of IT equipment is found in 250.6, Arrangment to Prevent Objectionable Current. The general requirement is that grounding conductors be installed to prevent *objectionable current* through the grounding path(s). The *NEC* does not define or describe what objectionable current is but, certainly, any current through grounding conductors that would in any way impair the performance of their intended function would be objectionable.

Section 250.6(B) outlines the following steps that can be taken to stop objectionable current:

- Discontinue one or more but not all of such grounding connections.
- Change the locations of the grounding connections.
- Interrupt the continuity of the conductor or conductive path causing the objectionable current
- Take other suitable remedial and approved action.[3]

Section 250.6(D), however, specifically applies to IT equipment or more broadly to all electronic equipment. It reads, "The provisions of this section shall not be considered as permitting electronic equipment from being operated on ac systems or branch circuits that are not connected to an equipment grounding conductor as required by this article. Currents that introduce noise or data errors in electronic equipment shall not be considered the objectionable currents addressed in this section." [4] This requirement was added after reports that some IT manufacturers or installers were not installing the equipment grounding conductors with the branch circuit to the equipment but were connecting IT equipment only to a local grounding point such as a driven ground rod.

Another general requirement, the importance of which cannot be overstated, is that the neutral or grounded system conductor not be grounded past the service disconnecting means (see figure 17-1, 250.24(A)(5), and photo 17-2). Grounding of the neutral conductor or system grounded conductor on the load side of the service creates parallel paths for neutral current over each path that is established such as over conduit, piping systems, cable shields and cable trays. An exception to this rule is that separately derived systems must be grounded if the system produced can be grounded under the conditions given in 250.20 (see chapter twelve for additional information on grounding and bonding requirements for separately derived systems).

Photo 17-2. Neutral conductors and equipment grounding conductors isolated from one another as required by 250.24(A)(5). Isolated grounding terminal bar shown.

Purposes of Grounding Electrical Equipment

A broad, general rule is provided in 250.4(A)(2) on grounding of electrical equipment. This rule applies to all electrical equipment, including information technology equipment, as there is no modification of this rule in Article 645. The rule states, "Normally non-current-carrying conductive materials enclosing electrical conductors or equipment, or forming part of such equipment, shall be connected to earth so as to limit the voltage to ground on these materials." [5]

Section 250.4(A)(5) contains another important performance rule that also clearly applies to information technology equipment. It reads, "Electrical equipment and wiring and other electrically conductive materials likely to become energized shall be installed in a manner that creates a permanent, low-impedance circuit…capable of safely carrying the maximum ground-fault current likely to be imposed on it from any point on the wiring system where a ground fault may occur to the electrical supply source."

The electrical supply source is either the transformer by the electric utility for the service or a separately derived electrical system. Electrical systems all follow the basic laws of physics: a complete circuit is required before current can exist, and current will always return to its source. The task, then, is to provide the "permanent, low-impedance circuit capable of safely carrying the maximum ground-fault current likely to be imposed on it." [6]

This section goes on to require that "the earth shall not be [used] as an effective ground-fault current path." [7]

Safety Grounds and Clean Grounds

The *NEC* does not use the terms *safety ground* or *clean ground*. These terms are sometimes used in the information-technology industry to differentiate between the equipment grounding conductor required in the branch circuit to the equipment and the isolated equipment grounding conductor sometimes installed for IT equipment. The reference to the safety ground in IT jargon is to the equipment grounding conductor as defined in Article 100. This safety ground is often the conduit for the branch circuit and is sometimes referred to as the *dirty ground* because it is considered to possibly carry unwanted electrical noise from a variety of sources (see figure 17-2). Sometimes unsafe actions are taken to interrupt the equipment grounding conductors(s) (safety or dirty ground) and install grounding means near the IT equipment. Such practice is not permitted by any safety code and can create shock or electrocution hazards.

As mentioned above, 645.15 requires that IT equipment be grounded in accordance with Article 250. Section 250.86 applies generally and requires, "except as permitted by 250.112(I), metal enclosures and raceways for other than service conductors shall be connected to the equipment grounding conductor." The three exceptions to this rule do not apply to IT equipment, nor does the rule in 250.112(I).

However, 250.110 provides conditions under which "exposed non-current-carrying metal parts of fixed equipment likely to become energized shall be connected to the equipment grounding conductor." Conditions from this section that might apply to IT equipment and require grounding include:

"(1) Where within 2.5 m (8 ft) vertically or 1.5 m (5 ft) horizontally of ground or grounded metal objects and subject to contact by persons

"(3) Where in electrical contact with metal

"(5) Where supplied by a metal-clad, metal-sheathed, metal-raceway, or other wiring method that provides an equipment ground…" [8]

It is safe to say that metal enclosures for IT equipment must be grounded by connection to an equipment grounding conductor. This equipment grounding conductor is required to be either the metal raceway acceptable in 250.118 as an equipment grounding conductor or an equipment grounding conductor contained within the wiring method to the IT equipment. Under no circumstances should the effective fault-current path to the source of power be interrupted and an equipment grounding connection be made only to a local grounding electrode such as water pipe, building steel or driven electrodes. To do so violates all the principles of safety. This connection certainly will provide a high-impedance path and will constitute ineffective grounding. Serious shock hazards can occur in this arrangement since the overcurrent device that supplies the equipment will not clear the fault because there will be little current due to the high-impedance ground-fault path.

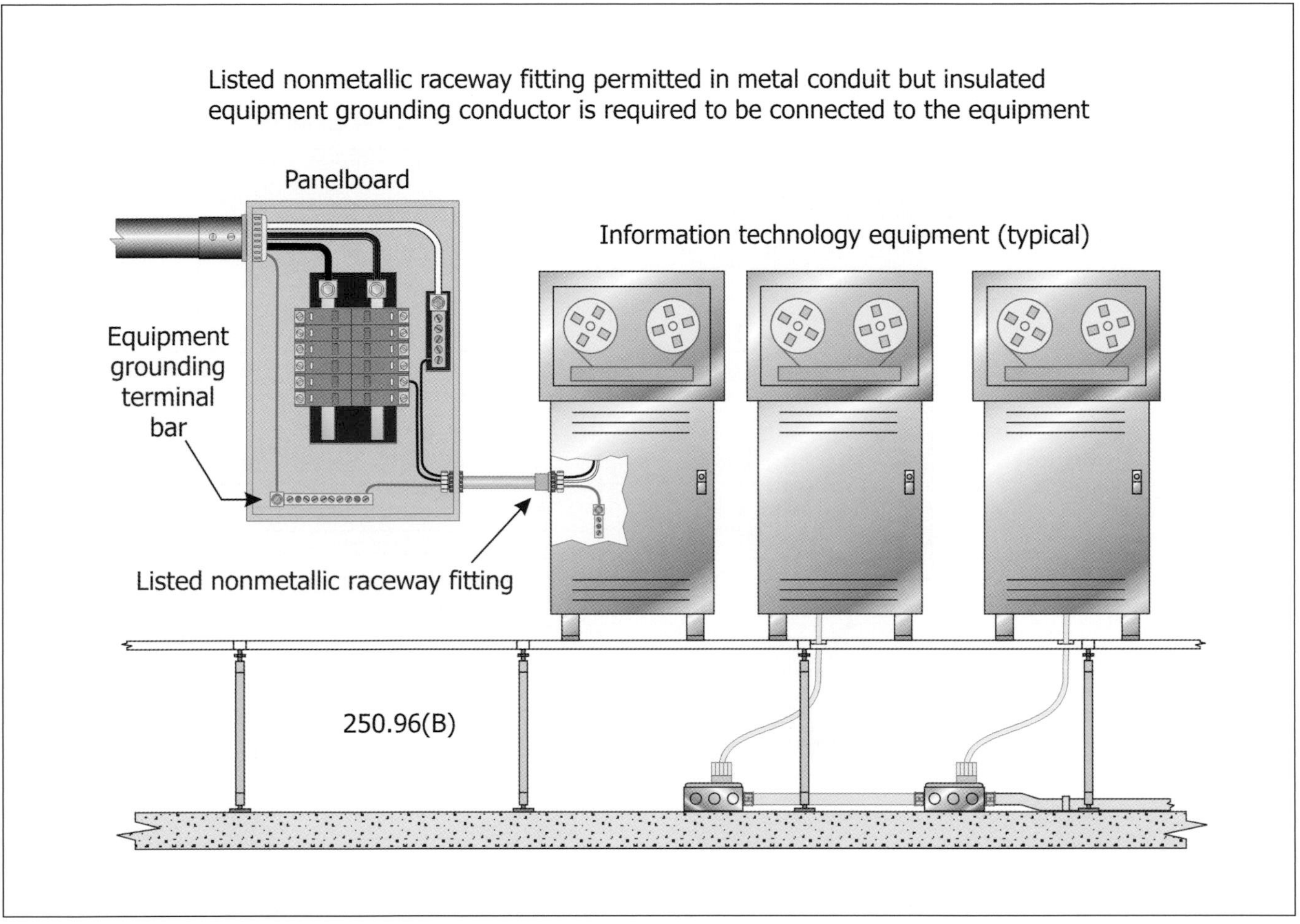

Figure 17-2. Isolated grounding connections for equipment

As mentioned, the term *clean ground* often refers to the isolated equipment grounding conductor sometimes specified for IT equipment or receptacle outlets by the equipment manufacturer or owner. This isolated equipment grounding conductor is permitted in 250.96(B). However, the section begins by stating, "Where installed for the reduction of electrical noise (electromagnetic interference) on the grounding circuit…." The *NEC* does not address the issue of who makes the decision about what the circumstances are that would justify the use of isolated equipment grounding. Usually, the owner, design engineer, or equipment manufacturer specifies the circuit to have isolated equipment grounding conductor(s).

Section 250.96(B) goes on to state, "an equipment enclosure supplied by a branch circuit shall be permitted to be isolated from a raceway containing circuits supplying only that equipment by one or more listed nonmetallic raceway fitting(s) located at the point of attachment of the raceway to the equipment enclosure" (see figure 17-2).

"The metal raceway shall comply with provisions of this article and shall be supplemented by an internal insulated equipment grounding conductor installed in accordance with 250.146(D) to ground the equipment enclosure." Note that this section does not require the use of a metal raceway to supply the IT equipment. If PVC conduit were used, the special insulating connector between the metal raceway and the IT equipment would not be required. Of course, the insulated equipment grounding conductor must be installed inside the PVC conduit to ground the IT equipment (see figure 17-2).

The reference here to installing the insulated equipment grounding conductor in accordance with 250.146(D) permits the equipment grounding conductor to pass through one or more panelboards, boxes, wireways, or other enclosures without being connected

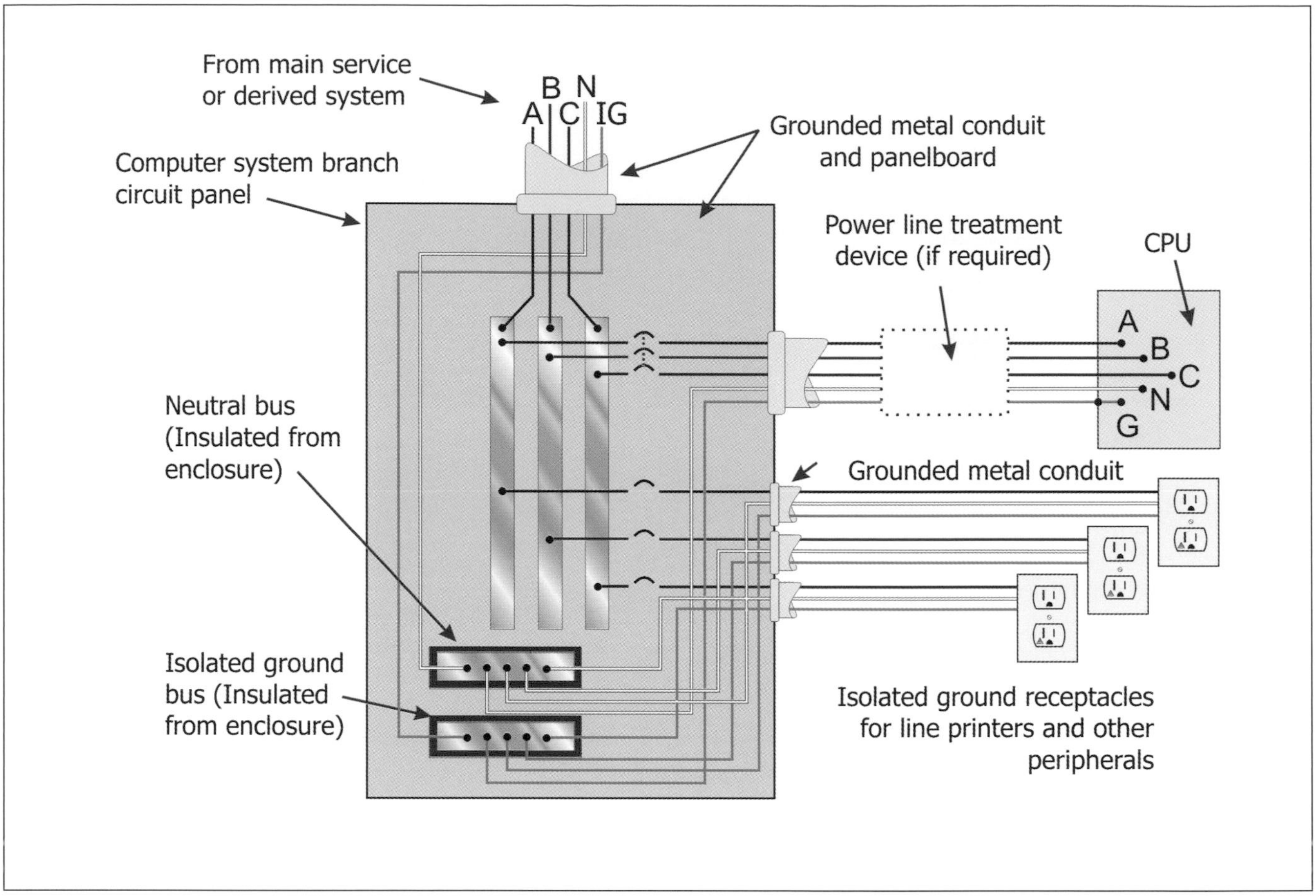

Figure 17-3. Isolated grounding receptacle connections

to equipment grounding terminal bars in the panelboard. The insulated equipment grounding conductor must be connected to the first of the following points to provide the low-impedance path required:

- Source of the separately derived system if it supplies the branch circuit
- Building disconnecting means
- Service equipment

Note that if a metal raceway is used, the raceway must be installed in a manner to ensure an adequate return fault-current path even though an insulated equipment grounding conductor is installed through the raceway for grounding the IT equipment. This ensures that the raceway does not become a shock hazard if a line-to-ground fault to, or in, the raceway occurs.

Proper Grounding Methods for a Data Processing System

To comply with the safety requirements of Article 250, an equipment grounding conductor of a type recognized in 250.118 must be installed as either the wiring method itself, such as a metal conduit or tubing, or an equipment grounding conductor of the wire type inside the conduit or tubing. The equipment grounding means is required to be connected to the metal enclosure for the IT equipment. The equipment grounding conductor, where it is a wire, is sized from Table 250.122 based on the rating of the overcurrent protective device ahead of the branch circuit. For example, a 10 AWG copper wire is permitted as the equipment grounding conductor in branch circuits having from 30-ampere through 60-ampere overcurrent protection. This will ensure the effective ground-fault return path will meet the rules in 250.4(A)(5) in being a "permanent, low-impedance circuit, capable of safely carrying the maximum ground-fault current likely to be imposed on it." This rule applies regardless of whether the branch circuits that supply the IT equipment originate in the electrical system provided from the electric utility service or in

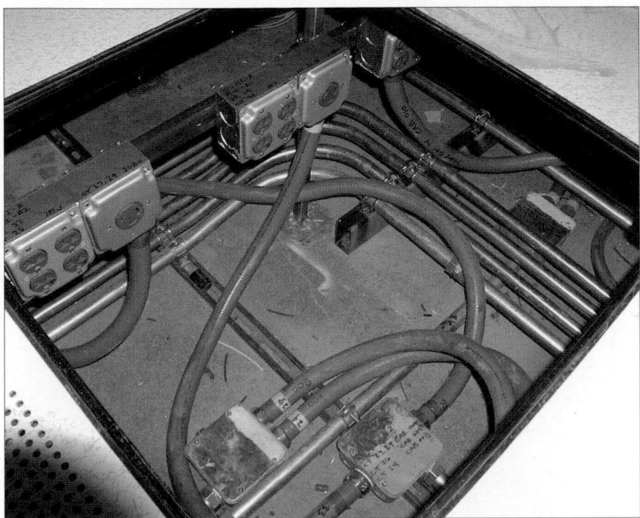

Photo 17-3. Isolated grounding receptacles installed under a raised floor in an information technology (IT) room

a dedicated power distribution unit (PDU) for the IT equipment.

As mentioned above, under no conditions should the effective ground-fault return path be interrupted and be connected to some other grounding means such as a local grounding electrode. Doing so can create a shock or electrocution hazard and aggravate the very problem such action is intended to solve.

Reducing Ground Loop Effects

The term *ground loops* is not defined in the *NEC*. It is used in the IT industry to refer to installations where ground current exists through multiple paths, and it is felt this ground current or electrical noise travels in a circular fashion, hence the term ground loops. Connecting computers and terminals on different

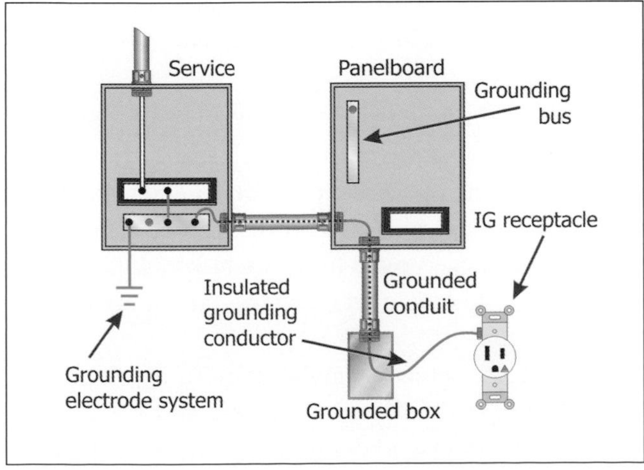

Figure 17-4. Isolated grounding receptacle connections

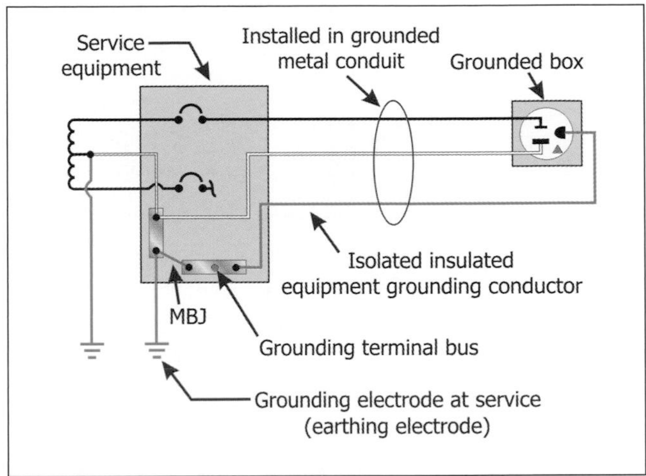

Figure 17-5. Isolated grounding receptacle connections

circuits that are grounded to building steel while the shielded communications cable between the computer and terminal completes the ground loop can cause this situation. Other connections through equipment grounding paths can add additional ground loops. Differences in ground potential will cause *noise current* and can couple with data signals. Many computer data signals operate at 5 V or less, which makes the system more susceptible to acquiring electrical noise from power circuits if cables are run closely together.

One method for solving this problem is to supply all equipment from the same power supply. The equipment grounding means of the branch-circuit wiring will serve to keep the ground (earth) potential the same. This solution can be impractical where the computer is located a long distance from peripherals and is supplied by different power supplies or different panelboards that, in turn, are supplied by different feeders. Known solutions to the problem of ground loops include:

1. Single point grounding. This solution might not be suitable at very high frequencies or where equipment is located a long distance from another and is connected by shielded cables.

2. A *balun coil*, which is a coil of insulated wire on a nonmetallic core. This can be installed in the power branch circuit or the data circuit. Manufacturers sometimes build this into equipment. Baluns are not effective at all frequencies.

3. Modems which are normally used as interfaces with telephone circuits.

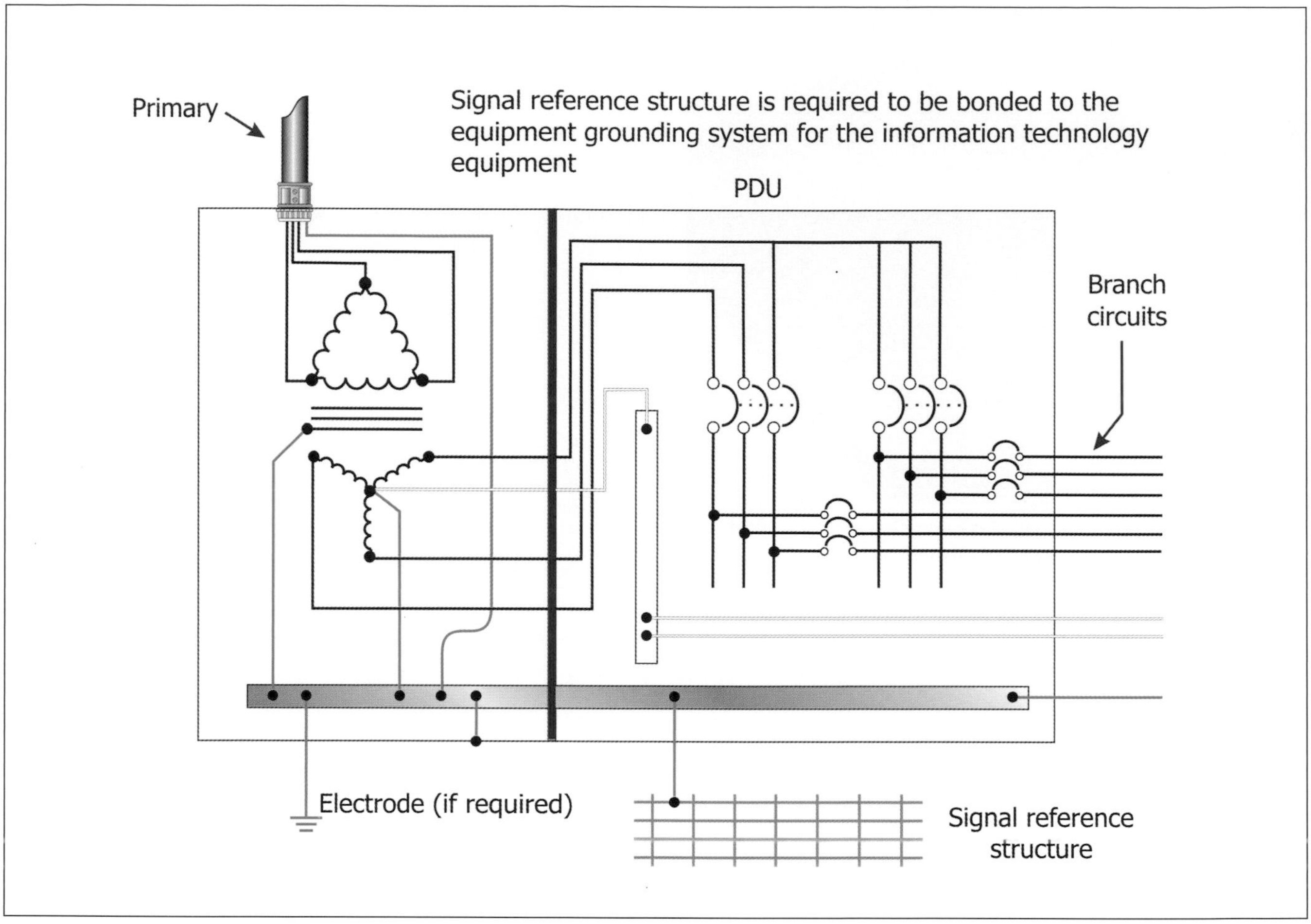

Primary

Signal reference structure is required to be bonded to the equipment grounding system for the information technology equipment

PDU

Branch circuits

Electrode (if required)

Signal reference structure

Figure 17-6. Signal reference grid is required to be bonded to equipment grounding conductor of the supply system(s)

4. Fiber optic transmission over completely non-conducting paths.

5. Optical isolators.

6. Interface devices such as surge arresters and surge protective devices (SPDs).

Equalizing Potential in a Computer Room

For the purposes of grounding IT equipment, an equipotential plane is described as, "a mass (or masses) of conducting material that, when bonded together, provide a uniformly low impedance to current over a large range of frequencies" [9] (see figures 17-7, and 17-8). The *signal reference structure* is defined as "a system of conductive paths among interconnected equipment that reduces noise-induced voltages to levels that minimize improper operation. Common configurations include grids and planes."[10] Keeping all electrical equipment, including IT equipment, at the same potential is necessary in an attempt to minimize disruptions in the operation of IT equipment.

Advantages of an equipotential plane include:

- Low impedance return path for radio frequency (RF) noise currents
- Containment of electromagnetic (noise) fields between their source (cables and similar) and the plane
- Increasing filtering effectiveness of contained electromagnetic fields, and
- Shielding of adjacent circuits or equipment

Equipotential plane structures can be established in several forms. Typical forms include:

- A conductive grid embedded in, or attached to, a concrete floor
- Metallic screen or metal strips under floor tile
- Ceiling grid above equipment, and
- Supporting grid of raised access flooring typical in computer rooms

Purpose of a Signal Reference Grid

The term *signal reference grid* refers to a structure such as a computer floor or a copper mesh grid installed under the computer floor. Information technology equipment is connected to the signal reference grid to equalize potential between components. All equipment in the room, including equipment that is wall-mounted, should be connected to the grid. The *NEC* is silent on this subject, but, as with other bonding or grounding methods used for IT equipment, such bonding cannot substitute for the equipment grounding conductor required to be installed with the branch-circuit conductors. The signal reference grid can be thought of as overlaying the equipment grounding conductors (sometimes referred to as safety grounds) that are required.

The signal reference grid serves as a signal reference plane over a broad range of frequencies. They are correctly referred to as *broadband grounding systems.* An effort is being made to reduce, eliminate or control the tendency of conductors connected to computers to resonate at higher frequencies. A grid provides multiple parallel conducting paths between its metal parts. If one path is a high-impedance path because of full or partial resonance, other paths of different lengths will be able to provide a lower impedance path. A signal reference grid could be constructed of continuous sheet copper or aluminum, zinc-plated steel, or any number of pure or composite metals with good surface conductivity (see photo 17-4). However, this type of construction is not only expensive but difficult to install in a computer room where other services have been or will be installed.

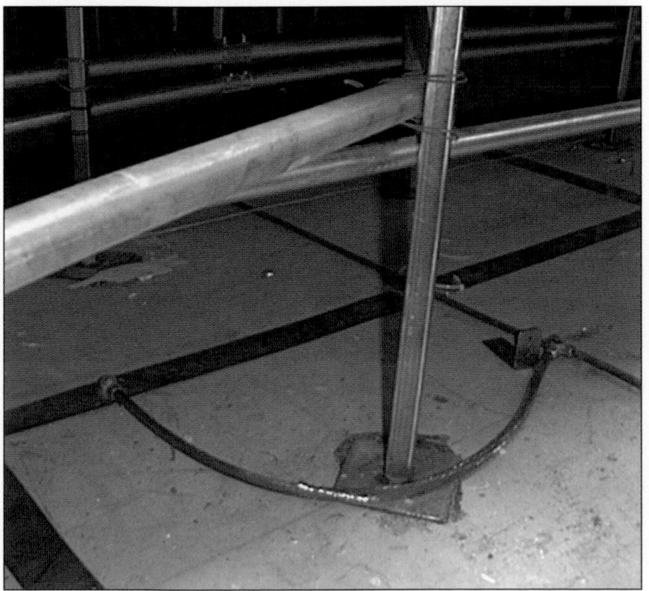

Photo 17-4. Signal reference grid connections independent of floor supports using 2" wide copper strips in a 2' x 2' pattern

Photo 17-5. Raised floor platform structure used as a signal reference grid

Grids of copper or aluminum strips are sometimes installed under a computer-raised floor. This grid provides a satisfactory constant potential reference network over a broad range of frequencies from dc to higher than 30 MHz. Typically, these have been formed of 4 AWG copper or aluminum conductors which have been electrically joined at their intersections, or by copper straps approximately 0.254 mm (0.010 in.) thick by 76 mm to 102 mm (3 in. to 4 in.) wide, also joined at their intersections. These grids typically lie directly upon the subfloor under the IT room raised floor. Cables and conduits under the floor would normally lie below the raised floor but above the grid.

As mentioned, the metal structure for the raised floor is sometimes used for the signal reference grid (photos 17-4 and 17-5). Typically, the floor grid is 600 mm x 600 mm (2 ft x 2 ft) as this size fits with standard cellular raised floor systems. The two essential requirements are bolted-down *stringers* (the lateral supporting braces installed between supporting pedestals) and suitably plated (tin or zinc) members so that low-resistance pressure connections can be made. It is important that all metal components of the floor structure be in good electrical contact with each other.

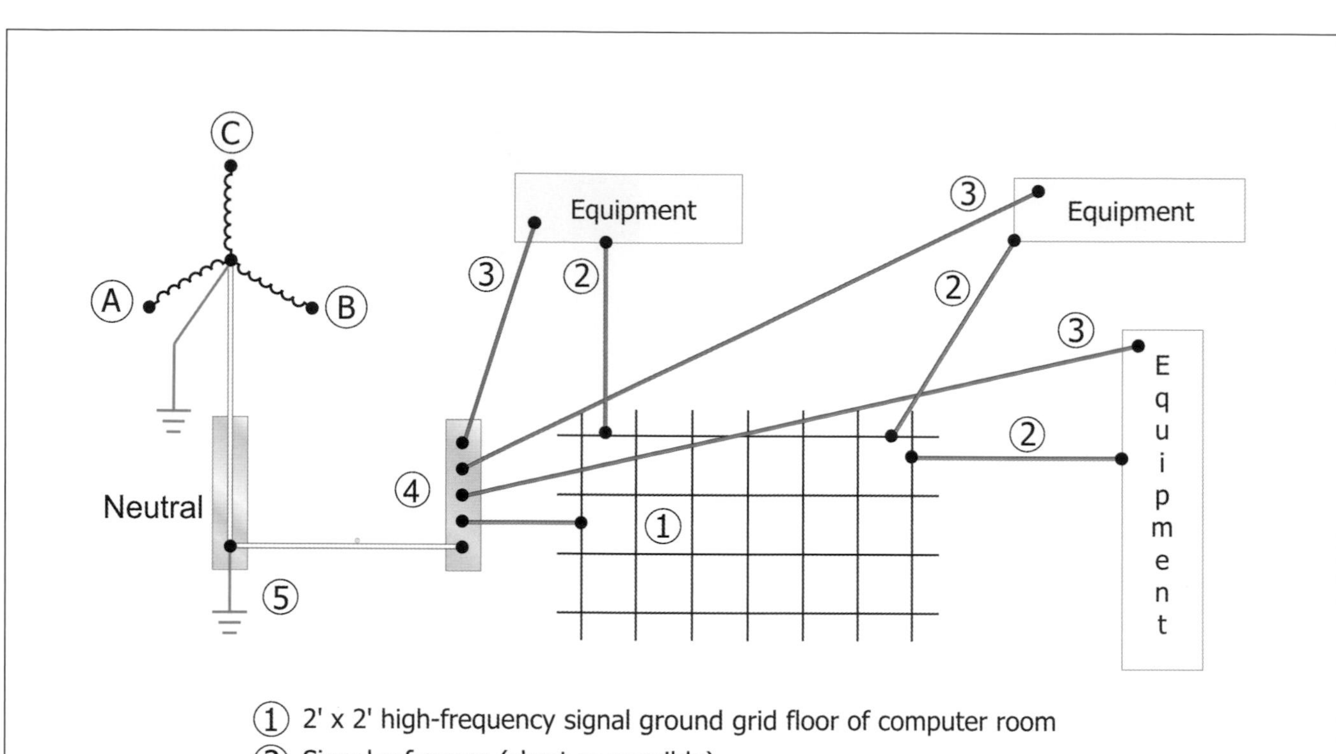

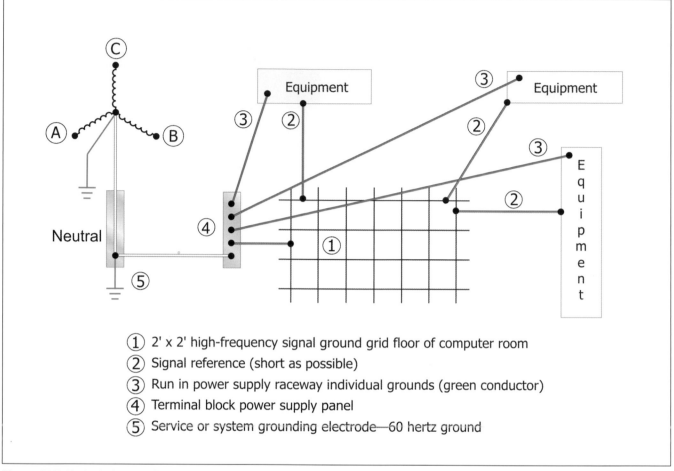

Figure 17-7. Typical signal reference grid connections showing the grid and the required equipment grounding conductors to equipment

This might require using bolts and clamps along with a copper or aluminum bonding conductor.

Some supporting structures are marketed in which the removable floor tiles lock onto the supports by gravity but it has not been demonstrated that the contact has a low enough resistance and is free from intermittent contact as people move about or loads are moved across the floor. The bolted down horizontal stringers (braces) with pressure-type spring washers or springs in the assembly have been shown to be highly suitable for the purpose.

Experience has shown that it is unnecessary that lift-out floor panels make a low-resistance contact with the supporting network of stringers. The plastic or synthetic rubber cushions or molded edging upon which the panels rest have been found to be adequate to drain static electricity from the panel if any should accumulate. A resistance as high as 20,000 megohms under 20 percent relative humidity or higher will satisfy this requirement.

Once the signal reference grid has been established in the IT room, the various IT units can each be connected to the grid by a flat braided copper strap. The connection should be made from each IT unit, preferably at a point near where the equipment grounding conductor is connected, to the nearest intersection of the signal reference grid. The strap should be no longer than necessary and should have few bends and very little loop or sag to minimize the impedance at high frequency. In addition, the effectiveness of the reference grid is improved if it is solidly connected to the power supply for the IT equipment by a very short strap.

If the IT equipment manufacturer does not subscribe to the philosophy of connecting the IT equipment to the grid at multiple locations, the stray capacitance between the IT equipment and

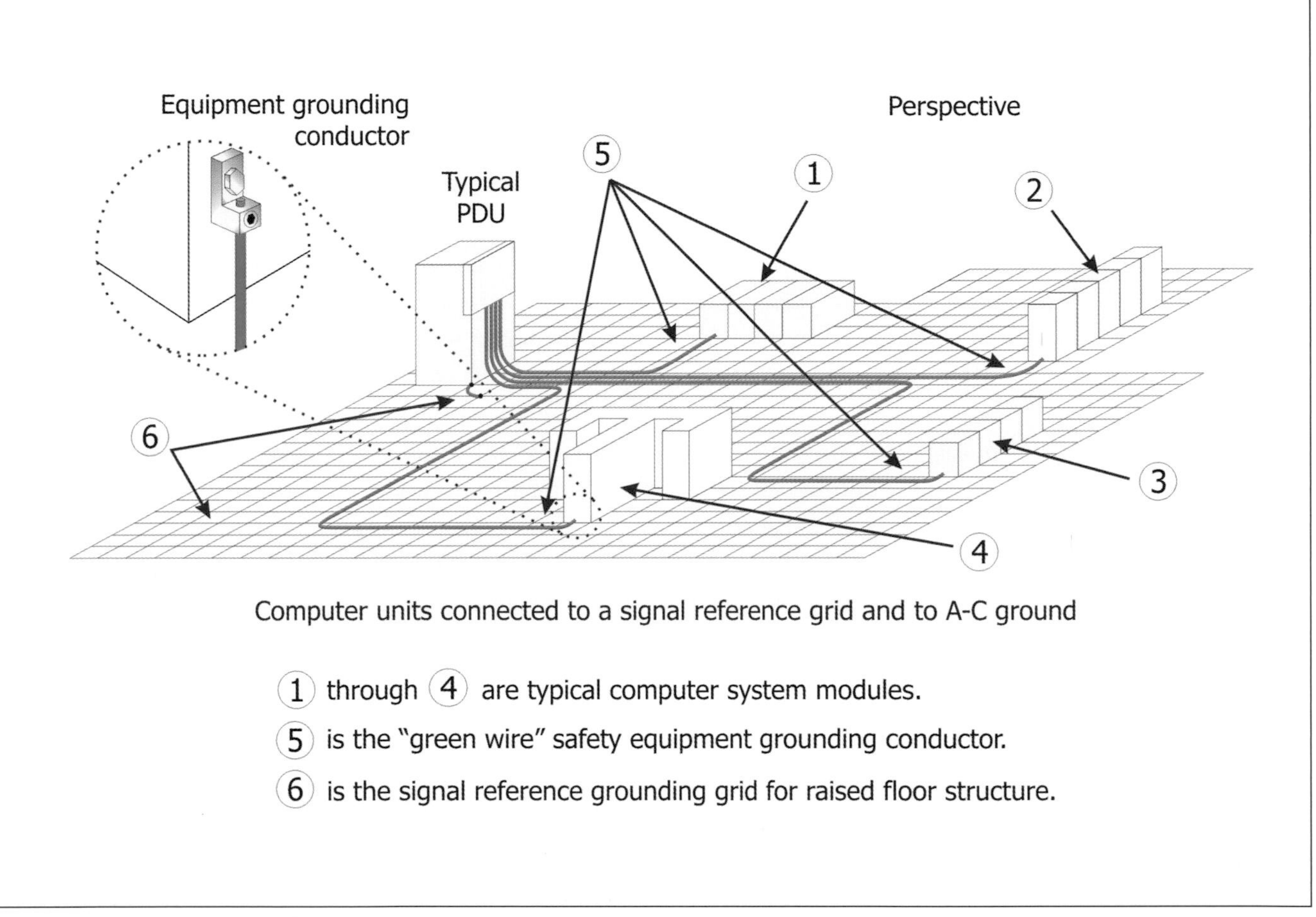
Equipment grounding conductor
Perspective
Typical PDU
5
1
2
3
4
6
Computer units connected to a signal reference grid and to A-C ground
1 through 4 are typical computer system modules.
5 is the "green wire" safety equipment grounding conductor.
6 is the signal reference grounding grid for raised floor structure.

Figure 17-8. Signal reference grid and equipment grounding conductors connected back to a single point at the source, usually a power distribution unit (PDU)

the raised floor reference grid will still help reduce voltage differences between grounded parts of the IT equipment.

High Frequency Effects in Grounding Conductors

Avoiding resonances at radio frequencies has become more important as the frequencies of digital signal circuits have increased beyond the 3 to 10 MHz clock repetition rate. Resonance occurs when the length of a conductor and the frequency of the alternating current are in tune. This is the principle of tuning a radio transmitting tower or antenna for maximum resonance and maximum radiation.

At frequencies slightly below and above resonance, the partial resonance still increases the impedance so it is ineffective as a constant potential conductor such as that needed as a good ground reference. Good engineering practice requires that a conductor any longer than 1/20 of a wavelength cannot be counted upon to equalize voltages between its ends. This amounts to only 1.34 m (4.4 ft) at 10 MHz.

The significance of this is that unless conductors can be limited to less than 1.22 m to 1.52 m (4 to 5 ft) in length, conventional grounding techniques with single-point grounds might not be effective for signal and noise frequencies up to 10 MHz. At higher frequencies and longer length of conductors, other techniques are needed to avoid resonance.

The use of multipoint grounding and short conductors appears to be the simplest and most reliable method for dealing with signals over 10 MHz in frequency (see figure 17-7). If bypass capacitors are installed instead of bonding straps, it is possible to make a conductor shield appear to be grounded at frequencies where the capacitors have very low impedance. At low frequencies, they carry very little current so the same system can have the characteristics of a single-point ground at power frequencies and other low frequencies. A signal reference grid is an effective way to ground systems and equipment installed in computer rooms.

Establishing a Single-Point Ground Reference

Deciding whether to use single-point grounding or multipoint grounding typically depends on the frequency range of equipment. Single-point grounding is typically applied to analog circuits with signal frequencies up to 300 kHz, while digital circuits with signal frequencies in the MHz range should utilize multipoint grounding.

The concept of single-point grounding is similar to the concept of equipotential bonding used in swimming pools (see figures 17-8). The intent is to maintain potential (voltage) differences among various IT equipment as low as possible. It should be noted that single-point grounding includes a set of grounding conductors that are in addition to the required equipment grounding conductors (sometimes referred to as *safety grounds*).

Single-point grounding should never be installed instead of the required equipment grounding conductors.

Single-point-grounding conductors are often connected from all equipment to an auxiliary (enhanced) grounding electrode in the vicinity of IT equipment. Requirements for auxiliary grounding electrodes are provided in 250.54. However, a bonding conductor must be installed from this auxiliary grounding electrode to the grounding electrode system for the electrical system. The auxiliary grounding electrode must never be isolated from the electrical system grounding electrode or be permitted to serve in place of the equipment grounding conductors (safety grounds). This clearly violates the installation instructions for equipment that is listed by a qualified electrical testing laboratory.

Surge Protection

The term *surge arrester* is defined in Article 100 as "a protective device for limiting surge voltages by discharging or bypassing surge current; it also prevents continued flow of follow current while remaining capable of repeating these functions." Rules for installing surge arrestors over 1 kV can be found in Article 280. They are permitted to be installed either at the service equipment or on the load side of the service equipment. Surge arresters installed on circuits of more than 1000 volts must be listed for the purpose as required in 280.6. Surge arresters must be made inaccessible to other than qualified persons unless listed for installation in accessible locations.

Surge arresters are of several basic types with different characteristics serving different functions in different ways. When used in combination in a coordinated manner, they provide protection by limiting overvoltage, reducing stress and unnecessary failures in components and reducing component failures in equipment caused by surges.

Two types of surge arresters become conductive when their threshold voltage is exceeded. The first type becomes and stays conductive until the current is reduced to zero such as when an alternating current passes through zero as it changes in amplitude and direction. The second type becomes conductive when its voltage threshold is exceeded but returns to its open circuit state as soon as the voltage drops below its threshold.

Pellet type and gas discharge type arresters are typical of the first type of surge arresters. They are usually capable of handling very high currents because the voltage drop through them becomes very low when they are conducting. However, if a weak, short duration impulse of a millisecond triggers the arrester, the shunting action would essentially short circuit the line for at least half of a cycle or possibly longer. The transient caused by shorting the power conductor can be greater than the event that triggered it.

The second type, typical of nonlinear insulating material, becomes conductive when voltage gradients exceed their threshold values. Some semiconductors in an avalanche mode exhibit similar characteristics not unlike Zener diodes. Metal oxide varistors or MOVs are marketed and widely used for arresting surges. These are constructed of an amorphous oxide material and exhibit these characteristics. Other types of surge arrestors of a silicon solid-state device are marketed which perform similar functions.

Surge arresters divert surge currents rather than absorb them. They are connected from ungrounded terminals in service equipment or panelboard to the

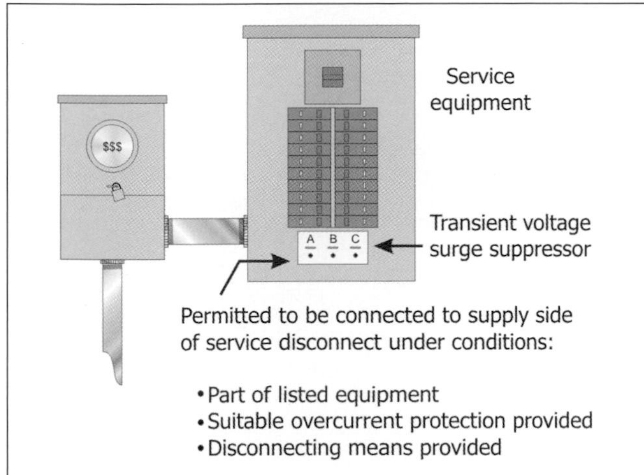

Figure 17-9. Surge protective device (SPD) component of a panelboard

ground terminal. Other common connection points include power conditioning equipment, the line and load side of transformers, motor generators and uninterruptible power supplies (UPS) systems.

There are risks associated with these devices, and they must be installed in compliance with manufacturer's installation instructions. Some devices can discharge flaming materials when they fail and must be mounted in suitable enclosures. The devices should be connected to the protected equipment with short conductors not more than .304 mm (1 ft) or so in length.

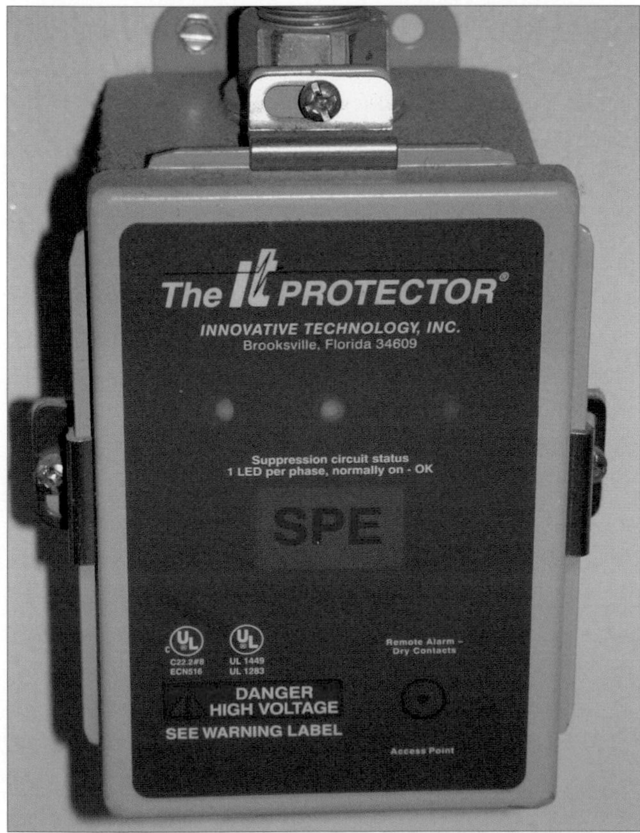

Photo 17-6. Typical three-phase surge protective device (SPD)

Some depth of technical understanding of surge voltages, the nature of traveling pulse waves, their reflection and diversion is needed for proper selection of surge arresters and the locations where they are most likely needed and can be most effective. In general, they are most likely needed wherever current in an inductive circuit element can be interrupted. They also might be useful where long cable leads interconnect widely separated systems as in different buildings or parts of a system within a very large building. Surge arresters are often installed at entrances of power and communications cables into a building to divert lightning and power switching surges before they reach IT equipment, power conditioning equipment or air-conditioning equipment.

Surge Protective Devices

Surge protective devices (SPDs) are available as standalone equipment or as a recognized component of listed equipment, such as switchboards or panelboards (see figure 17-9). Surge arresters rated 1 kV or less and surge protective devices are covered in Article 285 (see photo 17-6). The term *surge protective devices (SPDs)* is defined in Article 100 as "A protective device for limiting transient voltages by diverting or limiting surge current; it also prevents continued flow of follow current while remaining capable of repeating these functions and is designated as follows:

Type 1: Permanently connected SPDs intended for installation between the secondary of the service transformer and the line side of the service disconnect overcurrent device.

Type 2: Permanently converted SPDs intended for installation on the load side of the service disconnect overcurrent device, including SPDs located at the branch panel.

Type 3: Point of utilization SPDs

Type 4: Component SPDs, including discrete components, as well as assemblies.

Photo 17-7. Surge protective device (SPD) labels

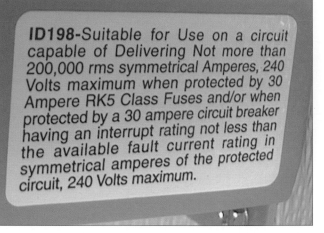

Photo 17-8. Surge protective device (SPD) labels showing short-circuit rating

Unlike surge arresters, surge protective devices (SPDs) are required to be listed per 285.5. They are not permitted on ungrounded systems, impedance grounded systems, or corner-grounded systems, unless listed for this use. According to 285.4, "where used at a point on a circuit, the SPD (surge arrester or TVSS) shall be connected to each ungrounded conductor." For other than receptacle-type devices, SPD (surge arrester or TVSS) must be marked with their short-circuit current rating and are not permitted to be installed at any point on the electrical system where the available fault current is in excess of that rating (see photos 17-6 through 17-8).

Installation requirements for SPDs are located in Part II of Article 285 and *Code* rules for connecting SPDs are provided in Part III of the article.

Harmonic Currents

Harmonics in the power source voltage to IT equipment can interfere with the proper operation of some IT equipment internal regulators. It can also create extra losses, excessive ground leakage current in line filters, and couple unwanted signals into low signal level conductors.

Harmonic currents are produced by equipment classed as being a nonlinear load. The term *nonlinear load* is defined in Article 100 as, "A load where the wave shape of the steady-state current does not follow the wave shape of the applied voltage." A fine print note following the definition indicates that "electronic equipment, electronic/electric-discharge lighting, ad-justable-speed drive systems, and similar equipment may be nonlinear loads." The power supplies in information technology equipment often are switching mode power supplies and are nonlinear loads. This nonlinear load can cause several problems in the electrical supply system including overloading neutral conductors and creating excessive heat in transformers.

Harmonic currents appear as reflected waveforms that distort the basic sine wave produced and provided by the electric utility. These reflected waves are multiples of the fundamental root frequency (60 Hz) sine wave. The odd multiples, such as the 3rd (180 Hz), 5th (300 Hz), 7th (420 Hz) add to the distortion of the sine wave and will appear as a single distorted waveform on an oscilloscope.

When the load current of IT units is distorted and contains harmonics, they interact with the power source impedance to create voltage drops at the harmonic currents that can be present. Line voltage distortion from this cause is the result of IT load characteristics interacting with the power source impedance. It is not correctable at the power source except to the extent that the source impedance can be reduced at each harmonic frequency by filters. Capacitors in parallel with the load can help but can cause unwanted resonance at specific frequencies.

Engineers will often specify K-rated transformers for IT installations. These transformers are designed to handle, without failure, the extra heat produced in the windings by the harmonic currents. K-rated transformers are available with ratings from

K-4 through K-20. In addition, design engineers often specify oversize neutrals to handle the extra nonlinear load without excessive voltage drop or heating.

[1] NFPA 70, *National Electrical Code* 2008, 645.4 (National Fire Protection Association, Quincy, MA, 2007, p. 70-544.

[2] NFPA 70, 645.15, p. 545.

[3] NFPA 70, 250.6(B), p. 70-97.

[4] NFPA 70, 250.4(A)(2), p. 95.

[5] NFPA 70, 250.4(A)(5), p.95.

[6] NFPA 70, 250.4(A)(5), p.95.

[7] NFPA 70, 250.110, p. 108–113.

[8] IEEE Std 1100-1999, *IEEE Recommended Practice for Powering and Grounding Electronic Equipment.*

For more detailed information on Grounding and Bonding requirements for Information Technology Equipment see FIPS 94 *Guidelines on Electrical Power for ADP Installations* and FIPS *Federal Information Processing Standards Publication*

[9] IEEE Std 1100-1999, *IEEE Recommended Practice for Powering and Grounding Electronic Equipment.*

[10] Burke, Thomas M., "Listing Requirements for ITE Installed in Computer Rooms," *IAEI News*, Volume 73, Number 1, January/February 2001, p. 11.

[11] Jones, Robert A., "Disconnecting Means—Requirements as Stated in 645.20 & 11," *IAEI News*, Volume 73, Number 1, January/February 2001, p. 22.

[12] Johnston, Michael J., "Installations and Inspections of Information Technology Equipment," *IAEI News*, Volume 73, Number 1, January/February 2001, p. 28.

17 Review Questions

The questions included here were developed using material included in this chapter. The answers can be found by reviewing the text. It is also important that students make use of *NEC*-2008, where many answers can be found.

1. Which of the following Articles in the *NEC* covers electrical requirements for Information Technology Equipment specifically?
   a. Article 250
   b. Article 645
   c. Article 725
   d. Article 800

2. The grounding requirements for information technology equipment in Article 250 apply
   a. To the entire installation
   b. Unless amended in Article 90
   c. Unless amended in Article 250
   d. Unless amended in Article 645

3. The wiring methods permitted in Article 645 for Information Technology equipment are permitted to be used
   a. Only where all the conditions of 645.4 are complied with
   b. For computer room equipment located in dedicated spaces
   c. For interconnection of peripheral equipment in open office spaces
   d. Provided a disconnecting means is located at exit doors

4. Power systems derived within listed information technology equipment that supply information technology systems
   a. Are considered separately derived systems and must be grounded
   b. Are considered separately derived systems but are not required to be grounded
   c. Are not considered separately derived systems
   d. Are considered Uninterruptible Power Systems (UPS)

5. Currents that introduce noise or data errors in electronic equipment
   a. Are considered a design flaw and are corrected as desired
   b. Are considered an unavoidable event in computer operation
   c. Are not considered objectionable currents
   d. Are not considered in design of systems

6. Which of the following sections in Article 250 generally requires that the grounded (often a neutral) conductor not be connected to ground past the service?
   a. 250.20
   b. 250.24(A)(5)
   c. 250.30(B)(2)
   d. 250.66

7. Which of the following statements most accurately describes an effective ground-fault current path?
   a. earth is not considered an effective path
   b. low-impedance circuit
   c. adequate current-carrying capacity
   d. all of the above

8. The terms "safety ground" and "clean ground"
   a. are defined in NFPA 70
   b. are defined in NFPA 75
   c. are IT jargon
   d. all of the above

9. Grounding IT equipment using a single connection only to "local" grounding means such as a ground rod rather than using the "safety" equipment grounding conductor
   a. is never permitted
   b. is a design choice
   c. is permitted to solve "dirty ground" problems
   d. is not likely to create a shock or electrocution risk

10. Which of the following statements is true regarding isolated grounding receptacles or equipment?
    a. They are permitted to be installed only where necessary.
    b. An insulated equipment grounding conductor is required.
    c. The equipment grounding conductor can pass through one or more panelboards, wireways, boxes, or other enclosures
    d. All of the above.

11. The insulated equipment grounding conductor for isolated grounding receptacles or equipment is permitted to terminate at which of the following locations?
    a. Source of the separately derived system if it supplies the branch circuit.
    b. The computer room disconnecting means.

c. The local grounding electrode near the computer room.
d. None of the above.

12. To reduce "ground loops" which of the following actions is sometimes taken?
a. Single point grounding.
b. Insert a balun coil.
c. Fiber optic cables.
d. All of the above.

13. Advantages of establishing an equipotential plane in IT equipment rooms include which of the following?
a. Low impedance return path for RF (radio frequency) noise currents.
b. Containment of electromagnetic (noise) fields between their source (cables and similar) and the plane.
c. Increasing filtering effectiveness of contained electromagnetic fields.
d. All of the above.

14. A "signal reference grid" can be thought of or is permitted to
a. Replace the equipment grounding conductor of the branch circuit.
b. Overlay the equipment grounding conductor of the branch circuit.
c. Be used in addition to the equipment grounding conductor of the branch circuit.
d. Any of the above.

15. Signal reference grids are typically constructed of which of the following materials or in which method?
a. Copper or aluminum strips.
b. 4 AWG copper or aluminum conductors.
c. The metal structure for the raised floor.
d. All of the above.

16. Which of the following statements about high frequencies in computer room circuits is or are true?
a. Digital signal circuits have increased beyond the 8 to 12 MHz clock repetition rate.
b. Resonance occurs when the length of a conductor and the frequency of the alternating current are in tune.
c. Good engineering practice requires that that a conductor any longer than 1/20th of a wavelength cannot be counted upon to equalize voltages between its ends.
d. All of the above.

17. Which of the following statements about single point grounding in computer rooms is or are true?
a. The intent is to maintain potential (voltage) differences among different IT equipment as low as possible.
b. Single-point grounding is typically applied to digital circuits with signal frequencies up to 300 kHz
c. Single point grounding includes a set of grounding conductors that replace the equipment grounding conductors.
d. Supplementary grounding electrodes are permitted to be isolated from the electrical system grounding electrode.

18. Which of the following statements about surge arrestors and surge protective devises (SPDs) is NOT true?
a. Surge arresters protect equipment by absorbing excessive voltages.
b. Surge arresters installed on circuits of more than 1000 volts are required to be listed.
c. Surge protective devices are required to be listed.
d. There are 4 types of surge protective devices (SPDs).

19. Which of the following statements about harmonic currents and nonlinear loads is NOT true?
a. Nonlinear loads can cause several problems in the electrical supply system including overloading neutral conductors and creating excessive heat in transformers.
b. Harmonic currents appear as reflected waveforms that distort the basic sine wave produced and provided by the electric utility.
c. The even multiples of the sine wave, such as the 2nd (120 Hz), 4th (240 Hz), 6th (360 Hz) are the most troublesome harmonic currents.
d. Engineers often specify oversize neutrals to handle the extra nonlinear load without excessive voltage drop or heating.

20. Power distribution units (PDUs) that are used for information technology equipment shall be permitted to contain multiple panelboards provided that each panelboard within the PDU contains no more than __________ overcurrent devices.
a. six
b. ten
c. fourty-two
d. two

Objectives

To understand...

- Grounding requirements for low-voltage circuits and systems less than 50 volts
- Low-voltage motor control and signaling circuits
- Requirements for intrinsically safe systems
- Purpose of an intersystem bonding termination
- Acceptable points on an existing building or structure grounding electrode system where a grounding electrode connection can be made
- Cable grounding requirements
- Grounding conductor points for a radio or TV antenna system

Chapter 18

Low-Voltage and Intersystem Grounding and Bonding

Grounding requirements for low-voltage circuits and systems less than 50 volts are covered in 250.20(A). Typical low-voltage systems falling into this category can include motor control circuits, Class 1, 2, or 3 circuits, and remote control or signaling circuits, such as low-voltage circuits for electric door release equipment or doorbell circuits. The *Code* requires these systems to be grounded under any of the following conditions:

Photo 18-1. Grounded secondary of control circuit transformer in motor controller

• "Where supplied by transformers, if the transformer supply system exceeds 150 volts to ground." [1] A low-voltage ac system must be grounded if it is supplied by a transformer secondary that is supplied on the primary side with a voltage exceeding 150 volts from the ungrounded conductors to ground. The primary circuits for these transformers are supplied from a grounded system, so there will be measurable voltages from any of the phase (ungrounded) conductors to ground (earth) or grounded metal parts.

• "Where supplied by transformers, if the transformer supply system is ungrounded." [2] A secondary system must also be grounded where the transformers for these circuits or systems are supplied on the primary side by an ungrounded system, such as by an ungrounded 480-volt or 240-volt power system, which is not required to be grounded by 250.20(B).

• "Where installed as overhead conductors outside of buildings." [3] When the conductors are installed in exterior locations as overhead conductors outside of buildings, the previously mentioned low-voltage systems or circuits must be grounded. These would include, but are not limited to, signal circuits, fire alarm, and other control circuits or systems.

Where low-voltage circuits or systems do not fall into the requirements of 250.20(A) as outlined above, grounding these systsems is optional. In other words, these types of systems or circuits could be operated ungrounded with no conductor of the system or circuit supplied by the secondary, if the transformer is intentionally grounded.

Low-Voltage Motor Control Circuits

Low-voltage motor control circuits are required to be grounded if any of the conditions in 250.20(A)(1) or 250.20(A)(2) exist.

Motor controllers and combination motor control units are often supplied with control-circuit transformers that are factory grounded on the secondary of the low-voltage side by the manufacturer (see photo 18-1). If the motor control circuits are field-installed and wired, then grounding may or may not be required, depending on the voltage or characteristics of the circuit on the primary side of the control circuit transformer. The determining factors for grounding a low-voltage motor control circuit on the secondary of the transformer are if it is supplied on the primary by a circuit derived from an ungrounded system, or if the primary is supplied by a circuit that exceeds 150 volts between any of the primary conductors to ground. The requirements for motor control circuits are located in Part VI of Article 430.

The motor control circuit is defined in 430.2 as "the circuit of a control apparatus or system that carries the electric signals directing the performance of the controller but does not carry the main power current. The requirements for motor control circuits include overcurrent protection, physical protection, disconnecting means, and control circuit transformers located in enclosures. Section 430.72(C)(1) includes the overcurrent protection requirements for the control circuit transformers. Motor control circuit transformers supplying "a Class 1 power-limited circuit, Class 2, or Class 3 remote-control circuit" complying with the requirements of Article 725, protection shall comply with Article 725."

Other Signaling Circuits

Low-voltage signaling circuits are required to be grounded circuits or systems if either 250.20(A)(1) or (2) applies. For example, the circuit is supplied by a transformer that has a primary voltage of 480-volts single-phase and steps the voltage down for the signal circuit to 24-volts AC for a horn or bell annunciation. Low-voltage signaling circuits are commonly used for doorbell installaions in dwelling units. The voltage for these door chime circuits is usually less than 24 VAC and the systems are ungrounded.

Intrinsically Safe Systems and Circuits

Branch circuits that supply intrinsically safe systems must include an equipment grounding conductor as covered in 250.118. The equipment grounding conductor is used for grounding the metal enclosure(s) and other metal parts and equipment of the system.

Article 504 of the *Code* includes the requirements for intrinsically safe systems, and 504.50 and 504.60 cover the grounding and bonding requirements for these systems.

Section 504.50(A) requires "intrinsically safe apparatus, associated apparatus, cable shields, enclosures, and raceways, if of metal, shall be connected to the equipment grounding conductor." A control drawing is required for these systems to provide specific information and instructions. These control drawings include grounding and bonding information critical to the integrity of the intrinsically safe system or circuit(s). Supplementary bonding to the grounding electrode may be needed for some associated apparatus (zener diode barriers, for example) if specified in the control drawing. Additional information relative to these systems can be found in ANSI/ISA RP 12.6.01-2003, *Wiring Practices for Hazardous (Classified) Locations Instrumentation Part 1: Intrinsic Safety.*

It is not uncommon for the required control drawing(s) to specify a grounding electrode conductor connection for these systems in addition to connection to an equipment grounding conductor. Usually, terminals for both of these conductors are located within the system enclosure or control panel. If a grounding electrode is required, then one of the electrodes in 250.52(A) must be used. The choice of which grounding electrode(s) must be used is governed by the same provisions specified in 250.30(A)(7) for separately derived systems. The electrode must be "as close as practicable and preferably in the same area as the grounding electrode conductor connection to the system." The grounding electrode shall be the nearest one of the following grounding electrodes in no particular order:

A structural metal grounding electrode of the structure as specified in 250.52(A)(1), (A)(2), (A)(3), and (A)(4) and shall comoply with 250.30(A)(7).

The size of this grounding electrode conductor is generally specified by the manufacturer's control drawing. Where shielded conductors or cables are used, shields shall be grounded in accordance with the required control drawing." The bonding requirements for metallic enclosures and raceways enclosing intrinsically safe systems or circuits in hazardous locations must be in accordance with any of the methods that are suitable for the bonding on the line side of a service.

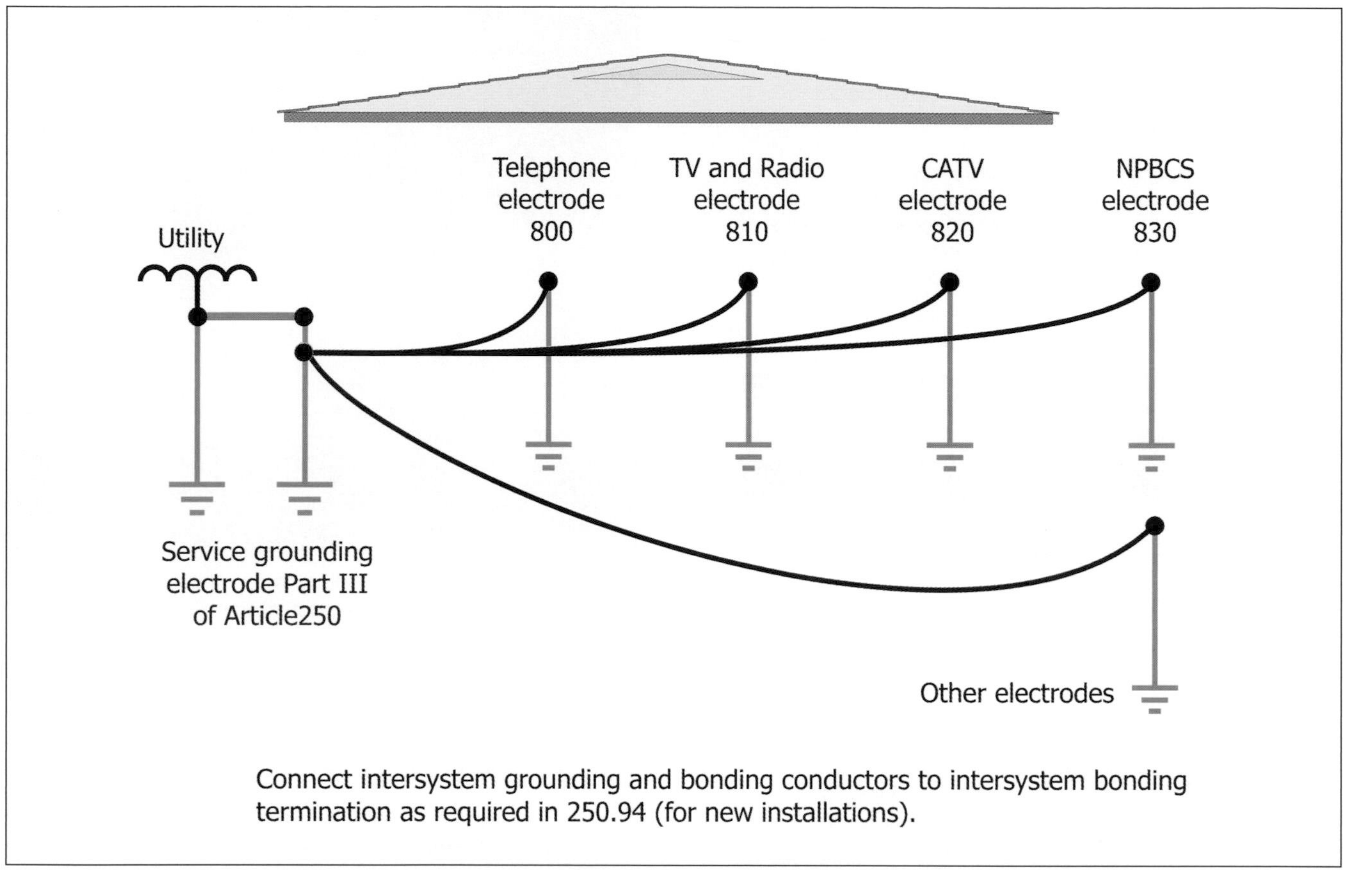

Figure 18-1. Intersystem grounding electrodes (bond together and connect to intersystem bonding termination)

Section 250.100 indicates that "regardless of the voltage of the electrical system, the electrical continuity of non-current-carrying metal parts of equipment, raceways, and other enclosures in any hazardous (classified) location as defined in 500.5 shall be ensured by any of the bonding methods specified in 250.92(B) (2) through (B)(4)." Note that these grounding methods are required even if an equipment grounding conductor of the wire type is installed.

Sections 501.30, 502.30, and 503.30 also provide information relative to restrictive grounding and bonding requirements in hazardous locations. Bonding must be extended and maintained for all intervening raceways and enclosures from the hazardous location all the way to the applicable service or separately derived system (see chapter sixteen for more detailed information on bonding requirements in hazardous locations).

The Purpose of Intersystem Grounding and Bonding

Intersystem grounding and bonding requirements serve a few important purposes [see figure 18-1]. The systems and circuits covered in *NEC* chapter 8 must be grounded (to earth) and bonded to the electrical power distribution systems for the building or structure.

By providing the connection to a grounding electrode, the systems and circuits are afforded reasonable protection from spike and surge currents and also brief elevated potentials due to lightning strikes. Bonding the electrodes of the systems listed in *NEC* chapter 8 to the grounding electrode system for the electrical system(s) of the building or structure limits the potential differences during normal operation and during surge or spike events and the effects of lightning discharges into the earth at close proximities. Bonding the electrodes of the two different systems together limits the potential differences and shock hazards that could result from isolated grounding (earthing) electrode connections.

The *NEC* is specific in Articles 800, 810, 820, and 830 about grounding electrodes to be used with these systems. Each article requires an effective

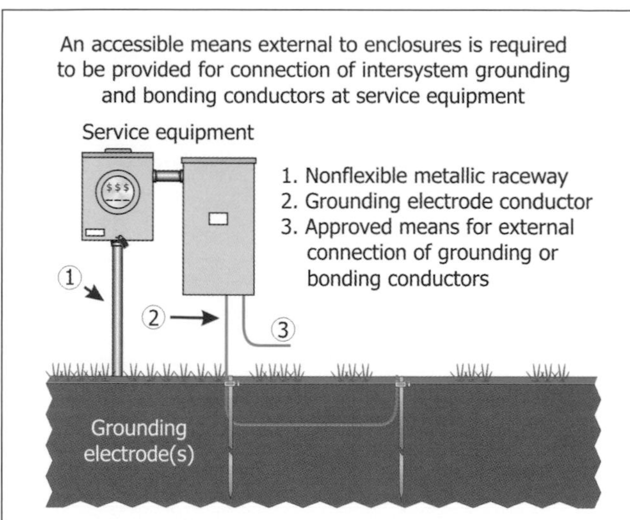

Figure 18-2. Bonding for other systems (250.94)

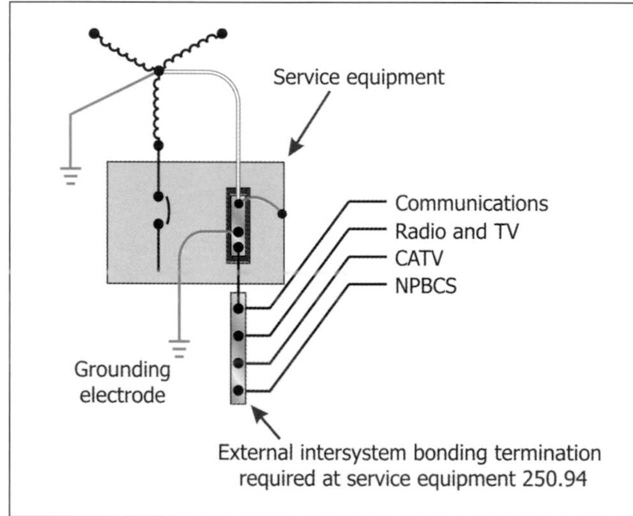

Figure 18-3. Intersystem bonding and grounding

bonding connection to the electrical power grounding electrode(s) system with a minimum 6 AWG copper conductor. This means that where items such as satellite dishes, TV antennas, radio antennas, and the like are installed at any occupancy, electrodes are required; and the electrode must be the same electrode used by the electrical service or system. All grounding electrodes that are installed must be bonded to the building grounding electrode system.

The grounding and bonding for these systems, although referred to as intersystem grounding and bonding, is usually not the grounding of the particular system itself. Part II of Article 250 is specifically titled Circuit and System Grounding. Section 250.20 specifies systems required to be grounded, systems

Photo 18-2. Communications system grounding conductor connected to electric service grounding electrode conductor

permitted to be grounded, and systems not permitted to be grounded.

Intersystem Bonding Termination

The requirements for safety of communications systems, antenna systems, CATV and radio systems, and network-powered broadband communications systems are located in *NEC* chapter 8 entitled Communications Systems. Section 250.94 includes the requirements for intersystem bonding and grounding (see figures 18-2 and 18-3).

An intersystem bonding termination is required to be provided external to service equipment and at disconnecting means for additional buildings or

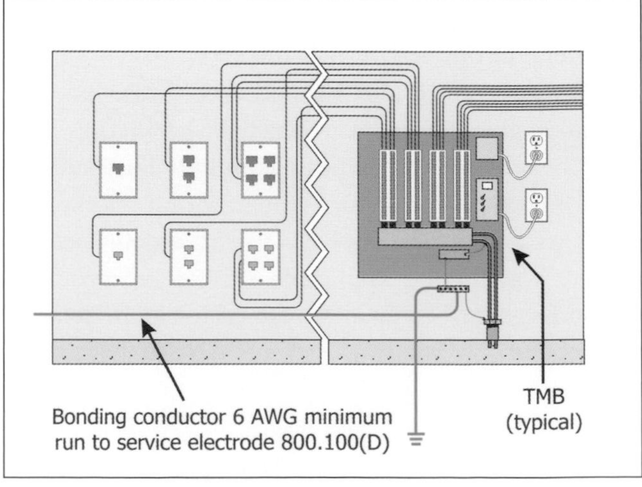

Figure 18-4. Communications systems grounding

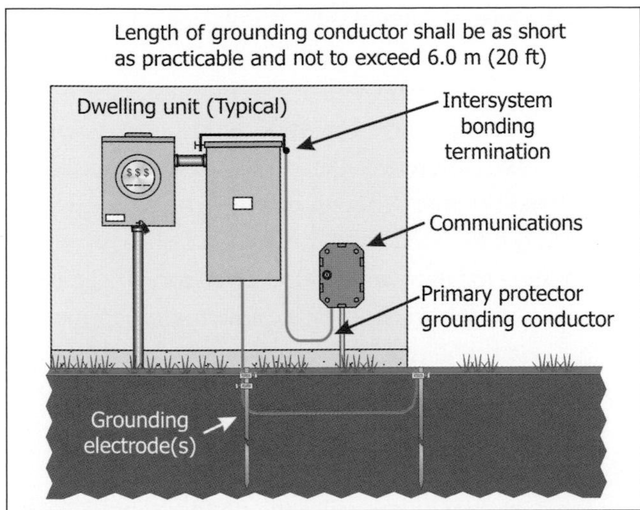

Figure 18-5. Length of grounding conductor for cable and primary protector (dwellings)

structures supplied by a feeder or branch circuit. The intersystem bonding termination is required to have provisions for connecting not less than three intersystem bonding conductors and shall be so installed that it does not interfere with opening service or metering equipment enclosures [250.94] (see figure 18-3).

The systems that the *Code* points to here are low-voltage (limited energy) systems and frequency signals or communications circuits and systems.

These systems are not normally grounded systems as covered in 250.20, where one conductor of the system or service is intentionally grounded. The grounding and bonding requirements for the systems provide safety from shock hazards and fire hazards and, in addition, provide important equipment and system/circuit protection from damaging surges or dips in the normal electrical supply. The following definitions are applicable to the installations of these voice, data, and video (VDV) systems.

Definitions

Grounding conductor. "A conductor used to connect equipment or the grounded circuit of a wiring system to a grounding electrode or electrodes." [5]

Bonded (Bonding). "Connected to establish electrical continuity and conductivity." [6]

Bonding jumper. "A reliable conductor to ensure the required electrical conductivity between metal parts required to be electrically connected." [7]

Intersystem bonding termination. "A device that provides a means for connecting communications system(s) grounding conductor(s) and bonding conductor(s) at the service equipment or at the disconnecting means for buildings or strutures supplied by a feeder or branch circuit."

Article 800, Communications Circuits, covers telephone, telegraph (except radio), outside wiring for fire alarm and burglar alarm, and similar central station systems; and telephone systems not connected to a central station system but using similar types of equipment, methods of installation, and maintenance for such circuits and systems. Grounding and bonding requirements for communications systems are found in Part IV and specifically in 800.100 and 800.106.

- "Insulation. The grounding conductor shall be insulated and shall be listed as suitable for the purpose.
- "Material. The grounding conductor shall be copper or other corrosion-resistant conductive material, stranded or solid.
- "Size. The grounding conductor shall not be smaller than 14 AWG.
- "Length. The primary protector grounding conductor shall be as short as practicable. In one- and two-family dwellings, the primary protector grounding conductor shall be as short as practicable, not to exceed 6.0 m (20 ft) in length[8] (see figures 18-5 and 18-6).

This grounding also serves as a path of discharge for lightning strikes and other events that work to raise potentials on these systems. By keeping the length of this grounding conductor as short as practicable, the effectiveness of the grounding conductor to provide a low-impedance path to ground to dissipate lightning and other surge events is established.

The exception to 800.100(A)(4) reads, "In one- and two-family dwellings where it is not practicable to achieve an overall maximum primary protector grounding conductor length of 6.0 m (20 ft), a separate communications ground rod meeting the minimum dimensional criteria of 800.100(B)(2)(2) shall be driven, the primary protector shall be connected to the communications ground rod in accordance with 800.100(C), and the communications ground rod shall be connected to the power grounding electrode system in accordance with 800.100(D)." In this case, the electrode for the communications system is to be installed and bonded to the service grounding electrode system with a bonding jumper sized not smaller than 6 AWG copper (see figures 18-5 and 18-6).

Section 800.100(A)(5) also requires that "the grounding conductor shall be run to the grounding electrode in as straight a line as practicable." If the grounding conductor is installed in a metal raceway, both ends of the grounding conductor must be bonded to the grounding conductor.

The grounding conductor shall be protected where exposed to physical damage.

Electrodes

For communications systems, the goal is to establish a connection to a grounding electrode. Section 800.100(B)(1) requires an intersystem bonding termination to which the communications system(s) grounding conductors must be connected. See Article 100 for the definition of intersystem bonding termination. The *NEC* also specifies several acceptable points on an existing building or structure grounding electrode system where this connection can be made (see figure 18-2 and 18-4). This preferred method eliminates distances and potential differences between the earthing circuits of the two systems. Where a building or structure has no intersystem bonding termination as required by 800.100(B)(1) for new installations, Section 800.100(B)(2) requires that the grounding conductor be connected "to the nearest accessible location on the following grounding electrodes:

"(1) The building or structure grounding electrode system as covered in 250.50

"(2) The grounded interior metal water piping system, within 1.5 m (5 ft) from its point of entrance to the building, as covered in 250.52

"(3) The power service accessible means external to enclosures as covered in 250.94

"(4) The metallic power service raceway

"(5) The service equipment enclosure

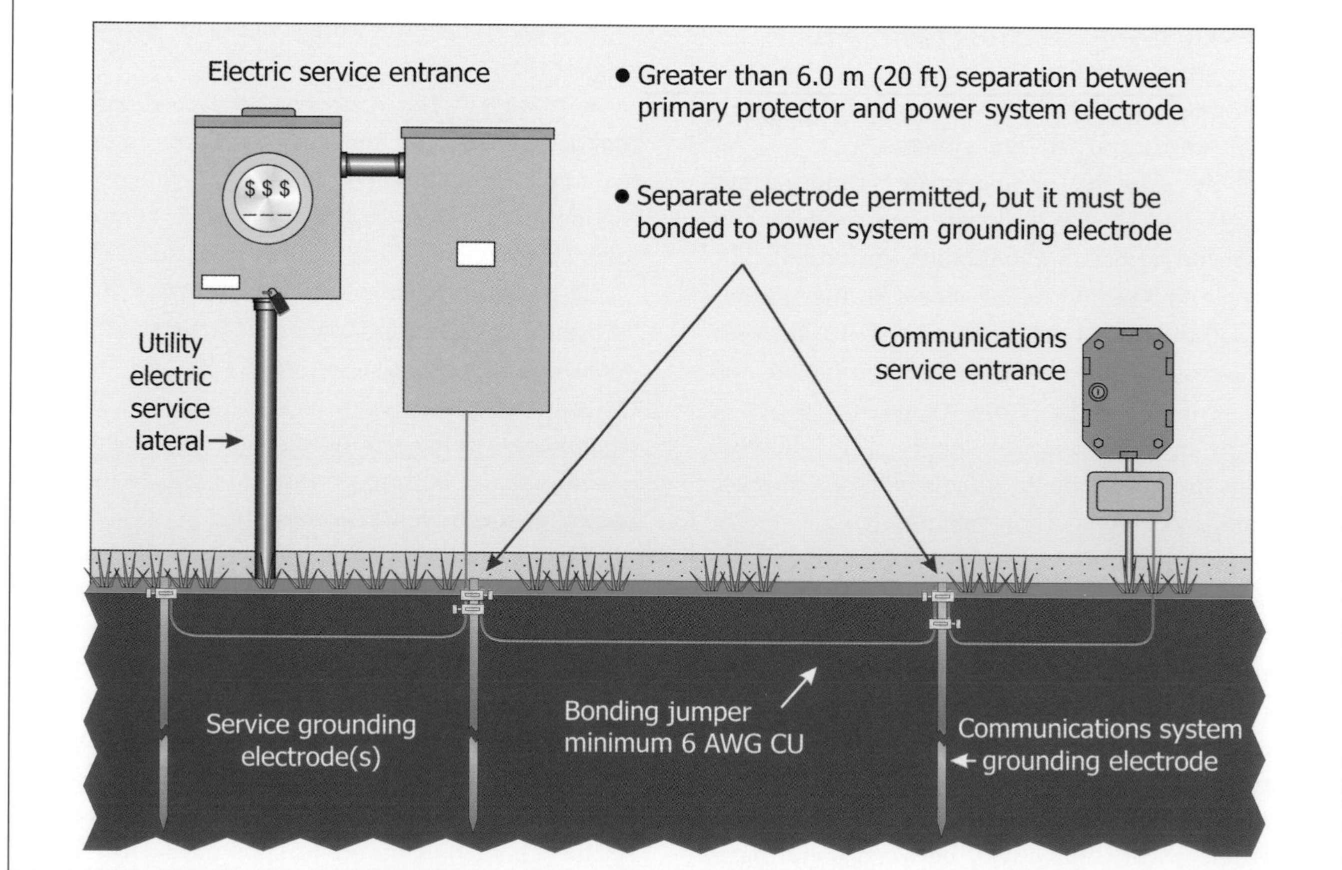

Figure 18-6. Length of grounding conductor for cable and primary protector (dwellings). The exception permits bonding of separate electrodes together

"(6) The grounding electrode conductor or the grounding electrode conductor metal enclosure

"(7) The grounding conductor or the grounding electrode of a building or structure disconnecting means that is grounded to an electrode as covered in 250.32."

For purposes of this section, the mobile home service equipment or the mobile home disconnecting means, as described in 800.90(B), shall be considered accessible.[9]

In buildings or structures that have no intersystem bonding termination or grounding electrode or grounding electrode system, an electrode is required to be installed. The connections of grounding conductors to grounding electrodes shall comply with the provisions in 250.70. "Connectors, clamps, fittings, or lugs used to attach grounding conductor and bonding jumpers to grounding electrodes or to each other that are to be concrete-encased or buried in the earth shall be suitable for this specific application." [10]

Intersystem Grounding and Bonding at Mobile Homes

The grounding for communications systems on a mobile home generally follows the same requirements as for permanent buildings or structures. Electrically, the objectives are similar. The goal is to establish a grounding (earth) connection to eliminate differences of potential, and to provide a path into the earth to dissipate lightning strikes.

"Where there is no mobile home service equipment located within 9.0 m (30 ft) of the exterior of the mobile home it serves, the primary protector ground shall be connected to a grounding conductor in accordance with 800.100(B)(2). The bonding requirements are a bit more specific for the mobile home applications. [11]

"(B) Bonding. The primary protector grounding terminal or grounding electrode shall be connected to the metal frame or available grounding terminal of the mobile home with a copper conductor not smaller than 12 AWG under any of the following conditions:

"(1) Where there is no mobile home service equipment or disconnecting means as in 800.106(A)

"(2) Where the mobile home is supplied by cord and plug" [12]

Articles 810, 820, and 830 include the grounding and bonding requirements for these particular installations although the system itself is usually ungrounded.

As indicated by 90.3, *NEC* chapter 8 is not subject to the other chapters and stands alone unless specifically referenced from therein. The term *grounding conductor* as used in chapter 8 in any of these articles is referring to the conductor that is connected to an electrode that functions similar to a grounding electrode conductor. The *Code* permits the minimum sizes of these grounding (earth) conductors to be smaller than normally is allowed by chapter 2 in Article 250. The communications system grounding conductor is permitted to be as small as a 14 AWG. However, the bonding conductor required to connect the grounding electrode for the power distribution system to the grounding electrode for the communication system is always required to be at least a minimum size of 6 AWG copper.

Communications Cable Grounding Requirements

The metallic sheaths of communications cables entering buildings shall be grounded as close as practicable to the point of entrance or shall be interrupted by an insulating joint or equivalent device located as close as practicable to the point of entrance. If the metallic shields of the cables are grounded, they are required to be grounded by a conductor that is insulated and suitable for the purpose. This conductor is required to be copper or other corrosion resistant conductor and can be solid or stranded not smaller than 14 AWG [800.100(A)(3)]. It is important that the routing and orientation of the conductor be installed so that it will be in as straight a line as practicable. When functioning to dissipate the effects of lightning, it is important that the conductor be installed with straight lines and gradual bends. The grounding conductor is required to be installed in a manner so as to be protected from physical damage. This is often accomplished by installing it in a raceway. If the grounding conductor is installed in a metal raceway, both ends of the raceway shall be bonded to the grounding conductor or the same terminal or electrode to which the grounding conductor is connected [800.100(A)(6)]. The grounding electrode to be used shall be electrode or point of grounding connection as indicated in 800.100(B):

"(B) Electrode. The grounding conductor shall be connected in accordance with 800.100(B)(1), (B)(2), or (B)(3):

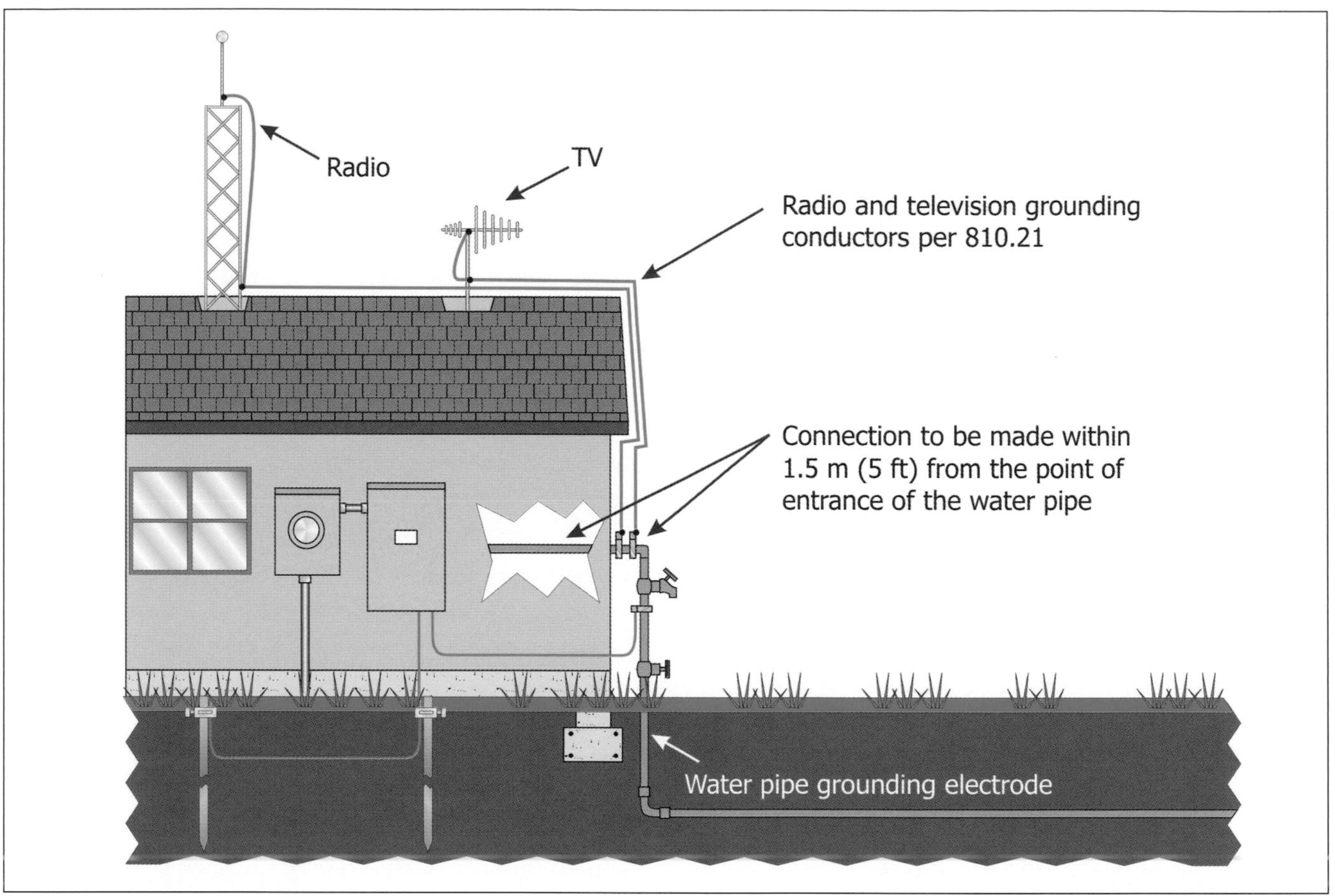

Figure 18-7. Interior metal water pipe electrode connections to be made within the first 1.52 m (5 ft) of the piping system entrance to the building or structure

"(1) In Buildings or Structures with an Intersystem Bonding Termination. The grounding conductor is required to be connected to the intersystem bonding termination. If the building or structure has no intersystem bonding termination, the grounding conductor is required to be connected to the nearest accessible location on the following:

"(1) The building or structure grounding electrode system as covered in 250.50

"(2) The grounded interior metal water piping system, within 1.5 m (5 ft) from its point of entrance to the building, as covered in 250.52

"(3) The power service accessible means external to enclosures as covered in 250.94

"(4) The metallic power service raceway

"(5) The service equipment enclosure

"(6) The grounding electrode conductor or the grounding electrode conductor metal enclosure

"(7) The grounding conductor or the grounding electrode of a building or structure disconnecting means that is grounded to an electrode as covered in 250.32." [13]

Differences of Potential

Bonding together all separate electrodes will limit potential differences between them and between their associated wiring systems. Grounding electrodes of different systems or circuits that are not bonded together can present hazards for property and persons. It is important for eliminating shock hazards to bond these intersystem electrodes to the power distribution system grounding electrode system.

Bonding Electrodes of Different Systems

Section 800.100(D) states that "a bonding jumper not smaller than 6 AWG copper or equivalent shall be connected between the communications grounding electrode and power grounding electrode system at the building or structure served where separate electrodes are used." Bonding together of all separate electrodes shall be permitted.

Radio and TV Antennas

Grounding and Discharge Equipment

Masts and metal structures supporting antennas shall be grounded in accordance with 810.21. Where required, each conductor of a lead-in from an outdoor antenna shall be provided with a listed antenna discharge unit. The antenna discharge unit shall be grounded in a manner specified by 810.21.

Grounding Conductor Installations

The grounding conductor for a radio or TV antenna system is required to be connected to an electrode or other suitable grounding connection point as specified by 810.21(F)(1) through (F)(3). These requirements are the same as the requirements for communications systems in Article 800. The size of the grounding conductors is a bit different, however, than the size required in Article 800.

"(A) Material. The grounding conductor shall be of copper, aluminum, copper-clad steel, bronze, or similar corrosion-resistant material. Aluminum or copper-clad aluminum grounding conductors shall not be used where in direct contact with masonry or the earth or where subject to corrosive conditions. Where used outside, aluminum or copper-clad aluminum shall not be installed within 450 mm (18 in.) of the earth.

"(B) Insulation. Insulation on grounding conductors shall not be required.

"(C) Supports. The grounding conductors shall be securely fastened in place and shall be permitted to be directly attached to the surface wired over without the use of insulating supports.

"(D) Mechanical Protection. The grounding conductor shall be protected where exposed to physical damage, or the size of the grounding conductors shall

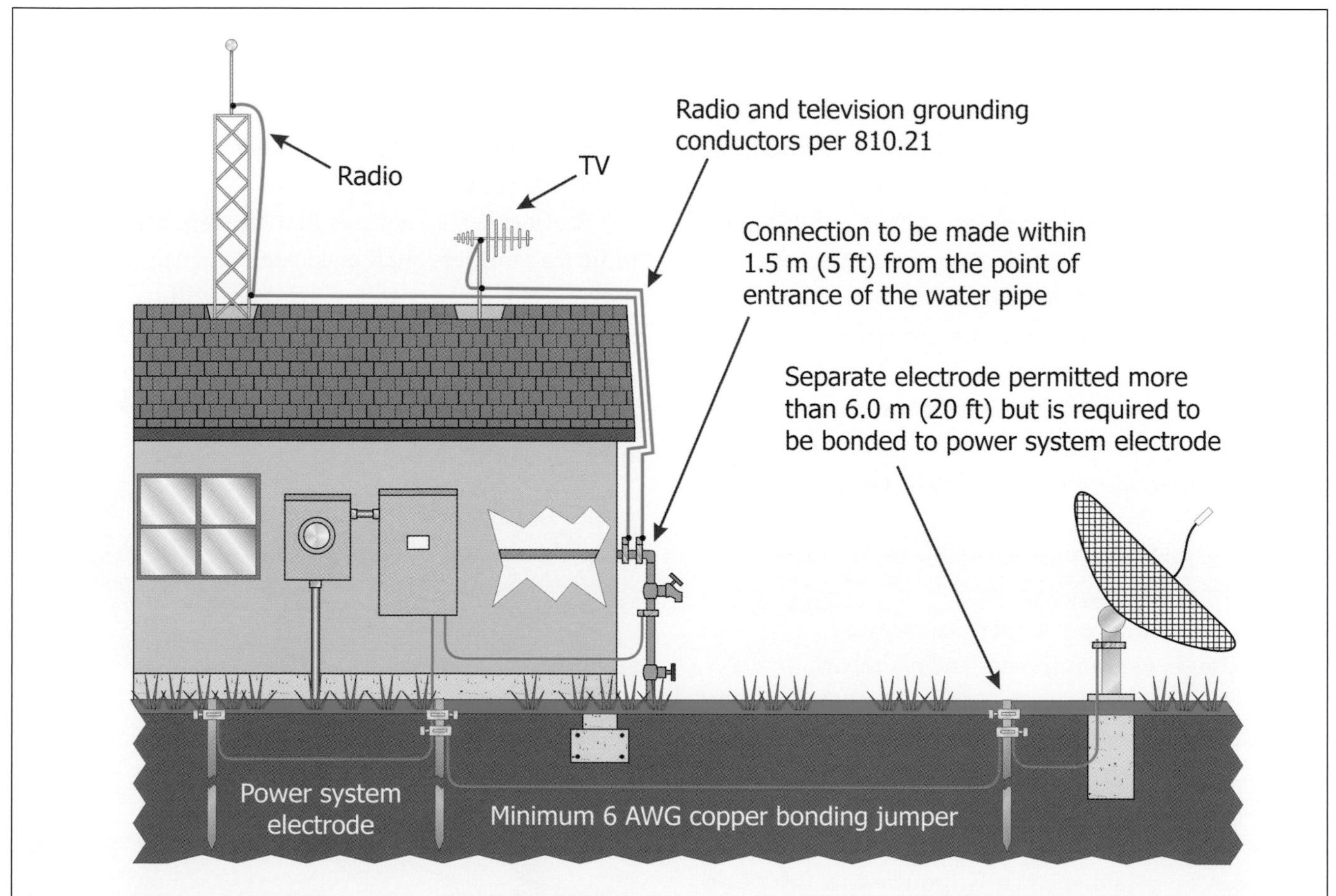

Figure 18-8. Interior metal water pipe electrode connections are to be made within the first 1.52 m (5 ft) of the piping system entrance to the building or structure. Bond separate grounding electrodes to power system or service electrode(s).

be increased proportionately to compensate for the lack of protection.

"(E) Run in Straight Line. The grounding conductor for an antenna mast or antenna discharge unit shall be run in as straight a line as practicable from the mast or discharge unit to the grounding electrode.

"(G) Inside or Outside Building. The grounding conductor shall be permitted to run either inside or outside the building.

"(H) Size. The grounding conductor shall not be smaller than 10 AWG copper, 8 AWG aluminum, or 17 AWG copper-clad steel or bronze.

"(I) Common Ground. A single grounding conductor shall be permitted for both protective and operating purposes." [14]

Electrode

The *Code* specifies several acceptable connection points on an existing building or structure grounding electrode system. This would be the preferred method, and eliminates distances and potential differences between the grounding (earthing) circuits of the two systems. Section 810.21(F)(1) requires the radio and television equipment grounding conductors(s) to be converted to an intersystem bonding termination. Where no intersystem bonding termination is present for use, Section 810.21(F)(2) requires the grounding conductor to be connected to the nearest accessible location on the following:

"(1) The building or structure grounding electrode system as covered in 250.50

"(2) The grounded interior metal water piping systems, within 1.52 m (5 ft) from its point of entrance to the building, as covered in 250.52 (see figures 18-7 and 18-8)

"(3) The power service accessible means external to the building, as covered in 250.94

"(4) The metallic power service raceway

"(5) The service equipment enclosure, or

"(6) The grounding electrode conductor or the grounding electrode conductor metal enclosures…" [15]

If the building or structure has no intersystem bonding termination or grounding electrode system as described in 810.21(F)(3), any of the electrodes in 250.52 can be established. It is required to install an electrode if any of the electrodes in 250.50 are not present.

According to 810.21(G) and (H), "the grounding conductor is permitted to run either inside or outside the building," and "shall not be smaller than 10 AWG copper, 8 AWG aluminum, or 17 AWG copper-clad steel or bronze." The connection of the grounding conductor shall be made to the grounding electrode and serves both as an operational grounding connection and as a protective earthing connection for lightning and other events affecting the system and potential differences.

The grounding electrode for a radio or television antenna is required to be bonded to the electrode system for the building. "A bonding jumper not smaller than 6 AWG copper or equivalent shall be connected between the radio and television equipment grounding electrode and the power grounding electrode system at the building or structure served if separate electrodes are used" [16]

The connection(s) to grounding electrode(s) is required to meet the requirements of 250.70 (see figure 18-8).

Amateur Transmission and Receiving Stations

Lead-in antenna conductors shall be installed in such a manner that affords a degree of protection against accidental contact by personnel.

Section 810.56 requires that "lead-in conductors to radio transmitters shall be located or installed so as to make accidental contact with them difficult." Where transmitting or receiving stations are installed, "each

Photo 18-3. Coaxial cable shields grounded at point of entrance to building or structure

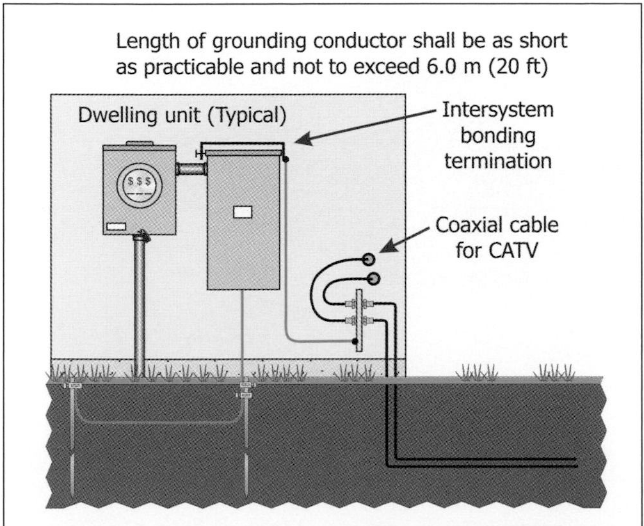

Figure 18-9. Length of grounding conductor for coaxial cable systems installations. The exception permits separate electrodes but they must be bonded to the power system electrode.

conductor of a lead-in for outdoor antennas shall be provided with an antenna discharge unit or other suitable means that will drain static charges from the antenna system.

"Exception No. 1: Where the lead-in protected by a continuous metallic shield that is permanently and effectively grounded with a conductor in accordance with 810.58, an antenna discharge unit or other suitable means shall not be required.

"Exception No. 2: Where the antenna is permanently grounded with a conductor in accordance with 810.58, an antenna discharge unit or other suitable means shall not be required." [17]

There are two grounding conductors required for this specific equipment: an operating grounding conductor and a protective grounding conductor. These grounding conductors for amateur transmitting

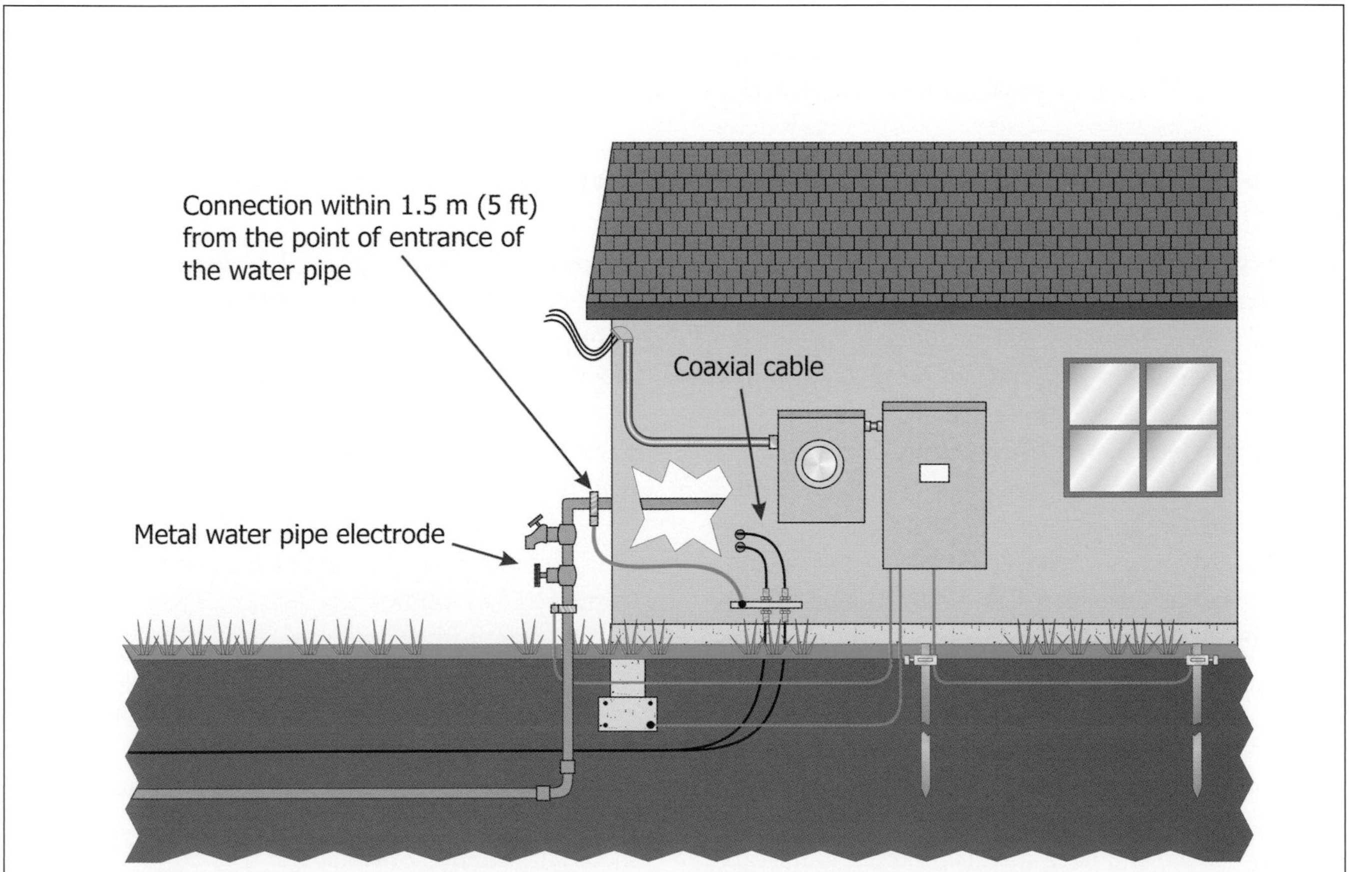

Figure 18-10. Interior metal water pipe electrode connections are to be made within the first 1.52 m (5 ft) of the piping system entrance to the building or structure.

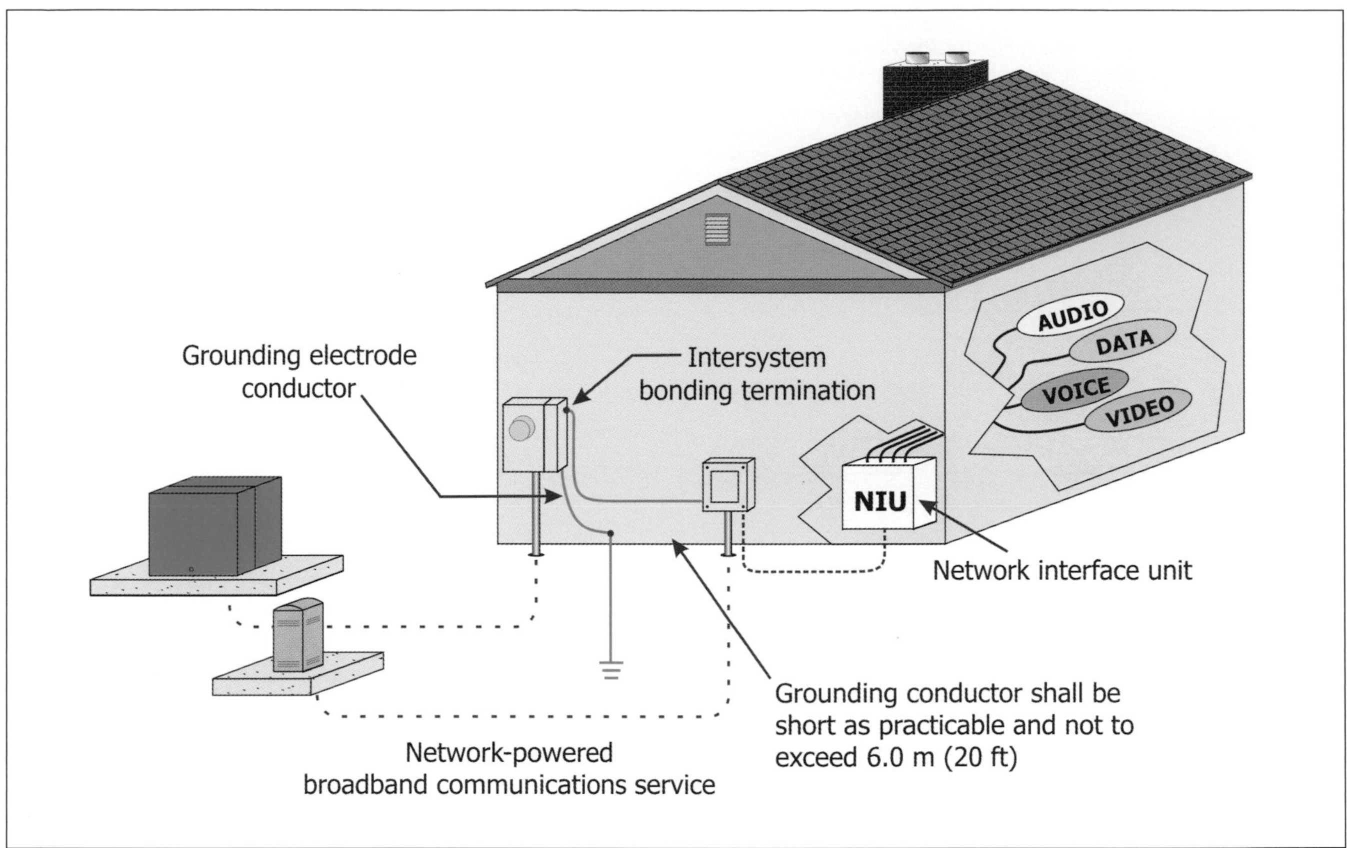

Figure 18-11. Cable network interface unit and primary protector grounding conductors connected to same electrode as power system or service.

and receiving stations are required to be installed in accordance with 810.58(A), (B), or (C) as follows:

"(A) Other Sections. All grounding conductors for amateur transmitting and receiving stations shall comply with 810.21(A) through (J).

"(B) Size of Protective Grounding Conductor. The protective grounding conductor for transmitting stations shall be as large as the lead-in but not smaller than 10 AWG copper, bronze, or copper-clad steel.

"(C) Size of Operating Grounding Conductor. The operating grounding conductor for transmitting stations shall not be less than 14 AWG copper or its equivalent." [18]

Community Antenna Television and Radio Distribution (CATV) Systems

CATV systems are usually wired with conductors that are coaxial and have shields inherent to the cable assembly. "The outer conductive shield of the coaxial cable shall be grounded at the building premises as close to the point of cable entrance or attachment as practicable."[19] Where these shields are required to be grounded, selecting a grounding location to achieve the shortest practicable grounding conductor will help limit potential differences between CATV and other metallic systems. "Where the outer conductive shield of a coaxial cable is grounded, no other protective devices shall be required" [*NEC* 820.93(A) through (D)] [20] (see photo 18-3).

Where the shielded cables are grounded, they shall be grounded in a manner specified in 820.100(A) through (D) as follows:

"(A) Grounding Conductor.

"(1) Insulation. The grounding conductor shall be insulated and shall be listed as suitable for the purpose.

"(2) Material. The grounding conductor shall be copper or other corrosion-resistant conductive material, stranded or solid.

"(3) Size. The grounding conductor shall not be smaller than 14 AWG. It shall have a current-carrying capacity approximately equal to that of the outer conductor of the coaxial cable…

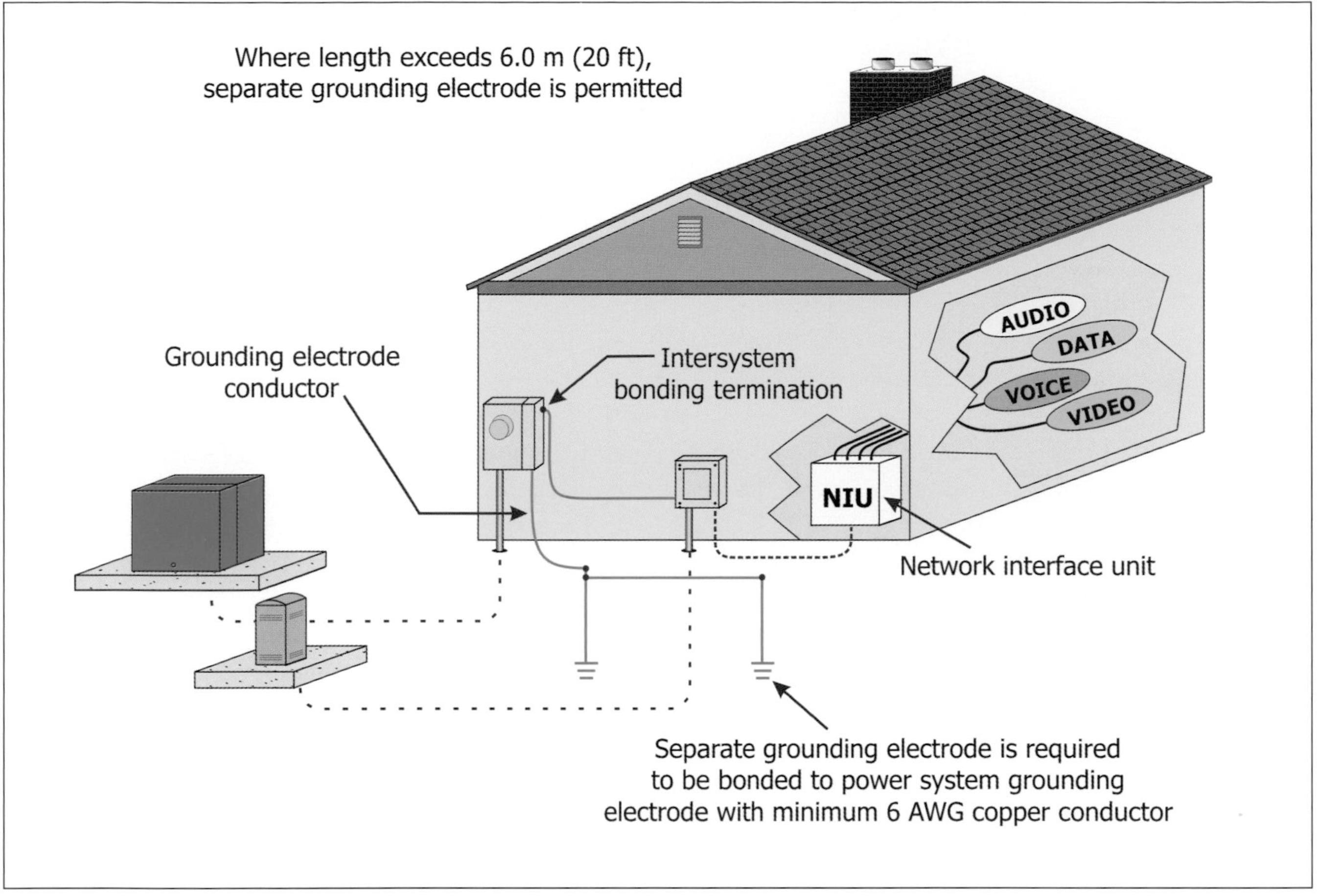

Figure 18-12. Cable network interface unit and primary protector grounding connections to separate electrode and bonded to power system grounding electrode as required.

"(4) Length. The grounding conductor shall be as short as practicable. In one- and two-family dwellings, the grounding conductor shall be as short as practicable, not to exceed 6.0 m (20 ft) in length." [An exception permits the grounding conductor to be connected to a separately installed electrode where the maximum distance specified cannot be achieved. In this case the separate electrode is to be installed and the electrode for the electrical service and the CATV electrode are required to be bonded together with a minimum 6 AWG copper bonding jumper as specified in 820.100(D)] (see figures 18-9 and 18-10).

"(5) Run in Straight Line. The grounding conductor shall be run to the grounding electrode in as straight a line as practicable.

"(6) Physical Protection. Where exposed to physical damage, the grounding conductor shall be protected. Where the grounding conductor is run in a metal raceway, both ends of the raceway shall be bonded to the grounding conductor or the same terminal or electrode to which the grounding conductor is connected. [820.100(A)(6)].

"(B) Electrode. The grounding conductor shall be connected in accordance with 820.100(B)(1), (B)(2), or (B)(3).

"(1) In Buildings or Structures with an intersystem bonding termination. Where the building or structure has an intersystem bonding termination as required by 250.94, the grounding conductor for community antenna television and radio distribution systems shall be connected to the intersystem bonding termination. If the building or structure does not include an intersystem bonding termination, the grounding conductor shall be connected to the nearest accessible location on the following:

"(1) The building or structure grounding electrode system as covered in 250.50

"(2) The grounded interior metal water piping

system, within 1.5 m (5 ft) from its point of entrance to the building, as covered in 250.52

"(3) The power service accessible means external to enclosures as covered in 250.94

"(4) The metallic power service raceway

"(5) The service equipment enclosure

"(6) The grounding electrode conductor or the grounding electrode conductor metal enclosure, or

"(7) To the grounding conductor or to the grounding electrode of a building or structure disconnecting means that is connected to an electrode as covered in 250.32.

"(2) In Buildings or Structures Without Grounding Means. If the building or structure served has no intersystem bonding termination or grounding means, as described in 820.100(B)(2), the grounding conductor shall be connected to either of the following:

"(1) To any one of the individual electrodes described in 250.52 (A)(1), (A)(2), (A)(3), (A)(4); or,

"(2) If the building or structure served has no intersystem bonding termination or grounding means as described in 820.100(B)(2) or (B)(3)(1), to an effectively grounded metal structure or to any of the individual electrodes described in 250.52(A)(5), (A)(7), and (A) (8)." [21]

Where the CATV system and or shielding of coaxial cables are connected to grounding electrodes, they are required to be connected in accordance with 820.100(C) and (D). The "connections to grounding electrodes must also comply with applicable requirements of 250.70." [22]

The electrode(s) of the CATV system must be bonded to the electrical service or system grounding electrode by means of a bonding jumper. Section 820.100(D) requires that "a bonding jumper not smaller than 6 AWG copper or equivalent shall be connected between the community antenna television system's grounding electrode and the power grounding electrode system at the building or structure served where separate electrodes are used." Selecting a grounding location to achieve the shortest practicable grounding conductor will help limit potential differences between CATV and other metallic systems.

Network-Powered Broadband Communications Systems

A typical basic network-powered broadband communications system configuration "includes a cable supplying power and broadband signal to a network interface unit that converts the broadband signal to the component signals (see figures 18-11 and 18-12). Typical cables are coaxial cable with both broadband signal and power on the center conductor, composite metallic cable with a coaxial member for the broadband signal and a twisted pair for power, and composite optical fiber cable with a pair of conductors for power. Larger systems may also include network components such as amplifiers that require network power." [23]

"Network interface units containing protectors, NIUs with metallic enclosures, primary protectors, and the metallic members of the network-powered broadband communications cable that are intended to be grounded shall be grounded as specified in 830.100(A) through (D)." [24] These requirements are similar to those in Articles 800, 810, and 820. The purpose of these grounding and bonding connections is also the same.

The size of the grounding conductor for the NPBCS is required to be not "smaller than 14 AWG" copper "and shall have a current-carrying capacity approximately equal to that of the grounded metallic member(s) and protected conductor(s) of the network-powered broadband communications cable. The grounding conductor shall not be required to exceed 6 AWG." [25]

"Connectors, clamps, fittings, or lugs used to attach grounding conductors and bonding jumpers to grounding electrodes or to each other that are to be concrete encased or buried in the earth shall be suitable for its application." [26] The grounding clamps and connectors suitable for use in accordance with 250.64 are suitable as well as listed clamps for use specifically with these low-voltage/current systems.

"(B) Electrode. The grounding conductor shall be connected as follows.

"(1) In Buildings or Structures with an Intersystem Bonding Termination. Where a building or structure has an intersystem bonding termination, the network-powered broadband communication systems grounding conductor shall be connected to the intersystem bonding termination. Where a building or structure does not include an intersystem bonding termination, the NPBCS grounding conductor shall be connected to the nearest accessible location on the following:

"(1) The building or structure grounding electrode system as covered in 250.50

"(2) The grounded interior metal water piping system, within 1.5 m (5 ft) from its point of entrance to the building, as covered in 250.52

"(3) The power service accessible means external to enclosures as covered in 250.94

"(4) The metallic power service raceway

"(5) The service equipment enclosure

"(6) The grounding electrode conductor or the grounding electrode conductor metal enclosure, or

"(7) To the grounding conductor or to the grounding electrode of a building or structure disconnecting means that is grounded to an electrode as covered in 250.32.

"(2) In Buildings or Structures Without an Intersystem Bonding Termination or Grounding Means. If the building or structure served has no intersystem bonding termination or grounding means, as described in (B)(2)…

"(1) To any one of the individual electrodes described in 250.52 (A)(1), (A)(2), (A)(3), or (A)(4) "(2) If the building or structure served has no grounding means as described in 830.100(B)(2) or (B)(3)(1), to any one of the individual electrodes described in 250.52(A)(7) and (A)(8) or to a ground rod or pipe not less than 1.5 m (5 ft) in length and 12.7 mm (½ in.) in diameter, driven, where practicable, into permanently damp earth and separated from lightning conductors as covered in 800.53 and at least 1.8 m (6 ft) from electrodes of other systems. Steam or hot water pipes or lightning-rod conductors shall not be employed as electrodes for protectors, NIUs, with integral protection, grounded metallic members, NIUs with metallic enclosures, and other equipment.[27]

"Connections to grounding electrodes are required to be in accordance with 250.70."

Lightning and Other Hazards

"On network-powered broadband communications conductors not exposed to lightning or accidental contact with power conductors, providing primary electrical protection in accordance with [830.90(A)] helps protect against other hazards, such as ground potential rise caused by power fault currents, and above-normal voltages induced by fault currents on power circuits in proximity to the network-powered broadband communications conductors.

"FPN No. 2: Network-powered broadband communications circuits are considered to have a lightning exposure unless one or more of the following conditions exist:

"(1) Circuits in large metropolitan areas where buildings are close together and sufficiently high to intercept lightning.

"(2) Areas having an average of five or fewer thunderstorm days per year and earth resistivity of less than 100 ohm-meters. Such areas are found along the Pacific coast." [28]

Bonding of Electrode Systems

It is also required to bond the grounding electrode for a network-powered broadband communications system to the power system grounding electrode or electrode system for the building or structure. "A bonding jumper not smaller than 6 AWG copper or equivalent must be connected between the radio and television equipment grounding electrode and the power grounding electrode system at the building or structure served where separate electrodes are used." [29]

The grounding and bonding requirements for the systems covered in *NEC* chapter 8 are important for protection against electric shock and fire, in addition to providing a level of protection from surges, spikes, dips, and lightning strikes. It is important to bond grounding electrodes of different systems together for safety and to comply with the minimum *NEC* rules. Effective grounding and bonding for these systems includes a connection of an equipment grounding conductor with the supply circuit for grounding equipment.

The exception to Section 250.94 covers the requirements for intersystem grounding and bonding conductor connections at existing buildings or structures where any intersystem bonding and grounding conductors required by 770.93, 800.100(B), 810.21(F), 820.100(B), 830.100(B) and an intersystem bonding termination is not installed. This exception allows "an accessible means external to enclosures for connecting intersystem bonding and grounding conductors shall be provided at the service equipment and at the disconnecting means for any additional buildings or structures by at least one of the following means:

"(1) Exposed nonflexible metallic raceways

"(2) Exposed grounding electrode conductor

"(3) Approved means for the external connection of a copper or other corrosion-resistant bonding or grounding conductor to the grounded raceway or equipment." [30]

Grounding clamps are available that are suitable for these grounding connections, but these grounding clamps are not permitted to be used for power distribution system grounding electrode connections unless so listed. An example would be the perforated metal-strap type grounding clamp used to connect grounding conductors of these systems to the nonflexible metallic service raceways. See the Guide Information for Electrical Equipment (White Book), under category (KDSH).

The authority having jurisdiction would have the responsibility of approval of any provided external connecting or bonding means. The fine print note indicates that "a 6 AWG copper conductor with one end bonded to the grounded nonflexible metallic raceway or equipment and with 150 mm (6 in.) or more of the other end made accessible on the outside wall is an example of the approved means covered in 250.94, Exception (3)." [31]

Summary

The grounding and bonding requirements for voice, data, and video systems covered by *NEC* chapter 8 are minimum requirements for safety. The *Code* includes grounding and bonding requirements for systems, circuits, and enclosures in chapter 2, Wiring and Protection. Chapter 8 is independent from the requirements of chapter 2, unless specifically referenced from the particular article in chapter 8. There are requirements specified in the articles of chapter 8 and specific requirements for grounding and bonding required for safety for these special systems. Proper grounding and bonding is essential for safety and operation of these systems. One of the most important requirements of the *Code* is to bond the electrodes or electrodes of different systems together. Failure to do so can result in shock hazards, fire hazards, and damage to electronic components of these systems. The *Code* includes the minimum requirements for personnel and building safety, this simply means that one must do at least that much to comply. It is not uncommon to see the minimum requirements of the *NEC* be exceeded relative to the grounding and bonding for these systems, but the minimum requirements must be met. For information on enhanced grounding electrodes and grounding and bonding for sensitive electronic equipment, see chapter seventeen of this text.

See 250.60 for use of air terminals (lightning rods), and NFPA 780, Lightning Protection Systems.

For more information relative to the applicable requirements for a listed secondary protector refer to UL 497A-1996, the *Standard for Secondary Protectors for Communications Circuits.*

For more information relative to determining acceptable installation practices for telecommunications systems, circuits, and equipment refer to nationally recognized standards such as ANSI/EIA/TIA 568-A-2001, *Standard for Installing Commercial Building Telecommunications Cabling* ; ANSI/EIA/TIA 569-1990, *Commercial Building Standard for Telecommunications Pathways and Spaces*; and ANSI/EIA/TIA 570-1991, *Residential and Light Commercial Telecommunications Wiring Standard.*

[1] NFPA 70, *National Electrical Code* 2008 250.20(A), (National Fire Protection Association, Quincy, MA, 2007), p. 70-97.

2 NFPA 70, 250.20(A), p. 70-97.

3 NFPA 70, 250.20(A), p. 70-97.

4 NFPA 70, 250.30(A)(7), p. 70-100.

5 NFPA 70, Article 100, p. 70-27.

6 NFPA 70, Article 100, p. 70-24.

7 NFPA 70, Article 100, p. 70-24.

8 NFPA 70, 800.100(A), p. 70-644.

9 NFPA 70, 800.100(B)(1), p70-644.

10 NFPA 70, 800.100(C), p. 70-645.

11 NFPA 70, 800.106(A), p. 70-645.

12 NFPA 70, 800.106(B), p. 70-645.

13 NFPA 70, 800.100(B)(1), p. 70-644.

14 NFPA 70, 810.21, p. 70-651 & 70-652.

15 NFPA 70, 810.21(F), p. 70-651.

16 NFPA 70, 810.21(J), p. 70-652.

17 NFPA 70, 810.57, p. 70-653.

18 NFPA 70, 810.58, p. 70-653.

19 NFPA 70, 820.93, p. 70-655.

20 NFPA 70, 820.93(A), p. 70-656.

21 NFPA 70, 820.100, p. 70-656.

22 NFPA 70, 820.100(C), p. 70-657.

23 NFPA 70, 830.1 FPN, p. 70-660.
24 NFPA 70, 830.100, p. 70-666.
25 NFPA 70, 830.100(A)(3), p. 70-666.
26 NFPA 70, 830.100(C), p. 70-667
27 NFPA 70, 830.100, p. 70-656.
28 NFPA 70, 830.90 FPN No. 1 and No. 2, p. 70-664.
29 NFPA 70, 810.21(H), p. 70-651.
30 NFPA 70, 250.94, p. 70-110.
31 NFPA 70, 250.94, p. 70-110.

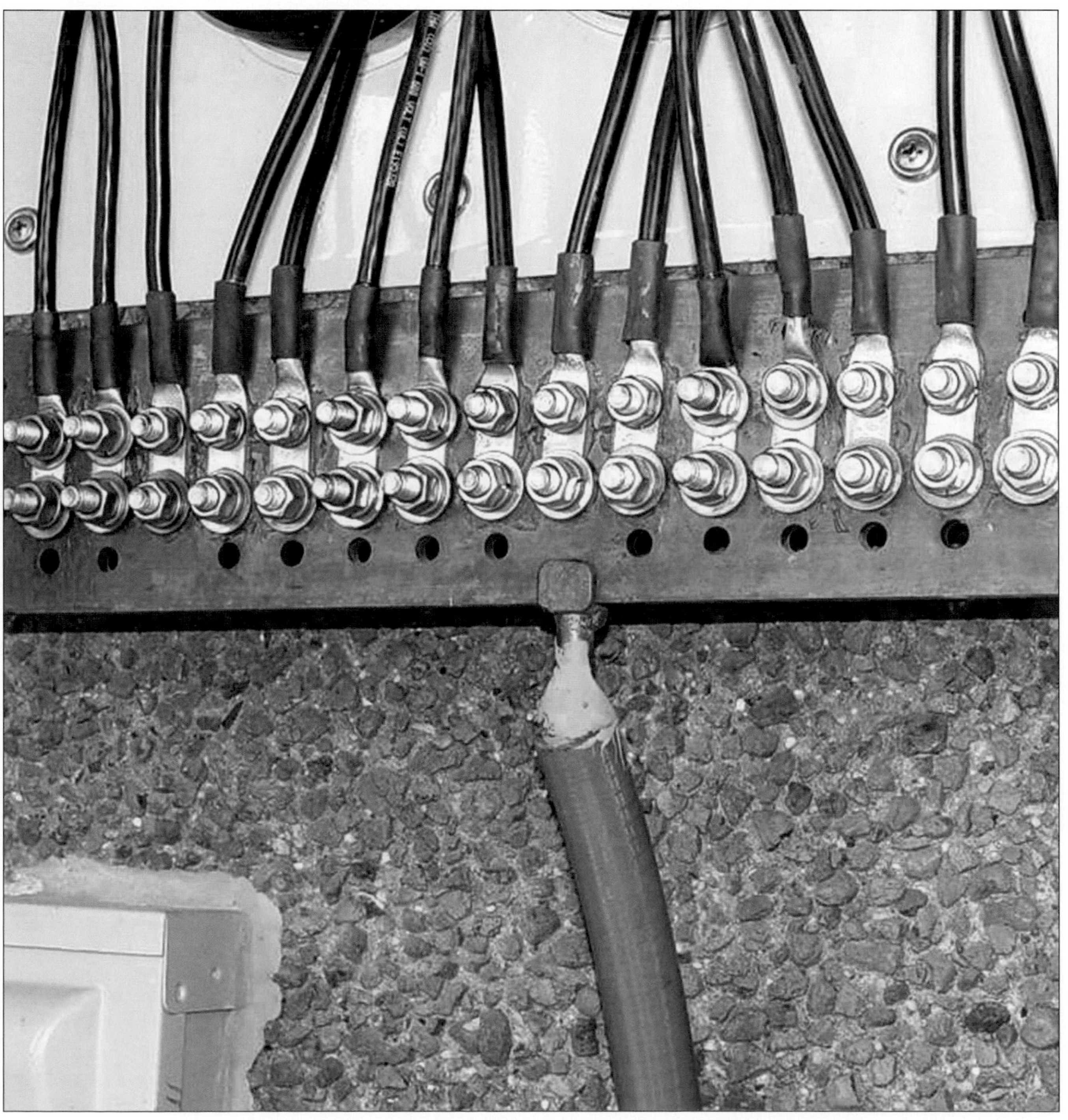

18 Review Questions

The questions included here were developed using material included in this chapter. The answers can be found by reviewing the text. It is also important that students make use of *NEC-2008*, where many answers can be found.

1. The conductor used to connect the equipment or grounded circuit conductor of a wiring system to a grounding electrode or to a point on the grounding electrode system is a ____________________.
   a. equipment grounding conductor
   b. grounding electrode conductor
   c. neutral conductor
   d. grounding conductor

2. An intersystem bonding termination for connecting intersystem bonding and grounding conductors shall be provided at ____________________.
   a. a separately derived system
   b. the service equipment or disconnecting means for additional buildings
   c. the telephone equipment mounting board
   d. The point of entry of a cable to the building

3. The minimum size copper grounding conductor for a receiving station shall be not less than_______.
   a. 12 AWG copper
   b. 10 AWG copper
   c. 6 AWG aluminum
   d. 6 AWG copper

4. For a one-family dwelling a grounding conductor for a coaxial cable shall generally be as short as practicable and not exceed ____________.
   a. 3.0 m (10 ft)
   b. 9.0 m (30 ft)
   c. 6.0 m (20 ft)
   d. 1.5 m (5 ft)

5. The shield of a coaxial cable shall be grounded with a grounding conductor not less than _______.
   a. 4 AWG copper
   b. 6 AWG copper
   c. 14 AWG copper or other corrosion resistant material
   d. 12 AWG copper

6. Where there is a grounding electrode installed for a coaxial cable antenna system that is greater than 6.0 m (20 ft) from the power grounding electrode system, there shall be a __________ bonding jumper installed between the two electrodes
   a. 4 AWG copper or equivalent
   b. 8 AWG copper or equivalent
   c. 6 AWG copper or equivalent
   d. 12 AWG copper or equivalent

7. A grounding conductor for a coaxial cable system installed on a building or structure that does not have an intersystem bonding termination shall be permitted to be connected to which of the following:
   a. the building or structure grounding electrode system as covered in 250.50
   b. the metallic power service raceway
   c. the service equipment enclosure
   d. all of the above

8. The minimum size bonding conductor connected to a primary protector grounding terminal for a network-powered broadband communications system at a mobile home that is cord-and-plug-connected shall be not less than _________.
   a. 2 AWG copper
   b. 4 AWG aluminum
   c. 12 AWG copper
   d. 10 AWG bronze

9. Where a grounding conductor for a communications system is installed in a metal raceway, the raceway shall be __________
   a. continuous
   b. bonded to both ends of the grounding conductor or to the same terminal to which the grounding conductor is connected.
   c. rigid metal conduit
   d. electrical metallic tubing

10. Where a grounding conductor for network-powered broadband communications systems is connected to concrete-encased or buried electrodes the connections shall be__________.
    a. accessible
    b. stainless steel
    c. suitable for its application
    d. irreversible compression-type

11. Grounding conductors for network-powered broadband communications systems shall be spaced at least _____________ from lightning conductors where practicable.
   a. 6 m (20 ft)
   b. 1.8 m (6 ft)
   c. 15 m (50 ft)
   d. 900 mm (3 ft)

12. A 24-volt ac system is required to be grounded where __________.
   a. supplied by a transformer with a supply syste of 120 volts
   b. where installed as overhead conductors outside of buildings
   c. where the supply system is grounded
   d. where ground detectors are not installed

13. Where an intrinsically safe system is required to be connected to a grounding electrode, the electrode shall be which of the following:
   a. a ground rod
   b. a pipe electrode
   c. the electrodes specified by 250.52(A)(1),(2),(3), and (4) where present and shall comply with 250.30(A)(7)
   d. a plate electrode

14. Where shields are used with intrinsically safe systems, and the shield is part of the intrinsically safe circuit, the shield is _____________.
   a. not required to be grounded
   b. required to be grounded
   c. required to be isolated
   d. required to be insulated

15. At a mobile home that is supplied by a cord-and-plug-connection, the primary protector bonding conductor for a communications system is required to be connected to the____________.
   a. water pipe at the sink location
   b. metal frame or available grounding terminal of the mobile home
   c. a ground rod
   d. the panelboard enclosure in the mobile home

16. "The conductor used to connect equipment or the grounded circuit of a wiring system to a grounding electrode or electrodes" best defines which of the following:
   a. grounded conductor
   b. grounding conductor
   c. grounding electrode conductor
   d. main bonding jumper

17. "A device that provides a means for connecting communications system(s) grounding conductor(s) and bonding conductor(s) at the service equipment or at the disconnecting means for buildings or structures supplied by a feeder or branch circuit" best describes which of the following:
   a. a grounding screw
   b. a grounding clip
   c. a grounding terminal lug
   d. an intersystem bonding termination

18. An intersystem bonding termination shall include provision for not less than _____intersystem bonding conductors.
   a. 4
   b. 5
   c. 3
   d. 2

Grounding of Systems or Circuits of 1kV and Over

The *NEC* indicates that where systems or circuits of 1000 volts (1 kV) and over are grounded, they shall comply with Article 250 plus specific rules in Part X of Article 250 (see figure 19-2). The requirements in Part X supplement or modify other rules in Article 250.

Objectives

To understand...

- Grounding rules for medium and high voltage (1 kV and over)
- Methods of grounding
- Use of surge arrestors
- Grounding of outdoor substations
- Stress reduction for cables

Chapter 19

Most medium-voltage systems in the 2.4 to 15 kV range are either low-resistance grounded or are high-resistance grounded. The only difference between low and high resistance grounding is the value of the resistor that, in turn, controls the amount of ground-fault current permitted during a ground-fault event. The other common method is to solidly ground the system, especially if it is exposed to lightning.

Medium voltage systems above 15 kV are typically either solidly grounded or ungrounded. For the ungrounded system, even though there is not an effective ground provided there is still a relationship to earth through the surge arresters installed where outdoor lines are open and commonly subjected to lightning surges and transient overvoltages (see figure 19-1 and photo 19-1).

Systems Rated 2400 Volts to 13,800 Volts

Grounding may be achieved through solid connections to earth or connection through a grounding resistor purposely installed in the equipment-grounding path at the source. The choice depends on available ground-fault current, the size of the system, tolerance for outages and tolerance for damage from ground faults. It is common in industrial systems to ground the neutral of systems rated 2400 volts and above through a resistor (see figure 19-2).

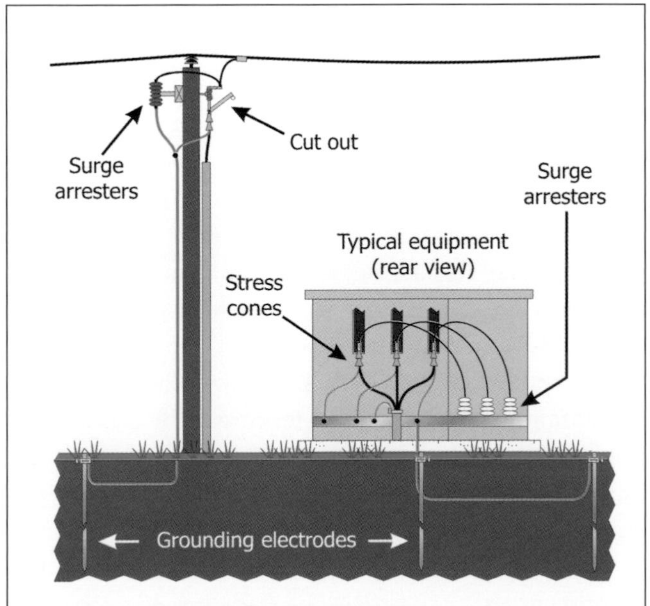

Figure 19-1. Effective grounding through surge arresters

Photo 19-1. High voltage lines typically connected to surge arresters at pole locations

Reactance grounding (use of a reactor or grounding transformer) in this voltage range is preferred if the circuits are overhead and, thus, subject to lightning exposure, if they serve rotating machinery, and if excessive ground-fault current will not develop.

Because of the higher voltage of these systems, compared to systems of 600 volts and less, the ground-fault current levels through the earth and through the grounding electrode and grounding electrode conductor are much higher. This current is sufficiently high that a current transformer can be placed around the grounding electrode conductor or system bonding jumper to give a positive indication that a ground fault exists and then used to clear the circuit in the event the ground fault persists over a predetermined time. Alternatively, a more typical protection scheme uses the residual current (*zero-sequence current*) from the phase current transformers to the protective relays or has a zero-sequence current transformer with a ground-fault protective relay. Compare this with a system of 600 volts or less. There, the ground-fault current sensor cannot be placed satisfactorily in the grounding electrode conductor, for little ground-fault current will exist in that conductor.

Systems Rated 15,000 Volts or More

Such systems are nearly always outdoors and have no rotating equipment served by them. Usually they are solidly grounded, which permits the use of grounded-neutral type surge arresters that cost less and provide

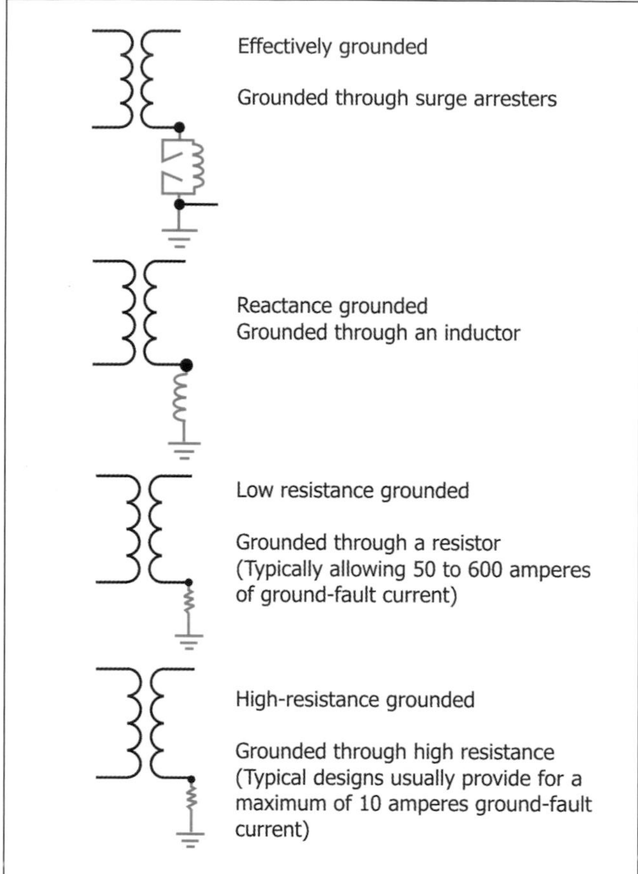

Figure 19-2. Typical methods of system grounding over 600 volts

better protection from overvoltage. Just as in systems over 600 volts, there will be different voltage levels for various utilization purposes. Additionally, some of the medium voltage levels are only for distribution to allow greater levels of power to be sent without having to use large conductors. Grounding must be reestablished by some means at each voltage level as was previously described for systems of 600 volts or less.

Grounding Outdoor Industrial Substations

The grounding of outdoor industrial substations involves not only the electrical system neutral and immediate electrical equipment enclosures but also includes grounding the fence and other supporting structures in the area.

A ground bus or grid should be established extending about 900 mm (3 ft) outside the periphery of the fence. The grid should be connected to many ground rods around its periphery and, in addition, should be connected to a metallic underground water piping system or other underground metallic structures where available. The ground grid should be a minimum size of 4/0 AWG copper and be approximately 25 percent of the capacity of the system. Conductor rating may be sized based on the capacity of bare conductors in free air. Copper bus is rated based on 1200 amperes per square inch. Connections from the fence and from all equipment within the fence, that is, transformer cases, steel structures, switchgear including operating mechanisms for gang-operated disconnects, and so forth, should be not less than 4/0 AWG copper nor less than 25 percent of the capacity of the secondary conductors. To be sure of having a good, permanent neutral ground, it is best to connect the neutral to two points on the ground bus.

A grounding bus and connections are effective only if the mechanical construction is sound and as permanent as possible. No connection should be soldered. It is preferable to properly braze or weld all connections and to protect all cable from mechanical injury. Common practices use exothermic welding of these connections or compression connections using special tools and connectors. If metallic enclosures are used for the mechanical protection of these conductors, the enclosures should be bonded to the enclosed conductor so both are connected in parallel at both ends of the enclosure.

There is some difference of opinion as to whether the fence earth grounding should be separated from the ground grid used for the substation. If the fence is connected to the station grid and a fault occurs, the fence will be elevated above earth potential by the IZ drop from the system and ground impedance. The potential gradient formed in the earth near the fence drops off very rapidly, typically within the first few feet. Anyone contacting the fence can therefore be subjected to a hazardous step-voltage or touch voltage potentials under fault conditions.

If the fence grounding is isolated from the substation ground, then the fence will be elevated in potential from the substation ground grid, and under fault conditions anyone contacting the fence and the equipment will be subjected to that hazard.

In most industrial substations, the fence and equipment are close proximity; therefore, in such cases, it is best to connect the fence grounds and the substation

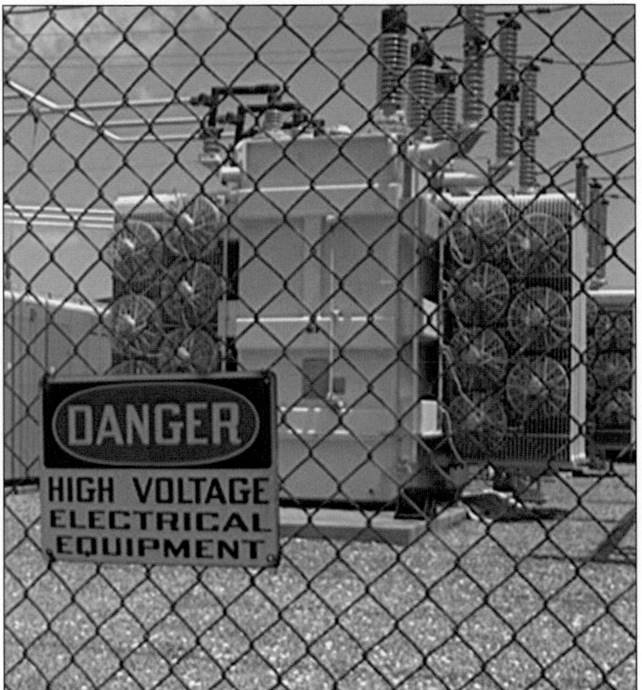

Photo 19-2. Substation surrounded by a grounded chain-linked fence

ground grid together. The shock hazard when using a common ground for the fence and the equipment can be kept to a minimum by having the ground connection resistance as low as possible (see photo 19-2).

Derived Neutral Systems

Section 250.182 permits "a system neutral point derived from grounding transformer…to be used for grounding high-voltage systems." The application is for *ungrounded* systems where there is still a need to detect and clear single phase-to-ground faults. These grounding type transformers include zigzag, wye-delta, or T-connected also known as *Scott T* (see figures 19-3 and 19-6). Ground-fault protection may be provided through the transformer and positive tripping can be accomplished with relatively low magnitudes of ground-fault current (see figure 19-3).

Solidly Grounded Neutral Systems

Section 250.184 also permits solidly grounded neutral systems and provides for two options, *single-point grounding* or *multiple-point grounding*. The neutral conductor of solidly grounded systems is generally permitted to have an insulation level of not less than 600 volts or to be bare.

Single-Point Neutral Grounded Systems

Single-point grounded neutral systems are permitted for systems of 1000 volts and greater, although they are less common. Where a single-point grounded neutral system is used, it is permitted to be supplied from either a separately derived system or a multigrounded neutral system provided that the equipment grounding conductor of the single-point grounded system is connected to the grounded neutral point at the source (see figure 19-4).

Single-point grounded neutral systems shall be installed as follows:

- They are required to have a grounding electrode that is connected to the system neutral conductor by means of a properly sized grounding electrode conductor.
- A system bonding jumper shall be installed from the system neutral to the source enclosure as well as to the grounding electrode conductor.
- An equipment grounding conductor is required at each building, structure, and equipment enclosure.
- A neutral conductor is only required where phase-to-neutral loads are supplied. The system neutral conductor is required to be insulated from the earth, except at one location (the point of grounding).

Equipment grounding conductor(s) are required to be run with the phase conductors and must meet the following conditions. They

- Are not permitted to carry continuous load
- May be bare or insulated
- Must have sufficient ampacity for fault current duty

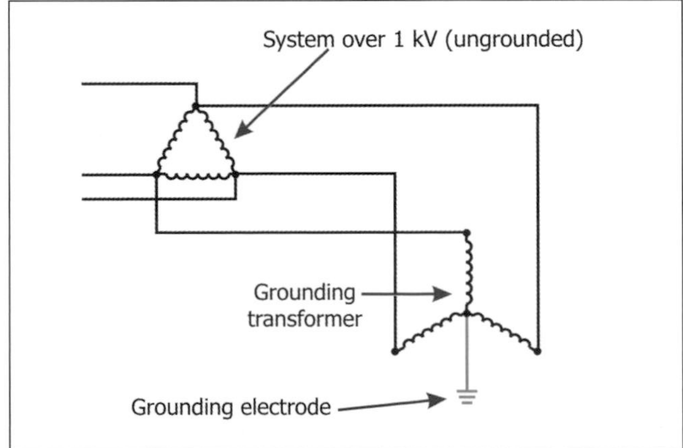

Figure 19-3. System grounding using grounding-type transformers

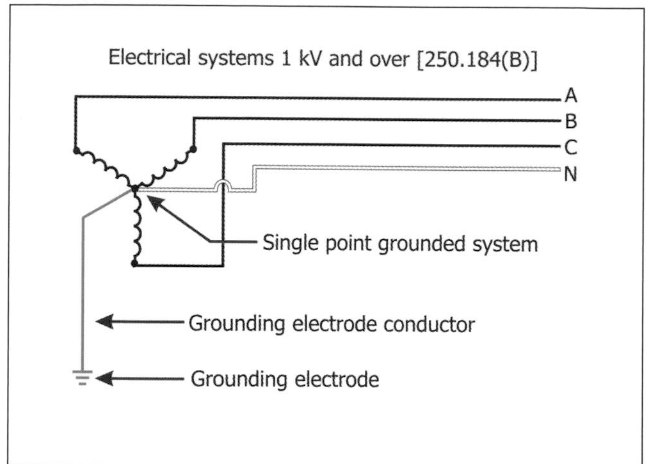

Figure 19-4. Single-point grounded neutral system

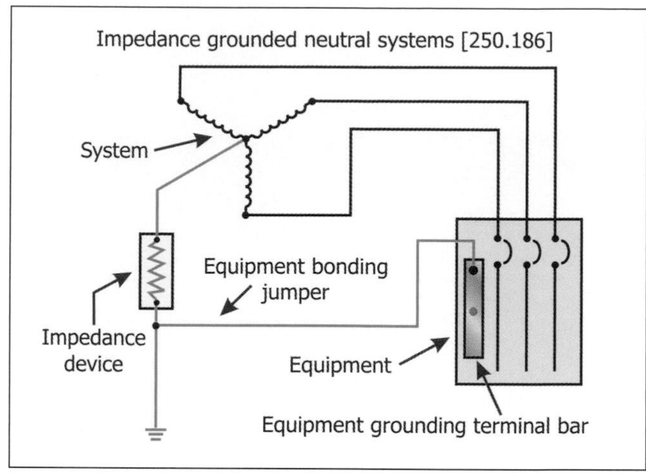

Figure 19-6. Impedance grounded neutral system

Multigrounded Neutral Systems

Multiple-point grounding, grounding the neutral conductor at the source and at additional points along the system, is permitted for transformer feeders supplying conductors to a building or other structure, direct buried portions of feeders employing a bare copper neutral and for overhead portions of a system installed outdoors (see figures 19-5).

Where multiple grounded neutral systems are used, the following requirements apply:

"Grounding shall be permitted at one or more of the following locations:

"a. Transformers supplying conductors to a building or other structure

"b. Underground circuits where the neutral is exposed

"c. Overhead circuits installed outdoors" [1]

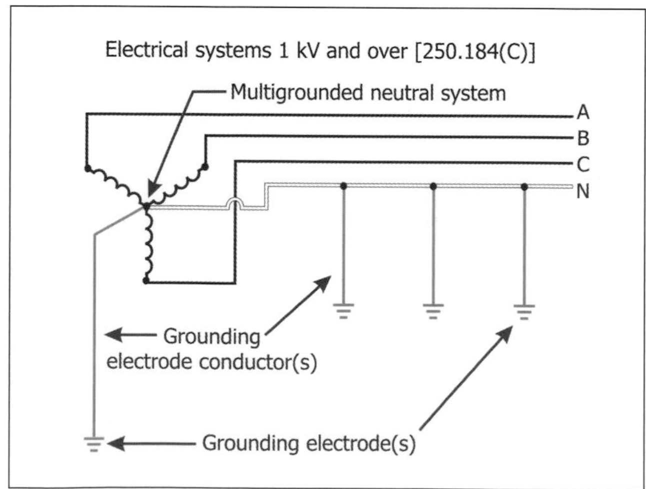

Figure 19-5. Multigrounded neutral systems

1. "The neutral conductor of a solidly grounded neutral system shall be permitted to be grounded at more than one point [250.184(C)(1)].

2. "The multigrounded neutral conductor shall be grounded at each transformer and at other additional locations by connections to a grounding electrode.

3. "At least one grounding electrode shall be installed and connected to the multigrounded neutral circuit conductor every 400 m (1300 ft).

4. "The maximum distance between any two adjacent electrodes shall not be more than 400 m (1300 ft).

5. "In a multigrounded shielded cable system, the shielding shall be grounded at each cable joint that is exposed to personnel contact." [2]

Impedance Grounded Neutral Systems

Section 250.186 permits impedance grounded neutral systems (see figure 19-6). Impedance grounding of these systems can be accomplished by reactance grounding, low-resistance grounding or high-resistance grounding.

Grounding of Systems Supplying Portable or Mobile Equipment

Special requirements apply to electrical systems that supply portable or mobile high-voltage equipment, other than substations installed on a temporary basis. The requirements are found in 250.188.

"(A) **Portable or Mobile Equipment.** Portable or mobile high-voltage equipment shall be supplied from a system having its neutral conductor grounded through an impedance. Where a delta-connected high-voltage system is used to supply portable or mobile equipment, a system neutral point and associated neutral conductor shall be derived.

"(B) **Exposed Non-Current-Carrying Metal Parts.** Exposed non-current-carrying metal parts of portable or mobile equipment shall be connected by an equipment-grounding conductor to the point at which the system neutral impedance is grounded.

"(C) **Ground-Fault Current.** The voltage developed between the portable or mobile equipment frame and ground by the flow of maximum level of ground-fault current shall not exceed 100 volts. [Editor's note: This somewhat limits the shock hazard.]

"(D) **Ground-Fault Detection and Relaying.** Ground-fault detection and relaying shall be provided to automatically de-energize any high-voltage system component that has developed a ground fault. The continuity of the equipment-grounding conductor shall be continuously monitored to de-energize automatically the high-voltage circuit to the portable or mobile equipment upon loss of continuity of the equipment-grounding conductor.

"(E) **Isolation.** The grounding electrode to which the portable or mobile equipment system neutral impedance is connected shall be isolated from and separated in the ground at least 6.0 m (20 ft) from any other system or equipment grounding electrode, and there shall be no direct connection between the grounding electrodes, such as buried pipe and fence, and so forth.

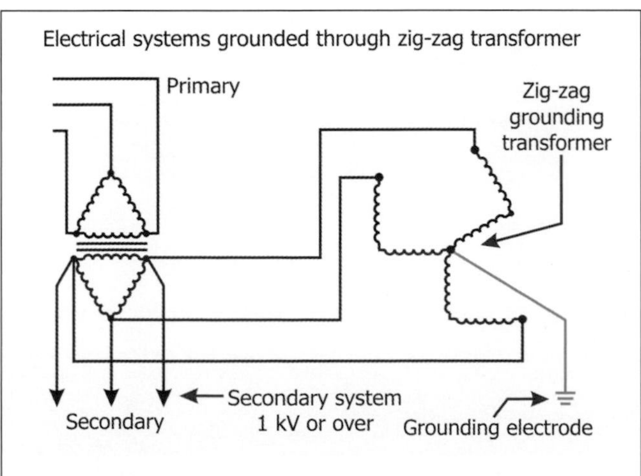

Figure 19-7 illustrates use of a zigzag grounding transformer for establishing a grounded system from an ungrounded system.

"(F) **Trailing Cable and Couplers.** High-voltage trailing cable and couplers for interconnection of portable or mobile equipment shall meet the requirements of Part III of Article 400 for cables and 490.55 for couplers." [3]

Grounding of High-Voltage Equipment

Section 250.190 generally requires grounding of all non-current-carrying metal parts of high-voltage fixed, portable, or mobile equipment including fences, housings, enclosures, and supporting structures.

The exception to this section permits an installation to be ungrounded if the equipment is isolated from ground and located to prevent any person who can make contact with the ground from contacting such metal parts when the equipment is energized.

Surge Arresters

Where surge arresters are used on secondary systems of less than 1000 volts, it is required that the connections to the service grounded conductor and to the grounding conductor be as short as practicable.

Line and ground connecting conductors shall not be smaller than 14 AWG copper or 12 AWG aluminum. The arrestor grounding conductor shall be connected to one of the following:

- Grounded service conductor
- Grounding electrode conductor
- Grounding electrode for the service
- Equipment grounding terminal in the service equipment.[4]

The use of method (3) is practical only on an ungrounded system.

If a surge arrester protects the primary of a transformer which supplies a secondary distribution system, the surge arrester grounding conductor may be interconnected to the secondary neutral, provided that in addition to the direct grounding connection at the arrester, the grounded conductor of the secondary has elsewhere a grounding connection to a continuous metallic underground water piping system. However, where there are not less than four secondary connections to any underground metallic water piping system per mile, the direct grounding connection at the arrester may be omitted.

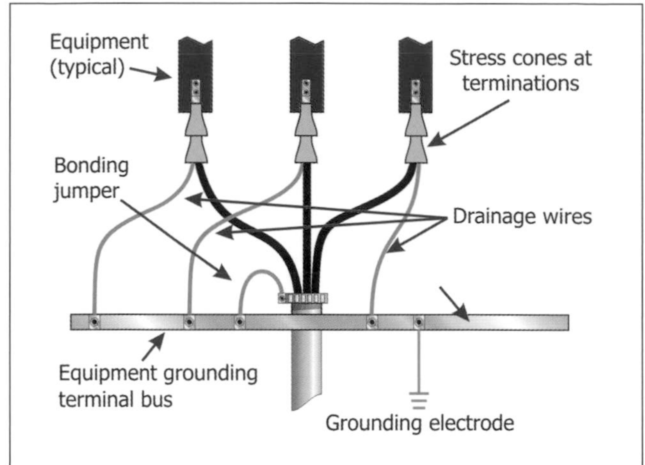

Figure 19-8. Stress reduction mean (grounding of cable shielding required)

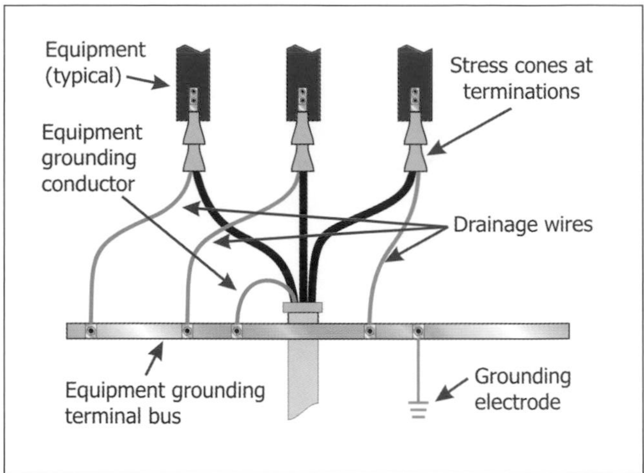

Figure 19-9. Stress reduction means (grounding of cable shielding required)

The grounding electrode conductor of a lightning arrester also may be connected to the secondary neutral. This is true if, in addition to the direct grounding connection at the arrester, the grounding conductor of the secondary system is part of a multi-grounded neutral system of which the primary neutral has at least four grounding connections in each mile of line, as well as a ground at each service.

Where the secondary is not grounded to a metallic water system but uses other available electrodes, an interconnection between surge arrester grounding conductor and the secondary neutral shall be made through a *spark gap*. The spark gap must have a breakdown voltage of at least twice the primary circuit voltage but not necessarily more than 10 kV. There shall be at least one other ground on the grounded conductor of the secondary not less than 6.0 m (20 ft) away from the surge arrester grounding electrode. No other connection between a surge arrester ground and a secondary neutral is allowed by the *Code,* except by special permission.

Stress Reduction Means and Cable Shielding

Medium and high-voltage cables above 2000 volts generally are required to be shielded. This shielding is usually in the form of either a conductive tape or stranded shield conductors. The purpose of the shielding is to evenly distribute voltage stresses through the insulation and bleed off to ground any capacitive voltage built up at the termination points (see photo 19-3). Section 310.6 addresses shielding requirements for insulated conductors operating at over 2000 volts. These conductors are required to have an ozone resistant insulation and be shielded. All metallic shields are required to be connected to a grounding electrode

Photo 19-3. 15,000-volt cables terminated at a switch showing stress reduction means and bleed wires connected to ground (earth)

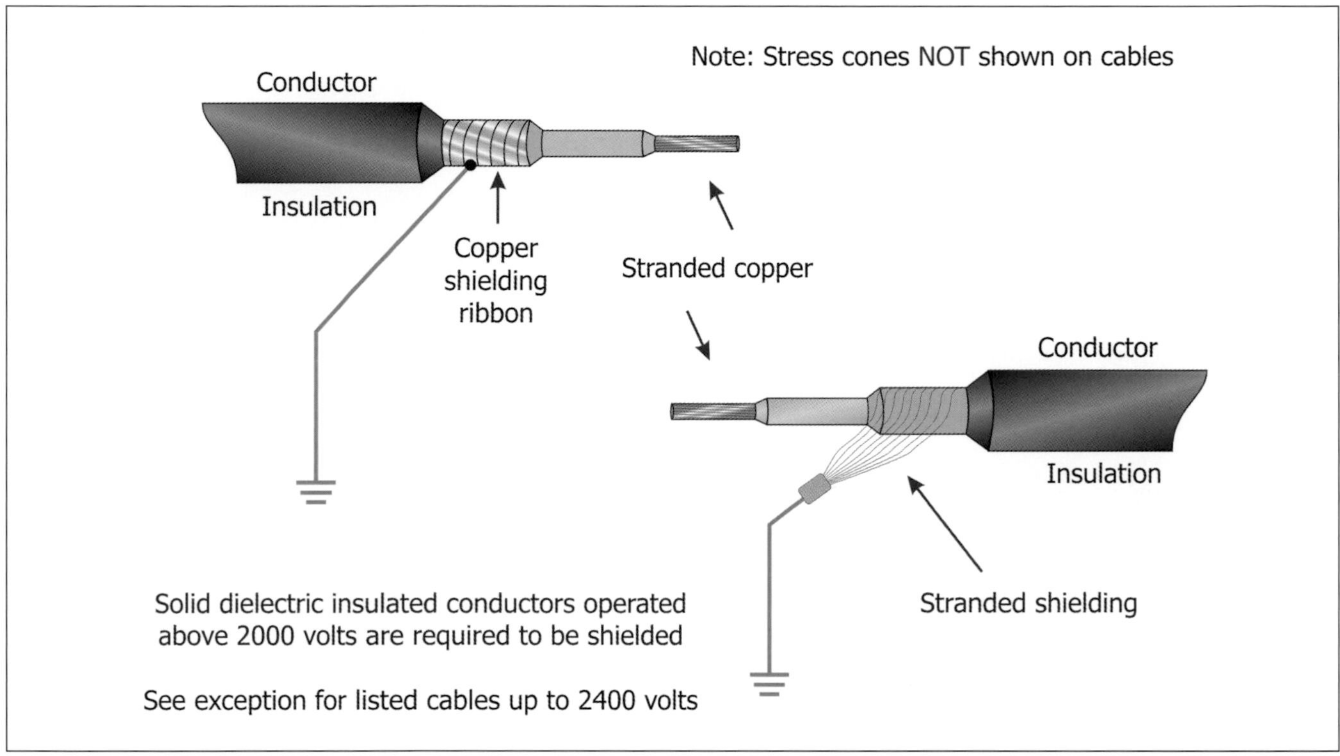

Figure 19-10. Cable shields are generally required for cables over 2000 volts. Cable shields are required to be grounded per 310.6.

conductor, grounding busbar, or grounding electrode [310.6] (see figures 19-8, 19-9 and 19-10).

"*Exception No. 1: Nonshielded insulated conductors listed by a qualified testing laboratory shall be permitted for use as field wiring up to 2000 volts under the following conditions:*

"*(a) Conductors shall have insulation resistant to electric discharge and surface tracking, or the insulated conductor(s) shall be covered with a material resistant to ozone, electric discharge, and surface tracking.*

"*(b) Where used in wet locations, the insulated conductor(s) shall have an overall nonmetallic jacket or a continuous metallic sheath.*

"*(c) Insulation and jacket thicknesses shall be in accordance with Table 310.13(D).*" [5]

"*Exception No. 2: Where permitted in 310.7, Exception No. 2.*"

Internal medium voltage conductors in equipment, even over 2000 volts, such as jumpers to potential transformers, some interconnecting cables, etc., are not shielded and are investigated for proper use when the equipment is tested and evaluated to the applicable ANSI/IEEE and UL standards.

Article 490 includes grounding requirements for equipment over 600 volts, nominal. Generally, the frames of switchgear and control assemblies are required to be grounded. "The metal cases or frames, such as instruments, relays, meters, and instrument and control transformers, located in or on switchgear and control, shall be connected to an equipment grounding conductor or, where permitted, the grounded conductor." [6] This can be accomplished by the mounting means to the metal enclosure or may have specific bonding jumpers installed to the equipment grounding bus.

[1] NFPA 70, *National Electrical Code* 2008, 250.184(C)(1), (National Fire Protection Association, Quincy, MA, 2007), p. 70-122.

[2] NFPA 70, 250.184(C), p. 70-122.

[3] NFPA 70, 250.188, p. 70-122.

[4] NFPA 70, 280.21, p. 70-124.

[5] NFPA 70, 310.6, p. 70-139.

[6] NFPA 70, 490.37, p. 70-351.

19 Review Questions

The questions included here were developed using material included in this chapter. The answers can be found by reviewing the text. It is also important that students make use of *NEC-2008*, where many answers can be found.

1. Most medium-voltage systems ____ to ____ are resistance grounded while others are high resistance grounded.
   a. 5.4 - 20 kV
   b. 2.4 - 15 kV
   c. 3.3 - 17 kV
   d. 4.6 - 18 kV

2. Most high-voltage systems ____ are effectively grounded utilizing surge arresters because the transmission lines are open and commonly subjected to lightning surges and transient overvoltages.
   a. above 15 kV
   b. above 12 kV
   c. above 10 kV
   d. above 11 kV

3. It is common in industrial systems to ground the neutral of systems rated ____ volts and above through a resistor.
   a. 2000
   b. 2400
   c. 1800
   d. 1380

4. Systems rated ____ volts or more are nearly always outdoors and have no rotating equipment served by them.
   a. 10,000
   b. 12,000
   c. 15,000
   d. 13,000

5. In systems over ____ volts, there will be different voltage levels for various utilization purposes.
   a. 1000
   b. 2000
   c. 600
   d. 1500

6. The grounding of outdoor industrial substations is concerned with grounding the fence and other supporting structures, all equipment and conductor enclosures and the grounding of the neutral. A ground bus or grid should be established extending about ____ feet outside the periphery of the fence.
   a. 3
   b. 4
   c. 6
   d. 8

7. Cable can be sized on the basis of the capacity of bare conductor in free air and copper bus on the basis of ____ amperes per square inch.
   a. 1250
   b. 1000
   c. 1200
   d. 1300

8. Connections from the fence and from all equipment within the fence, that is, transformer cases, steel structures, switchgear including operating mechanisms for gang-operated disconnects, etc., should be not less than . ____AWG nor less than ____ percent of the capacity of the secondary conductors.
   a. 3/0 - 50
   b. 4/0 - 25
   c. 2/0 - 40
   d. 1/0 - 30

9. To be sure of having a good, permanent neutral grounding connection, it is best to connect the neutral to ____ points on the ground bus.
   a. four
   b. three
   c. two
   d. five

10. Impedance-grounded neutral systems are permitted and can be accomplished by all but one of the following methods:
   a. reactance grounding
   b. resistance grounding
   c. high-resistance grounding
   d. low-resistance grounding

11. Without exception, where surge arresters are installed either indoors or outdoors they are required to be ____ to unqualified persons unless otherwise listed for installation in accessable locations.
   a. inaccessible
   b. accessible
   c. readily accessible
   d. available

12. Where there are not less than four secondary connections to any underground metallic water piping system per each ____, the direct grounding connection at the arrester may be omitted.

a. 1/4 mile
b. mile
c. 1/2 mile
d. 3/4 mile

13. A single point grounded system shall be permitted to be supplied from which of the following?

a. a separately derived system
b. an ungrounded system
c. a high resistance grounded neutral system
d. None of the above

14. Generally, solid dielectric insulated conductors operating above _____ volts are required to have ozone-resistant insulation and are required to be shielded.

a. 5000
b. 2400
c. 2000
d. all of the above

Fundamentals of Lightning Protection

Lightning is an atmospheric electrical discharge which may occur within a cloud, between clouds, between a cloud and earth (or items located on the earth), and sometimes in the case of very tall structures between items on or attached to the earth and the clouds. The four basic types of cloud-to-ground lightning discharges are: (1) downward negative lightning, (2) upward negative lightning, (3) downward positive lightning, and (4) upward positive lightning. Upward negative and upward positive discharges are generally only associated with very tall structures (greater than 300 feet tall) or objects of moderate heights located in higher elevations.[1]

Objectives

To understand...

- Fundamentals of lightning protection systems
- Industry standards related to lightning protection systems
- Components of lightning protection systems
- Conductor sizes and installations
- Grounding network and equipotenial bonding
- Qualifty control programs for lightning protection systems

Chapter 20

The Lightning Discharge

The details of the process by which a cloud becomes electrically charged are not yet fully understood, but they are generally associated with charge separation resulting from the updraft of air in the center of a thunderstorm cell and the velocity of particles falling through the cloud. The growing consensus is that the dominant electrification mechanism for cumulonimbus clouds is the graupel-ice mechanism.[1] Hydrometeors, which are liquid or frozen water particles in the atmosphere, are classified according to the effect of gravity on the particle. Those particles which fall at speeds of 0.3 meters per second or greater are referred to as precipitation particles. Those which fall at lower speeds are referred to as cloud particles (typically ice crystals). The electrification of individual particles occurs as a result of collisions between precipitation and cloud particles in the presence of water droplets. In general, the polarity and amplitude of charge separated during the collisions depend on several factors such as temperature, cloud water content, ice crystal size, relative velocity of the collisions, chemical contaminants in the water, and variations in size of supercooled water droplets.

The resulting charge separation process will typically yield a net positive charge near the top of the cloud, a net negative charge below the layer of positive charge, and, often, an additional positive charge pocket at the bottom of the cloud. However, it is generally agreed that charges of both polarities often exist in any region of the cloud.

It is speculated that the lightning discharge is initiated by the emission of positive corona from precipitation particles (such as raindrops) which have been deformed by strong electric fields. When the ambient electric field exceeds the electric field required to support the propagation of corona streamers, a streamer will be developed. Should a number of these streamers occur sequentially, a field enhancement sufficient for breakdown, and perhaps the formation of a stepped leader, may occur. The stepped leader will be faint and typically heavily branched and generally begins in steps of approximately 60 feet. These steps generally lengthen as the stepped leader approaches the earth.[2]

An electrostatic charge will be induced on the earth and items located on the earth due to the charge in the cloud and the charge in the stepped leader. As the stepped leader approaches the earth, the electric field between the tip of the leader and the earth is increased. As the electric field increases, point discharge currents are produced from items on the earth, such as blades of grass, trees, poles, and structures. When the critical electric field strength is exceeded, a streamer is produced, usually from a location where the electric field is concentrated. When the stepped leader attaches to the streamer, a return stroke is produced. The return stroke is the intensely luminous part of the strike where the peak transfer of current occurs. This process may be repeated as one or more subsequent return strokes in a single lightning flash. In subsequent return strokes, the subsequent leader (called a dart leader) is much faster and contains much less branching. The leader – return stroke process may occur once in a flash or may occur up to 26 times in a single flash[3].

The final step of the stepped leader is often referred to as the striking distance. The striking distance is a function of the charge in the leader channel. It is often characterized by the following relationship:

$$D = 10\, I^{0.65}$$

where:

D is the striking distance in meters and

I is the peak current in the strike in kilo-amperes.

Photo 1 is a popular picture that was first published in *National Geographic* magazine. It is forwarded here as a good example of the branching of the initial stepped leader and it provides a good example of the relative luminosity of the return stroke. The picture also provides visual documentation of the upward streamers produced by objects on the earth as the channel nears its final step. Finally, one can clearly see the striking distance in this flash as the channel changes from a nearly vertical path to one in a direction approximately 5 o'clock from the origin of the last step.

Purpose of a Lightning Protection System

In general, the purpose of a lightning protection system is to provide a high probability attachment point for

Photo 1. Lightning striking a tree showing branching of stepped leader, streamers from objects in vicinity of the flash, luminosity of the return stoke, and the striking distance associated with the flash Courtesy of *National Geographic.*

a lightning discharge which may strike a structure, and to provide a low impedance path to a grounding electrode system that is capable of dissipating the energy in the strike swiftly and safely into the earth without significant internal arcing that may result in fires or significant overvoltages that may damage internal equipment. Lightning protection systems meet this objective through the installation of the following subsystems: (1) strike termination network, (2) down conductor network, (3) grounding electrode network, (4) equipotential bonding network, and (5) surge protection. These subsystems will be discussed in greater detail later in this chapter. Lightning protection systems cannot prevent the lightning event from occurring nor do they "attract" lightning from greater distances than the conventional calculated attractive area of a structure.

The purpose of lightning protection standards vary depending upon application. For risk-based standards, such as IEC 62305,[4] the designer is allowed to develop a lightning protection system design that focuses on the protection of the specific threat(s) identified by the risk assessment. For instance, if the risk assessment identified that the only threat of concern is the threat against physical damage or fire, a system may be designed to address only that specific threat. Conversely, if the assessment identified that internal electrical systems at the site are susceptible but the structure is not susceptible to a direct strike, the lightning protection system design need only address the protection of the specific susceptible electrical equipment.

Few, if any, United States lightning protection standards can be considered risk-based. United States lightning protection standards generally describe a minimum-acceptable set of requirements which may be supplemented as necessary by the lightning protection system designer or authority having jurisdiction. The primary United States lightning protection standard, NFPA 780,[5] defines its purpose "to provide for the safeguarding of persons and property from hazards arising from exposure to lightning." It meets this purpose by providing specific minimum-acceptable requirements for a number of applications, such as ordinary structures, structures containing flammable vapors, heavy-duty stacks, and watercraft. A detailed discussion on the application of NFPA 780 is provided in the following section.

NFPA 780

As stated earlier, the principle lightning protection standard in the United States is NFPA 780. It is the primary US implementing document for the IEC 62305 series of documents and is referenced as providing baseline requirements in numerous specialized lightning protection documents such as DOD,[6,7] DOE,[8] NASA,[9] and FAA[10] lightning protection standards. Prior to the development of the IEC 62305 series of standards, NFPA 780 was used by many countries across the world as their own national lightning protection standard. The scope of the document covers traditional lightning-protection-system installation requirements for: (1) ordinary structures, (2) miscellaneous structures and special occupancies, (3) heavy-duty stacks, (4) watercraft, and

(5) structures containing flammable vapors, flammable gases, or liquids that give off flammable vapors. The document is not applicable to lightning-protection-system installation requirements for (1) explosives manufacturing buildings and magazines, and (2) electric generating, transmission, and distribution systems. However, informative annex information on the protection of structures housing explosives is included in the document and recommendations on the protection of wind turbines have been included in the 2008 edition as an informative annex. This annex on wind turbines addresses the protection of the structure but not the electrical generating, transmission, and distribution systems which may be associated with the structure.

The layout of the document is arranged such that the administrative information is included in chapter 1. This administrative information includes the scope, purpose, the use of listed components, mechanical execution of work, maintenance requirements, and the use of metric units of measurements.

Chapter 2 provides a listing of referenced publications.

Chapter 3 contains definitions of terms used in the standard. The chapter includes definitions of official NFPA terms and terms specific to the standard. Where terms are not defined, definitions provided in the NFPA Glossary or common usages of the terms apply.

Chapter 4 provides the minimum requirements for the protection of ordinary structures. It provides the core of requirements on which all remaining chapters are based. It is the most important chapter in the standard as it addresses all of the key elements of a lightning protection system.

Chapter 5 provides requirements for the protection of miscellaneous structures and structures with special occupancies. Included in this chapter are protection of masts, spires, and flagpoles, metal tanks and towers, concrete tanks and silos, air-inflated structures, and guyed structures. The chapter also cautions that provisions need to be made for settling and rising of wood-framed elevators when protecting grain-, coal-, and coke-handling and processing structures.

Chapter 6 provides protection requirements for heavy-duty stacks. The chapter addresses the protection of masonry, steel-reinforced concrete and metal stacks and provides requirements above and beyond those in chapter 4 in the area of corrosion protection, strike termination device locations and materials, and the installation of fasteners.

Chapter 7 is also a principal chapter in the standard as it addresses key parameters associated with the protection of structures which may contain flammable vapors. This chapter introduces details on the concept of the use of masts and overhead ground wire lightning protection systems, even though these types of systems are listed as approved in chapter 4. It also introduces a level of protection efficiency which exceeds that described in chapter 4 for ordinary structures. In the case of the zone-of-protection for structures covered by chapter 7, the rolling sphere radius is reduced from the 150-foot striking distance specified in chapter 4 to a 100-foot striking distance (the rolling sphere model will be discussed later in the following section). The basic requirements of this chapter are also referenced as applicable for the baseline requirements in the informative annex on the protection of structures housing explosives.

Chapter 8 deals with the protection of watercraft. The basic principles associated with the protection of structures are applicable although the size and location of components are modified. Corrosion issues and the use of dissimilar metals are key issues. The zone of protection used in this application specifies the use of a 100-foot striking distance versus the 150-foot striking distance used for ordinary structures.

NFPA 780 also contains 14 informative (non-mandatory) annexes. Annex A provides explanatory material expanding on some of the specific text in the body of the standard. Annex N provides a listing of informational references. Four annexes provide explanations or discussions on items applicable to lightning protection systems. These are annexes on the principles of lightning protection systems, explanation of bonding principles, a discussion on inspection and maintenance procedures, and a discussion on grounding system measurement techniques. Two annexes provide guides, one on lightning risk assessment and one on personal safety. The remaining annexes address specific protection applications. These are protection of trees, protection for picnic

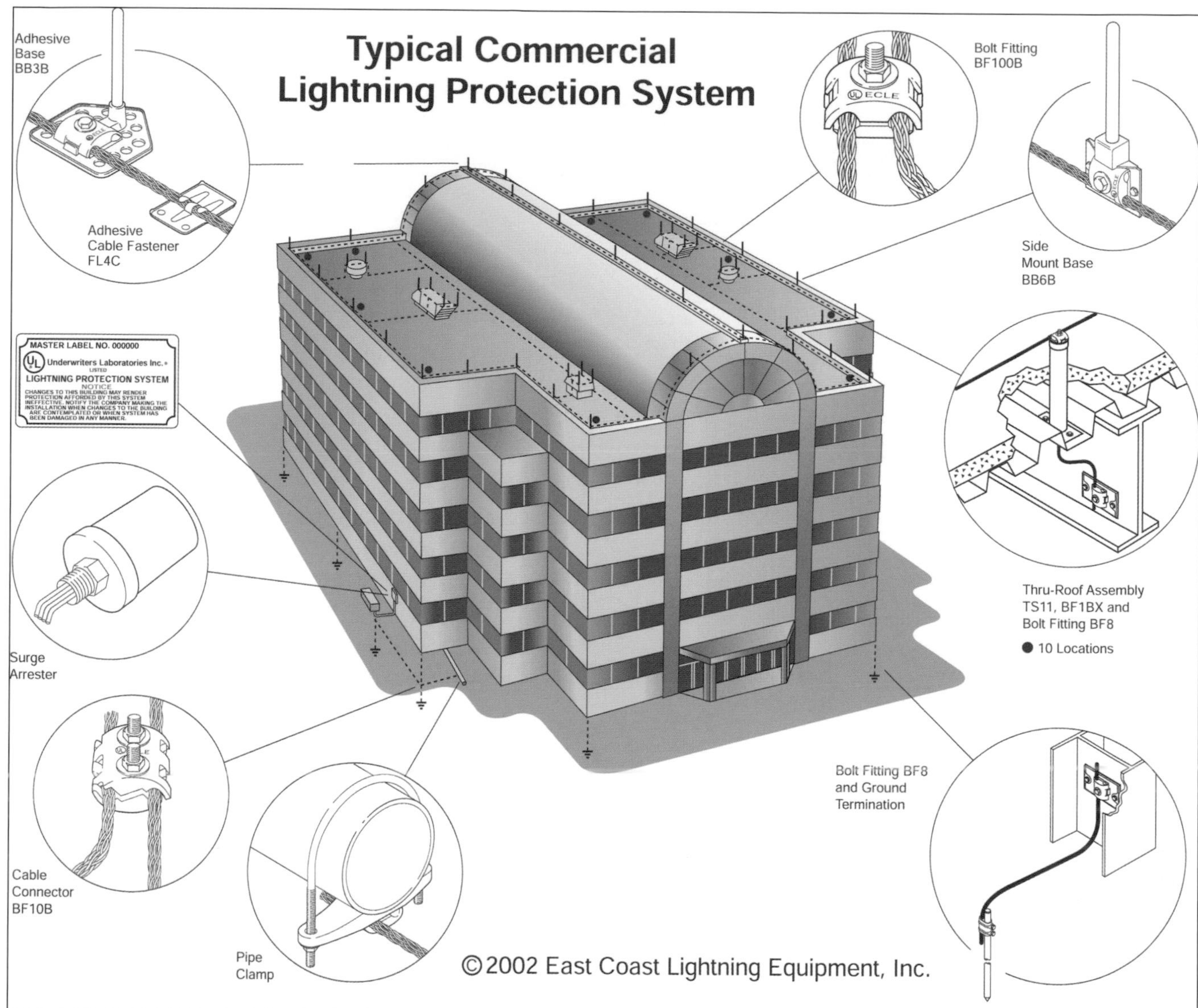

Figure 1. Example of commercial lightning protection system layout Courtesy of East Coast Lightning Equipment of Winsted, CT

grounds and open spaces, protection for livestock in fields, protection for parked aircraft, protection of wind turbines, and protection of structures housing explosives materials.

The Lightning Protection System

The primary components of a lightning protection system are the strike termination network, down conductor network, grounding electrode network, equipotential bonding network, and surge protection. Figure 1 illustrates the relationship each component has with respect to a complete lightning protection system. Each of these primary components is discussed below in this section.

Strike Termination Network

Strike Termination Device Placement

We should begin by discussing the difference between the terms *strike termination device* and *air terminal*. NFPA 780 defines a strike termination device as a component of a lightning protection system that is intended to intercept lightning flashes and connect them to a path to ground. Strike termination devices include air terminals, metal masts, and permanent metal parts of structures with a thickness equivalent to no less than 3/16-inch thick steel and overhead ground wires. An air terminal is defined as a strike termination device that is a receptor for attachment of flashes to the lightning protection system and is listed for the purpose. The key distinction between an air terminal and other strike termination devices is that the air terminal is listed in accordance with UL 96[11] for the purpose of providing lightning strike termination. Photo 2 provides an example of various configurations and materials of air terminals available to the lightning protection system designer.

All parts of a protected structure must be provided with strike termination devices at locations as required to ensure vulnerable parts of the structure are within a protected area or a zone of protection. NFPA 780 identifies three methods by which the placement of strike termination devices can be determined. The oldest method used for the determination of the protective area of a lightning-protection strike termination device is the protective angle method. The concept for this method is commonly attributed to Benjamin Franklin, but it was not until W. H. Preece[12] published the results of his experiments measuring the electric field around a vertical air terminal in 1880 that an accurate protective zone was proposed. The 1-to-1 (45 degree) protective angle proposed by Preece is still valid today for structures up to 50 feet tall. NFPA 780 allows a 2-to-1 (60 degree) protective angle for strike termination devices at heights of 25 feet or less.

The most universally used method for determining the protected area of a lightning protection system is the rolling sphere method. In this method, the final striking distance of the stepped leader is modeled to serve as the radius of a sphere. It is assumed that lightning can strike any point on the face of the sphere. Where the sphere is tangent to earth and resting against a strike termination device, all space in the vertical plane between the two points of contact and under the sphere is considered to be in the zone of protection. A zone of protection is also provided in the space between strike termination devices and in the vertical plane under the sphere when the sphere is resting on two or more strike termination devices (see figure 2). A striking distance of 150 feet is used in United States lightning protection standards for ordinary applications. The striking distance is reduced to 100 feet for structures requiring increased protection efficiency, such as structures housing flammable vapors and gases and structures housing explosives materials.

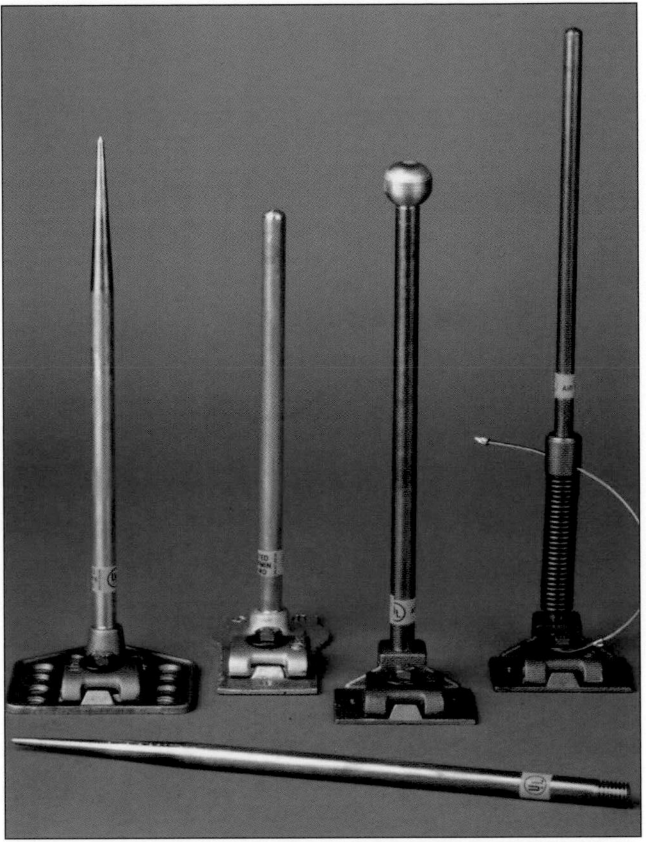

Photo 2. Examples of available air terminal configurations

Courtesy of East Coast Lightning Equipment of Winsted, CT

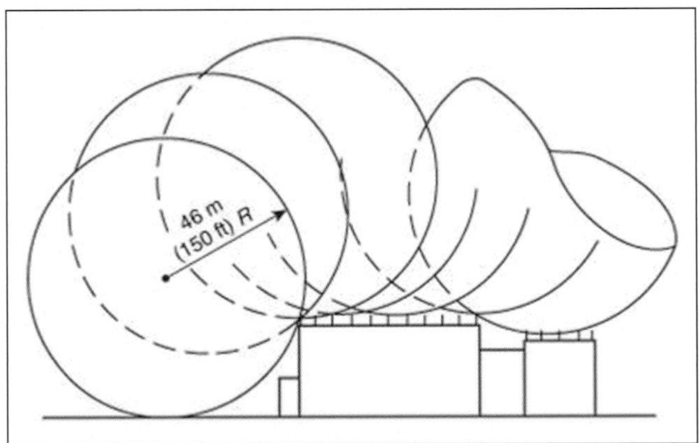

Figure 2. Zone of protection as determined by the rolling sphere method [from NFPA 780-2008, Figure 4.7.3.1(B)]

United States lightning protection standards do not recognize the mesh method of lightning protection for placement of strike termination devices, recognized by IEC 62305-3[13] and well established in many European standards. It is the opinion of U.S. standards-making organizations that the installation of air terminals or other strike termination devices enhances the probability of a successful termination by

Photo 3. Measurement to ensure air terminal is no greater than two feet from corner of roof Courtesy of Bonded Lightning Protection of Argyle, TX

Photo 4. Example of air terminal spacing on roof showing air terminal spacing within two feet of the edge and at a spacing of 20 feet around the perimeter Courtesy of Lightning Prevention Systems of West Berlin, NJ

providing a protuberance in the electric field gradient that will generate a successful streamer. United States lightning protection standards do recognize what will be described in this chapter as a *modified mesh method* that requires air terminals a minimum of 10-inches above the protected structure to be located along the perimeter of flat and gently sloping roofs and along the ridges of pitched roofs with roof conductors running along the perimeters or along the ridge as required to interconnect the air terminals. Air terminals are required to be located within 2 feet (0.6 m) of the ridge

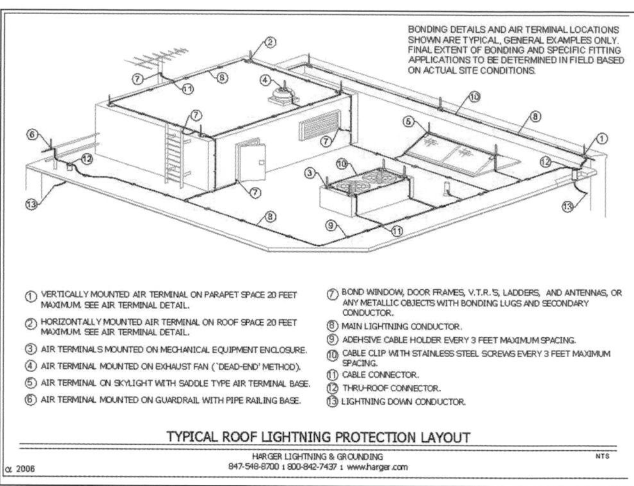

Figure 3. Strike termination device layout for flat roof with typical protrusions Courtesy of Harger Lightning Protection of Grayslake, IL

ends on pitched roofs or at edges and outside corners of flat or gently sloping roofs (see photo 3). The spacing between the air terminals on ridges and around the perimeter of a flat roof shall not exceed 20 feet (see photo 4); with the exception that air terminals projecting 2 feet or more above the roof of the structure may be spaced at 25 feet intervals. For flat or gently sloping roofs exceeding 50 feet in length or width, additional strike termination devices shall be provided in the interior of the roof at intervals not exceeding 50 feet. This relaxed spacing for the center of the flat roof is due to the relaxation of the electric field in the center of a large flat area.

Discussions in IEC TC81 (Lightning Protection) working group meetings have confirmed there is scientific justification for the relaxation of the air terminal placement requirements for the center of a large flat roof due to the reduction of the electric field gradient over that which will exist at corners and edges of a structure. The key to the effectiveness of this modified mesh arrangement is the location of strike termination devices where there will be greatest electric field concentration: the corners, edges, and ridges. It must be stressed that any protrusions on a flat roof (such as a vent, HVAC unit, etc.) must be provided with a strike termination device if it is not within the zone of protection of another device. Figure 3 provides an example drawing of a strike-termination device layout for a flat roof with typical protrusions in accordance with NFPA 780.

Photo 5. Guyed wooden mast for an overhead wire lightning protection system Courtesy of Harger Lightning Protection of Grayslake, IL

Approved Strike Termination Devices

The strike termination network may consist of masts (remote of or attached to the structure), overhead ground wires, or an arrangement of air terminals and interconnecting roof conductors located around the perimeter of flat or gently sloping roofs or along the ridges of peaked roofs. Masts may be metallic or wooden poles or they may be in the form of a metal tower. Wooden poles must be provided with an air terminal and a minimum of one down conductor. While both NFPA 780-2008 and UL 96A,[14] 12th edition, allow a single down conductor, a second down conductor is recommended. Masts will often extend at least 40 feet and the inductance of a single conductor can lead to a significant voltage drop at high frequencies over the length of the conductor. In addition, the second conductor adds a level or reliability since it is possible that any single conductor can be damaged over its normal lifetime. Photo 5 provides an example of a wooden pole used as a terminating mast for an overhead wire lightning protection system. In addition, metal parts of the structure having a thickness greater than 3/16-inch (4.8 mm) may serve as strike termination devices as long as they are properly connected to the down conductor system. As discussed in the previous paragraph, many international countries and international lightning protection standard IEC 62305-3 also allow a mesh arrangement of main-sized conductors installed on a structure to serve as an approved strike termination system. However, this mesh-type system without the use of supplementing air terminals is not recognized by United States lightning protection standards as an approved strike termination technique.

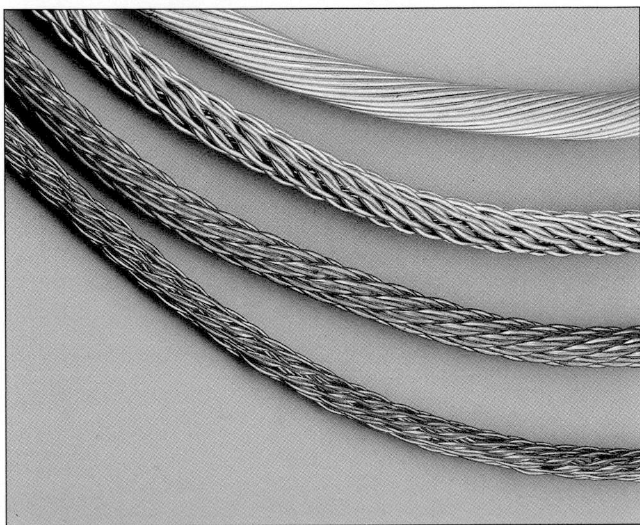

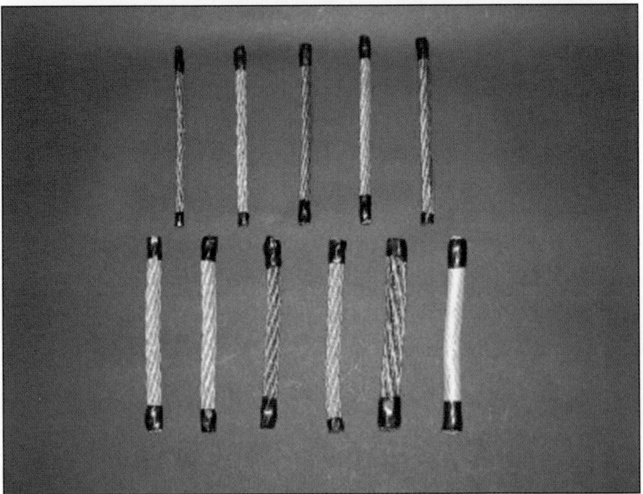

Photos 6 and 7. Examples of main-sized lightning protection conductors Courtesy of East Coast Lightning Equipment of Winsted, CT (photo 6) and Harger Lightning Protection of Grayslake, IL (photo 7).

Main Conductor Network

General

The material used for main conductors may be either copper or aluminum, depending upon the application. Care must be taken to minimize the probability of corrosion and contact with dissimilar metals. In no case shall a copper conductor be installed on an aluminum surface, nor an aluminum conductor be installed on a copper surface. In addition, aluminum conductors must not be used where they may come in contact with the earth. NFPA 780 defines a main-sized conductor by type of material, cross section area, weight per length, size of each strand, and height of the structure on which it is installed. Table 1 provides a summary of the NFPA 780-2008 requirements for a main conductor. Photos 6 and 7 show main-sized conductors that meet the requirements of Table 1. The weaves used in the production of the cables differ from standard electrical cable to make it more flexible to fit the approved connectors and fittings better. Conductors used in a lightning protection system should be listed for the purpose in accordance with UL 96.

Roof conductors

Each strike termination device must be provided with a minimum of two paths to ground. Both NFPA 780 and UL 96A allow a single exception to this rule for the case where a strike termination device is located below the main roof level and the length of the *dead-end run* is less than 16 feet (4.9 m). The dead-end run must maintain a horizontal or downward coursing to the nearby roof conductor or down conductor. The application of this exception is most often found in the protection of dormers.

A main-size conductor shall be used to interconnect the strike termination devices (roof conductors) and connect them to the grounding electrode network (down conductors). The coursing of the main-size conductors shall be such that the paths from the strike termination devices are downward, horizontal, or rising at no more than ¼ pitch to connections with their respective grounding electrodes. No bend of a conductor shall form an included angle of less than 90 degrees, nor shall it have a radius of bend less than 8 inches (203 mm). Additionally, the development of "U" or "V" (down and up) pockets must be avoided, as sharp bends in the primary current conductor

Main Conductor	Thickness	Cross section area	Weight per unit length	Size of each strand
Copper cable Structure ≤ 75 ft		($29 mm^2$) 57,400 circular mils	(278 g/m) 187 lb per 1000 ft	17 AWG
Aluminum cable Structure ≤ 75 ft		($50 mm^2$) 98,600 circular mils	(141 g/m) 95 lb per 1000 ft	14 AWG
Copper solid strip Structure ≤ 75 ft	(1.30 mm) 0.051 inches	($29 mm^2$) 57,400 circular mils		
Aluminum solid strip Structure ≤ 75 ft	(1.63 mm) 0.064 inches	($50 mm^2$) 98,600 circular mils		
Copper cable Structure > 75 ft		($58 mm^2$) 115,000 circular mils	(558 g/m) 375 lb per 1000 ft	15 AWG
Aluminum cable Structure > 75 ft		($97 mm^2$) 192,000 circular mils	(283 g/m) 190 lb per 1000 ft	13 AWG
Copper solid strip Structure > 75 ft	(1.63 mm) 0.064 inches	($58 mm^2$) 115,000 circular mils		
Aluminum solid strip Structure > 75 ft	(2.61 mm) 0.1026 inches	($97 mm^2$) 192,000 circular mils		

Table 1. NFPA 780 requirements for minimum size of main conductors

Photo 8. Example of proper bend radius in roof level main conductors Courtesy of Harger Lightning Protection of Grayslake, IL

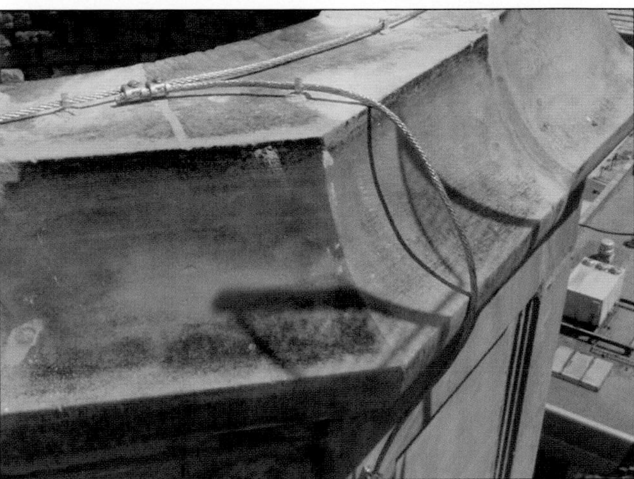

Photo 9. Example of a proper down conductor bend radius Courtesy of Guardian Equipment Company of Novi, MI

will add unnecessary inductance (impedance) to the conductor circuit. Photo 8 provides an example of a proper gradual bend in the roof conductor providing a two-way path for the corner air terminal and in the conductor used to connect the roof conductor to the through-roof connector. The through-roof connector provides the mechanism by which lightning currents can be transferred to the structural steel frame of the structure. The use of structural steel as a down conductor is discussed later in this section. It should also be noted that the air terminal in photo 8 is within 2 feet of the corner of the roof as required by NFPA 780-2008, 4.8.2.

Down conductors

A low impedance path from the strike termination network to the grounding electrode network is critical to the efficient operation of a lightning protection system. The function of the down conductor network is to conduct the majority of the lightning current from the strike termination network to the grounding electrode network. Even though the current will typically be divided among the down conductors, currents on the order of tens of thousands of amperes on a single down conductor would not be uncommon for small structures where the number of down conductors may be limited. For even a very small down conductor impedance, a significant voltage could be developed from the top of the down conductor (roof level) to the bottom of the down conductor (grounding electrode network). For this reason, significant care should be taken to minimize the impedance of the down conductor network. This is achieved by coursing the down conductor as direct as possible minimizing bends where practicable. Any bends that must be included in the down conductor coursing should be as gradual as possible and should never include a bend radius of less than 8 inches (203 mm). Photo 9 provides a good example of down conductor coursing where a bend in the down conductor is necessary. Increasing the number of parallel conductors will act to both decrease the overall impedance of the down conductor network as well as decrease the current density on any one conductor.

A minimum of two down conductors shall be provided for any application. Additional down conductors are required for structures exceeding 250 feet (76 m) in perimeter. Additional down conductors should be added as necessary to ensure that the average distance between down conductors does not exceed 100 feet (30 m).

Down conductors must be as widely separated as practicable. It is also recommended that they be located close to corners of the structure where practicable as corners provide a higher probability of streamer development and therefore lightning strike attachment. Another consideration for down conductor placement is the proximity to human traffic. Locating the down conductors in high traffic density areas is not recommended due to the higher probability of step and touch potential risks. Where conductors are located in public areas they should be protected

Photo 10. Example of interior connections to the steel frame from the through-roof connector Courtesy of Guardian Equipment Company of Novi, MI

against physical damage and consideration should also be given to utilizing a protection technique that will limit the probability of touch potential. Other considerations influencing the specific location of down conductors are the location of strike termination devices, the direct coursing of the conductors, local earth conditions, security against displacement, the location of large metallic bodies, and the location of underground metallic piping systems.

Photo 11. Attachment to structural steel using a bonding plate
Courtesy of East Coast Lightning Equipment of Winsted, CT

The use of the frame of a structural steel framed building as the down conductor network is encouraged. It will generally provide lower impedance than will conventional main-sized conductors due to the physical size of the steel frame members and the number of members that will act as parallel conductors. In this case, the strike- termination network components will typically be connected to the frame of the structure using a through-roof connector such as the device shown in photo 8. The connection at the top and bottom of the steel frame of the structure shall be made at a contact point where the frame is cleaned to base metal using a bonding plate having a contact area of not less than 8 square inches (5200 mm^2). Photo 10 provides an example of the interior connections to the steel frame from the through-roof connector. Methods allowed for contact to the structural steel are clamping, bolting or threading and tapping. Welding or brazing of the conductor directly to the structural steel is also allowed. Photo 11 provides an example of the installation of a bonding plate on a structural steel member at the earth level for the purpose of providing a connection between the structural steel and the grounding electrode(s).

It should be noted that many international standards allow the use of electrically continuous concrete reinforcing bars to serve as lightning protection system down conductors (see IEC 62305-3, 5.3.5). However, IEC 62305-3, E.5.3.5 recommends the use of natural components such as the concrete reinforcing bars because the large number of parallel conductors would result in a considerable drop in the maximum voltage across the down conductor network. However, this clause also states, "If electrical continuity of the natural down-conductors cannot be guaranteed, conventional down-conductors should be installed." It is for this reason that the use of concrete reinforcing bars is not allowed in either NFPA 780 or UL 96A as a substitute for the required down conductor. These standards do require the interconnection of the concrete reinforcing bars with the lightning protection system at the top and bottom for the purpose of potential equalization.

Grounding Network

General

The purpose of a lightning-protection grounding network is to provide a low impedance connection with the earth in order to dissipate lightning currents in such a manner as to minimize the peak voltage that will appear on the system. The grounding electrode network must also be designed to reduce to the extent possible the risk due to touch and step potentials. For a properly designed and installed lightning protection system, the largest contributor to the overall voltage that will appear on the lightning protection system is the grounding electrode component. Therefore, a low-impedance grounding electrode network will lower the threat due to touch potentials, internal side flashes, and the threat to internal electrical equipment.

All grounding electrodes, as well as conductive items in contact with the earth, exhibit specific impedances with respect to remote earth. These impedances are typically characterized in terms of resistance per unit of distance. As current is injected into the grounding electrode, a voltage gradient related to the injected current and earth resistance gradient is developed. This voltage gradient is often referred to as step voltage or step potential. In homogeneous earth, the step voltage is usually the highest where the current passing from the grounding electrode to the earth is strongest. This is typically at the outer extremity of the grounding electrode. The maximum potential in the soil decreases with increased area covered by grounding system. Vertical grounding electrodes, such as ground rods, affect only a small surface area of ground surrounding them. A grounding mesh network, radials, and ground ring electrode can greatly reduce the grounding electrode potential over that of a driven ground rod.

Many factors come into play in the design of the optimum lightning protection grounding network. The specific design of the lightning protection grounding network may be driven by local conditions (temperature variations, soil moisture and type, local earth resistivity, etc.) as well as the vulnerability of the structure or its contents to step potentials, ground potential rise, and the high frequency components of the lightning discharge.

Requirements

Grounding electrodes may be made of copper, copper-clad steel, or stainless steel. Each down conductor must be terminated by a grounding electrode dedicated to the lightning protection system. Electrical or telecommunication system grounding electrodes shall not be utilized as a lightning-protection grounding electrode. However, these grounding electrodes, along with all incoming or exiting metallic piping must be bonded to the lightning protection grounding system. Connections between the down conductors and grounding electrodes may be made by bolting, brazing, welding, or high compression connectors listed for the purpose. Clamps suitable for direct burial may also be used. Grounding electrodes shall be installed below the frost line where possible.

Approved grounding electrodes

Approved grounding electrodes for use in lightning protection installations are ground rods, ground plate electrodes, concrete-encased electrodes, radials, and ground ring electrodes. Combinations of these devices may be used. The construction and installation requirements for each of these electrodes are similar to that given in the *National Electrical Code*[15] (NFPA 70), Article 250.52.

Ground rods shall be not less than ½-inch (12.7 mm) in diameter and 8 feet (2.4 m) long. The ground rods shall extend vertically not less than 10 feet (3 m) into the earth (see Figure 4). The earth shall be compacted and made tight against the length of the conductor or ground rod. Ground rods shall be free of paint or other nonconductive coatings. Where multiple connected

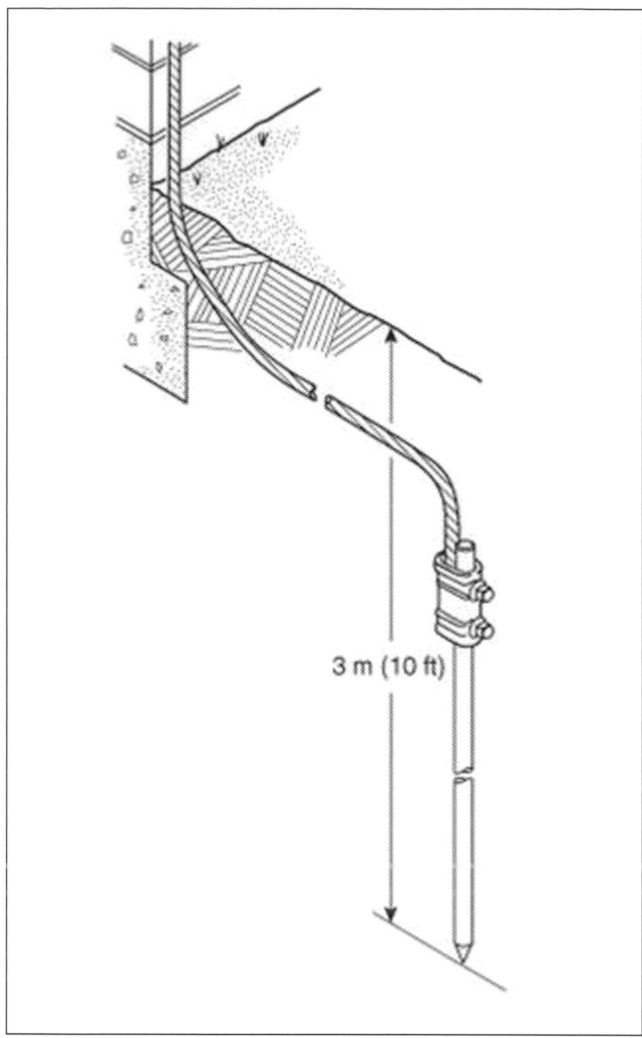

Figure 4. Typical single ground rod installation [from NFPA 780-2008, Figure 4.13.2.3(B)]

ground rods are used, the separation between any two ground rods shall be at least the sum of their lengths where practicable.

Ground plate electrodes shall have a minimum thickness of 0.032 inch (0.8 mm) and a minimum surface area of 2 ft^2 (0.18 m^2). The plate shall be buried not less than 18 inches (460 mm) below grade. Photo 12 provides an example of a ground plate electrode being installed.

Concrete-encased electrodes must be limited to new construction because it is necessary to ensure that the electrode was properly constructed. The electrode must be located near the bottom of a concrete foundation or footing that is in direct contact with the earth and shall be encased by not less than 2 inches (50 mm) of concrete. The encased electrode

Photo 12. Example of a buried copper plate grounding electrode Courtesy of Bonded Lightning Protection of Argyle, TX

shall consist of either not less than 20 feet (6 m) of bare copper main-size conductor or at least 20 feet (6 m) of one or more steel reinforcing bars or rods not less than ½ inch (12.7 mm) in diameter that have been effectively bonded together by either welding or overlapping 20 diameters and wire-tying. UL 96A, 10.5.1 allows concrete-encased electrodes only when utilized in conjunction with other grounding electrodes, such as rods, plates, or ground ring electrodes. NFPA 780 has not such a restriction. International lightning protection standards (IEC 62305-3, 5.4.4) indicate that concrete-encased electrodes are preferred over other grounding electrodes with the exception of the ground ring electrode.

A radial electrode system shall consist of one or more main-size conductors, each in a separate trench extending outward from the location of each down conductor. Each radial electrode shall be not less than 12 feet (3.6 m) in length and not less than 18 inches (460 mm) below grade and shall diverge at an angle no greater than 90 degrees.

A ground ring electrode is a main-sized lightning conductor encircling the structure. It shall be in direct contact with earth or it may be encased in a concrete footing at a depth below the frost line where possible but not less than 18 inches (460 mm). Photo 13 shows a trench being prepared for the installation of a ground ring electrode. The ground ring electrode is referred to in IEC 62305-3 as a Type B grounding system and is the preferred grounding system for all applications. Only Type B grounding systems are recommended in the international standard for installations on bare solid rock and for structures with extensive electronic systems or with high risk of fire.

Ground potential rise

When an impulse current is injected onto a grounding electrode, an impulse voltage is developed. This voltage (known as a *ground potential rise*) will exist as long as the current is flowing. Should independently grounded systems or systems utilizing metallic piping in contact with the earth exist in a structure, this ground potential rise could create a sufficient potential to cause breakdowns through unintended paths. This could include arcing through the air, through the earth, or through internal systems. The result could be fire, equipment damage, and possibly human injury.

As an example, consider the situation where a lightning protection system is installed with four grounding electrodes having a resistance to earth of 25 ohms which are interconnected at the ground level. Using the IEC 62305-1 model for current division, one would assume that the current associated with a direct strike would divide equally between the lightning protection grounding electrodes and the services. For an average 40 kA lightning strike, the current in each lightning protection grounding electrode will be 5 kA (half of the total current will flow through the lightning protection system and this will be equally divided between grounding electrodes because they are of equal resistance). This will result in an estimated ground potential rise of 125 kV. Let us also assume that an isolated gas line is installed in the structure. This lightning current and resulting voltage associated with the lightning protection grounding system will be sufficient to cause significant damage should an arc occur between the lightning protection grounding electrode (which will be at 125 kV) and the incoming gas line, which would be at 0 volts. An alternative failure mode would be for the current path to be through the utilization hardware via the electrical service ground. Either case could create a serious problem. The solution is to ensure that all incoming and exiting conductors be bonded to a common grounding system for the structure. In this case, all of the services would share the same potential but the majority of the current would be dissipated through the lower impedance lightning protection grounding system. NFPA 780, 4.14 requires that all grounding media in or on a structure be interconnected to provide a common ground potential. This interconnection shall include lightning protection, electric service, telephone, and antenna system grounds, as well as underground metallic piping systems. Where corrosion is an issue, the bond between the incoming piping system and the lightning protection grounding system may be made through a spark gap.

Photo 13. Installation of a ground ring electrode Courtesy of East Coast Lightning Equipment of Winsted, CT

Equipotential bonding

The previous discussion on ground potential rise justified the need to have a ground-level equipotential bonding system. An earlier section discussed the roof level conductors which would provide a roof-level equipotential bonding system. There may be some conditions where additional bonding is necessary in the structure.

When lightning currents flow through a down conductor, a voltage is developed between the top and bottom of the down conductor as a result of the impedance of the cable. This voltage will produce an electric field. The current flowing though the conductor will also create a magnetic field around the conductor. These electromagnetic fields can result in coupling in adjacent or nearby conductors. If the voltage between a lightning protection main conductor and an adjacent conductor exceed the breakdown voltage of air, a side flash will occur. The probability of a side flash is increased for independently grounded items interconnected only at the ground level where one of the conductors, such as conduit or piping, meanders greatly or where there is a significant inductance in the circuit.

NFPA 780, 4.21.2 provides a formula in which it simplifies the calculations required to determine the probability of side flash. The formula requires a number of assumptions to get to its simplified state, but it can be accurately used to determine the need for bonding above the ground level bonds required. When grounded metal bodies are located within the following calculated bonding distance, they shall be bonded to the lightning protection system:

$D = h (k_m) / 6n$

where:

D = calculated bonding distance

h = vertical distance between the bond being considered and the nearest lightning protection system bond

k_m = 1 if the flashover is through air, or 0.50 if through dense material such as concrete, brick, wood, etc.

n = a value related to the number of down conductors that are spaced at least 25 feet (7.6 m) apart; located within a zone of 100 feet (30 m) from the bond in question; and where bonding is required within 60 feet (18 m) from the top of any structure.

The value of n is given as:

n = 1 where there is only one down conductor in this zone;

n = 1.5 where there are only two down conductors in this zone;

n = 2.25 where there are three or more down conductors in this zone.

(When the bonding calculation is performed for a location below a level 60 feet (18 m) from the top of a structure, n is the total number of down conductors in the lightning protection system.)

For structures of 40 feet (12 m) in height or less, the value of h may be taken to be either the height of the building or the vertical distance between the nearest bonding connection of the grounded metal body to the lightning protection system and the point on the down conductor where the bonding connection is being considered.

These bonding equations are also valid for determining side flash distances from overhead ground wires or spacing of masts from a structure. For ungrounded metallic bodies, bonding is applicable only if the isolated body reduces the spacing between two grounded items (such as a window frame between a lightning protection system conductor and a water pipe).

Surge protection

Surges can occur from a number of sources. Among these are direct strikes to incoming lines, strikes near an incoming line, and ground potential rises as a result of a strike to or near the structure. Magnetic and capacitive coupling from a lightning strike occurring up to 1500 feet (500 m) can induce a transient capable of damaging electrical and electronic systems.

The protection of internal electrical and electronic systems will require that measures be taken for the protection against lightning electromagnetic pulses (LEMP). Potential equalization measures are required for incoming electrical services (including data and communication services). This is normally implemented using spatial shielding and surge protective devices (SPDs). All lightning protection standards require protection against incoming surges at a minimum. In most cases, this leads to requirements for equipotential bonding SPDs.

The susceptibility of electrical and electronic systems can be reduced in some cases by establishing lightning protection zones. In some cases, this may be accomplished by taking advantage of natural shielding procedures such as that provided by locating the equipment in a steel framed or steel reinforced concrete structure, by utilizing the shielding provided by the equipment's enclosure, or locating it in an equipment rack. The boundary point of the initial lightning protection zone (LPZ 1) also requires the installation of SPDs to protect the equipment. Additional lightning protection zones may be established as necessary, depending upon the sensitivity of the electronic equipment.

Prior to September 2006, UL required only a surge arrestor or a "TVSS marked for LPS" on the incoming power service to the structure in order to obtain Master Label certification. UL 1449, edition 3[16] now incorporates surge arrestors and transient voltage surge suppressors into a single document. The standard classifies as a Type 1 SPD permanently connected SPDs intended for installation between the secondary of the service transformer and the line side of the service-equipment overcurrent device, and without an external overcurrent protective device. Type 2 SPDs are permanently connected SPDs intended for installation on the load side of the service-equipment overcurrent device; including SPDs located at the branch panel.

United States lightning protection standards agree on the surge protective device requirements to be installed at the service entrances. They primarily address common mode surges as a minimum, but indicate differential mode surges may need to be considered in some applications. In the case of electrical service entrances, line-to-ground and line-to-neutral modes must be protected but line-to-line protection may also be provided. For each electrical service entrance of less than 1 kV, the following is acceptable:

- a surge protective device with a rated maximum discharge current of 40 kA or more complying with the Standard for Transient Voltage Surge Suppressors, UL 1449 and IEEE C62.11, Standard for Metal – Oxide Surge Arresters for AC Power Circuits installed on the supply or load side of the service disconnect overcurrent protection installed in accordance with NFPA 70, Article 285 (Type 1 SPD),
- a transient voltage surge suppressor marked for LPS application with a rated maximum discharge current of 40 kA or more, installed on the load side of the service disconnect overcurrent protection in accordance with NFPA 70, Article 285 (Type 2 SPD)
- Type 1 or Type 2 surge protective devices (SPDs) rated 20 kA or more nominal discharge current in accordance with the Standard for Surge Protective Devices, UL 1449, edition 3, installed in accordance with NFPA 70, Article 285.

For circuits greater than 1 kV, a surge arrester with a rated maximum discharge current of 40 kA or more meeting the requirements of IEEE C62.11, Standard for Metal – Oxide Surge Arresters for AC Power Circuits or IEEE C62.1, Standard for Gapped Carbide Surge Arresters for AC Power Circuits may be used provided it is installed in accordance with the requirements of NFPA 70, Article 280.

Surge protection for all conductive signal, data, and communication lines shall be provided with a rated maximum discharge current of 10 kA or more at the point of entrance. The protection provided shall be installed in accordance with Articles 800, 810, 820, or 830 (as applicable) of the *National Electrical Code* and shall comply with the Standard for Antenna-Discharge Units, UL 452, the Standard for Protectors for Paired-Conductor Communications Circuits, UL 497, and the Standard for Protectors for Coaxial Communications Circuits, UL 497C.

Surge protective devices should be installed in such a manner that they can be tested or monitored (as applicable). Photo 14 provides an example of a surge protective device installed on a service entrance and photo 15 forwards an example of an installation for communication lines.

Quality Control Programs

There are two well-recognized quality control programs in the United States for lightning protection system installations. These are the Underwriters Laboratories (UL) Master Label program and the Lightning Protection Institute (LPI) Certified System program. Each of these programs requires the use of components qualified to UL 96.[11]

UL Master Label Program

The best known of all lightning protection quality control programs is the Underwriters Laboratories Inc. (UL) Master Label program. It is well respected by numerous organizations such as many federal agencies and insurance underwriters as a program that will ensure that proper materials and workmanship were provided when the lightning protection system

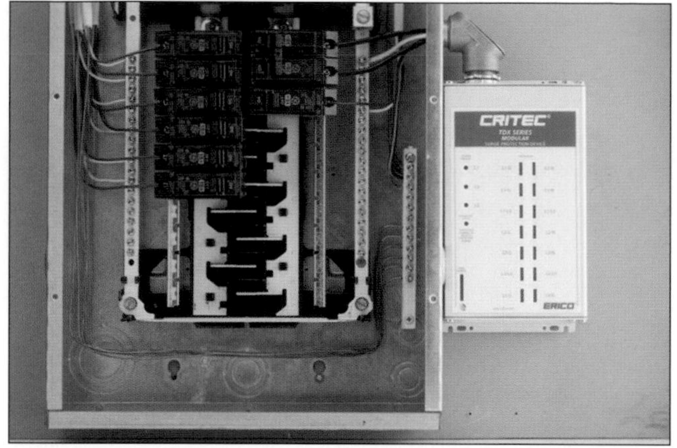
Photo 14. Installation of surge protection on incoming service
Courtesy of Erico

Photo 15. Installation of surge protection on communication lines at the service entry

was installed. Underwriters Laboratories has been testing and certifying lightning protection equipment since 1908. In 1923, UL began issuing Master Labels for lightning protection system installations. UL engineer Karl Klock, in a 1938 radio address, reported that in the first 15 years the service was in operation, approximately 78,000 buildings received a Master Label and that the minimal reports of failure reflected a protection efficiency of 99.9%.

UL issues Master Labels for systems by inspecting system components and checking completed installations. Components are labeled in accordance with UL 96 while installations are required to comply with UL's internationally recognized Standard UL 96A.[14]

The program requires that installers demonstrate competence in the installation of lightning protection systems by becoming a UL Listed installer. Once the UL Listed installer completes an installation, they submit the Master Label certification application. A UL field representative then inspects the installation and instantly communicates the results electronically to UL and the installer. If necessary, a letter detailing any variances is issued to the installer. After variances are corrected, the installer resubmits the application for re-inspection. In some instances, system designs and variance corrections can be reviewed electronically. The UL Listed installer forwards the certificate to the premise owner/operator, and posts the certificate on the UL web site, providing proof that the lightning protection system complies with the applicable standard.[17]

Master Label certificates must be renewed every five years, whenever the building changes structurally, or when modifications are made to the system. The UL Listed installer can repair or modify the system and arrange to have it re-evaluated by UL to determine its continued compliance with UL 96A.

LPI Certified System

The Lightning Protection Institute (LPI) also offers a certification program for lightning protection installations. Although not as well known as the UL Master Label program, the LPI Certified System program offers an additional layer of quality control and accountability. The LPI Certified System program requires that essential inspections be performed once the grounding system is installed, once conductors are installed (prior to being concealed if applicable), and once the roof system is completed. These inspections will be made by a LPI certified master installer and witnessed by a representative of the owner. Upon completion, an independent inspection on the completed system will be performed either by a LPI certified inspector or by a UL inspector and a final report on findings will be provided. When an owner receives an LPI certification for the installation, they can be ensured that all aspects of the installation have been systematically scrutinized for compliance with LPI 175, NFPA 780, and UL 96A, as applicable.

NEC Versus NFPA 780 Grounding and Bonding Requirements

Little difference exists between permitted grounding electrodes identified in NFPA 70, 250.52 and those identified in NFPA 780, 4.13; with the exception, that NFPA 780 allows the use of radials. The issues of the higher frequency components and higher current densities in the lightning threat warrant the need to provide a larger area covered by the grounding system to decrease the maximum potential developed under discharge conditions.

The primary difference between the grounding requirements identified in NFPA 70 Article 250 and NFPA 780 is in the definition of the term *grounding electrode* and the resulting conflict between NFPA standards as to the bonding of gas lines to the lightning protection grounding system. NFPA 70, 250.52(B), NFPA 780, and NFPA 54, 7.13.2 all agree that gas piping shall not be used as a grounding electrode.

The *National Electrical Code* defines *grounding electrode* as "a conducting object through which a direct connection to the earth is established." NFPA 780 defines the term as "the portion of a lightning protection system, such as a ground rod, ground plate, or ground conductor that is installed for the purpose of providing electrical contact with the earth." The *for the purpose* clause is an important distinction in the lightning protection community, as the intent is that all metallic conductors entering or exiting the structure must be interconnected with the lightning protection grounding system to eliminate the dangerous effects of ground potential rises; but the lightning protection grounding system is designed to conduct the majority of the current into the earth.

By the definition given in the *NEC*, an underground gas line is a grounding electrode because it establishes a connection to earth the same as an underground water pipe. The key point in Article 250.52(B) is that it shall not be used for that purpose. NFPA 70-2008, 250-104(B) acknowledges that gas piping "that is likely to be energized" shall be bonded to the service equipment enclosure. However, it is not clear to an electrical inspector from this text that during a direct or nearby lightning strike the gas line *is* likely to become energized (see the earlier discussion on ground potential rise). It is also questionable whether the bond at the service equipment enclosure is the best place to make this bond. For the purpose of lightning protection, this bond is best made at the service entrance.

Summary

This chapter provides a general overview of the principles of lightning protection. It focuses on the installation requirements of NFPA 780 and UL 96A. The installation of a lightning protection system is much different from the installation of electrical service wiring due to the high current densities, high rates of current rise (di/dt), and resulting mechanical forces. For these reasons, specialized material such as that specified in NFPA 780 and UL 96 is required and the system should be installed by qualified personnel trained and certified in the installation of lightning protection systems (such as LPI-certified and UL listed installers).

The chapter identifies the key elements of a lightning protection system. These are the strike termination network, down conductor network, grounding electrode network, equipotential bonding network, and surge protection. It provides a description of each of these elements and provides some minimum–acceptable parameters. It also discusses two quality control programs that can be utilized to ensure a lightning protection system is installed in accordance with established lightning protection standards.

Key principles discussed in this chapter are the differences in the definition of *grounding electrodes* between the *National Electrical Code* and national lightning protection standards such as NFPA 780 and UL 96A. The differences in the definition are minor but the application can lead to issues as to whether a bond to an underground pipe would constitute its use as a grounding electrode. The interconnection of all incoming conductors to the lightning protection system is critical for the safe and efficient operation of the system.

References

[1] Rakov, Vladimir A., and Uman, Martin A., *Lightning: Physics and Effects*, Cambridge University Press, Cambridge, UK, 2003.

[2] Malan, D. J., *Physics of Lightning*, The English University Press Ltd., London, 1963.

[3] Workman, E. J., Brook, M., Kitagawa, N., "Lightning and charge storage," *Journal of Geophysical Research*, 65, 1513-1517, 1960.

[4] International Electrotechnical Commission, IEC 62305 Parts 1 - 4:2006, "Protection against lightning," 1st ed. (Geneva: International Electrotechnical Commission, January 2006).

[5] National Fire Protection Association, NFPA 780-2008, "Standard

for the Installation of Lightning Protection Systems," (Quincy, MA: National Fire Protection Association, 15 August 2007).

[6] Department of Defense, DOD 6055.9-STD, "Ammunition and Explosives Safety Standards," Chapter 7 (Washington, D.C.: Department of Defense, 5 October 2004).

[7] Naval Sea Systems Command, NAVSEA OP-5, "Ammunition and Explosives Ashore," vol. 1, 6th rev., change 2, chapter 6, (Washington, D.C.: Naval Sea Systems Command, March 1995).

[8] Department of Energy, DOE M 440.1-1, "DOE Explosives Safety Manual," (Washington, D.C.: Department of Energy, 2002).

[9] National Aeronautics and Space Administration, NASA E-0012E, "Standard for Facility Grounding and Lightning Protection," (Kennedy Space Center, FL: National Aeronautics and Space Administration, 2001).

[10] Federal Aviation Administration, FAA-STD-109, "Lightning and Surge Protection, Grounding, Bonding, and Shielding Requirements for Facilities and Electronic Equipment," rev. D, (Washington, D.C.: Federal Aviation Administration, 9 August 2002).

[11] Underwriters Laboratories, UL 96, "UL Standard for Safety for Lightning Protection Components," 5th ed., (Northbrook, IL: Underwriters Laboratories, 12 May 2005).

[12] Preece, W. H., "On the space protected by a lightning conductor," *Phil. Magazine*, 9, pp.427-430.

[13] International Electrotechnical Commission, IEC 62305-3:2006, "Protection against lightning – Part 3: Physical damage to structures and life hazard," 1st ed., (Geneva: International Electrotechnical Commission, January 2006.

[14] Underwriters Laboratories, UL 96A, "UL Standard for Safety for Installation Requirements for Lightning Protection Systems," 12th ed., (Northbrook, IL: Underwriters Laboratories, 23 May 2007).

[15] National Fire Protection Association, NFPA 70, *National Electrical Code*, (Quincy, MA : National Fire Protection Association, 2008).

[16] Underwriters Laboratories, UL 1449, "UL Standard for Safety for Surge Protective Devices," 3rd ed., (Northbrook, IL: Underwriters Laboratories, 29 September 2006).

[17] http://www.ul.com/lightning

Review Questions

The questions included here were developed using material included in this chapter. The answers can be found by reviewing the text. It is also important that students make use of *NEC-2008*, where many answers can be found.

1. Between which of the following are lightning events likely to occur?
   a. cloud-to-ground
   b. cloud-to-cloud
   c. between charge centers within a cloud
   d. all of the above

GIVEN: The striking distance is the final step of a stepped leader for cloud-to-ground lightning. It is the distance the final step travels when making attachment with the object on the ground. The relationship between *striking distance* and *peak current* is given in the body of the chapter. By relating the probability of occurrence of a given peak lightning current to its associated striking distance, a protection efficiency can be determined for a given striking distance. This striking distance can be used in conjunction with the rolling sphere method for strike termination device placement in lightning protection system design.

2. The term *striking distance* refers to
   a. the distance within which an item must be bonded to the lightning protection system
   b. the height of a building or structure
   c. the final step of a stepped leader
   d. all of the above

GIVEN: A lightning protection system cannot prevent lightning from occurring. It can only provide a mechanism by which the energy can be controlled until it can be safely dissipated into the earth.

3. Which of the following is not the purpose of a lightning protection system?
   a. provide a high probability attachment point for a lightning strike
   b. prevent lightning from occurring in the vicinity of the strike
   c. provide a low impedance path to earth for the dissipation of lightning currents
   d. limit internal arcing and overvoltages on electrical/electronic equipment

4. Which of the following is not a part of the lightning protection system?
   a. strike termination network
   b. down conductor network
   c. grounding network
   d. potential equalization network
   e. all of the above

5. Which of the following lightning protection standards allow a risk-based solution to lightning protection?
   a. NFPA 780
   b. UL 96A
   c. IEC 62305
   d. all of the above

6. Which of the following chapters of NFPA 780 provide the baseline requirements on which the remainder of the document is built?
   a. Chapter 2
   b. Chapter 3
   c. Chapter 4
   d. Chapter 5

7. NFPA 780 allows the following to act as a strike termination device:
   a. air terminal
   b. metal mast
   c. structural component equivalent to 3/16-inch thick steel
   d. all of the above

8. Which of the following standards provides the criteria for the listing of lightning protection components:
   a. NFPA 780
   b. UL 96A
   c. UL 96
   d. NFPA 70

9. A main-sized lightning protection conductor is used to:
a. interconnect air terminals

b. connect the air terminal network to the grounding electrodes
c. serve as the ground ring electrode
d. all of the above

10. The minimum bend radius of a main conductor may be:
a. 12 inches
b. 8 inches
c. 4 inches
d. None. There shall be no bends in main conductors

11. A lightning protection mast may be in the form of which of the following:
a. a metal pole with 3/16-inch cap thickness
b. wooden pole with an air terminal and two down conductors
c. metal tower
d. all of the above

12. Which of the following is not allowed by NFPA 780 as a method for determining the placement of strike termination devices?
a. rolling sphere method
b. protective angle
c. electric field intensification method
d. "modified-mesh" method

GIVEN: The thickness of a HVAC enclosure is generally not thick enough to meet the criteria for strike termination devices.

13. Which of the following is not allowed to be used as a strike termination device?
a. air terminals
b. HVAC
c. metallic structural components with a thickness of 3/16-inch or greater
d. all of the above

14. Which of the following is not allowed to be used as a lightning protection grounding electrode (as defined by NFPA 780)?
a. ground rod
b. radial electrodes
c. concrete-encased electrodes
d. underground metallic piping (such as gas or water lines)

15. Which of the following is required to be bonded to the lightning protection grounding system?
a. electrical service ground
b. metallic water pipe
c. gas piping
d. all of the above

16. Which of the following materials is not allowed by NFPA 780 to be used as a ground rod?
a. galvanized ground rod
b. copper-clad steel ground rod
c. copper ground rod
d. stainless steel ground rod

17. Which of the following are not allowed by US lightning protection standards to be used as a lightning protection system down conductor?
a. main-sized conductor
b. concrete reinforcing steel
c. structural steel
d. all of the above are allowed to be used as a down conductor

18. Which of the following methods is not used by NFPA 780 to define a main-sized lightning protection conductor?
a. cross section area
b. weight per unit length
c. size of each strand
d. diameter

Tables

Included in this chapter are several tables that are vital or useful in studying the subject of grounding and correctly applying the *NEC* rules in that regard. In addition, several tables are included to assist the installer or inspector of electrical systems in properly designing, installing and maintaining these systems. By carefully utilizing the information in these tables, the electrical system will benefit from additional safety and should serve the owner for years to come.

Chapter 21

21 Soares Book on Grounding and Bonding

The following tables are included:

Table 21-1. Table 250.66, Grounding Electrode Conductor for AC Systems from the *NEC*.

Table 21-2. Analysis of *NEC* Table 250.66 Grounding Electrode Conductor and Service Conductor Compared as to Relative Size.

Table 21-3. Comparison of Copper Grounding Electrode Conductor per *NEC* Table 250.66 when Used with and without Steel Conduit for Physical Protection.

Table 21-4. Comparison of Aluminum Grounding Electrode Conductors per *NEC* Table 250.66 when Used with and without Aluminum Conduit for Protection.

Table 21-5. Rating of Grounding Electrode Conductors Specified in *NEC* Table 250.66 and Voltage Drop under Maximum Short-time Rating.

Table 21-6. *NEC* Table 250.122 Minimum Size Equipment Grounding Conductors for Grounding Raceways and Equipment.

Table 21-7. Analysis of Table 250.122 of the *National Electrical Code*.

Table 21-8. Table 8 from *NEC* Chapter 9 Conductor Properties.

Table 21-11. Maximum length of electrical metallic tubing, intermediate metal conduit and rigid steel conduit that may safely be used as an equipment grounding circuit conductor.

Table 21-14. Aluminum conduit used as an equipment grounding conductor (EGC) compared with equipment grounding conductors as specified in Table 250.122 of the *National Electrical Code*.

Table 21-15. Sizes of conductor (Copper Bar) for use in the Equipment Grounding Strap for overcurrent devices from 1600 amperes through 5000 amperes.

Note: Tables **21**-16 through **21**-19 were derived from software (SCA) and Testing developed at the Georgia Institute of Technology and sponsored by the producers of steel EMT, IMC, and rigid steel conduit. Additional information about the research and testing of steel conduit and tubing is available at www.steelconduit.org and the computer based modeling GEMI software is available online at http://www.steelconduit.org/gemi.htm.

Table 21-16. Maximum length of equipment grounding conductor that may safely be used as an equipment grounding conductor based on ground-fault current of 400% of the overcurrent device rating. Circuit 120 volts to ground; 40 volts drop at the point of fault. Ambient temperature 25°C.

Table 21-17. Maximum length of EMT that may safely be used as an equipment grounding conductor based on ground-fault current of 400% of the overcurrent device rating. Circuit 120 volts to ground; 40 volts drop at the point of fault. Ambient temperature 25°C.

Table 21-18. Maximum length of IMC that may safely be used as an equipment grounding conductor based on ground-fault current of 400% of the overcurrent device rating. Circuit 120 volts to ground; 40 volts drop at the point of fault. Ambient temperature 25°C.

Table 21-19. Maximum length of RMC that may safely be used as an equipment grounding conductor based on ground-fault current of 400% of the overcurrent device rating. Circuit 120 volts to ground; 40 volts drop at the point of fault. Ambient temperature 25°C.

Table 250.66 Grounding Electrode Conductor for Alternating-Current Systems			
Size of Largest Ungrounded Service-Entrance Conductor or Equivalent Area for (AWG/kcmil) Conductors[a]		Size of Grounding Electrode Parallel Conductor (AWG/kcmil)	
Copper	Aluminum or Copper-Clad Aluminum	Copper	Aluminum or Copper-Clad Aluminum[b]
2 or smaller	1/0 or smaller	8	6
1 or 1/0	2/0 or 3/0	6	4
2/0 or 3/0	4/0 or 250	4	2
Over 3/0 thru 350	Over 250 thru 500	2	1/0
Over 350 thru 600	Over 500 thru 900	1/0	3/0
Over 600 thru 1100	Over 900 thru 1750	2/0	4/0
Over 1100	Over 1750	3/0	250 kcmil

Notes:

1. Where multiple sets of service-entrance conductors are used as permitted in Section 230.40, Exception No. 2. the equivalent size of the largest service-entrance conductor shall be determined by the largest sum of the areas of the corresponding conductors of each set.

2. Where there are no service-entrance conductors, the grounding electrode conductor size shall be determined by the equivalent size of the largest service-entrance conductor required for the load to be served.

[a] This table also applies to the derived conductors of separately derived ac systems.

[b] See installation restrictions in 250.64(A).

Table 21-1.

Size of Copper Service Conductor AWG	Continuous Rating of Service Conductor 75°C Wire Ampacity	Size of Copper Grounding Electrode Conductor AWG	Continuous Rating Grounding Electrode Conductor 75°C Wire Ampacity	Grounding Electrode Conductor Expressed as a percent of Service Conductor
Analysis of Table 250.66 *Grounding Electrode Conductor and Service Conductor Compared As To Relative Size				
2 or smaller	115	8	50	43%
0	150	6	65	43%
3/0	200	4	85	43%
350 kcmil	310	2	115	37%
600 kcmil	420	1/0	150	36%
1000 kcmil	545	2/0	175	32%
1250 kcmil (over 1100 kcmil)	590	3/0	200	34%

* This grounding Electrode Conductor is that part of the Grounding System which is the sole connection between the grounding electrode and the grounded system conductor.

Table 21-2.

Comparison of Copper Grounding Electrode Conductors per Table 250.66 When Used With and Without Steel Conduit for Physical Protection.					
Copper Wire Size. AWG	DC Resistance of Bare Wire. Ohms/100'	Trade Size of conduit to enclose conductor. Inch	*Approximate DC Resistance of conduit. Ohms/100'	*Approximate Impedance of Conductor in Conduit. Ohms/100' at 500 Amps Sq. In. density	
				Conduit bonded at both ends	Conduit NOT bonded
8	0.07780	1/2	0.0321	0.105	0.210
6	0.04910	3/4	0.0242	0.078	0.156
4	0.03080	3/4	0.0242	0.078	0.156
2	0.01940	1	0.0154	0.056	0.112
0	0.01220	1	0.0154	0.056	0.112
00	0.00967	1	0.0154	0.056	0.112
000	0.00766	1 1/4	0.0120	0.044	0.088

* These values are approximate but sufficiently accurate for all practical purposes. Manufacturing tolerances for conduit, although quite satisfactory for its physical requirements, can vary the resistance up to about 15 percent compared to standard dimensions.

Table 21-3.

Comparison of Aluminum Grounding Electrode Conductors per Table 250.66 When Used With and Without Aluminum Conduit for Physical Protection.			
Aluminum Wire Size	DC Resistance Ohms/100'	Trade Size of Conduit to Enclose Conductor In.	Approximate DC Resistance of Aluminum Conduit Ohms/100'
6	0.08080	1/2	0.00800
4	0.05080	1/2	0.00800
2	0.03190	3/4	0.00605
1/0	0.02010	1	0.00385
3/0	0.01260	1	0.00385
4/0	0.00100	1 1/4	0.00300
250 kcmil	0.00847	1 1/4	0.00300

*Must be installed to comply with 250.64, that is, made electrically continuous from the point of attachment to cabinets or equipment to the grounding electrode and shall be securely fastened to the ground clamp.

Table 21-4.

Rating of Grounding Electrode Conductors Specified in Table 250.66 and Voltage Drop under Maximum Short Time Rating.

Grounding Conductor Size AWG	DC Resistance per 100 ft. Copper Conductor	Ampacity 75° C Wire Continuous Rating	Circular Mils	*Short-Time Rating Amps	Ratio Short time Rating to Continuous Rating	Voltage Drop per 30 m (100 ft.) at Short time Rating
8	0.077800	50	16510	391	7.8	30.4
6	0.049100	65	26240	621	9.6	30.5
4	0.030800	85	41740	988	11.6	30.4
3**	0.024500	100	52620	1245	12.5	30.5
2	0.019400	115	66360	1571	13.7	30.5
1**	0.015400	130	83690	1981	15.2	30.5
1/0	0.012200	150	105600	2499	16.7	30.5
2/0	0.009670	175	133100	3150	18.0	30.5
3/0	0.007660	200	167800	3972	19.9	30.4
4/0**	0.006080	230	211600	5008	21.8	30.5
kcmil						
250**	0.005150	255	250000	5917	23.2	30.5
300**	0.004290	285	300000	7101	24.9	30.5
350**	0.003670	310	350000	8284	26.7	30.4
400**	0.003210	335	400000	9467	28.3	30.4
500**	0.002580	380	500000	11,834	31.1	30.5

* Based on 1 ampere per 42.25 of circular mil area for five seconds.

** Not shown in Table 250.66 but listed here to assist in sizing grounding conductor lengths over 30 m (100 ft.) long.

Design Recommendation. If length grounding conductor exceeds 30 m (100 ft.), the conductor should be increased in size so that voltage drop based on short-time current rating of the conductor specified should not exceed 40 volts. In other words, the resistance of the grounding conductor used when over 30 m (100 ft) long should not exceed the resistance of 30 m (100 ft) of the conductor specified in the table.

Table 21-5.

Table 250.122. Minimum Size Equipment Grounding Conductors for Grounding Raceway and Equipment.

Rating or Setting of Automatic Overcurrent Device in Circuit Ahead of Equipment, Conduit, etc., Not Exceeding (Amperes)	Size (AWG or kcmil)	
	Copper	Aluminum or Copper-Clad Aluminum*
15	14	12
20	12	10
30	10	8
40	10	8
60	10	8
100	8	6
200	6	4
300	4	2
400	3	1
500	2	1/0
600	1	2/0
800	1/0	3/0
1000	2/0	4/0
1200	3/0	250
1600	4/0	350
2000	250	400
2500	350	600
3000	400	600
4000	500	800
5000	700	1200
6000	800	1200

Note: Where necessary to comply with 250.4(A)(5) or 250.4(B)(4), the equipment grounding conductor shall be sized larger than this table.

* See installation restrictions in Section 250.120.

Table 21-6.

Analysis of Table 250.122						
Rating of O. C. Device in Amperes	EGC Copper Conductor AWG	EGC Copper Conductor Cir. Mils.	*Short-Time Rating of EGC Conductor in Amps	** K-factor	Percent Rating of EGC Conductor to OC Device	*** Ampacity per Table 310.16
20	12	6530	155	7.7	125%	25
30	10	10,380	246	8.2	117%	35
40	10	10,380	246	6.1	88%	35
60	10	10,380	246	4.1	58%	35
100	8	16,510	391	3.9	50%	50
200	6	26,240	621	3.1	33%	65
400	3	52,620	1,245	3.1	25%	100
600	1	83,690	1,981	3.3	22%	130
800	1/0	105,600	2,499	3.1	19%	150
1000	2/0	133,100	3,150	3.2	18%	175
1200	3/0	167,800	3,972	3.3	17%	200
1600	4/0	211,600	5,008	3.1	14%	230
2000	250 kcmil	250,000	5,914	3.0	13%	255
2500	350 kcmil	350,000	8,285	3.3	12%	310
3000	400 kcmil	400,000	9,467	3.2	11%	335
4000	500 kcmil	500,000	11,834	3.0	10%	380
5000	700 kcmil	700,000	16,568	3.3	9%	460
6000	800 kcmil	800,000	18,935	3.2	8%	490

* One ampere per 42.25 circular mils for five seconds.

** K factor. Ampere rating of overcurrent device times K equals short-time rating of E.G.C. conductor in amperes.

*** Based on 75°C copper wire, Table 310.16.

Table 21-7.

Conductor Properties

			Conductors							DC Resistance at 75°C (167°F)					
			Stranding			Overall				Copper				Aluminum	
	Area			Diameter		Diameter		Area		Uncoated		Coated			
Size (AWG/ kcmil)	mm²	Circular mils	Quantity	mm	in.	mm	in.	mm²	in.²	ohm/ km	ohm/ kFT	ohm/ km	ohm/ kFT	km	kFT
18	0.823	1620	1	—	—	1.02	0.040	.823	0.001	25.5	7.77	26.5	8.08	42.0	12.8
18	0.823	1620	7	0.39	0.015	1.16	0.046	1.06	0.002	26.1	7.95	27.7	8.45	42.8	13.1
16	1.31	2580	1	—	—	1.29	0.051	1.31	0.002	16.0	4.89	16.7	5.08	26.4	8.05
16	1.31	2580	7	0.49	0.019	1.46	0.058	1.68	0.003	16.4	4.99	17.3	5.29	26.9	8.21
14	2.08	4110	1	—	—	1.63	0.064	2.08	0.003	10.1	3.07	10.4	3.19	16.6	5.06
14	2.08	4110	7		0.024	1.85	0.073	2.68	0.004	10.3	3.14	10.7	3.26	16.9	5.17
12	3.31	6530	1	—	—	2.05	0.081	3.31	0.005	6.34	1.93	6.57	2.01	10.45	3.18
12	3.31	6530	7	0.78	0.030	2.32	0.092	4.25	0.006	6.50	1.98	6.73	2.05	10.69	3.25
10	5.261	10,380	1	—	—	2.588	0.102	5.26	0.008	3.984	1.21	4.148	1.26	6.561	2.00
10	5.261	10,380	7	0.98	0.038	2.95	0.116	6.76	0.011	4.070	1.24	4.226	1.29	6.679	2.04
8	8.367	16,510	1	—	—	3.264	0.128	8.37	0.013	2.506	0.764	2.579	0.786	4.125	1.26
8	8.367	16,510	7	1.23	0.049	3.71	0.146	10.76	0.017	2.551	0.778	2.653	0.809	4.204	1.28
6	13.30	26,240	7	1.56	0.061	4.67	0.184	17.09	0.027	1.608	0.491	1.671	0.510	2.652	0.808
4	21.15	41,740	7	1.96	0.077	5.89	0.232	27.19	0.042	1.010	0.308	1.053	0.321	1.666	0.508
3	26.67	52,620	7	2.20	0.087	6.60	0.260	34.28	0.053	0.802	0.245	0.833	0.254	1.320	0.403
2	33.62	66,360	7	2.47	0.097	7.42	0.292	43.23	0.067	0.634	0.194	0.661	0.201	1.045	0.319
1	42.41	83,690	19	1.69	0.066	8.43	0.332	55.80	0.087	0.505	0.154	0.524	0.160	0.829	0.253
1/0	53.49	105,600	19	1.89	0.074	9.45	0.372	70.41	0.109	0.399	0.122	0.415	0.127	0.660	0.201
2/0	67.43	133,100	19	2.13	0.084	10.62	0.418	88.74	0.137	0.3170	0.0967	0.329	0.101	0.523	0.159
3/0	85.01	167,800	19	2.39	0.094	11.94	0.470	111.9	0.173	0.2512	0.0766	0.2610	0.0797	0.413	0.126
4/0	107.2	211,600	19	2.68	0.106	13.41	0.528	141.1	0.219	0.1996	0.0608	0.2050	0.0626	0.328	0.100
250		—	37	2.09	0.082	14.61	0.575	168	0.260	0.1687	0.0515	0.1753	0.0535	0.2778	0.0847
300		—	37	2.29	0.090	16.00	0.630	201	0.312	0.1409	0.0429	0.1463	0.0446	0.2318	0.0707
350		—	37	2.47	0.097	17.30	0.681	235	0.364	0.1205	0.0367	0.1252	0.0382	0.1984	0.0605
400		—	37	2.64	0.104	18.49	0.728	268	0.416	0.1053	0.0321	0.1084	0.0331	0.1737	0.0529
500		—	37	2.95	0.116	20.65	0.813	336	0.519	0.0845	0.0258	0.0869	0.0265	0.1391	0.0424
600		—	61	2.52	0.099	22.68	0.893	404	0.626	0.0704	0.0214	0.0732	0.0223	0.1159	0.0353
700		—	61	2.72	0.107	24.49	0.964	471	0.730	0.0603	0.0184	0.0622	0.0189	0.0994	0.0303
750		—	61	2.82	0.111	25.35	0.998	505	0.782	0.0563	0.0171	0.0579	0.0176	0.0927	0.0282
800		—	61	2.91	0.114	26.16	1.030	538	0.834	0.0528	0.0161	0.0544	0.0166	0.0868	0.0265
900		—	61	3.09	0.122	27.79	1.094	606	0.940	0.0470	0.0143	0.0481	0.0147	0.0770	0.0235
1000		—	61	3.25	0.128	29.26	1.152	673	1.042	0.0423	0.0129	0.0434	0.0132	0.0695	0.0212
1250		—	91	2.98	0.117	32.74	1.289	842	1.305	0.0338	0.0103	0.0347	0.0106	0.0554	0.0169
1500		—	91	3.26	0.128	35.86	1.412	1011	1.566	0.02814	0.00858	0.02814	0.00883	0.0464	0.0141
1750		—	127	2.98	0.117	38.76	1.526	1180	1.829	0.02410	0.00735	0.02410	0.00756	0.0397	0.0121
2000		—	127	3.19	0.126	41.45	1.632	1349	2.092	0.02109	0.00643	0.02109	0.00662	0.0348	0.0106

1. These resistance values are valid ONLY for the parameters as given. Using conductors having coated strands, different stranding type, and especially, other temperatures, change the resistance.

2. Formula for temperature change: $R_2 = R_1\ [1+a(T_2\text{-}75)]$ where: $a_{cu} = 0.00323, a_{AL} = 0.00330$, at 75°C.

3. Conductors with compact and compressed stranding have about 9 percent and 3 percent, respectively, smaller bare conductor diameters than those shown. See Table 5A for actual compact cable dimensions.

4. The IACS conductivities used: bare copper = 100% aluminum = 61%.

5. Class B stranding is listed as well as solid for some sizes. Its overall diameter and area is that of its circumscribing circle.

(FPN): The construction information is per NEMA WC8-1992. The resistance is calculated per National Bureau of Standards Handbook 100, dated 1966, and Handbook 109, dated 1972.

Table 21-8. *NEC* Table 8, Chapter 9

Maximum length of steel conduit or tubing that may safely be used as an equipment grounding circuit conductor.						
Conduit Trade Size In.	AWG/ kcmil	OC 75°C	Fault 500%	EMT *Length (ft)	IMC *Length (ft)	RGC *Length (ft)
1/2	14	15	75	231	232	227
	12	20	100	246	255	247
	10	30	150	226	250	240
3/4	12	20	100	266	262	257
	10	30	150	255	259	250
	8	40	200	261	275	264
	8	50	250	211	229	222
1	8	40	200	288	284	275
	8	50	250	236	237	229
	6	60	300	263	270	257
	6	70	350	228	240	228
	4	80	400	254	274	259
	4	90	450	228	251	238
	3	100	500	227	256	244
1 1/4	3	100	500	268	274	258
	2	110	550	273	283	265
	2	125	625	244	257	243
	1	125	625	266	281	263
	1	150	750	226	245	234
1 1/2	1	125	625	285	292	273
	1	150	750	243	256	241
	1/0	150	750	264	279	260
2	2/0	175	875	280	287	268
	3/0	200	1000	268	281	261
	4/0	225	1125	257	277	257
	4/0	250	1250	234	257	240
2 1/2	250	250	1250	289	269	261
	250	300	1500	249	238	229
	300	250	1250	302	280	271

Based on a clearing ground-fault current of 500 percent of overcurrent device rating; circuit 120 volts to ground; 50 volts drop at the arc; 30° C ambient temperature; 75°C conductor temperature. Calculated with "Steel Conduit Analysis Vs 1.2" by Georgia Institute of Technology.

* Measurement in feet

Table 21-11. **(continued on next page)**

Maximum length of steel conduit or tubing that may safely be used as an equipment grounding circuit conductor.						
Conduit Trade Size In.	AWG/ kcmil	OC 75°C	Fault 500%	EMT *Length (ft)	IMC *Length (ft)	RGC *Length (ft)
2 1/2	400	300	1500	278	263	253
	400	350	1750	244	239	227
	500	350	1750	255	249	236
	500	400	2000	227	230	216
3	500	350	1750	280	262	253
	500	400	2000	251	240	231
	600	400	2000	260	248	238
	600	450	2250	235	231	220
	700	450	2250	242	237	225
	700	500	2500	221	222	209
3 1/2	600	400	2000	280	256	248
	600	450	2250	257	237	229
	700	450	2250	264	244	235
	700	500	2500	243	228	218
	750	450	2250	267	246	237
	750	500	2500	246	230	221
	900	500	2500	254	237	227
	900	600	3000	220	212	200
4	750	450	2250	277	253	245
	750	500	2500	255	236	228
	900	500	2500	264	243	234
	900	600	3000	229	216	206
	1000	500	2500	269	247	238
	1000	600	3000	233	219	210

Based on a clearing ground-fault current of 500 percent of overcurrent device rating; circuit 120 volts to ground; 50 volts drop at the arc; 30° C ambient temperature; 75°C conductor temperature. Calculated with "Steel Conduit Analysis Vs 1.2" by Georgia Institute of Technology.

* Measurement in feet

Table 21-11 (continued).

Aluminum Conduit used as an equipment grounding circuit conductor (EGC) compared with equipment grounding circuit (EGC) conductors as specified in Table 250.122 of the *National Electrical Code.*						
					Nearest Wire AWG Equivalent to Conduit	
Size of Trade Conduit In.	Area of Conduit Wall Sq. In.	Area of Conduit Wall Based on NEMA Standards Sq. In.	DC Resistance Ohms/M ft.	Largest EGC Conductor as specified in Table 250.122 Aluminum Wire	Aluminum Wire	Copper Wire
1/2	0.254	0.254	0.08000	12 AWG	4/0 AWG	2/0 AWG
3/4	0.337	0.337	0.06050	8	300 kcmil	4/0
1	*0.530	0.500	0.03850	6	500 kcmil	300 kcmil
1 1/4	*0.680	0.670	0.03000	4	600 kcmil	350 kcmil
1 1/2	*0.790	0.790	0.02580	4	700 kcmil	500 kcmil
2	*1.030	1.080	0.01980	1	900 kcmil	600 kcmil
2 1/2	1.710	1.710	0.01190	1	1500 kcmil	1000 kcmil
3	*2.310	2.250	0.00880	2/0	2000 kcmil	1250 kcmil
3 1/2	2.700	2.700	0.00752	2/0	over	1500 kcmil
4	*3.400	3.180	0.00598	2/0	2000 kcmil	2000 kcmil
5	4.300	4.300	0.00473	3/0	over 2000 kcmil	over 2000 kcmil

*Values as measured from a sample.

Column 3 gives values as calculated on basis of NEMA standards.

Table 21-14.

Sizes of conductor (Copper Bar) for use in the equipment grounding circuit and for the Equipment Grounding Strap for overcurrent devices from 1600 amperes through 5000 amperes.

Rating of OC Device in Amps.	Copper EG Conductor: Copper Bar - Inch	Copper EG Conductor: Cir. Mils.	* Conductor Rating Amps	** Short-Time Rating of EGC Conductor in Amps	*** K factor	**** Percent Rating of EGC Conductor to OC Device
1600	1/4 x 1	318,300	366	10,610	5.3 (6.6)	22.9
2000	1/4 x 1 1/4	397,900	457	13,260	6.0 (6.6)	22.9
2500	1/4 x 1 1/2	477,450	549	15,920	6.0 (6.4)	21.9
3000	1/4 x 2	636,600	647	21,220	6.0 (7.1)	21.6
4000	1/4 x 2 1/2	795,750	809	26,525	6.3 (6.6)	20.2
5000	1/4 x 4	1,273,000	1,220	42,430	6.8 (8.5)	24.4

* Based on 30°C rise, 40°C ambient.

** One ampere per 30 circular mils for 5 seconds. Use this value ONLY where insulated, current-carrying conductors will NOT come in contact with the bare equipment grounding bus bars. Where there is likelihood of contact, then recalculate the 5 second withstand rating on the basis of one ampere...for five seconds...for every 42.25 circular mils of copper conductor. This is discussed in chapter eleven of this text.

*** K-factor. Ampere rating of O. C. device times K-factor equals short-time rating of EGC conductor in amperes. Figures in brackets are based on the actual EGC conductor used.

**** Based on 30°C rise, 40°C ambient for EGC conductor.

Table 21-15.

Maximum length of equipment-grounding conductor that may safely be used as an equipment-grounding circuit conductor. Based on a ground-fault current of 400% of the overcurrent device rating. Circuit 120 volts to ground; 40 volts drop at the point of fault. Ambient temperature 25°C.							
						For Copper and Aluminum	
Copper Equipment Grounding Conductor AWG Size***	Copper Circuit AWG Conductors	Maximum Length of Run (in feet) using Copper Equipment Ground Conductor	Aluminum Equipment Grounding Conductor AWG Size***	Aluminum Circuit AWG Conductors	Maximum Length of Run (in feet) using Aluminum Equipment Grounding Conductor	Overcurrent Device Rating Amperes 75°C**	Fault Clearing Current 400% O.C. Device Rating Amperes
14	14	253	12	12	244	15	60
12	12	300	10	12	226	20	80
10	10	319	8	8	310	30	120
10	8	294	8	8	232	40	160
10	6	228	8	4	221	60	240
8	3	229	6	1	222	100	400
6	3/0	201	4	250 kcm	195	200	800
4	350 kcm	210	2	500 kcm	204	300	1200
3	600 kcm	195	1	900 kcm	192	400	1600
2	2-4/0	160	1/0	2-400 kcm	163	500	2000
1	2-300 kcm	160	2/0	2-500 kcm	161	600	2400
1/0	3-300 kcm	134	3/0	3-400 kcm	131	800	3200
2/0	4-250 kcm	114	4/0	4-400 kcm	115	1000	4000
3/0	4-300 kcm	106	250 kcm	4-500 kcm	107	1200	4800
4/0	4-600 kcm	93	350 kcm	4-900 kcm	97	1600	6400
250 kcm	5-600 kcm	78	400 kcm	5-800 kcm	79	2000	8000
350 kcm	6-600 kcm	*	600 kcm	6-900 kcm	*	2500	10,000
400 kcm	8-500 kcm	*	600 kcm	8-750 kcm	*	3000	15,000
500 kcm	8-1000 kcm	*	800 kcm	8-1500 kcm	*	4000	16,000
700 kcm	10-1000 kcm	*	1200 kcm	10-1500 kcm	*	5000	20,000
800 kcm	12-1000 kcm	*	1200 kcm	12-1500 kcm	*	6000	24,000

* Calculations necessary
** 60°C for 20- and 30-ampere devices
*** Based on *NEC* Chapter 9, Table 8

Table 21-16. Table derived from Software (SCA) and Testing developed at Georgia Institute of Technology and sponsored by the producers of Steel EMT, IMC, and rigid steel conduit.

Maximum length of electrical metallic tubing that may safely be used as an equipment-grounding circuit conductor. Based on a ground-fault current of 400% of the overcurrent device rating. Circuit 120 volts to ground; 40 volts drop at the point of fault. Ambient temperature 25°C.

EMT Trade Size	Conductors AWG	Overcurrent Device Rating Amperes 75°C*	Fault Clearing Current 400% O.C. Device Rating Amperes	Maximum Length of EMT Run in Feet
1/2 (16)	3-12	20	80	395
	4-10	30	120	358
3/4 (21)	4-10	30	120	404
	4-8	50	200	332
1 (27)	4-8	50	200	370
	3-4	85	340	365
1 1/4 (35)	3-2	115	460	391
1 1/2 (41)	3-1	130	520	407
	3-2/0	175	700	364
2 (53)	3-3/0	200	800	390
	3-4/0	230	920	367
2 1/2 (63)	3-250 kcm	255	1020	406
3 (78)	3-350 kcm	310	1240	404
	3-500 kcm	380	1520	370
	3-600 kcm	420	1680	353
4 (103)	3-900 kcm	520	2080	353
	3-1000 kcm	545	2180	347

* 60°C for 20- and 30-ampere devices

Table 21-17. Table derived from Software (SCA) and Testing developed at Georgia Institute of Technology and sponsored by the producers of Steel EMT, IMC, and rigid steel conduit.

Maximum length of intermediate metal conduit that may safely be used as an equipment-grounding circuit conductor. Based on a ground-fault current of 400% of the overcurrent device rating. Circuit 120 volts to ground; 40 volts drop at the point of fault. Ambient temperature 25°C.

IMC Trade Size	Conductors AWG	Overcurrent Device Rating Amperes 75°C*	Fault Clearing Current 400% O.C. Device Rating Amperes	Maximum Length of IMC Run in Feet
1/2 (16)	3-12	20	80	398
	4-10	30	120	383
3/4 (21)	4-10	30	120	399
	4-8	50	200	350
1 (27)	4-8	50	200	362
	3-4	85	340	382
1 1/4 (35)	3-2	115	460	392
1 1/2 (41)	3-1	130	520	402
	3-2/0	175	700	377
2 (53)	3-3/0	200	800	389
	3-4/0	230	920	375
2 1/2 (63)	3-250 kcm	255	1020	368
3 (78)	3-350 kcm	310	1240	367
	3-500 kcm	380	1520	338
	3-600 kcm	420	1680	325
4 (103)	3-900 kcm	520	2080	320
	3-1000 kcm	545	2180	314

* 60°C for 20- and 30-ampere devices

Table 21-18. Table derived from Software (SCA) and Testing developed at Georgia Institute of Technology and sponsored by the producers of Steel EMT, IMC, and rigid steel conduit.

Maximum length of galvanized rigid conduit that may safely be used as an equipment-grounding circuit conductor. Based on a ground-fault current of 400% of the overcurrent device rating. Circuit 120 volts to ground; 40 volts drop at the point of fault. Ambient temperature 25°C.

GRC Trade Size	Conductors AWG	Overcurrent Device Rating Amperes 75°C*	Fault Clearing Current 400% O.C. Device Rating Amperes	Maximum Length of GRC Run in Feet
1/2 (16)	3-12	20	80	384
	4-10	30	120	364
3/4 (21)	4-10	30	120	386
	4-8	50	200	334
1 (27)	4-8	50	200	350
	3-4	85	340	357
1 1/4 (35)	3-2	115	460	365
1 1/2 (41)	3-1	130	520	377
	3-2/0	175	700	348
2 (53)	3-3/0	200	800	363
	3-4/0	230	920	347
2 1/2 (63)	3-250 kcm	255	1020	356
3 (78)	3-350 kcm	310	1240	355
	3-500 kcm	380	1520	327
	3-600 kcm	420	1680	314
4 (103)	3-900 kcm	520	2080	310
	3-1000 kcm	545	2180	304

* 60°C for 20- and 30-ampere devices

Table 21-19. Table derived from Software (SCA) and Testing developed at Georgia Institute of Technology and sponsored by the producers of Steel EMT, IMC, and rigid steel conduit.

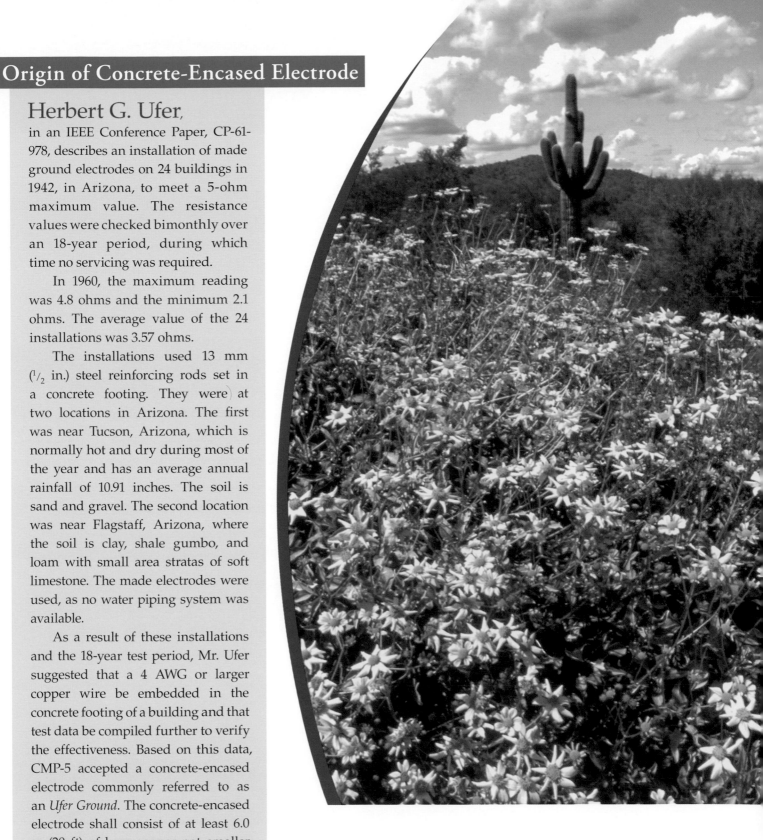

Origin of Concrete-Encased Electrode

Herbert G. Ufer,

in an IEEE Conference Paper, CP-61-978, describes an installation of made ground electrodes on 24 buildings in 1942, in Arizona, to meet a 5-ohm maximum value. The resistance values were checked bimonthly over an 18-year period, during which time no servicing was required.

In 1960, the maximum reading was 4.8 ohms and the minimum 2.1 ohms. The average value of the 24 installations was 3.57 ohms.

The installations used 13 mm ($^1/_2$ in.) steel reinforcing rods set in a concrete footing. They were at two locations in Arizona. The first was near Tucson, Arizona, which is normally hot and dry during most of the year and has an average annual rainfall of 10.91 inches. The soil is sand and gravel. The second location was near Flagstaff, Arizona, where the soil is clay, shale gumbo, and loam with small area stratas of soft limestone. The made electrodes were used, as no water piping system was available.

As a result of these installations and the 18-year test period, Mr. Ufer suggested that a 4 AWG or larger copper wire be embedded in the concrete footing of a building and that test data be compiled further to verify the effectiveness. Based on this data, CMP-5 accepted a concrete-encased electrode commonly referred to as an *Ufer Ground*. The concrete-encased electrode shall consist of at least 6.0 m (20 ft) of bare copper not smaller than 4 AWG encased in 50 mm (2 in.) of concrete near the bottom of the footing or foundation.

[See *The History and Mystery of Grounding* online at http://www.iaei.org/products/pdfs/historyground.pdf]

Appendix A

Appendix B

National Electrical Grounding Research Project

Beginning in 1995, the National Electrical Grounding Research Project (NEGRP) initiated a study of effectiveness of various types of buried grounding electrodes in differing geographies around the U.S. The NEGRP adopted a similar and ongoing research originated by the International Association of Electrical Inspectors, Southern Nevada Chapter (the IAEI /SNC grounding study) in 1992. During the term of the project, electrode resistance, earth resistivity, soil temperature, soils moisture and other measurements were recorded for more than twenty types of buried grounding electrodes in ten sites, situated in six geographic locations around the continental U.S. After being buried for extended periods, some electrodes at selected sites were exhumed for corrosion analysis, and observation.

One major goal of this research was to produce data and other information that would be useful to those in industry who require this to aid in the design of new systems, in the analysis of existing system performance, and in the explanation of failures of grounding electrodes.

In a report such as this, it would be preferable to state results simply and conclude facts about the features

Site Names			
SITE NAME	SITE LOCATION	INSTALLATION DATE	REMOVAL DATE
PA	Pawnee, Las Vegas, NV	May 1992	2003
LM	Lone Mountain, Las Vegas, NV	Jun 1992	2004
BA	Balboa, Henderson, NV	Aug 1992	2001
PE	Pecos, Las Vegas, NV	Dec 1992	2004
CH	Charleston, Las Vegas, NV	Dec 1992	NA
VA	Staunton, VA	Jun 1997	NA
TX	Dallas, TX	Jun 1998	NA
NY	Hibernia, NY	Sep 1998	2006
IL	Northbrook, IL	Sep 1998	NA
CA	Moffet Field, Mountain View, CA	Jan 2002	NA

Table 1. Site Names

of the project; however, owing to the long-term nature of the project and the number of details evaluated, it is important to state the features of the project as many individual observations. A large amount of information was produced and many details of the project have not been fully evaluated or explored. Much can still be determined in a review of the information produced by the project.

Conductance of earth varies by location

The nature of soil varies greatly in different locations. New York, Balboa, Las Vegas and Virginia sites produced higher earth resistivity and electrode resistance results than did other sites. The earth resistivity and electrode resistance readings from the Illinois site are significantly lower in comparison to other sites. The Illinois site was situated in a water runoff area and was usually wet. Although the New York site and Nasa sites also showed evidence of water for long periods, the values of earth resistivity and electrode resistance in these two sites were always higher than Illinois.

Table 2 showing earth resistivity values covers ten years of data for each site and is arranged by minimums first, sorted in descending order.

Earth resistivity proves to be a valuable indicator of electrode performance

An apparent correlation to electrode resistance values is observed in the recorded earth resistivity data. Test pin spacing (depth of measurement) for earth resistivity proves to be a prime indicator that can be indicative of the prospective performance of electrode resistance values. Care should be taken to observe earth resistivity at pin spacings (depths) relative to the proposed electrode depths. In this project 10-foot and 20-foot pin spacings were used to obtain earth resistivity values, yet these test pin spacings produced values consistent with the reported electrode resistance. This close correlation is possibly due to the homogenous nature of the soil at lesser depths.

Earth resistivity and electrode resistance appear to be inverse to temperature

The Illinois site exhibited a clearly inverse relationship between temperature and earth resistivity (and electrode resistance) where increases in soil temperature are effectively mirrored by reductions in earth resistivity when viewed over time. In most other sites, the earth resistivity and electrode resistance similarly showed linkage to seasonal temperature variations. This factor

Earth Resistivity

SITES	EARTH RESISTIVITY 10 ft. Depth			EARTH RESISTIVITY 20 ft. Depth		
	MIN	MAX	AVG	MIN	MAX	AVG
NY	15378	24014	19527	11011	23363	20409
VA	7603	19533	9930	8235	15090	10269
BA	3217	31597	8736	1915	25278	11474
IL	2313	3497	2821	1896	2306	2098
TX	1563	4577	3319	1655	8158	3242
PE	766	12868	1734	306	14056	1651
CH	708	9575	2084	575	13175	2079
CA	517	4577	2763	460	9000	2118
LM	153	9000	2527	766	9506	4638
PA	77	9575	1362	613	9958	3158

Table 2. Earth Resistivity

could be important in the design and initial installation. For example, an electrode installed in summer that resulted in a 25-ohm initial value could potentially exceed this value for much of the other seasons. Except for other factors, it might also be concluded that an electrode installed in the winter that meets the ohmic design criteria would typically not venture above this ohmic value in the summer.

Of the 18 electrodes in the Balboa test site, 14 of these exceeded 25 ohms for 90% or more of the readings; electrode Types E, L and R were the exceptions. Similarly, for electrodes in the New York test site, 100% of the readings exceeded 25 ohms for 13 of the 15 electrodes in the site; the exceptions being electrodes Type L and R.

It is noted that the percentage of readings for electrode Type B, (*NEC* Section 250.52) two exceeded 25 ohms for 74% or more of the readings in four of the 10 test sites.

Electrode Type K readings exceeded 25 ohms for 75% or more of the readings at nine of the 10 test sites, with Texas readings exceeding 25 ohms for 21% of the readings.

The Illinois site had the fewest number of electrodes with readings exceeding 25 ohms. Electrodes H and K had 11% and 79% respectively that exceeded 25 ohms. This could be due to generally wet soil conditions based upon site location in a drainage field.

In the New York site, 10 of 15 electrodes exceeded 50 ohms more than 50% of the time. In the Balboa site, nine of 16 electrodes exceeded 50 ohms 50% of the time, as compared to the Illinois site where no electrodes exceed 50 ohms except for the K electrode (18%).

Although the Las Vegas Pecos and Balboa sites were in the same geographical area, the results differed greatly. In the Balboa site, nine of 18 electrodes exceeded 50 ohms more than 50% of the time; whereas in the Pecos site, only the K electrode exceeded 50 ohms 50% of the time.

It should be noted that an electrode that exceeded 50 ohms also exceeded the *NEC* 25-ohm requirement by a factor of two.

The data and observations documented under this research project support the following findings.

Observations

Considering the national sites, the electrodes in the Illinois site exhibited the lowest resistance readings; where the New York site exhibited the highest readings. The majority of readings in New York did not meet the 25-ohm requirement of *NEC* Section

Site Descriptions

SITE	LOCATION	SOIL
PA	Pawnee Las Vegas, NV Central Valley	Water table ranging from 10 to 20 ft. generally clayey silt
LM	Lone Mountain Las Vegas, NV North Valley	Normally dry, silty, sandy clays
BA	Balboa Henderson, NV South Valley	Normally dry, rock and gravel
PE	Pecos Las Vegas, NV North Valley	Normally dry , silt and sand to 6 ft., silty clays 6 to 11 ft.
CH	Charleston Las Vegas, NV East Valley	Normally dry, sand, and gravel, surface froth
VA	Staunton, VA	Sand and silt to a depth of approximately 6 ft; silty clays 6 to 11 ft.
TX	Dallas, TX	Normally dry, sandy soil to a depth of 5 to 6 ft.; increasingly stiff sandstone to 11 ft.
NY	Hibernia, NY	Water table at 10 ft to 11 ft.; gravel, sand and clays with occasional cobbles to 11 ft. depth
IL	Northbrook, IL Illinois	Periodically covered with rainwater runoff, contains silts and clays to a depth of 11 ft.
CA	Moffet Field, Mountain View, CA	Normally dry, sand, silt with inorganic and organic clays

Table 3. Site Descriptions

250.52. Comparing all electrodes in each national site, E, L and R electrodes had the best results. Types S, W and X electrodes did relatively well and for most sites; where the K and G electrodes were poor performers.

Analysis limited to horizontal and vertical rod type electrodes for the Las Vegas sites showed that the vertical electrodes had lower resistance values than their horizontal counterparts. Similarly, the vertical electrodes in national sites exhibited lower resistance values, with exception of the Illinois site. For most sites, this feature appeared to be consistent.

When yearly mean electrode resistance readings were compared for the Las Vegas sites, in almost every case for every electrode, the Balboa site had the highest yearly mean readings; whereas, the Pawnee site had the lowest values. For the national sites, the yearly mean readings for all electrodes were highest in the New York site, and lowest in the Illinois and Texas sites. The B, E, F, G, H and L electrodes were occurring in Las Vegas and national sites.

The worst case results electrode resistance values were in the New York and Balboa sites. The best-case results were generally with Illinois, Pawnee, and Texas sites. Electrode resistance values are required to be lower than 25 ohms for individual rod, pipe and plate electrodes to meet *NEC* 250.56(3) requirement.

Although concrete-encased electrodes per *NEC* 250.52 are not required to be lower than the 25-ohm

Electrodes

TYPE	MATERIAL	LENGTH	INSTALLATION
A	No. 2 AWG stranded copper wire	50 ft	Centered in 12 in. of sand at 36 in. below grade measured to the conductor. Buried horizontally in sand backfill
B	No. 4 uncoated steel reinforcing bar	20 ft.	Within, and near the bottom of a concrete foundation consisting of 12 in. by 12 in. of 2500 psi concrete. The top of the concrete was located 6 in. below grade.
C	No. 4 solid copper wire	25 ft.	Centered in 6 in. by 6 in. of ERICO® Ground Enhancement Material, (GEM™) located at 20 in., to the bottom of the concrete. Installed horizontally
D	No. 4 solid copper wire	25 ft.	Centered in 6 in. by 6 in. of 2500 psi concrete. Located at 20 in. to the bottom of the concrete. This electrode is designed to represent the thickened edge of post tensioned concrete construction and is installed horizontally
E	Copper bonded steel rod	8 ft.	5/8 in. diameter, centered in a 9 in. diameter, 9 ft. deep boring in earth, encased in ERICO® (GEM™) backfill, installed vertically.
F	Copper bonded steel rod	8 ft.	5/8 in. diameter, installed horizontally centered in a trench, encased in ERICO® (GEM™) backfill, 24 in. to the bottom of the (GEM™) backfill material.
G	Copper bonded steel rod	8 ft.	5/8 in. diameter, directly buried in a trench 36 in. deep, installed horizontally.
H	Copper bonded steel rod	8 ft.	5/8 in. diameter, driven vertically in earth.
I	Galvanized steel rod	10 ft.	3/4 in. diameter, driven vertically in earth.
J	Galvanized steel rod	10 ft.	3/4 in. diameter 10 ft. directly buried horizontally at a depth of 36 in.
K	Copper grounding "pole" plate	NA	For Las Vegas sites 30 in. depth., T&B® Blackburn™ Model PBH, .025 in. thickness 7 in. wide by 7 3/8 in. long, with connection capable of up to 4 AWG stranded wire For national sites, buried at 36in., T&B® Blackburn™ GP-114, 14 in. diameter, .025 in. thickness.
L	Lyncole® XIT™, copper tube	10	Vertical chemically charged electrode assembly in an 11 ft. deep, 9 in. diameter boring.
M	Steel and concrete	NA	Arrangement designed to represent a light pole base with six each, 2 ft. long No. 4 reinforcing steel vertical, tied with 3 each No. 2 steel horizontal hoops separated 12 in., vertically in a 2 ft., deep 36 in. diameter excavation, encased in 2500 psi concrete.
N	4 AWG solid copper wire	20 Coil	20 ft. of wire rolled into a coil approximately 18 in. diameter, installed in a 2 ft. round by 2 ft. deep excavation in concrete 2500 psi.

Table 4a. Electrodes

Electrodes

TYPE	MATERIAL	LENGTH	INSTALLATION
P	Wood pole with copper grounding pole plate to 6 AWG solid copper wire	8	Approximately 18 in. diameter, with Blackburn GP-114 copper grounding pole plate attached at bottom using 6 AWG solid copper wire wrapped in a spiral for 6 ft., at approximately 6 in. spacing between wraps.
Q	Copper bonded steel rod	8	1/2 in. diameter, rod installed horizontally, directly buried at 30 in. depth.
R	LEC® Chemrod	10	Vertically charged electrode assembly in an 11 ft. deep 9 in. diameter boring encased in Ground Augmentation Fill (GAF™).
S	Lyncole® XIT™	10	Horizontal chemically charged electrode assembly installed at a depth of 36 in.
T	Galvanized steel water pipe	8	3/4 in. diameter galvanized (water) pipe, driven vertically in earth.
V	4/0 AWG 7-strand copper wire	20	4/0 AWG 7-strand bare copper cable, directly buried horizontally 36 in. deep.
W	Assembly of copper bonded rods	NA	Dominion Virginia Power Co., ground cage, (proprietary) using multiple 8 ft. long, 5/8 in. diameter copper bonded rods installed vertically in a mortar backfill.
X	Stainless steel rod	8	5/8 in. diameter stainless steel ground rod driven vertically in earth.
Y	LEC® Chemrod	10	LEC® Chemrod™ charged electrode assembly installed at a depth of 36 in., encased in GAF™ backfill.
Z	Copper plated steel mesh	8	Wire mesh consisting of No. 6 copper plated steel on 4 in. centers measuring 2 ft. wide. Directly buried in earth.
AA	Copper tube	10	ERICO® Horizontal Model Chemical Electrode. Installed with backfill (GEM™)
AB	Copper Tube	11	Harger® Vertical Model Chemical Electrode, installed in backfill (Ultrafill™)
AC	Copper tube	10	ERICO® Vertical Model Electrode. Installed in backfill (GEM™)
AD	Copper tube	10	Lyncole® Vertical Model Sectional Chemical. Electrode, installed in Lynconite II backfill.
AE	Copper tube	10	Lyncole® Horizontal Model Sectional Electrode, installed horizontally in Lynconite II backfill.
DC1	Rod	8	DC1A, B, and C are 5/8 in. copper bonded steel rod, installed vertically.
DC2	Pipe	8	DC2A, B, and C are ¾ in. by 8 ft. galvanized steel pipe.
DM	Rod and pipe	8	DM1A, 2A, and 3A are 5/8 in. copper bonded steel rods, installed vertically. DM1B, 2B, and 3B are ¾ in. galvanized steel pipe, installed vertically.

Table 4b. Electrodes

PERCENTAGE OF MEASUREMENTS 25 OHMS OR GREATER										
TYPE	PA	LM	BA	PE	CH	VA	TX	NY	IL	CA
A	1%	7%	100%	38%	72%					
B	1%	4%	93%	5%	3%	99%	74%	100%		31%
C	50%		52%		59%					
D	5%	3%	95%	67%	87%					
E	1%	2%					2%	100%		
F	25%	2%	93%	2%		52%	1%	100%		8%
G	6%	92%	100%	52%	87%	99%	22%	100%		33%
H	3%		99%			75%		100%	11%	28%
I	1%		98%		1%					
J		2%	100%		28%					
K	80%	94%	100%	95%	97%	100%	21%	100%	79%	100%
L	40%	29%	13%		13%	15%		74%		
M	7%	68%	100%	60%	3%					
N	12%	11%	100%	88%	6%					
O	3%	12%	100%	55%						
P	3%	17%	100%							
Q	48%	44%	100%	48%	90%					
R			3%					42%		
S	2%			14%		19%		100%		
T						97%		100%		35%
V						99%	5%	100%		52%
W						4%		100%		
X						62%		100%		
Y						8%		100%		
Z						76%		100%		39%
AA										
AB										8%
AC										
AD										
AE										

ELECTRODE NOT INSTALLED IN THIS SITE

LESS THAN 25 OHMS

Table 5. Electrodes Exceeding 25 Ohms

value and may be installed without testing, the concrete-encased electrode, (Type B) exceeded the 25-ohm value in the Texas, Virginia and New York sites for the term of the project.

Temperature measurements were taken in the soil at three different depths per site. Reliability was shown in the data for all sites since there was less fluctuation in soil temperature as depth increased.

It is apparent in comparisons of data from many sites that an inverse relationship exists between earth resistivity and soil temperature. Distinct patterns can be seen in data for many of the sites. These patterns appear to be sympathetic to seasonal variations. Knowledge of this relationship could be useful in determining the suitability of an electrode.

The resistance reading for the national sites showed that Illinois had only two electrodes that exceeded 25 ohms during the term of the study. This could be due to the Illinois site being located in moist soil conditions. Most of the electrodes in the Dallas site for the term of the study were less than 25 ohms. The difference in resistance readings between the sites can be attributed primarily to earth resistivity. Soil chemistry, moisture and temperature play a part in modifying earth resistivity, thus electrode resistance, whereas corrosion and other mechanical conditions can modify electrode performance also.

Data from this research can be used to make inferences for appropriateness of electrodes to be used in similar soil conditions.

PERCENTAGE OF MEASUREMENTS 50 OHMS OR GREATER										
TYPE	PA	LM	BA	PE	CH	VA	TX	NY	IL	CA
A			61%	5%	13%					
B			7%			6%	1%	87%		
C	46%		7%		29%					
D	4%		11%	9%	77%			32%		
E	1%	1%						32%		
F	18%		8%			2%		95%		
G	3%	76%	97%	6%	51%	86%	1%	100%		8%
H	2%		49%			1%		100%		9%
I			1%		1%					
J			100%		5%					
K	78%	77%	100%	84%	39%	100%	3%	100%	18%	91%
L	25%	2%								
M		25%	98%	2%	3%					
N			95%	3%	3%					
O		8%	65%							
P		1%	100%							
Q	1%	4%	97%	3%	13%					
R			2%							
S								66%		
T						5%		100%		
V						17%	1%	100%		
W										
X						4%	7%	100%		
Y								13%		
Z						1%		92%		
AA										
AB										
AC										
AD										
AE										

ELECTRODE NOT INSTALLED IN THIS SITE

LESS THAN 50 OHMS

Table 6. Electrodes Exceeding 50 Ohms

DC experiments were included in the research as a means to evaluate the effects of corrosion due to the presence of direct currents. With many of the DC electrodes in the national sites, the data showed a sharp decrease in current with time. It is assumed that this decrease is due to the corrosive effects of the DC current possibly resulting in progressively higher resistance levels between the electrode and earth ground.

Dissimilar metals experiments were included to evaluate the possibility of corrosion of electrodes caused by different metals (Zn, Cu) installed in close proximity to one another. Data was developed showing changes of resistance, mainly caused by dissimilar metals in an earth-coupled cell that could possibly produce a direct current flow and lead to accelerated corrosion. In most cases, the trendline showed a decrease in current over time, which indicates an increase of resistance between electrodes and earth due to corrosion caused by DC current from dissimilar metals located in an electrolyte (earth).

Results for the CDA® sponsored benign corrosion experiment showed that the 4/0 AWG and 500 KCM bare copper conductors backfilled with earth, Bentonite or GEM,™ received a CR of no worse than 0/1— slight superficial occurrences of corrosion, not over all surfaces. The CR rating is a method of evaluating the corrosion of electrodes exclusively in this research project.

Earth resistivity values are used to predict properly the performance of electrodes. Standardized testing methods such as those referenced in IEEE 81-1983 were used to determine these values. Care was taken to ensure reliability and reproducibility of readings. Even though this was done, there were a few abnormalities in the data. Nonetheless, the data for most sites show reliability through the term of the project, except for California, which showed some variability. It is known that brackish water exists below the surface of the California site. Earth resistivity was varied by location. Resistivity values for the national sites ranged from 517 to 24,014 ohm-cm. In the Las Vegas sites, the range was 30 ohm-cm to 31,597 ohm-cm. Data support the concept that a direct relationship exists between earth resistivity and electrode resistance.

Low earth resistivity or low electrode resistance values are not necessarily a product of moist soil. Each of the national sites was in different geographic locations, and, therefore, subjected to different local weather climates and soil content. Soil moisture values for the Virginia and New York sites indicate mostly wet soil for the majority of readings while average resistivity values were higher, in the range of 9,817 and 20,424 ohm-cm for 10 ft. spacing respectively. Electrode-resistance average values were higher in the range of 39.6 ohms in Virginia to 94.6 ohms in New York.

An independent corrosion analysis was conducted for some of the grounding electrodes exhumed from sites. The analysis indicated that the majority of the grounding electrode materials performed well over the approximately ten-year exposure test except for the following:

a. loss of zinc on galvanized steel rods resulted in excessive corrosion,

b. copper-bonded steel ground rods showed minimal corrosion; however, the exposed steel at the unplated end of the ground rod was particularly vulnerable to corrosion although the average loss was minimal,

c. some electrodes filled with salts and encased in Bentonite corroded at the end of the electrode and around the weep holes by contrast to others that displayed minimal corrosion,

d. vertical electrodes in GEM™ displayed minimal corrosion in contrast to their horizontal counterparts, and

e. there was generally insignificant corrosion of the three types of connectors—mechanical, compression, and exothermic—during the term of the study.

The project concluded in 2006 and the final report was published in mid-2007. The complete report is available through the Fire Protection Research Foundation, Quincy, Massachusetts.

Information provided by Travis Lindsey and updated to *NEC*-2008.

Index

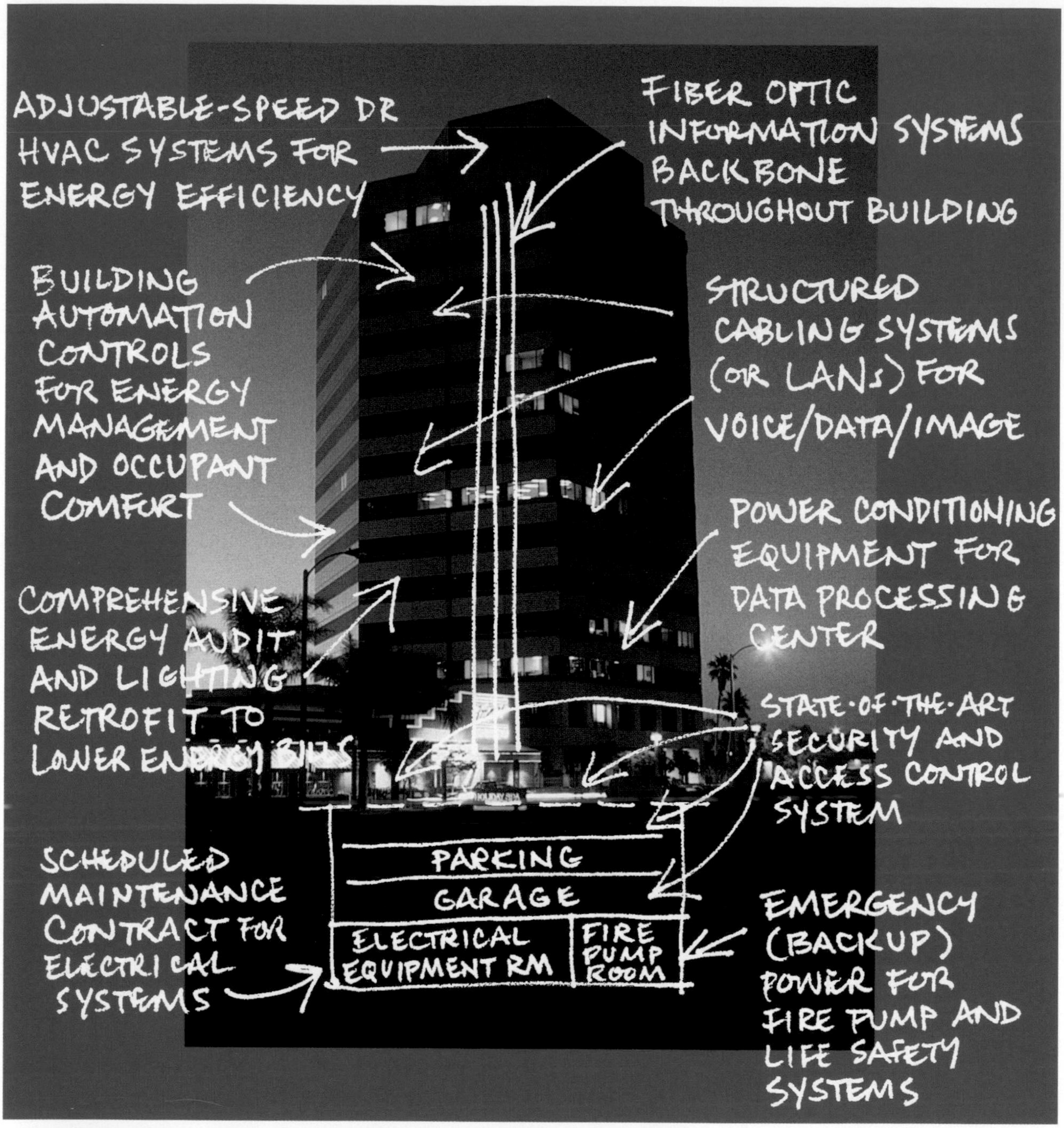

Soares Book on Grounding and Bonding
Tenth Edition

Cover design: John Watson

Contributors: Michael J. Johnston
Paul A. Dobrowsky
Charles F. Mello
Travis Lindsey
Mitchell Guthrie

Technical Edit and Review:
Michael J. Johnston
Paul A. Dobrowsky
Michael K. Weitzel

Technical Drawings: Michael J. Johnston

Graphic Production: John Watson
Laura Hildreth

Production Manager: Laura Hildreth

Project Manager: Michael J. Johnston

Research Editor: Laura Hildreth

Technical Editor: Michael J. Johnston

Creative Director: John Watson

Managing Editor: Kathryn Ingley

Editor in Chief: James W. Carpenter

Photos / Illustration Reprints:
AFC Cable Systems, New Bedford, MA
AVO International, Biddle Instruments, Dallas, TX
Electro-Test Inc., Milwaukie, OR
Cooper Industries, Bussman Division, St. Louis, MO
Eagle Electric Company, Long Island, NY
Insulated Cable Engineers Association, South Yarmouth, MA
International Electronic and Electrical Engineers, Piscataway, PA
National Electrical Manufacturers Association, Rosslyn, VA
National Fire Protection Association, Quincy, MA
NFPA Research Foundation, Quincy, MA
Post Glover Resistors, Inc., Erlanger, KY
Square D Company Palatine, IL
Steel Conduit Section, NEMA , Rosslyn, VA
Thomas & Betts, Memphis, TN
Underwriters Laboratories, Northbrook, IL
Greaves Corporation, Guilford, CT
Lyncole XIT, Torrance, CA
Harger, Grayslake, IL
Megger Group Limited, Dallas, TX
Galvan Industries, Inc., Harrisburg, NC
Charles F. Mello, Electro-test Inc.
Brady Davis
Michael J. Johnston
IAEI Archives

Composed at: International Association of Electrical Inspectors in Palatino LT Std, Arial, and Garamond
Printed at: Mosaic Printing, Cheverly, Maryland, on Anthem Matte text and bound in C1S White Tango cover